Nonlinear Analysis and Applications:
To V. Lakshmikantham on his 80th Birthday
Volume 2

Nonlinear Analysis and Applications: To V. Lakshmikantham on his 80th Birthday

Volume 2

Edited by

Ravi P. Agarwal

Department of Mathematical Sciences,
Florida Institute of Technology,
Melbourne, Florida, U.S.A.

and

Donal O'Regan

Department of Mathematics,
National University of Ireland,
Galway, Ireland

Springer Science+Business Media, B.V.

A C.I.P. Catalogue record for this book is available from the Library of Congress.

ISBN 978-94-010-4001-3 ISBN 978-94-010-0035-2 (eBook)
DOI 10.1007/978-94-010-0035-2

Printed on acid-free paper

Contents

VOLUME I

VOLUME II

Operator Equations in Ordered Function Spaces

S. Heikkilä

University of Oulu, Department of Mathematical Sciences
Box 3000, FIN-90014 University of Oulu, Finland.
e-mail: sheikki@cc.oulu.fi

Abstract. In this paper we present analysis in ordered function spaces, and apply it and results derived for operator equations in posets to prove existence and comparison results for operator equations in ordered function spaces.

AMS Subject Classification. 47H07, 47H15

Key words. Operator equation, poset, ordered function space, weak convergence, least, greatest, maximal, minimal, discontinuity.

1 Introduction

In this paper we shall first present recently proved existence and comparison results for the operator equation $Lu = Nu$, where the operators L and N are defined on a poset U, and their values are in another poset P. Our main purpose is to modify these results to the case where P is contained in one of the function spaces:

- $L^p(\Omega, E)$, $1 \leq p \leq \infty$, where $\Omega = (\Omega, \mathcal{A}, \mu)$ is a σ-finite measure space and E is an ordered Banach space;

- $C(X, E)$, where X is a topological space and E is an ordered normed space.

We assume that the space E has the following property.

(E0) Bounded and increasing sequences of E have weak limits.

In the case when P is an order interval of $L^p(\Omega, E)$, equipped with a.e. pointwise ordering, we prove existence results for least and greatest solutions of $Lu = Nu$, and study their dependence on N. Further existence and comparison results will be proved when

$$P = \{u \in L^p(\Omega, E) \mid \|u(t)\| \leq r(t) \text{ for a.e. } t \in \Omega\},$$

where $r : \Omega \to \mathbf{R}$, and E is lattice-ordered with properties (E0) and

(E1) the mapping $E \ni x \mapsto x^+ := \sup\{0, x\}$ is demicontinuous, and $\|x^+\| \leq \|x\|$ for all $x \in E$.

When $L^p(\Omega, E)$ is replaced by $C(X, E)$ we have to replace demicontinuity by continuity in (E1), and "for a.e." by "for all" in the definition of P.

Finally, we present some fixed point theorems in the case when L is the identity operator.

Main features of this paper are:

– No continuity assumptions are imposed on the operators L and N.

– Many hypotheses common in papers dealing with equations in ordered Banach spaces, such as normality, (full) regularity and/or solidity of their order cones, are not assumed in our main theorems.

2 Existence and comparison results for equations in posets

Let $P = (P, \leq)$ be a partially ordered set (poset). When $a, b \in P$, $a \leq b$, we denote $[a) = \{u \in P \mid a \leq u\}$, $(a] = \{u \in P \mid u \leq a\}$ and $[a, b] = [a) \cap (b]$. Recall that a subset C of P is *well-ordered* (resp. *inversely well-ordered*) if each nonempty subset of C has the least (resp. greatest) element. In both cases C is a chain.

A nonempty subset A of a poset P is called *relatively well-order complete* in P if each nonempty well-ordered or inversely well-ordered chain of A has supremums and infimums in P. If $A = P$, we say that P is *well-order complete.*

Our first existence and comparison result deals with the existence of least and greatest solutions of equation $Lu = Nu$, and their dependence on N.

Proposition 2.1. *Given posets U and P, mappings $L, N : U \to P$, and elements a, b of P, $a \leq b$, assume that the following hypotheses are valid.*

(I) *If $u, v \in U$, $u \leq v$ and $Lu \leq Lv$, then $a \leq Nu \leq Nv \leq b$.*

(II) *Equation $Lu = x$ has for each $x \in [a, b]$ least and greatest solutions, and they are increasing with respect to x.*

(III) *$N[U]$ is relatively well-order complete in P.*

Then equation $Lu = Nu$ has least and greatest solutions, and they are increasing with respect to N.

Hint to the proof. The given hypotheses ensure that the relation

$$Gx = Nu, \quad \text{where } u \text{ is the least solution of } \; Lu = x \in [a, b],$$

defines an increasing mapping $G : [a, b] \to [a, b]$ whose range $G[a, b]$ is contained in $N[U]$, and is by the hypothesis (III) relatively well-order complete in P, and hence also in $[a, b]$. Thus G has the least fixed point x_* by [10, Theorem 1.2.1]. As for the proof that the least solution u_* of $Lu = x_*$ is the least solution of $Lu = Nu$, and that u_* is increasing with respect to N, see the proof of [2, Theorem 1.1.2]. □

The hypothesis (I) of Proposition 2.1 implies that the range $N[U]$ of N is contained in the order interval $[a, b]$. In what follows, this property is replaced by the following property.

We say that an element c of a poset P is an *order-center of* P if $\sup\{c, v\}$ and $\inf\{c, v\}$ exist and belong to P for all $v \in P$.

For instance, a closed disk of $\mathbf{R}^2$, ordered coordinatewise, is well-order complete, and its center is also an order center.
The following result is proved in [8].

Proposition 2.2. (cf. [8, Proposition 2.2]) *Let U and P be posets, let c be an order center of P, and let L, $N : U \to P$ satisfy the following hypotheses.*

(H0) *L is a bijection, and both L^{-1} and $N \circ L^{-1}$ are increasing.*

(H1) *$N[U]$ is a relatively well-order complete in P.*

Then the following results hold.

(a) Equation $Lu = \inf\{c, Nu\}$ has the greatest solution a_c in $(L^{-1}c]$.

(b) Equation $Lu = \sup\{c, Nu\}$ has the least solution b_c in $[L^{-1}c)$.

(c) Equation $Lu = Nu$ has least and greatest solutions u_c and u^c in $[a_c, b_c]$.

(d) The solutions u_c and u^c are increasing with respect to N.

Hint to the proof. Defining $G = N \circ L^{-1}$ one obtains an increasing mapping $G : P \to P$, whose range $G[P]$ equals to $N[U]$, and is therefore relatively well-order complete in P by the hypothesis (H1). The asserted results follow then from [8, Proposition 2.1]. □

The hypotheses of Proposition 2.2. don't ensure the existence of least and greatest solutions of $Lu = Nu$ in P. However, one can prove the following result (cf. [9]).

Proposition 2.3. *Let U and P be posets, let c be an order center of P, and let L, $N : U \to P$ satisfy the following hypotheses.*

(H1) *$N[U]$ is relatively well-order complete in P.*

(H2) *L is a bijection, and both L and $N \circ L^{-1}$ are increasing.*

Then equation $Lu = Nu$ has a minimal solution u_- and a maximal solution u_+ in the sense that if $u \in P$ and $Lu = Nu$, then $u \le u_-$ implies $u = u_-$, and $u_+ \le u$ implies $u = u_+$.

Remarks 2.1. The proofs of the above stated results are based on chain methods introduced in [10]. In the special case when the range of N is finite the chain methods used in the proof of Proposition 2.2 can be reduced to the algorithmic methods (cf. [6, 8]). In concrete examples the finiteness of the ranges of a real valued mapping N is often ensured by approximating its values within the accuracy of given number of digits. In fact, such approximations are necessary if one wants to use computers to approximate solutions of equations which contain transcendental functions.

Since partial orderings are the only structures needed in the proofs of the above results, they can be applied to problems where only ordinal scales are available.

The Axiom of Choice or its variant, Zorn's Lemma, are not needed in the proofs of Propositions 2.1 and 2.2.

3 Analysis in ordered spaces

We shall first present some basic results on the integration theory of Banach-valued functions. Let $\Omega = (\Omega, \mathcal{A}, \mu)$ be a σ-finite measure space and $E = (E, \|\cdot\|)$ a Banach space.

If a sequence (x_n) of a Banach space E converges weakly to x, denote $x_n \rightharpoonup x$, or ${}^w\lim_{n\to\infty} x_n = x$. Recall (cf. e.g., [19]) that in such a case (x_n) is bounded, i.e. $\sup_n \|x_n\| < \infty$, and

$$\|x\| \leq \liminf_{n\to\infty} \|x_n\|. \tag{3.1}$$

We say that a function $u : \Omega \to E$ is *μ-measurable* if u is almost everywhere (a.e.) pointwise limit of a sequence of *step functions* of the form $t \mapsto \sum_{i=1}^n \chi_{A_i}(t)u_i$, where χ_{A_i} denotes the characteristic function of $A_i \in \mathcal{A}$ with $\mu(A_i) < \infty$, and $u_i \in E$, $i = 1, \ldots, n$. By a result due to Pettis (cf. e.g., [19]) a function $u : \Omega \to E$ is μ- measurable if and only if it is weakly μ- measurable and almost separably-valued. Applying this property one can prove the following result.

Lemma 3.1. *If (u_n) is a sequence of μ- measurable functions $u_n : \Omega \to E$, and if $u_n(t) \rightharpoonup u(t)$ for almost every (a.e.) $t \in \Omega$, then u is μ- measurable.*

Proof. Since each u_n is weakly μ- measurable, then the limit u is also weakly μ- measurable. Moreover, for each $n \in \mathbf{N}$ there is a μ- null set Z_n in Ω such that a set $\{u_n(t) \mid t \in \Omega \setminus Z_n\}$ is separable. Denoting $Z = \bigcup_{n=0}^\infty Z_n$, then Z is a μ- null-set and the set $D = \overline{co} \bigcup_{n=0}^\infty \{u_n(t) \mid t \in \Omega \setminus Z\}$ is separable. As a closed and convex set D is weakly closed, so that $u(t) \in D$ for a.e. $t \in \Omega$. Thus u is also almost separably-valued, and hence μ- measurable by Pettis Theorem. □

Recall that a μ- measurable function $u : \Omega \to E$ is *μ- integrable* if and only if the function $t \mapsto \|u(t)\|$ is μ- integrable. The integral of a step function $u = \sum_{i=1}^n \chi_{A_i}(t)u_i$ over Ω is defined by

$$\int_\Omega u(s)\, d\mu(s) = \sum_{i=1}^n \mu(A_i)u_i.$$

If u is μ- integrable, define

$$\int_\Omega u(s)\, d\mu(s) = \lim_{n\to\infty} \int_\Omega u_n(s)\, d\mu(s), \tag{3.2}$$

where $(u_n)_{n=1}^\infty$ is any sequence step functions satisfying $\lim_{n\to\infty} u_n(t) = u(t)$ for a.e. $t \in \Omega$. It is easy to see that

$$\|\int_\Omega u(s)\, d\mu(s)\| \leq \int_\Omega \|u(s)\|\, d\mu(s).$$

If $A \in \mathcal{A}$, define

$$\int_A u(s)\,d\mu(s) = \int_\Omega \chi_A(s)u(s)\,d\mu(s).$$

Let E and V be Banach spaces. We say that a function $g : V \to E$ is *demicontinuous* if $x_n \to x$ in V implies $g(x_n) \rightharpoonup g(x)$ in E.

Proposition 3.1. *Let E and V be Banach spaces and Ω a real interval. If $g : V \to E$ is demicontinuous, and if $u : \Omega \to V$ is μ- measurable, then $g \circ u : \Omega \to E$ is μ- measurable.*

Proof. Let $u : \Omega \to V$ be μ- measurable, and let $(u_n)_{n=0}^\infty$ be a sequence of step functions which converges pointwise a.e. on Ω to u. If $g : V \to E$, then the function $g \circ u_n : \Omega \to E$ is for each $n \in \mathbf{N}$ a step function, and hence μ- measurable. Assuming that g is demicontinuous, then $g(u_n(t)) \rightharpoonup g(u(t))$ for a.e. $t \in \Omega$. This implies by Lemma 3.1 that $g \circ u$ is μ- measurable. □

Assuming that E is an ordered Banach space we have the following result, which is proved in [10].

Lemma 3.2. *Let E be an ordered Banach space. If $u, v : \Omega \to E$ are μ-integrable and $u(t) \le v(t)$ for a.e. $t \in \Omega$, then*

$$\int_\Omega u(s)\,d\mu(s) \le \int_\Omega v(s)\,d\mu(s). \tag{3.3}$$

Denote by $L^p(\Omega, E)$, $1 \le p \le \infty$, the space of all μ- measurable functions $u : \Omega \to E$ for which $t \mapsto \|u(t)\|$ belongs to $L^p(\Omega, \mathbf{R})$. If E is an ordered Banach space, and if we identify a.e. equal functions, then $L^p(\Omega, E)$ is an ordered Banach space with respect to the p-norm:

$$\|u\|_p = \left(\int_\Omega \|u(t)\|^p\,d\mu(t)\right)^{\frac{1}{p}},\ 1 \le p < \infty, \quad \|u\|_\infty = \operatorname{essup}\{\|u(t)\| \mid t \in \Omega\},$$

and the partial ordering

$$u \le v \text{ if and only if } u(t) \le v(t) \text{ for a.e. } t \in \Omega. \tag{3.4}$$

Recall that if E is lattice-ordered, then $x^- := \sup\{0, -x\}$ can be represented in the form $x^- = x^+ - x$. Consequently, if E has property (E1), given in the Introduction, then also the mapping $x \mapsto x^-$ is demicontinuous. Applying Proposition 3.1 and the definitions of the p-norms we obtain the following result.

Corollary 3.1. *Let E be a lattice-ordered Banach space with property (E1) given in the Introduction. If $u \in L^p(\Omega, E)$, $1 \le p \le \infty$, then the mappings $t \mapsto u(t)^\pm$ are also in $L^p(\Omega, E)$.*

The next result is a kind of weak monotone convergence theorem.

Proposition 3.2. *Given an ordered Banach space E and $p \in [1,\infty)$, assume that an increasing and bounded sequence $(u_n)_{n=0}^{\infty}$ of $L^p(\Omega, E)$ converges weakly a.e. pointwise to $u : \Omega \to E$. Then $u \in L^p(\Omega, E)$, and $u = \sup_n u_n$ if $(u_n)_{n=0}^{\infty}$ is increasing, and $u = \inf_n u_n$ if $(u_n)_{n=0}^{\infty}$ is decreasing.*

Proof. Let $(u_n)_{n=0}^{\infty}$ be an increasing and bounded sequence in $L^p(\Omega, E)$, $u : \Omega \to E$, and assume that

$$u(t) = {}^{w}\lim_{n\to\infty} u_n(t) = \sup_n u_n(t) \quad \text{for a.e.} \quad t \in \Omega. \tag{3.5}$$

The function u is measurable by Lemma 3.1. In view of (3.1) and (3.5) we obtain

$$\|u(t)\|^p \le \liminf_{n\to\infty} \|u_n(t)\|^p < \infty \quad \text{for a.e.} \quad t \in \Omega.$$

The above inequality, Fatou's Lemma and the boundedness of (u_n) in $L^p(\Omega, E)$ imply that

$$\int_\Omega \|u(t)\|^p d\mu(t) \le \int_\Omega \liminf_{n\to\infty} \|u_n(t)\|^p d\mu(t) \le \liminf_{n\to\infty} \int_\Omega \|u_n(t)\|^p d\mu(t) < \infty.$$

This proves that $t \mapsto \|u(t)\| \in L^p(\Omega, \mathbf{R})$, so that $u \in L^p(\Omega, E)$. Moreover, it follows from (3.5) that $u = \sup_n u_n$. If $(u_n)_{n=0}^{\infty}$ is decreasing, apply the above proof to the sequence $(-u_n)_{n=0}^{\infty}$. □

4 Well-order completeness in ordered function spaces

The existence and comparison results of Section 2 will be applied in Section 5 to the case when the poset P is contained in one of the function spaces

- $L^p(\Omega, E)$, $1 \le p \le \infty$, where $\Omega = (\Omega, \mathcal{A}, \mu)$ is a σ-finite measure space and E is an ordered Banach space;

- $C(X, E)$, where X is a topological space and E is an ordered normed space.

The main purpose of this section is to prove results which enable to replace the hypotheses (III) and (H1) concerning the relative well-order completeness of the range $N[U]$ of the operator N by more concrete hypotheses. In the proofs we need the follo wing Lemma which follows, e.g., from [2, Lemma A.3.1].

Lemma 4.1. *Let W be a well-ordered subset of an ordered normed space E, and assume that each increasing sequence of W has a weak limit in E. Then W contains an increasing sequence which converges weakly to $\sup W$.*

As an application of Proposition 3.2 and Lemma 4.1 we prove the following result.

Lemma 4.2. *Given an ordered Banach space E whose bounded and increasing sequences have weak limits, and $p \in [1,\infty)$, assume that C is a bounded and well ordered chain of $L^p(\Omega, E)$ whose increasing sequences are*

a.e. pointwise bounded. If $\mu(\Omega) < \infty$, then C contains an increasing sequence which converges weakly a.e. pointwise to $\sup C$.
The above result holds also when the boundedness hypotheses of C are replaced by the existence of a function $h \in L^p(\Omega, \mathbf{R})$ such that

$$\|u(t)\| \leq h(t) \quad \textit{for a.e. } t \in \Omega \textit{ and all } u \in C.$$

Proof. Let C be a bounded and well-ordered chain of $L^p(\Omega, E)$. Because $\mu(\Omega) < \infty$, then $L^p(\Omega, E)$ is continuously embedded in $L^1(\Omega, E)$, so that C is a bounded and well-ordered chain in $L^1(\Omega, E)$. If u, $v \in C$ and $u \leq v$, then $\int_\Omega u \leq \int_\Omega v$ by Lemma 3.2. Thus $\{\int_\Omega v\}_{v \in C}$ is a bounded and well-ordered subset of E whose bounded and increasing sequences have weak limits. By Lemma 4.1 there is an increasing sequence $(u_n)_{n=0}^\infty$ in C such that

$$(4.1) \qquad {}^w\lim_{n\to\infty} \int_\Omega u_n = \sup_n \int_\Omega u_n = \sup_{v \in C} \int_\Omega v.$$

Moreover, $(u_n)_{n=0}^\infty$ as an increasing sequence of C is a.e. pointwise bounded, by a hypothesis, so that

$$(4.2) \qquad u(t) = {}^w\lim_{n\to\infty} u_n(t) = \sup_n u_n(t)$$

exists for a.e. $t \in \Omega$ by a hypothesis, and $u \in L^p(\Omega, E)$ by Proposition 3.2.

To show that u is an upper bound of C, let $w \in C$ be given. Assume first that $u_n \leq w$ for each $n \in \mathbf{N}$. Then $u \leq w$ by (4.2). It follows from (4.1) and (4.2) that

$$(4.3) \qquad \int_\Omega w \leq \sup_{v \in C} \int_\Omega v = \sup_n \int_\Omega u_n \leq \int_\Omega u.$$

If A is a measurable subset of Ω, then $\int_A u \leq \int_A w$ and $\int_{\Omega \setminus A} u \leq \int_{\Omega \setminus A} w$ by Lemma 3.2. If $\int_A u < \int_A w$, then

$$\int_\Omega u = \int_A u + \int_{\Omega \setminus A} u < \int_A w + \int_{\Omega \setminus A} w = \int_\Omega w,$$

which contradicts with (4.3). Thus $\int_A u = \int_A w$ for each measurable subset A of Ω, so that $w = u$ by [11, VI, Corollary 5.16]. The above proof shows that

$$w = u, \quad \text{whenever } w \in C \text{ and } u_n \leq w \text{ for all } \ n \in \mathbf{N}.$$

If $w \leq u_n$ for some $n \in \mathbf{N}$, then $w \leq u$ by (4.2). Thus $w \leq u$ for each $w \in C$.

To show that $u = \sup C$, let $v \in L^p(\Omega, E)$ be an upper bound of C. Then $u_n(t) \leq v(t)$ for a.e. $t \in \Omega$ and for all $n \in \mathbf{N}$. This result and (4.2) imply that $u(t) = \sup_n u_n(t) \le v(t)$ for a.e. $t \in \Omega$, i.e., $u \le v$. Thus $u = \sup C$ in $L^p(\Omega, E)$.

To prove the last assertion assume that each $u \in C$ satisfies $\|u(t)\| \le h(t)$ for a.e. $t \in \Omega$, where $h \in L^p(\Omega, \mathbf{R})$. Then $\|u\|_p \le \|h\|_p$ for each $u \in C$ so that C is bounded and a.e. pointwise bounded. □

Consider next the case when $p = \infty$.

Lemma 4.3. *Let E be an ordered Banach space whose bounded and increasing sequences have weak limits. If $\mu(\Omega) < \infty$, then each bounded and well-ordered chain C of $L^\infty(\Omega, E)$ contains an increasing sequence which converges weakly a.e. pointwise to* $\sup C$.

Proof. If $u \in L^\infty(\Omega, E)$ then $\|u(t)\| \le \|u\|_\infty$ for a.e. $t \in \Omega$, which implies that $\|u\|_1 \le \mu(\Omega)\|u\|_\infty$. This shows that $L^\infty(\Omega, E)$ is continuously embedded in $L^1(\Omega, E)$. Let C be a bounded and well ordered chain in $L^\infty(\Omega, E)$. C is a bounded and well ordered also in $L^1(\Omega, E)$. Hence, if $(u_n)_{n=0}^\infty$ is an increasing sequen ce in C, there is $M > 0$ such that for each $n \in \mathbf{N}$,

$$\|u_n(t)\| \le \|u_n\|_\infty \le M \quad \text{for a.e.} \quad t \in \Omega.$$

Thus $(u_n)_{n=0}^\infty$ is also a.e. pointwise bounded. Hence, C has by the proof of Lemma 4.2 a supremum u in $L^1(\Omega, E)$, and there is an increasing sequence $(u_n)_{n=0}^\infty$ in C which converges weakly a.e. pointwise in Ω to u. Moreover, (3.1) and the above inequality imply that

$$\|u(t)\| \le \liminf_{n\to\infty} \|u_n(t)\| \le M \quad \text{for a.e.} \quad t \in \Omega.$$

This implies that u belongs to $L^\infty(\Omega, E)$ and it is easy to see that $u = \sup C$ also in $L^\infty(\Omega, E)$. □

The next result extends the results of Lemma 4.2 and Lemma 4.3 to the case when μ is σ-finite.

Proposition 4.1. *Let $(\Omega, \mathcal{A}, \mu)$ be a σ-finite measure space and E an ordered Banach space whose bounded and increasing sequences have weak limits. If C is a bounded and a.e. pointwise bounded well-ordered chain in $L^p(\Omega, E)$, $1 \le p \le \infty$, then C has the supremum u^* in $L^p(\Omega, E)$. Moreover, there exists an increasing sequence (u_n) of C such that $u_n(t) \rightharpoonup u^*(t)$ for a.e. $t \in \Omega$.*

Proof. Since Ω is σ-finite, then $\Omega = \bigcup_{n=0}^\infty \Omega_n$, where $\Omega_n \subseteq \Omega_{n+1}$ and $\mu(\Omega_n) < \infty$ for each $n \in \mathbf{N}$. Let C be a well-ordered, bounded and a.e. pointwise bounded chain in $L^p(\Omega, E)$, $1 \le p \le \infty$. The restriction $C|\Omega_n = \{u|\Omega_n \mid u \in C\}$ is for each $n \in \mathbf{N}$ a well-ordered, bounded and a.e. pointwise bounded chain in $L^p(\Omega_n, E)$. It follows from Lemma 4.2 and Lemma 4.3 that

$$v_n = \sup(C|\Omega_n)$$

exists in $L^p(\Omega_n, E)$. Defining $v_n(t) = 0$ for $t \in \Omega \setminus \Omega_n$, we obtain a sequence of μ-measurable functions $v_n : \Omega \to E$. This sequence is increasing, since

$\Omega_n \subseteq \Omega_{n+1}$. It is also a.e. pointwise bounded, whence

$$u^*(t) = {}^w\lim_{n\to\infty} v_n(t) = \sup_{n\in\mathbf{N}} v_n(t)$$

exists for a.e. $t \in \Omega$. Defining $u^*(t) = 0$ for the remaining $t \in \Omega$, we obtain by Proposition 3.2 a function $u^* \in L^p(\Omega, E)$. If $u \in C$, then $u|\Omega_n \leq v_n$, so that

$$u(t) \leq v_n(t) \leq u^*(t) \quad \text{for a.e. } t \in \Omega_n \text{ and for each } \ n \in \mathbf{N}.$$

Thus $u \leq u^*$ for each $u \in C$, so that u^* is an upper bound of C. If $v \in L^p(\Omega, E)$ is another upper bound of C, then

$$u(t) \leq v(t) \text{ for a.e. } t \in \Omega \text{ and for each } u \in C.$$

Thus $u|\Omega_n \leq v|\Omega_n$ for all $n \in \mathbf{N}$ and $u \in C$, whence

$$v_n(t) \leq v(t) \text{ for a.e. } t \in \Omega \text{ and for each } n \in \mathbf{N}.$$

This result and the definition of u^* imply that $u^* \leq v$, whence $u^* = \sup C$ in $L^p(\Omega, E)$.

To prove the last assertion, notice first that by the proofs of Lemmas 4.2 and 4.3 there exists for each $n \in \mathbf{N}$ an increasing sequence $(u_k^n)_{k=0}^\infty$ of C and a μ-null set $Z_n \subset \Omega_n$ such that

$$v_n(t) = {}^w\lim_{k\to\infty} u_k^n(t) = \sup_{k\in\mathbf{N}} u_k^n(t) \quad \text{for each } \ t \in \Omega_n \setminus Z_n.$$

Denoting

$$u_n = \max\{u_k^j \mid 0 \leq j,\, k \leq n\}, \quad n \in \mathbf{N},$$

we obtain an increasing sequence (u_n) of C, which satisfies

$$u_k^n(t) \leq u_n(t) \leq u^*(t) \quad \text{for each } k = 0, \ldots, n \text{ and } \ t \in \Omega_n \setminus Z_n.$$

Since C is a.e. pointwise bounded, we may assume that $(u_n(t))_{n=0}^\infty)$ is increasing and bounded for each $t \in \Omega \setminus Z$, where $Z = \cup_{n=0}^\infty Z_n$. Thus

$$u(t) = {}^w\lim_{k\to\infty} u_n(t) = \sup_{n\in\mathbf{N}} u_n(t)$$

exists for each $t \in \Omega \setminus Z$. Moreover, the definitions of u and v_n imply that

$$v_n(t) \leq u(t) \leq u^*(t) \quad \text{for each } \ t \in \Omega_n \setminus Z_n.$$

Thus

$$u^*(t) = {}^w\lim_{n\to\infty} v_n(t) \leq u(t) \leq u^*(t)$$

for a.e. $t \in \Omega$. This result implies that $u = u^*$, whence $u_n(t) \rightharpoonup u^*(t)$ for a.e. $t \in \Omega$. □

Next we study existence of supremums of well-ordered chains in the space $C(X,E)$ of continuous functions $u : X \to E$, where X is a topological space and E an ordered normed space. Define a partial ordering in $C(X,E)$ by

$$u \leq v \text{ if and only if } u(t) \leq v(t) \text{ for each } t \in X. \tag{4.4}$$

We say that a subset C of $C(X,E)$ is *equicontinuous* if for each $t \in X$ and for each $\epsilon > 0$ there exists a neighborhood U of t such that

$$\|u(s) - u(t)\| \leq \epsilon \text{ for all } u \in C \text{ and } s \in U. \tag{4.5}$$

Proposition 4.2. *Let E be an ordered normed space, X a topological space, and let C be an equicontinuous and well-ordered subset of $C(X,E)$, whose increasing sequences have weak pointwise limits. Then $v = \sup C$ exists in $C(X,E)$.*

Proof. The hypotheses given for C imply that for each $t \in E$ the set $\{u(t)\}_{u \in C}$ is a well-ordered subset of E whose increasing sequences have weak limits. This implies by Lemma 4.1 that

$$v(t) = \sup\{u(t)\}_{u \in C} \tag{4.6}$$

exists in E for each $t \in \Omega$. To prove that the so obtained function $v : X \to E$ is the supremum of C in $C(X,E)$ it suffices to show its continuity. Let $t \in X$ and $\epsilon > 0$ be given. By the equicontinuity hypothesis there is such a neighborhood U of t that

$$\|u(s) - u(t)\| \leq \epsilon \text{ whenever } s \in U \text{ and } u \in C.$$

Let $s \in U$ be fixed. By Lemma 4.1 there exists an increasing sequence $(v_n)_{n=0}^{\infty}$ in C such that $(v_n(s))_{n=0}^{\infty}$ converges weakly in E to $v(s)$, and an increasing sequence $(u_n)_{n=0}^{\infty}$ in C such that $(u_n(t))_{n=0}^{\infty}$ converges weakly in E to $v(t)$. Denoting $z_n = \max\{v_n, u_n\}$, $n \in \mathbf{N}$, we obtain an increasing sequence $(z_n)_{n=0}^{\infty}$ in C. By a hypothesis it converges weakly pointwise, whence $(z_n(s))_{k=0}^{\infty}$ converges weakly in E to $v(s)$, and $(z_n(t))_{k=0}^{\infty}$ converges weakly in E to $v(t)$. Thus $(z_n(s) - z_n(t))_{k=0}^{\infty}$ converges weakly in E to $v(s) - v(t)$, so that

$$\|v(s) - v(t)\| \leq \liminf_{n \to \infty} \|z_n(s) - z_n(t)\| \leq \epsilon.$$

This holds for each $s \in U$, which shows that v is continuous at t. Thus $v \in C(X,E)$, whence v is the supremum of C in $C(X,E)$.

If X is a separable topological space we have the following result.

Proposition 4.3. *Let E be an ordered normed space, X a separable topological space, and let C be an equicontinuous well-ordered subset of $C(X,E)$ whose increasing sequences have weak pointwise limits. Then $v = \sup C$ exists, and there is an increasing sequence (u_n) of C such that $u_n(t) \rightharpoonup v(t)$ for each $t \in X$.*

Proof. Let $D = \{t_j\}_{j\in \mathbf{N}}$ be a dense subset of X. It follows from Proposition 4.2 that $v = \sup C$ exists in $C(X,E)$ and satisfies (4.6). Moreover, the proof of Proposition 4.2 implies that the set $C \cup \{v\}$ is equicontinuous, and that for each $j \in \mathbf{N}$ there is a sequence $(u_k^j)_{k=0}^\infty$ in C which converges weakly to $v(t_j)$. Denote

$$u_n = \max\{u_k^j \mid 0 \le j,\, k \le n\}, \quad n \in \mathbf{N}.$$

The so obtained sequence (u_n) is increasing, is contained in C, and $u_n(t) \rightharpoonup v(t)$ for each $t \in D$. To prove this convergence also when t belongs to the complement of D, let $t \in X \setminus D$, $\epsilon > 0$ and $f \in E'$ be given. Choose a neighborhood U of t such that

$$\text{(4.7)} \qquad \|u(s) - u(t)\| \le \frac{\epsilon}{1 + 4\|f\|} \quad \text{for all } u \in C \cup \{x\} \text{ and } s \in U.$$

Since D is a dense subset of E, we can choose s in (4.7) so that it belongs to $U \cap D$. The sequence $(u_n(t))$ has by a hypothesis a weak limit z. We have to prove that $z = v(t)$. Since $u_n(t) \rightharpoonup z$ and $u_n(s) \rightharpoonup v(s)$, there is $n \in \mathbf{N}$ such that

$$\text{(4.8)} \qquad |f(u_n(t)) - f(z)| \le \frac{\epsilon}{4} \quad \text{and} \quad |f(u_n(s)) - f(v(s))| \le \frac{\epsilon}{4}.$$

Applying (4.7) and (4.8) we get

$$|f(z - v(t))| = |f(z) - f(v(t))| \le |f(z) - f(u_n(t))| + \|f\|\, \|u_n(t) - (u_n(s)\|$$
$$+ |f(u_n(s)) - f(v(s))| + \|f\|\, \|v(s)) - v(t)\| \le \epsilon.$$

This holds for each $f \in E'$ and for each $\epsilon > 0$, whence $z = v(t)$. Thus $u_n(t) \rightharpoonup v(t)$ also when $t \in X \setminus D$, which concludes the proof. □

Remarks 4.1. The results of this section generalize somewhat those derived in [5].

In Propositions 4.1 and 4.3 there is an increasing sequence of C which converges to $\sup C$. This implies by [10, Lemma 1.1.4] that C is countable.

The a.e. pointwise weak convergence of sequences in $L^p(\Omega, E)$ cannot be topologized, in general. Thus the results derived in ordered topological spaces are not sufficient for our purposes.

The results of this section hold also when weak convergence are replaced by strong convergence.

5 Existence and comparison results for operator equations in ordered function spaces

Throughout this section we assume that $\Omega = (\Omega, \mathcal{A}, \mu)$ is a σ-finite measure space, and that X is a topological space. The results of Section 4 will now be applied to modify the existence and comparison results introduced in Section 2 for equation $Lu = Nu$ to the cases when the values of L and N are in ordered function spaces $L^p(\Omega, E)$, $1 \le p \le \infty$, equippe d with a.e. pointwise ordering (3.4), or in $C(X, E)$, equipped with pointwise ordering (4.4). Our first result is a consequence of Proposition 2.1 and Proposition 4.1.

Theorem 5.1. *Let E be an ordered Banach space with property*

(E0) *Bounded and increasing sequences of E have weak limits.*

Given a poset U, mappings L, $N : U \to L^p(\Omega, E)$, $1 \le p \le \infty$, and elements a, b of $L^p(\Omega, E)$, $a \le b$, assume that the following hypotheses are valid.

(I) *If u, $v \in U$, $u \le v$ and $Lu \le Lv$, then $a \le Nu \le Nv \le b$.*

(II) *Equation $Lu = x$ has for each $x \in [a, b]$ least and greatest solutions, and they are increasing with respect to x.*

(IV) *$N[U]$ is a norm-bounded and a.e. pointwise bounded subset of $L^p(\Omega, E)$.*

Then equation $Lu = Nu$ has least and greatest solutions, and they are increasing with respect to N.

Proof. By Proposition 2.1 it suffices to show that

(III) $N[U]$ is relatively well-order complete in $P = [a, b]$.

Let C be a nonempty and well-ordered subset of $N[U]$. The hypothesis (IV) implies that C is norm-bounded and a.e. pointwise bounded. It then follows from Proposition 4.1 and from the hypothesis (E0) that $u = \sup C$ exists in $L^p(\Omega, E)$. Since $C \subset [a, b]$, then $\sup C \in [a, b]$. Moreover, $\inf C = \min C \in [a, b]$. If D is a nonempty and inversely well-ordered subset of $N[U]$, then $C = -D$ is a nonempty, well-ordered, norm-bounded and a.e. pointwise bounded subset of $[-b, -a]$, whence $\sup C$ exists in $[-b, -a]$ by the above proof. Thus $\inf D = -\sup C$ exists and belongs to $[a, b]$. Moreover, $\sup D = \max D \in [a, b]$.

The above proof shows that the hypothesis (III) is valid, which concludes the proof. □

The next result is an application of Proposition 2.2 and Corollary 3.1.

Theorem 5.2. *Assume that the hypotheses of Theorem 5.1, with (II) replaced by*

(II') *Equation $Lu = x$ has for each $x \in [a, b]$ a unique solution which is increasing with respect to x,*

are satisfied, and that E is lattice ordered and has properties (E0) and

(E1) *the mapping $E \ni x \mapsto x^+ := \sup\{0, x\}$ is demicontinuous, and $\|x^+\| \le \|x\|$ for all $x \in E$.*

Then for each $c \in [a, b]$ following results hold.

(a) Equation $Lu = \inf\{c, Nu\}$ has the greatest solution a_c in $L^{-1}[a, c]$.

(b) Equation $Lu = \sup\{c, Nu\}$ *has the least solution* b_c *in* $L^{-1}[c, b]$.

(c) Equation $Lu = Nu$ *has least and greatest solutions* u_c *and* u^c *in* $[a_c, b_c]$.

(d) The solutions $\{u_c,\ u^c \mid c \in [a, b]\}$ *are increasing with respect to* N.

Proof. The given hypotheses and property (E0) imply by the proof of Theorem 5.1 that $N[U]$ is relatively well-order complete in $[a, b]$. Thus the hypothesis (H1) of Proposition 2.2 holds when $P = [a, b]$. The hypotheses (I) and (II') ensure that also the hypothesis (H0) of Proposition 2.2 holds when $P = [a, b]$. Let $c,\ u \in [a, b]$ be given, and denote $v(t) = \sup\{c(t), u(t)\}$, $t \in \Omega$. Since E has property (E1), and since $v(t) = c(t) + (u(t) - c(t))^+$, then $v \in L^p(\Omega, E)$ by Corollary 3.1. This result and the definition of v imply that $v = \sup\{c, u\}$ in $L^p(\Omega, E)$. Moreover, since $a \le c,\ u \le b$, then $v \in [a, b]$. The proof that $\inf\{c, u\}$ exists in $L^p(\Omega, E)$ and belongs to $[a, b]$ is similar. Thus each element c of $[a, b]$ is its order center.

The above proof shows that all the hypotheses of Proposition 2.2 hold when $P = [a, b]$ and $c \in [a, b]$, which concludes the proof. □

In the case when the order cone E_+ of E is regular the hypotheses can be reduced as follows.

Corollary 5.1. *The results of Theorems 5.1 and 5.2 hold also when property (E0) and hypothesis (IV) are replaced by the assumption that the order cone* E_+ *is regular, or equivalently, that all order-bounded and increasing sequences of* E *converge strongly.*

Proof. Let $a,\ b \in L^p(\Omega, E)$ be as in the hypothesis (I) of Theorem 5.1. This hypothesis implies that $N[U] \subset [a, b]$. If E_+ is regular, it follows from [10, Proposition 5.8.8] that $[a, b]$ is well-order complete, so that the hypothesis (III) holds. The verification of this hypothesis was the only place where property (E0) and the hypothesis (IV) were needed in the proof of Theorem 5.1, which concludes the proof. □

Next we derive existence and comparison results for the operator equation $Lu = Nu$ when the values of the operators L and N are in the set

$$P = \{u \in L^p(\Omega, E) \mid \|u(t)\| \le r(t) \ \text{ for a.e. } \ t \in \Omega\}, \tag{5.1}$$

where $r : \Omega \to \mathbf{R}$ and $1 \le p \le \infty$. As a consequence of Propositions 2.2 and 4.1 we obtain the following result.

Theorem 5.3. *Let* E *be a lattice-ordered Banach space with properties (E0) and (E1), and that* $L,\ N : U \to P$, *where* P *is given by (5.1), satisfy the following hypotheses:*

(H0) L *is a bijection, and both* L^{-1} *and* $N \circ L^{-1}$ *are increasing;*

(H3) $N[U]$ *is a norm-bounded subset of* P, *or* $r \in L^p(\Omega, \mathbf{R})$ *in (5.1).*

Then equation $Lu = Nu$ *has least and greatest solutions* u_- *and* u_+ *in* $[a_-, b_+]$, *where* a_- *is the greatest solution of equation* $Lu = -(Nu)^-$ *in* $L^{-1}(P_-)$, *where* $P_- = \{u \in P \mid u \le 0\}$, *and* b_+ *is the least solution of equation* $Lu = (Nu)^+$ *in* $L^{-1}(P_+)$, *where* $P_+ = \{u \in P \mid u \geq 0\}$. *Moreover, the solutions* u_- *and* u_+ *are increasing with respect to* N.

Proof. We shall first show that

(H1) $N[U]$ is a relatively well-order complete subset of P.

If C is a well-ordered subset of $N[U]$, then C is norm-bounded by the hypothesis (H3), and a.e. pointwise bounded as a subset of P. Thus C contains by Proposition 4.1 an increasing sequence (u_n) which converges a.e. pointwise weakly to $u = \sup C$. It follows from (3.1) that

$$\|u(t)\| \leq \liminf_{n\to\infty} \|u_n(t)\| \leq r(t) \quad \text{for a.e.} \quad t \in \Omega,$$

whence $u = \sup C \in P$. Moreover, $\inf C = \min C \in P$. If D is a nonempty and inversely well-ordered subset of P, then $C = -D$ is nonempty, well-ordered, norm-bounded and a.e. pointwise bounded subset of $-P = P$, whence $\sup C$ exists in P by the above proof. Thus $\inf D = -\sup C$ exists and belongs to P. Moreover, $\sup D = \max D \in P$.

The above proof shows that the hypothesis (H1) is satisfied. Next we shall show that the zero function 0 is an order center of P. If $u \in P$, then $u^+ = t \mapsto \sup\{0, u(t)\}$ belongs to $L^p(\Omega, E)$ by the proof of Theorem 5.2. Moreover, it follows from (5.1) and (E1) that

$$\|u^+(t)\| = \|u(t)^+\| \leq \|u(t)\| \leq r(t) \qquad \text{for a.e.} \quad t \in \Omega.$$

Thus $u^+ = \sup\{0, u\} \in P$. The proof that $\inf\{0, u\}$ exists in $L^p(\Omega, E)$ and belongs to P is similar, whence 0 is an order center of P. Thus all the hypotheses of Proposition 2.2 hold with $c = 0$, whence the following results hold:

(a) Equation $Lu = \inf\{0, Nu\}$ has the greatest solution a_- in $L^{-1}[(0] \cap P]$.

(b) Equation $Lu = \sup\{0, Nu\}$ has the least solution b_+ in $L^{-1}[[0) \cap P]$.

(c) Equation $Lu = Nu$ has least and greatest solutions u_- and u_+ in $[a_-, b_+]$.

(d) The solutions u_- and u_+ are increasing with respect to N.

Noticing that $\inf\{0, Nu\} = -(Nu)^-$, $\sup\{0, Nu\} = (Nu)^+$, $(0] \cap P = \{u \in P \mid u \leq 0\}$ and $[0) \cap P = \{u \in P \mid u \geq 0\}$, the conclusions of Theorem follow from the above results (a)–(d). □

The next result is a consequence of Proposition 2.3.

Theorem 5.4. *Assume that* E *be a lattice-ordered Banach space with properties (E0) and (E1), and that* L, $N : U \to P$, *where* P *is given by (5.1), satisfy the following hypotheses:*

(H2) *L is a bijection, and both L and $N \circ L^{-1}$ are increasing;*

(H3) *$N[U]$ is a norm-bounded subset of P, or* $r \in L^p(\Omega, \mathbf{R})$ *in (5.1).*

Then equation $Lu = Nu$ *has maximal and minimal solutions.*

Proof. It follows from the proof of Theorem 5.3 that the hypothesis (H1) holds, and that 0 is an order center of P, defined by (5.1). The conclusion follows then from Proposition 2.3. □

Consider next the case when the values of the operators L, N are in $C(X,E)$. Applying Proposition 4.2 instead of Proposition 4.1 in the proof of Theorem 5.1 we obtain the following result.

Proposition 5.1. *Let X be a topological space and E an ordered normed space with property (E0). Given a poset U, mappings $L,\ N : U \to C(X,E)$, $1 \le p \le \infty$, and elements a, b of $C(X,E)$, $a \le b$, assume that the following hypotheses are valid.*

(I) *If $u,\ v \in U$, $u \le v$ and $Lu \le Lv$, then $a \le Nu \le Nv \le b$.*

(II) *Equation $Lu = x$ has for each $x \in [a,b]$ least and greatest solutions, and they are increasing with respect to x.*

(V) *$N[U]$ is a bounded and equicontinuous subset of $C(X,E)$.*

Then equation $Lu = Nu$ has least and greatest solutions, and they are increasing with respect to N.

Theorem 5.2 has the following counterpart when $L^p(\Omega,E)$ is replaced by $C(X,E)$.

Proposition 5.2. *Assume that the hypotheses of Proposition 5.1, with (II) replaced by*

(II') *Equation $Lu = x$ has for each $x \in [a,b]$ a unique solution which is increasing with respect to x,*

are satisfied, and that E is lattice ordered and has properties (E0) and

(E2) *the mapping $E \ni x \mapsto x^+ := \sup\{0,x\}$ is continuous, and $\|x^+\| \le \|x\|$ for all $x \in E$.*

Then for each $c \in [a,b]$ following results hold.

(a) Equation $Lu = \inf\{c, Nu\}$ has the greatest solution a_c in $L^{-1}[a,c]$.

(b) Equation $Lu = \sup\{c, Nu\}$ has the least solution b_c in $L^{-1}[c,b]$.

(c) Equation $Lu = Nu$ has least and greatest solutions u_c and u^c in $[a_c, b_c]$.

(d) The solutions $\{u_c,\ u^c \mid c \in [a,b]\}$ are increasing with respect to N.

Replacing P in Theorem 5.3 by

$$P = \{u \in C(X,E) \mid \|u(t)\| \le r(t) \ \text{ for all } \ t \in X\}, \tag{5.2}$$

where $r : X \to \mathbf{R}$, and applying Proposition 4.3, instead of Proposition 4.1, in the proof, we obtain the following result.

Proposition 5.3. *Let X be a separable topological space, E a lattice-ordered normed space with properties (E0) and (E2), and let $L,\ N : U \to P$, where P is given by (5.2), satisfy the following hypotheses:*

(H0) *L is a bijection, and both L^{-1} and $N \circ L^{-1}$ are increasing;*

(H4) *$N[U]$ is a norm-bounded and equicontinuous subset of P.*

Then the equation $Lu = Nu$ has least and greatest solutions u_- and u_+ in $[a_-, b_+]$, where a_- is the greatest solution of equation $Lu = -(Nu)^-$ in $L^{-1}(P_-)$, where $P_- = \{u \in P \mid u \leq 0\}$, and b_+ is the least solution of equation $Lu = (Nu)^+$ in $L^{-1}(P_+)$, where $P_+ = \{u \in P \mid u \geq 0\}$. Moreover, the solutions u_- and u_+ are increasing with respect to N.

When P is is given by (5.2) in Theorem 5.4 we obtain the following result.

Proposition 5.4. *Let X be a separable topological space, E a lattice-ordered normed space with properties (E0) and (E2), and let L, $N : U \to P$, where P is given by (5.2), satisfy the following hypotheses:*

(H2) *L is a bijection, and both L and $N \circ L^{-1}$ are increasing;*

(H4) *$N[U]$ is a norm-bounded and equicontinuous subset of P.*

Then the equation $Lu = Nu$ has maximal and minimal solutions.

6 Special cases

In this section we present some fixed point theorems which are special cases to results derived in Section 5. In our first result we assume that E is a *weakly complete Banach lattice*, i.e., E is lattice-ordered, its weak Cauchy sequences posses weak limits, and the norm $\|\cdot\|$ of E and its valuation $E \ni x \mapsto |x| = \sup\{x, -x\}$ satisfy

(E) $\|x\| \leq \|y\|$ whenever x, $y \in E$ and $|x| \leq |y|$.

In what follows, $\Omega = (\Omega, \mathcal{A}, \mu)$ is a σ-finite measure space. When $u : \Omega \to E$, denote $|u| = t \mapsto |u(t)|$.

Our first fixed point result is a consequence of Theorem 5.1.

Proposition 6.1. *Assume that E is a weakly complete Banach lattice, and $G : L^p(\Omega, E) \to L^p(\Omega, E)$, $1 \leq p \leq \infty$, an increasing mapping, which satisfies the following hypothesis.*

(G1) *There exists a bounded, linear and positive operator $T : L^p(\Omega, E) \to L^p(\Omega, E)$ with spectral radius $\rho(T) < 1$ and a function $v \in L^p(\Omega, E_+)$ such that*

$$|Gu(t)| \leq T|u|(t) + v(t) \quad \text{for a.e. } t \in \Omega \text{ and for all } u \in L^p(\Omega, E). \tag{6.1}$$

Then the operator G least and greatest fixed points u_* and u^* in $L^p(\Omega, E)$. Moreover, u_* and u^* are increasing with respect to G.

Proof. The properties given for T in the hypothesis (G1) ensure that the function $b = \sum_{n=0}^{\infty} T^n v$ is the unique solution of equation

$$b = v + Tb. \tag{6.2}$$

If $u \in [-b, b]$, or equivalently, if $|u|(t) := |u(t)| \leq b(t)$, for a.e. $t \in \Omega$, the hypothesis (G1) implies that

$$|Gu(t)| \leq T|u|(t) + v(t) \leq Tb(t) + v(t) = b(t) \quad \text{for a.e. } t \in \Omega.$$

Thus $Gu \in [-b, b]$, and $\|Gu(t)\| \leq \|b(t)\|$ for a.e. $t \in \Omega$ by (E). In particular, $\|Gu\|_p \leq \|b\|_p$ for each $u \in [-b, b]$, so that $G[-b, b]$ is a bounded subset of $[-b, b]$. Because all bounded and increasing sequences of a weakly complete Banach lattice are strongly convergent by [12, Theorem 1.c4], then E has property (E0).

The above proof shows that all the hypotheses of Theorem 5.1 hold when $L = I$, $G = N$ and $[a, b] = [-b, b]$, whence G has the least fixed point u_* and the greatest fixed point u^* in $[-b, b]$, and they are increasing with respect to N. To prove that u_* and u^* are least and greatest of all the fixed points of G in $L^p(\Omega, E)$ it is enough to show that all the fixed points of G are contained in $[-b, b]$. So, let $u \in L^p(\Omega, E)$ be a fixed point of G. Applying (G1) we get

$$|u(t)| \leq T|u|(t) + v(t) \quad \text{for a.e.} \quad t \in \Omega.$$

Thus the function $w = t \mapsto |u(t)|$ satisfies the inequality

$$w \leq Tw + v,$$

This inequality and the Abstract Gronwall Lemma (cf. [20, Proposition 7.15]) imply that $w \leq b$. Thus $|u(t)| \leq b(t)$ for a.e. $t \in \Omega$, so that $u \in [-b, b]$. □

Since a weakly complete Banach lattice has also property (E1), we obtain as a consequence of Theorem 5.2 the following fixed point result.

Proposition 6.2. *Let the hypotheses of Proposition 6.1 hold, and let $b \in L^p(\Omega, E)$ be the solution of (6.2). Then for each $c \in [-b, b]$ the operator G has the least fixed point u_c and the greatest fixed point u^c in the order interval $[a_c, b_c]$ of $L^p(\Omega, E)$, where a_c is the greatest solution of equation $u = \inf\{c, Gu\}$ and b_c is the least solution of equation $u = \sup\{c, Gu\}$.*

The next result is a special case of Theorems 5.3 and 5.4.

Proposition 6.3. *Assume that E is a lattice-ordered Banach space with properties (E0) and (E1), and that $G : L^p(\Omega, E) \to L^p(\Omega, E)$, $1 \leq p \leq \infty$, is an increasing mapping which satisfies the following hypothesis.*

(G2) *There exists a bounded, linear and positive operator $T : L^p(\Omega, \mathbf{R}) \to L^p(\Omega, \mathbf{R})$ with spectral radius $\rho(T) < 1$ and a function $v \in L^p(\Omega, \mathbf{R})$ such that (denoting $\|u\| = t \mapsto \|u(t\|$)*

$$\|Gu(t)\| \leq T\|u\|(t) + v(t) \quad \textit{for a.e. } t \in \Omega \textit{ and for all } \; u \in L^p(\Omega, E).$$

Then the operator G has

(a) least and greatest fixed points u_- and u_+ in $[a_-, b_+]$, where a_- is the greatest solution of equation $u = -(Gu)^-$, and b_+ is the least solution of equation $u = (Gu)^+$;

(b) minimal and maximal fixed points.

Moreover, the fixed points u_- and u_+ are increasing with respect to G.

Proof. Replacing the valuation of E by its norm in the first part of the proof of Proposition 6.1, we see that

(i) equation (6.2), has a unique solution $b \in L^p(\Omega, \mathbf{R})$;

(ii) if $\|u(t)\| \leq b(t)$ for a.e. $t \in \Omega$, then $\|Gu(t)\| \leq b(t)$ for a.e. $t \in \Omega$;

(iii) if u is a fixed point of G, then $\|u(t)\| \leq b(t)$ for a.e. $t \in \Omega$.

The result (ii) implies that $G[P] \subset P$, where

$$P = \{u \in L^p(\Omega, E) \mid \|u(t)\| \leq b(t) \text{ for a.e. } t \in \Omega\},$$

and that $\|Gu\|_p \leq \|b\|_p$ for all $u \in P$. Thus $G[P]$ is a bounded subset of P.

The above proof shows that the hypotheses of Theorem 5.3 are satisfied when $L = I$ and $N = G$, whence the assertions follow from Theorems 5.3 and 5.4 and from the result (iii) above. □

Consider next the case when G is a self-mapping of a pointwise ordered space $C(X, E)$, where X is a topological space and E is an ordered normed space. As a consequence of Propositions 5.1–5.4 we obtain the following results.

Proposition 6.4. *Assume that X is a topological space, that E is a weakly complete Banach lattice, and that $G : C(X, E) \to C(X, E)$ is an increasing mapping, which satisfies the following hypothesis.*

(G3) *There exists a bounded, linear and positive operator $T : C(X, E) \to C(X, E)$ with spectral radius $\rho(T) < 1$ and a function $v \in C(X, E_+)$ such that*

$$|Gu(t)| \leq T|u|(t) + v(t) \quad \text{for all } t \in X \text{ and } u \in C(X, E).$$

(G4) *$G[-b, b]$ is an equicontinuous subset of $C(X, E)$, where $b \in C(X, E_+)$ is the solution of (6.2).*

Then the operator G least and greatest fixed points u_* and u^* in $C(X, E)$. Moreover, u_* and u^* are increasing with respect to G.

Proposition 6.5. *Let the hypotheses of Proposition 6.4 hold, and let $b \in C(X, E)$ be the solution of (6.2). Then for each $c \in [-b, b]$ the operator G has the least fixed point u_c and the greatest fixed point u^c in the order interval $[a_c, b_c]$ of $C(X, E)$, where a_c is the greatest solution of equation $u = \inf\{c, Gu\}$ and b_c is the least solution of equation $u = \sup\{c, Gu\}$.*

Proposition 6.6. *Let X be a separable topological space, E a lattice-ordered normed space with properties (E0) and (E1), and $G : C(X, E) \to C(X, E)$ an increasing mapping which satisfies the following hypothesis.*

(G5) *There exists a bounded, linear and positive operator $T : C(X, \mathbf{R}) \to C(X, \mathbf{R})$ with spectral radius $\rho(T) < 1$ and a function $v \in C(X, \mathbf{R})$ such that (denoting $\|u\| = t \mapsto \|u(t\|$)*

$$\|Gu(t)\| \leq T\|u\|(t) + v(t) \quad \text{for all } t \in X \text{ and } u \in C(X, E).$$

Then the operator G has

(a) least and greatest fixed points u_- and u_+ in $[a_-, b_+]$, where a_- is the greatest solution of equation $u = -(Gu)^-$, and b_+ is the least solution of equation $u = (Gu)^+$;

(b) minimal and maximal fixed points.

Moreover, the fixed points u_- and u_+ are increasing with respect to G.

Remarks 6.1. Ordered reflexive Banach spaces have property (E0). Examples of such spaces are, for instance, the Sobolev spaces $W^{1,p}(\Omega)$ and $W^{1,p}(\Omega)$, ordered a.e. pointwise, where $1 < p < \infty$ and Ω is a bounded domain in $\mathbf{R}^m$. These spaces satisfy also the properties listed in (E2), and hence those of (E1) (cf. [2]).

All the properties imposed on the space E in this paper hold if E is a weakly complete Banach lattice. Examples of such spaces are, e.g., UMB-lattices defined in [1, XV,14], the spaces $\mathbf{R}^m$, $m = 1, 2, \ldots$, and l^p, $p \in [1, \infty)$, ordered coordinatewise and normed by p-norm, and spaces $L^p(\Omega, \mathbf{R})$, where $p \in [1, \infty)$ and $\Omega = (\Omega, \mathcal{A}, \mu)$ is a measure space, equipped with p-norm and a.e. pointwise ordering. In particular, we can choose E to be one of these spaces in the above considerations, and get existence and comparison results for finite and infinite systems of differential equations, and also random differential equations when μ above is a probability measure.

In [10, Section 5.8] there exists a number of further examples of ordered function spaces to which the theory presented in Section 2 could be applied.

As for other existence and comparison results for operator equations in abstract spaces, see, e.g., [2–7, 13–18].

The results of Proposition 2.2 are applied in [8] to prove existence and comparison results for solutions of implicit discontinuous functional differential problems, e.g., of the form

$$\begin{cases} L_1 u(t) = N_1 u(t) := f_2(t, u, u(t), L_1 u(t)) \ \text{ a.e. in } \ J = [t_0, t_1], \\ L_2 u(t) = N_2 u(t) := B_2(t, u, L_2 u(t)) \ \text{ in } \ J_0 = [t_0 - r, t_0], \end{cases}$$

where

$$\begin{cases} L_1 u(t) := \frac{d}{dt}(\varphi(t)u(t)) - f_1(t, u, u(t), \frac{d}{dt}(\varphi(t)u(t))), \quad t \in J, \\ L_2 u(t) := u(t) - B_1(t, u, u(t)), \quad t \in J_0, \end{cases}$$

and their dependence on E-valued functions f_i and B_i, $i = 1, 2$, by assuming that $\varphi \in C(X, (0, \infty))$, and that E satisfies the hypotheses (E0) and (E2). In these applications P is a subset of the product of the spaces $L^1(J, E)$, $J := [t_0, t_1]$, and the space $C(J_0, E)$, $J_0 := [t_0 - r, t_0]$. This application offers a hint how one can generalize the results of Sections 5 and 6.

References

[1] G. Birkhoff, Lattice Theory, *Amer. Math. Soc. Publ.* **XXV**, Rhode Island, 1940.

[2] S. Carl and S. Heikkilä, Nonlinear Differential Equations in Ordered Spaces, *Chapman & Hall/CRC*, London, 2000.

[3] S. Carl and S. Heikkilä, On discontinuous implicit evolution equations, *J. Math. Anal. Appl.*, **219** (1998), 455–471.

[4] S. Carl and S. Heikkilä, Operator and differential equations in ordered spaces, *J. Math. Anal. Appl.*, **234** (1999), 31–54.

[5] S. Heikkilä, On well-ordered sets in ordered function spaces, *Proceedings of FSDONA-99, Acad. Sci. Czech Repub., Prague* (2000), 125–132.

[6] S. Heikkilä, New algorithms to solve equations and systems in ordered spaces, *Neural, Parallel & Scientific Computations*, **9** (2001), 407–416.

[7] S. Heikkilä, New iterative methods to solve equations and systems in ordered spaces, *Nonlinear Anal.*, **51** (2002), 1233–1244.

[8] S. Heikkilä, Existence and comparison results for operator and differential equations in abstract spaces, *J. Math. Anal. Appl.*, **274**, 2 (2002), 586–607.

[9] S. Heikkilä, On the existence of minimal and maximal solutions of operator equations in posets, *Manuscript* (2002).

[10] S. Heikkilä and V. Lakshmikantham, Monotone Iterative Techniques for Discontinuous Nonlinear Differential Equations, *Marcel Dekker*, New York, 1994.

[11] S. Lang, Real and Functional Analysis, Springer-Verlag, Berlin, 1993.

[12] J. Lindenstrauss and L. Tzafriri, Classical Banach Spaces II, Function Spaces, *Springer-Verlag*, Berlin, 1979.

[13] E. Liz, Monotone iterative techniques in ordered Banach spaces, *Nonlinear Anal.*, **30** (1997), 5179–5190.

[14] E. Liz, Abstract monotone iterative techniques and applications to impulsive differential equations, *Dynam. Contin. Discrete Impuls. Systems*, **30** (1997), 443–452.

[15] E. Liz and J. J. Nieto, An abstract monotone iterative method and applications, *Dynam. Systems Appl.*, **7** (1998), 365–376.

[16] J. J. Nieto, An abstract monotone iterative technique, *Nonlinear Anal.*, **20**(2) (1997), 1923–1933.

[17] J. J. Nieto and A. Torres, Approximation of solutions for nonlinear problems with an application to the study of aneurysms of the circle of Willis, *Nonlinear Anal.*, **40** (2000), 513–521.

[18] V. Seda, Monotone iterative technique for decreasing mappings, *Nonlinear Anal.*, **40** (2000), 577–588.

[19] K. Yoshida, Functional Analysis, *Springer-Verlag*, Berlin, 1974.

[20] E. Zeidler, Nonlinear Functional Analysis and its Applications I, *Springer-Verlag*, Berlin, 1985.

Hereditary Symmetry of Resolving Systems for Nonlinear Equations with Fredholm Operators

B. Karasözen[1], I. Konopleva[2] and B. Loginov[2]

[1]Department of Mathematics, Middle East Technical University, 06531 Ankara, Turkey e–mail: bulent@metu.edui.tr

[2] Department of Mathematics, Ulyanovsk State Technical University, Ulyanovsk, 432027, Russia e–mail:i.konopleva@ulstu.ru, loginov@ulstu.ru

Abstract. The notion of the equivalent finite-dimensional resolving system, degeneralizing the bifurcation equation is introduced for nonlinear equations in Banach spaces with Fredholm operator at the derivative. The equivalence of Lyapounov and Schmidt resolving systems for stationary equations is established. For resolving systems the inheritance of intertwining and symmetry properties of the original nonlinear equation are proved. Applications of resolving systems theory to linear problems (Cauchy problem for linear differential equation and perturbation of the linear equation by a small linear term), analogs of Grobman-Hartman theorem for differential equations with a degenerate operator at the derivative and Andronov-Hopf bifurcation under cosymmetry conditions are given.

AMS Subject Classification. 58E09, 34G20, 37D10

Key words. Bifurcation theory, Lyapunov-Schmidt construction, resolving systems, branching equation in the root subspace, hereditary symmetry, Grobman-Hartman theorem, Andronov-Hopf bifurcation, cosymmetry.

1 Introduction

Let E_1, E_2 be Banach spaces and $A : E_1 \supset D_A \to E_2$ and $B : E_1 \supset D_B \to E_2$ be densely defined closed linear Fredholm operators. If $D_B \subset D_A$, then A is subordinated to B (i.e. $\| Ax \| \leq \| Bx \| + \| x \|$ on D_B), if $D_A \subset D_B$, B is subordinated to A (i.e. $\| Bx \| \leq \| Ax \| + \| x \|$ on D_A). We consider the differential equation

$$(1.1) \qquad A\frac{dx}{dt} = Bx + f(x,t), \quad \|f(x,t)\| = o(\|x\|), \quad \|x\| \to 0$$

with initial value $x(0) = 0$ or periodicity conditions ([6]-[9],[19]). The nontriviality of the zero-subspaces $\mathcal{N}(A) = \text{span}\{\phi_1, \dots, \phi_m\}$ or $\mathcal{N}(B) = \text{span}\{\varphi_1, \dots, \varphi_n\}$ and of the defect-subspaces $\mathcal{N}^*(A) = \text{span}\{\widehat{\psi}_1, \dots, \widehat{\psi}_m\}$, $\mathcal{N}^*(B) = \text{span}\{\psi_1, \dots, \psi_n\}$ is assumed with $\mathcal{N}(A) \cap \mathcal{N}(B) = \{0\}$. The corresponding bi-orthogonal systems are introduced $\{\vartheta_j\}_1^m \in E_1^*$, $\langle \phi_i, \vartheta_j \rangle = \delta_{ij}$, $\{\zeta_j\}_1^m \in E_2$, $\left\langle \zeta_i, \widehat{\psi}_j \right\rangle = \delta_{ij}$, $\{\gamma_j\}_1^n \in E_1^*$, $\langle \varphi_i, \gamma_j \rangle = \delta_{ij}$, $\{z_j\}_i^n \in$

E_2, $\langle z_i, \psi_j \rangle = \delta_{ij}$. It means that we study the nonstationary (dynamical) equation (1.1) with the degenerate Fredholm operator A at the derivative and the stationary equation $Bx + f(x) = 0$ with the degenerate Fredholm operator B with the aim to reduce these problems to finite-dimensional resolving systems (RS), which inherit the symmetry of the original problems. For the stationary case, the branching equation in the root subspace (BEqR) arises as the resolving system [7]-[11]. In the nonstationary case this approach allows the investigation of the Andronov-Hopf bifurcation [10] i.e., the periodic solutions of equation (1.1). We will make use of the results in [5]- [16] systematically in this communication. Together with the group symmetry for equation (1.1), the properties of intertwining operators, the so-called non-group symmetry [1], are also investigated. We study also some other aspects of equations of the type (1): the Grobman-Hartman theorem, the notion of center manifold and Andronov-Hopf bifurcation under cosymmetry conditions. Some parts of this work were presented at international conferences [12-13].

The authors are thankful to Professor V. S. Mokeychev for his useful advise concerning to operators subordinateness definition.

For the convenience of the reader we give here some auxiliary results from our previous works [5],[10].

Definition 1.1. *The elements* $\phi_i^{(s)}$, $s = 1, \ldots, q_i$, $\phi_i^{(1)} = \phi_i$, $i = 1, \ldots, m$ ($\varphi_i^{(s)}$, $s = 1, \ldots, p_i$, $\varphi_i^{(1)} = \varphi_i$, $i = 1, \ldots, n$) *form the complete canonical generalized Jordan set (GJS ≡ B-JS) relative to the operator-function* $A - \varepsilon B$ *(correspondingly* $B - \mu A$*) if*

$$\begin{aligned} &A\phi_i^{(s)} = B\phi_i^{(s-1)}, \ \left\langle \phi_i^{(s)}, \vartheta_j \right\rangle = 0, s = 2, \ldots q_i, \ i, j = 1, \ldots, m; \\ (1.2) \qquad &(B\varphi_i^{(s)} = A\varphi_i^{(s-1)}, \ \left\langle \varphi_i^{(s)}, \gamma_j \right\rangle = 0, \ s = 2, \ldots, p_i, i, j = 1, \ldots, n) \\ &D_q \equiv \det \left[\left\langle B\phi_i^{(q_i)}, \widehat{\psi}_j \right\rangle \right] \neq 0, \quad (D_p \equiv \det \left[\left\langle A\varphi_i^{(p_i)}, \psi_j \right\rangle \right] \neq 0). \end{aligned}$$

This GJS is called bi-canonical if the corresponding B^*-JS (A^*-JS) of the adjoint operator A^* (B^*) is also canonical.

The conditions in (1.2) determine the B-JS (A-JS) uniquely. Its elements are linearly independent and form a basis for the root-subspace $K(A;B)$ ($K(B;A)$) of the Fredholm point $\lambda = 0 \in \sigma_B(A)$ ($\mu = 0 \in \sigma_A(B)$) of the operator-function $A - \lambda B$ ($B - \mu A$), where $k_A = \dim K(A;B) = \sum_{i=1}^{m} q_i$ ($k_B = \dim K(B;A) = \sum_{i=1}^{n} p_i$) is called the root-number of the Fredholm point.

Lemma 1.1. *[11,16,19] Elements of B and B^*-Jordan sets (A-and A^*-Jordan sets) of the operator-functions $A - \lambda B$ and $A^* - \lambda B^*$ ($B - \mu A$ and $B^* - \mu A^*$)*

can be chosen so that the following bi-orthogonality conditions are satisfied

$$\begin{aligned}\left\langle \phi_i^{(j)}, \vartheta_k^{(l)} \right\rangle &= \delta_{ik}\delta_{jl},\ \left\langle \zeta_i^{(j)}, \widehat{\psi}_k^{(l)} \right\rangle = \delta_{ik}\delta_{jl},\ j(l) = 1, \dots, q_i(q_k), \\ \vartheta_k^{(l)} &= B^*\widehat{\psi}_k^{(q_k+1-l)},\ \zeta_i^{(j)} = B\phi_i^{(q_i+1-j)},\ i, k = 1, \dots, m\end{aligned} \tag{1.3}$$

$$\begin{aligned}\left\langle \varphi_i^{(j)}, \gamma_k^{(l)} \right\rangle &= \delta_{ik}\delta_{jl},\ \left\langle z_i^{(j)}, \psi_k^{(l)} \right\rangle = \delta_{ik}\delta_{jl},\ j(l) = 1, \dots, p_i(p_k), \\ \gamma_k^{(l)} &= A^*\psi_k^{(p_k+1-l)},\ z_i^{(j)} = A\varphi_i^{(p_i+1-j)},\ i, k = 1, \dots, n.\end{aligned} \tag{1.4}$$

The relations ((1.3)-(1.4)) allow us to introduce the projectors [7, 8]

$$\begin{aligned}\mathbf{p} &= \sum_{i=1}^{m}\sum_{j=1}^{q_i} \left\langle \cdot, \vartheta_i^{(j)} \right\rangle \phi_i^{(j)} = \langle \cdot, \vartheta \rangle \phi : E_1 \to E_1^{k_A} = K(A, B), \\ \mathbf{q} &= \sum_{i=1}^{m}\sum_{j=1}^{q_i} \left\langle \cdot, \widehat{\psi}_i^{(j)} \right\rangle \zeta_i^{(j)} = \left\langle \cdot, \widehat{\psi} \right\rangle \zeta : E_2 \to E_{2,k_A} = \operatorname{span}\{\zeta_i^{(j)}\},\end{aligned} \tag{1.5}$$

$$\begin{aligned}\mathbf{P} &= \sum_{i=1}^{n}\sum_{j=1}^{p_i} \left\langle \cdot, \gamma_i^{(j)} \right\rangle \varphi_i^{(j)} = \langle \cdot, \gamma \rangle \varphi : E_1 \to E_1^{k_B} = K(B; A), \\ \mathbf{Q} &= \sum_{i=1}^{n}\sum_{j=1}^{p_i} \left\langle \cdot, \psi_i^{(j)} \right\rangle z_i^{(j)} = \langle \cdot, \psi \rangle z : E_2 \to E_{2,k_B} = \operatorname{span}\{z_i^{(j)}\}\end{aligned} \tag{1.6}$$

(where $\phi = (\phi_1^{(1)}, \cdots, \phi_1^{(q_1)}, \cdots, \phi_m^{(1)}, \cdots, \phi_m^{(q_m)})$, and the vectors ϑ, $\widehat{\psi}$, ζ, φ, γ, ψ, z are defined analogously) generating the following direct sums expansions

$$E_1 = E_1^{k_A} \dot{+} E_1^{\infty - k_A}, \quad E_2 = E_{2,k_A} \dot{+} E_{\infty - k_A}, \tag{1.7}$$

$$E_1 = E_1^{k_B} \dot{+} E_1^{\infty - k_B}, \quad E_2 = E_{2,k_B} \dot{+} E_{2,\infty - k_B}. \tag{1.8}$$

The intertwining relations are realized

$$A\mathbf{p} = \mathbf{q}A \text{ on } D_A, \quad B\mathbf{p} = \mathbf{q}B \text{ on } D_B, \tag{1.9}$$

$$(B\mathbf{P} = \mathbf{Q}B \text{ on } D_B\ , A\mathbf{P} = \mathbf{Q}A \text{ on } D_A), \tag{1.10}$$

$$A\phi = \mathfrak{A}_A\zeta,\ B\phi = \mathfrak{A}_B\zeta,\ B^*\widehat{\psi} = \mathfrak{A}_B\vartheta, \tag{1.11}$$

$$(B\varphi = \mathcal{A}_B z,\ A\varphi = \mathcal{A}_A z,\ A^*\psi = \mathcal{A}_A\gamma), \tag{1.12}$$

with cell-diagonal matrices $\mathfrak{A}_A = (A_1, \ldots, A_m)$, $\mathfrak{A}_B = (B_1, \ldots, B_m)$ ($\mathcal{A}_B = (B^1, \ldots, B^n)$, $\mathcal{A}_A = (A^1, \ldots, A^n)$) where the $q_i \times q_i$-cells ($p_i \times p_i$-cells) have the forms

$$A_i = \begin{pmatrix} 0 & 0 & 0 & \ldots & 0 & 0 \\ 0 & 0 & 0 & \ldots & 0 & 1 \\ \vdots & \vdots & \vdots & \ddots & \vdots & \vdots \\ 0 & 0 & 1 & \ldots & 0 & 0 \\ 0 & 1 & 0 & \ldots & 0 & 0 \end{pmatrix}, \quad B_i = \begin{pmatrix} 0 & 0 & 0 & \ldots & 0 & 1 \\ 0 & 0 & 0 & \ldots & 1 & 0 \\ \vdots & \vdots & \vdots & \ddots & \vdots & \vdots \\ 0 & 1 & 0 & \ldots & 0 & 0 \\ 1 & 0 & 0 & \ldots & 0 & 0 \end{pmatrix}$$

(B^i's have the same form as the A_i's, correspondingly A^i's have also the same form as the B_i's). The following relations for the operators A and B are valid

$$\begin{gathered} \mathcal{N}(A) \subset E_1^{k_A},\ AE_1^{k_A} \subset E_{2,k_A},\ A(E_1^{\infty-k_A} \cap D_A) \subset E_{2,\infty-k_A}, \\ \mathcal{N}(B) \subset E_1^{\infty-k_A},\ BE_1^{k_A} \subset E_{2,k_A},\ B(E_1^{\infty-k_A} \cap D_B) \subset E_{2,\infty-k_A}. \end{gathered} \tag{1.13}$$

The mappings $B : E_1^{k_A} \to E_{2,k_A}$, $\overset{\sqcap}{A} = A : E_1^{\infty-k_A} \cap D_A \to E_{2,\infty-k_A}$ are one-to-one. Analogously, the operators B and A act as invariant pairs of subspaces $E_1^{k_B}$, E_{2,k_B} and $E_1^{\infty-k_B}$, $E_{2,\infty-k_B}$, further $\overset{\sqcap}{B} = B : D_B \cap E_1^{\infty-k_B} \to E_{2,\infty-k_B}$, $A : E_1^{k_B} \to E_{2,k_B}$ are isomorphisms.

In fact, the expansions ((1.7)-(1.8)) and the intertwining equalities ((1.9)-(1.12)) follow from the definitions ((1.5)-(1.6)) of the projectors **p** and **q** (**P** and **Q**) and the Lemma 1.1. The inclusions in (1.13) and their duals can be verified directly from the relations ((1.9)-(1.10)). The Jordan sets satisfying bi-orthogonality conditions ((1.3)-(1.4)) are called three-canonical sets. They possibly do not exist in the general case of analytical (polynomial) operator-functions with a spectral parameter [9]. However, in this communication all assertions obtained for linear operator-functions and connected with three-canonical GJSs remain true for the analytical case under the conditions of Jordan sets three-canonicity.

2 Construction of resolving systems

For the construction of the RSs we will use Lyapounov or Schmidt procedures [19]. Schmidt's approach in application to the dynamical case (1.1) was considered by N. A. Sidorov [17] and the problem (1.1) was reduced to a system of differential equations of infinite order or to a system of Volterra integral equations. Therefore we will start first with Lyapounov's approach [19].

Let us write the equation (1.1) as a projection on the subspaces (1.7) by setting $\xi_{is} = \left\langle x, \vartheta_i^{(s)} \right\rangle$ and $x = \xi \cdot \phi + w = \sum_{i=1}^{m} \sum_{s=1}^{q_i} \left\langle x, \vartheta_i^{(s)} \right\rangle \phi_i^{(s)} + w$,

$w \in E_1^{\infty-k_A}$. After the application of the projectors $(\mathbf{I}-\mathbf{q})$ and $\mathbf{q}$ the following system arises

$$\overset{\sqcap}{A}\frac{dw}{dt} = \overset{\sqcap}{B} w + (\mathbf{I}-\mathbf{q})f(\xi\cdot\phi + w, t), \tag{2.1}$$

$$A_i\frac{d\xi_i}{dt} = B_i\xi_i + \mathbf{q}_i f(\xi\cdot\phi + w, t), \tag{2.2}$$

$$\xi_i = (\xi_{i1},\dots,\xi_{iq_i}),\ \mathbf{q}_i = \sum_{j=1}^{q_i}\left\langle \cdot, \widehat{\psi}_i^{(j)}\right\rangle \zeta_i^{(j)},\ i=1,\dots,m.$$

The vectors $\xi = (\xi_{11},\dots,\xi_{1q_1},\dots,\xi_{m1},\dots,\xi_{mq_m})$ are considered as parameters, upon which depend the solution of the equation (2.1), i.e. $w = w(\xi,t)$. In this communication *the unique solvability question of the equation (2.1), is not discussed, it will be only assumed.* Substituting the solution $w = w(\xi,t)$ of the first equation (2.1) into the second one and setting $f_{ij}(\xi,t) \equiv < f(\xi\cdot\phi + w(\xi,t)), \widehat{\psi}_i^{(q_i-j+1)} >$, we can write the equation (2.2) in the form of the following system $\mathcal{F}(\xi,t)=0$:

$$0 = \xi_{iq_i} + f_{iq_i}(\xi,t),\ i=1,\dots,m, \tag{2.3}$$

$$\begin{cases} \dot{\xi}_{iq_i} = \xi_{iq_i-1} + f_{iq_i-1}(\xi,t), \\ \dots\dots\dots\dots\dots\dots\dots\dots\dots\dots \\ \dot{\xi}_{i2} = \xi_{i1} + f_{i1}(\xi,t), \quad i=1,\dots,m. \end{cases} \tag{2.4}$$

The differential-algebraic system ((1.16)-(1.17)) will be called the resolving system (RS) for the nonlinear problem (1.1).

Now we apply Lyapounov's and Schmidt's approaches to a stationary problem with a small parameter $\varepsilon \in \mathbb{C}^1$

$$Bx = \varepsilon Ax + R(x,\varepsilon), \quad R(0,\varepsilon)=0, \quad \|R(x,\varepsilon)\| = o(\|x\|). \tag{2.5}$$

Setting $x = u+v$, $v \in E_1^{k_B}$, $u \in E_1^{\infty-k_B}$ and taking into account that $\mathbf{Q}Bu = 0$, $u \in D_B$; $\mathbf{Q}Au = 0$, $u \in E_1^{\infty-k_B}\cap D_A$; $(I-\mathbf{Q})Av = 0$, one finds that the equation

$$(I-\mathbf{Q})Bu - \varepsilon(I-\mathbf{Q})Au = (I-\mathbf{Q})R(u+v,\varepsilon) \tag{2.6}$$

defines an isomorphism $u = u(v,\varepsilon)$ of a sufficiently small neighborhood ω of the point $v=0$, $\varepsilon=0$ from $E_1^{k_B}\dot{+}\mathbb{C}^1$ into small neighborhood $\Omega \subset E_1^{\infty-k_B}$ of the point $u=0$. After substituting $u(v,\varepsilon)$ into the $\mathbf{Q}$-projected equation (2.5), one obtains the BEqR as RS

$$\mathbf{Q}Bv = \varepsilon\mathbf{Q}Av + \mathbf{Q}R(u(v,\varepsilon)+v,\varepsilon) = 0 \tag{2.7}$$

and in the coordinate form as

$$\tau(\xi,\varepsilon) \equiv (\mathcal{A}_B - \varepsilon \mathcal{A}_A)\xi - \mathbf{Q}R(u(\xi,\varepsilon) + \xi\cdot\varphi,\varepsilon) = 0 \tag{2.8}$$

or

$$\begin{cases} 0 = \varepsilon\xi_{ip_i} + \left\langle R(u(\xi,\varepsilon)+\xi\cdot\varphi,\varepsilon), \psi_i^{(1)}\right\rangle, \\ \xi_{ip_i} = \varepsilon\xi_{ip_i-1} + \left\langle R(u(\xi,\varepsilon)+\xi\cdot\varphi,\varepsilon), \psi_i^{(2)}\right\rangle, \\ \cdots\cdots\cdots\cdots\cdots\cdots\cdots\cdots \\ \xi_{i3} = \varepsilon\xi_{i2} + \left\langle R(u(\xi,\varepsilon)+\xi\cdot\varphi,\varepsilon), \psi_i^{(p_i-1)}\right\rangle, \\ \xi_{i2} = \varepsilon\xi_{i1} + \left\langle R(u(\xi,\varepsilon)+\xi\cdot\varphi,\varepsilon), \psi_i^{(p_i)}\right\rangle. \end{cases} \tag{2.9}$$

Let us consider the BEqR based on Schmidt's lemma. According to Lemma 1.1 and generalized Schmidt's lemma in [19], there exists a bounded operator $\Gamma = \widehat{B}^{-1}$, where

$$\widehat{B} = \overset{\sqcap}{B} \dot{+} \sum_{i=1}^{n}\sum_{s=1}^{p_i}\left\langle \cdot, A^*\psi_i^{(p_i+1-s)}\right\rangle A\varphi_i^{(s-1)} = B + V = B + \sum_{i=1}^{n}\left\langle \cdot, \gamma_i^{(1)}\right\rangle z_i^{(1)},$$

and $\varphi_i^{(0)} \overset{\text{def}}{=} \varphi_i^{(p_i)}$, $\varphi_i^{(p_i+1)} = \varphi_i^{(1)}$, $A\varphi_i^{(0)} = A\varphi_i^{(p_i)} = z_i^{(1)}$, $\overset{\sqcap}{B} = B|_{E_1^{\infty-k_B}} :$ $E_1^{\infty-k_B} \to E_{2,\infty-k_B}$. Therefore the equation (1.1) may be represented again in the form

$$\widehat{B}\, x = \left[\overset{\sqcap}{B} \dot{+} \sum_{i=1}^{n}\sum_{s=1}^{p_i}\left\langle \cdot, \gamma_i^{(s)}\right\rangle z_i^{(p_i+2-s)}\right] x = \varepsilon A x + R(x,\varepsilon) + \sum_{i=1}^{n}\xi_{i1} z_i^{(1)}, \tag{2.10}$$

$$\xi_{is} = \left\langle x, \gamma_i^{(s)}\right\rangle, \quad s = 1,\dots,p_i,\ i = 1,\dots,n. \tag{2.11}$$

The first equation (2.10) of this system can be solved for $x = (I-\varepsilon\Gamma A)^{-1}\Gamma R(x,\varepsilon)+$ $+\sum\limits_{i=1}^{n} \xi_{i1}(I-\varepsilon\Gamma A)^{-1}\varphi_i^{(1)}$. After substituting $x = w + v = w + \xi\cdot\varphi$, it follows

$$\widehat{B}\, w + \sum_{i=1}^{n}\sum_{s=2}^{p_i} \xi_{is} z_i^{(p_i+2-s)} = \varepsilon A w + \varepsilon A(w + \varepsilon\cdot\varphi) + R(w + \xi\cdot\varphi,\varepsilon),$$

that defines the isomorphism $w = w(\xi,\varepsilon)$ of sufficiently small neighborhood $\omega(0,0) \subset E_1^{k_B} \dot{+} \mathbb{C}^1$ into small neighborhood $\Omega(0) \subset E_1$ of the point $w = 0$. Application of the formulae

(2.12)

$$\Gamma^*\gamma_j^{(s)} = \Gamma^* A^* \psi_j^{(p_j+1-s)} = \psi_j^{(p_j+2-s)},$$

$$\varphi_j^{(\sigma)} = (\Gamma A)^{\sigma-1}\varphi_j^{(1)} = \varphi_j^{\left(\sigma - \left[\frac{\sigma}{p_j}\right]p_j\right)}, \quad \psi_i^{(\sigma)} = (\Gamma^* A^*)^{\sigma-1}\psi_j^{(1)} = \psi_j^{\left(\sigma - \left[\frac{\sigma}{p_j}\right]p_j\right)}$$

and the substitution $w = w(\xi, \varepsilon)$ into the equalities (2.11) give Schmidt's BEqR as RS in the following form [9,12,14]

(2.13)

$$\begin{cases} t_{js}(\xi,\varepsilon) = \xi_{js} - \frac{\varepsilon^{s-1}}{1-\varepsilon^{p_i}}\xi_{i1} - \left\langle (I - \varepsilon A\Gamma)^{-1} R(w(\xi,\varepsilon) + \xi\cdot\varphi, \varepsilon), \psi_j^{(p_j+2-s)} \right\rangle = 0, \\ t_{j1}(\xi,\varepsilon) \equiv -\frac{\varepsilon^{p_j}}{1-\varepsilon^{p_j}}\xi_{j1} - \left\langle (I - \varepsilon A\Gamma)^{-1} R(w(\xi,\varepsilon) + \xi\cdot\varphi, \varepsilon), \psi_j^{(1)} \right\rangle = 0, \end{cases}$$

$$s = 2, \ldots, p_j, \; j = 1, \ldots, n.$$

In contrast to our previous works [9,14] we continue here with further simplification of the BEqR (2.13) based on the generalized Jordan chains, using the adjoint operator-function(formulae (2.12) for $\psi_j^{(\sigma)}$)

$$t_{j1}(\xi,\varepsilon) = -\frac{1}{1-\varepsilon_j^p}\left[\varepsilon^{p_j}\xi_{j1} + \left\langle R(w(\xi,\varepsilon) + \xi\cdot\varphi,\varepsilon), \psi_1^{(1)} + \varepsilon\psi_j^{(2)} + \cdots + \varepsilon^{p_j-1}\psi_j^{(p_j)} \right\rangle\right] = 0.$$

Analogously for $s = 2$ we obtain

$$t_{j2}(\xi,\varepsilon) = \xi_{j2} - \frac{1}{1-\varepsilon^{p_j}}\left[\varepsilon\xi_{j1} + \left\langle R(w(\xi,\varepsilon) + \xi\cdot\varphi,\varepsilon), \psi_j^{(p_j)} + \varepsilon\psi_j^{(1)} + \right.\right.$$
$$\left.\left. + \cdots + \varepsilon^{p_j-1}\psi_j^{(p_j-1)} \right\rangle\right] = 0$$

and so on. As a result the following system arises

(2.14)

$$\begin{cases} \varepsilon^{p_j}\xi_{j1} + \left\langle R(w(\xi,\varepsilon) + \xi\cdot\varphi,\varepsilon), \psi_j^{(1)} + \varepsilon\psi_j^{(2)} + \cdots + \varepsilon^{p_j}\psi_j^{(p_j)} \right\rangle = 0, \\ (1-\varepsilon^{p_j})\xi_{j2} - \varepsilon\xi_{j1} - \\ - \left\langle R(w(\xi,\varepsilon) + \xi\cdot\varphi,\varepsilon), \psi_j^{(p_j)} + \varepsilon\psi_j^{(1)} + \cdots + \varepsilon^{p_j-1}\psi_j^{(p_j-1)} \right\rangle = 0, \\ (1-\varepsilon^{p_j})\xi_{j3} - \varepsilon^2\xi_{j1} - \\ - \left\langle R(w(\xi,\varepsilon) + \xi\cdot\varphi,\varepsilon), \psi_j^{(p_j-1)} + \varepsilon\psi_j^{(p_j)} + \varepsilon^2\psi_j^{(1)} + \cdots + \varepsilon^{p_j-1}\psi_j^{(p_j-2)} \right\rangle = 0, \\ \cdots\cdots\cdots \\ (1-\varepsilon^{p_j})\xi_{jp_j} - \varepsilon^{p_j-1}\xi_{j1} - \\ - \left\langle R(w(\xi,\varepsilon) + \xi\cdot\varphi,\varepsilon), \psi_j^{(2)} + \varepsilon\psi_j^{(3)} + \cdots + \varepsilon^{p_j-2}\psi_j^{(p_j)} + \varepsilon^{p_j-1}\psi_j^{(1)} \right\rangle = 0. \end{cases}$$

Multiplying the first equation by ε and adding it to the second one, and multiplying the second equation by ε and subtracting it from the third one and so on. Then after crossing out the factor $(1 - \varepsilon^{p_j})$ one obtains the equivalent system

$$\varepsilon^{p_j}\xi_{j1} + \left\langle R(w(\xi,\varepsilon)+\xi\cdot\varphi,\varepsilon), \psi_j^{(1)} + \varepsilon\psi_j^{(2)} + \cdots + \varepsilon^{p_j-1}\psi_j^{(p_j)}\right\rangle = 0, \tag{2.15}$$

$$\left\{\begin{array}{l} \xi_{j2} - \varepsilon\xi_{j1} - \left\langle R(w(\xi,\varepsilon)+\xi\cdot\varphi,\varepsilon), \psi_j^{(p_j)}\right\rangle = 0, \\ \xi_{j3} - \varepsilon\xi_{j2} - \left\langle R(w(\xi,\varepsilon)+\xi\cdot\varphi,\varepsilon), \psi_j^{(p_j-1)}\right\rangle = 0, \\ \cdots\cdots\cdots\cdots\cdots\cdots\cdots\cdots\cdots\cdots \\ \xi_{jp_j-1} - \varepsilon\xi_{jp_j-2} - \left\langle R(w(\xi,\varepsilon)+\xi\cdot\varphi,\varepsilon), \psi_j^{(3)}\right\rangle = 0, \\ \xi_{jp_j} - \varepsilon\xi_{jp_j-1} - \left\langle R(w(\xi,\varepsilon)+\xi\cdot\varphi,\varepsilon), \psi_j^{(2)}\right\rangle = 0. \end{array}\right. \tag{2.16}$$

Multiplying the first equation (2.16) by ε^{p_j-1} and add it to (2.15)

$$\varepsilon^{p_j-1}\xi_{j2} + \left\langle R(w(\xi,\varepsilon)+\xi\cdot\varphi,\varepsilon), \psi_j^{(1)} + \varepsilon\psi_j^{(2)} + \varepsilon^{p_j-2}\psi_j^{(p_j-1)}\right\rangle = 0$$

and multiplying the second equation (2.16) by 1, the first one by ε and adding them and multiplying the resulting equation by ε^{p_j-2} and adding it to (2.15) we get

$$\varepsilon^{p_j-2}\xi_{j3} + \left\langle R(w(\xi,\varepsilon)+\xi\cdot\varphi,\varepsilon), \psi_j^{(1)} + \varepsilon\psi_j^{(2)} + \varepsilon^{p_j-3}\psi_j^{(p_j-2)}\right\rangle = 0$$

Multiplying the third equation (2.16) by 1, the second one by ε, the first one by ε^2 and adding them and multiplying the resulting equation by ε^{p_j-3} and adding to (2.15) we get

$$\varepsilon^{p_j-3}\xi_{j4} + \left\langle R(w(\xi,\varepsilon)+\xi\cdot\varphi,\varepsilon), \psi_j^{(1)} + \varepsilon\psi_j^{(2)} + \cdots + \varepsilon^{p_j-4}\psi_j^{(p_j-3)}\right\rangle = 0.$$

Continuing the analogous operations gives the equivalent system

(2.17)
$$\left\{\begin{array}{l} \varepsilon^{p_j}\xi_{j1} + \left\langle R(w(\xi,\varepsilon)+\xi\cdot\varphi,\varepsilon), \psi_j^{(1)} + \varepsilon\psi_j^{(2)} + \cdots + \varepsilon^{p_j-1}\psi_j^{(p_j)}\right\rangle = 0, \\ \varepsilon^{p_j-1}\xi_{j2} + \left\langle R(w(\xi,\varepsilon)+\xi\cdot\varphi,\varepsilon), \psi_j^{(1)} + \varepsilon\psi_j^{(2)} + \cdots + \varepsilon^{p_j-2}\psi_j^{(p_j-1)}\right\rangle = 0, \\ \varepsilon^{p_j-2}\xi_{j3} + \left\langle R(w(\xi,\varepsilon)+\xi\cdot\varphi,\varepsilon), \psi_j^{(1)} + \varepsilon\psi_j^{(2)} + \cdots + \varepsilon^{p_j-3}\psi_j^{(p_j-2)}\right\rangle = 0, \\ \cdots\cdots\cdots\cdots\cdots\cdots\cdots\cdots\cdots\cdots\cdots\cdots \\ \varepsilon^2\xi_{p_j-1} + \left\langle R(w(\xi,\varepsilon)+\xi\cdot\varphi,\varepsilon), \psi_j^{(1)} + \varepsilon\psi_j^{(2)}\right\rangle = 0, \\ \varepsilon\xi_{jp_j} + \left\langle R(w(\xi,\varepsilon)+\xi\cdot\varphi,\varepsilon), \psi_j^{(1)}\right\rangle = 0, \quad j = 1,\ldots,n. \end{array}\right.$$

From the system (2.15),(2.16) we can pass to (2.9), i. e. Lyapounov's BEqR. In fact, the system (2.16) coincides with the last $p_j - 1$ equations (2.9). The first equation (2.9) can be obtained by adding (2.15) to the first equation (2.16) multiplied by ε^{p_j-1}, the second one multiplied by ε^{p_j-2} and so on ..., the last $(p_j - 1)$-th equation multiplied by ε.

Remark 2.1. *The system (2.15),(2.16) obtained by equivalent transformations of the Schmidt's BEqR (2.13) is equivalent to Lyapounov's BEqR (2.9).*

Consequently, the BEqR's ((2.14)-(2.15)) and ((2.16)-(2.17)) as resolving systems for the stationary equation (2.5) are equivalent to the A.M. Lyapounov's RS (2.9), but in general they don't coincide.

3 Intertwining operators

Let's assume that there exist linear operators $L \in \mathcal{L}(E_1)$ and $K \in \mathcal{L}(E_2)$ intertwining operators A and B, i.e.

$$KA = AL, \quad KB = BL \tag{3.1}$$

on D_B (D_A) when A is subordinated to B (when B is subordinated to A).

These can be both separate operators, and in general not invertible, and some of their parametric families, in particular the representations L_g and K_g of a certain group G. We are interested in the transformation of the generalized Jordan chains under the actions of the operators L and K.

Let further everywhere the zero-subspaces $\mathcal{N}(A)$ and $\mathcal{N}(B)$ be invariant relative to the operator L and the ranges $\mathcal{R}(A)$ and $\mathcal{R}(B)$ be invariant subspaces relative to the operator K in previous assumptions on A and B. The last invariance implies the invariance of the zero-subspaces $\mathcal{N}(A^*)$ and $\mathcal{N}(B^*)$ relative to K^*.

Condition I *Assume there exist the direct supplements $E_1^{\infty-m}$ to $\mathcal{N}(A)$ and $E_1^{\infty-n}$ to $\mathcal{N}(B)$ invariant relative to L.*

According to Lemma 1.2 in [6] the Condition I can be replaced by the equivalent assumption that the bi-orthogonal systems $\{\vartheta_i\}_1^m \in E_1^*$ for the operator-function $A - \lambda B$ and $\{\gamma_i\}_1^n \in E_1^*$ for $B - \mu A$ form the bases of invariant subspaces relative to the operator L^*.

Let the transformation L in the invariant subspace $E_1^m = \operatorname{span}\{\phi_1, \ldots, \phi_m\}$ ($E_1^n = \operatorname{span}\{\varphi_1, \ldots, \varphi_n\}$) acts according to the formulae

$$\begin{aligned} L\phi_i &= \mathfrak{A}'\phi_i = \sum_{j=1}^{m} a_{ji}\phi_j, \ \mathfrak{A} = \|a_{ij}\|_{i,j}^m, \\ L\varphi_i &= \mathcal{A}'\varphi_i = \sum_{j=1}^{n} \alpha_{ji}\varphi_j, \ \mathcal{A} = \|\alpha_{ij}\|_{i,j=1}^n. \end{aligned} \tag{3.2}$$

Then for an arbitrary $\phi = \sum_{i=1}^{m} \xi_i\phi_i$ ($\varphi = \sum_{i=1}^{n} \eta_i\varphi_i$) the action of the operator

L on ϕ (φ) is equivalent to the transformation of its coordinates

$$\widetilde{\xi}_i = (\mathfrak{A}\xi)_i = \sum_{j=1}^{m} a_{ij}\xi_j \quad (\widetilde{\eta}_i = (\mathcal{A}\eta)_i = \sum_{j=1}^{n} \alpha_{ij}\eta_j). \tag{3.3}$$

Consequently Condition I implies

$$L\vartheta_i = \sum_{j=1}^{m} a_{ij}\vartheta_j \quad (L\gamma_i = \sum_{j=1}^{n} \alpha_{ij}\gamma_j). \tag{3.4}$$

Analogously the invariance of $\mathcal{R}(A)$ ($\mathcal{R}(B)$) relative to the operator K defines the matrices $\mathfrak{B}$ and $\mathcal{B}$

$$\mathfrak{B}\widehat{\psi}_i = \sum_{j=1}^{m} b_{ij}\widehat{\psi}_j \quad (\mathcal{B}\psi_i = \sum_{j=1}^{n} \beta_{ij}\psi_j). \tag{3.5}$$

Lemma 3.1. *Let the operators B and A in the equation (1.1) be intertwined ((3.1)) by the operators L and K, operator L be invertible and Condition I be fulfilled. Then the operator K is invertible in the ranges $\mathcal{R}(A)$ and $\mathcal{R}(B)$ of the operators A and B. Furthermore assume that there exists canonical B-JS relative to the operator-function $A - \lambda B$. Then the lengths of the B-Jordan chains of the elements ϕ, $L\phi \in \mathcal{N}(A)$ are equal (i.e. $q(L\phi) = q(\phi)$) and the operator L transforms the B-JCh of the element ϕ into B-JCh of $L\phi$. If the elements ϕ_i are enumerated in the JChs lengths in increasing order $q_i = \ldots = q_{i_1} < q_{i_1+1} = \ldots = q_{i_1+i_2} < \ldots < q_{i_1+\ldots+i_{k-1}+1} = \ldots = q_{i_1+\ldots+i_{k-1}+i_k}$ then for every s the B-JChs $\phi_i^{(s)}$ are transformed by the block-diagonal matrix, the blocks of which begin with the index i_s+1 in the (i_s+1)-th row.*

Proof: We prove first the invertibility of the operator K on $\mathcal{R}(A)$ and its invertibility on $\mathcal{R}(B)$ can be proved analogously. Let us assume that there exists $0 \neq y_0 \in \mathcal{R}(A)$ such that $Ky_0 = 0$. Then for some $x_0 \in D_A$ one has $Ax_0 = y_0$, $KAx_0 = 0$, whenever $ALx_0 = 0$ and $Lx_0 \in \mathcal{N}(A)$, which leads to a contradiction. □

Let's assume further, that element $\phi = \phi^{(1)}$ has B-JCh of the length $q(\phi)$. Then for the first step we get $A\phi^{(2)} = B\phi^{(1)}, \langle\phi^{(2)}, \vartheta_i\rangle = 0, i = 1, \ldots, m \Longrightarrow AL\phi^{(2)} = KA\phi^{(2)} = KB\phi^{(1)} = BL\phi^{(1)}$ and $\left\langle KB\phi^{(1)}, \widehat{\psi}_j\right\rangle = \left\langle B\phi^{(1)}, K^*\widehat{\psi}_j\right\rangle = 0$, $j = 1, \ldots, m$, since $\mathcal{N}^*(B)$ is invariant relative to K^*. Consequently $L\phi^{(2)} = (L\phi^{(1)})^{(2)}$. Analogously we obtain for the s-th step: $(L\phi)^k = L\phi^{(k)}$, $k = 2, \ldots, s$, i.e. $A(L\phi)^k = AL\phi^{(k)} = KA\phi^{(k)} = KB\phi^{(k-1)} = BL\phi^{(k-1)} = B(L\phi)^{(k-1)}$, $k = 2, \ldots, s$, and consequently the equation $Ax = B\phi^{(s)}$ has the unique solution $x = \phi^{(s+1)}$, $\langle\phi^{(s+1)}, \vartheta_k\rangle = 0$, $k = 1, \ldots, n$. Then

$$AL\phi^{(s+1)} = KA\phi^{(s+1)} = KB\phi^{(s)} = BL\phi^{(s)} = B(L\phi)^{(s)}$$

and also $\langle BL\phi^{(s)}, \vartheta_k \rangle = 0$, thus $L\phi^{(s+1)} = (L\phi)^{(s+1)}$. This Jordan chain breaks at the element $\phi^{(q)}$, i.e. $\left\langle B\phi^{(q)}, \widehat{\psi}_{j_0} \right\rangle \neq 0$ for some j_0, $\langle \phi^{(q)}, \vartheta_i \rangle = 0$, $i = 1, \ldots, m$ and the equation $Ax = B\phi^{(q)}$ has no solution. Since the operator K is invertible on $\mathcal{R}(B)$, then there exists an index j_0 such that $\left\langle BL\phi^{(q)}, \widehat{\psi}_{j_0} \right\rangle = \left\langle KB\phi^{(q)}, \widehat{\psi}_{j_0} \right\rangle \neq 0$, consequently $q(L\phi) = q(\phi)$.

The length of GJCh for an arbitrary element $\phi \in \mathcal{N}(A)$ is determined by the index of the first nonzero projection on the spans of the sets of the vectors ϕ_i with the equal GJChs lengths and the expansion of the $L\phi$ by the base $\{\phi_i\}_1^m$ is beginning with the indices of the corresponding set. Consequently the operator L defines on $K(A; B)$ of block-diagonal matrix of the indicated form. □

Corollary 3.1. *Under the conditions of Lemma 3.1 the transformation formulas for the coordinates ξ_{ij} have the form*

$$\widetilde{\xi}_{i_1+\ldots+i_\sigma+j,k} = \sum_{s=i_1+\ldots+i_\sigma+1}^{i_1+\ldots+i_\sigma+i_{\sigma+1}} a_{i_1+\ldots+i_\sigma+j,s}\xi_{sk}, \tag{3.6}$$
$$j = 1, \ldots, q_{\sigma+1}, \quad k = 1, \ldots, q_\sigma.$$

Remark 3.1. *The dual assertion to Lemma 3.1 relative to the operator-function $B - \mu A$ is also true.*

Lemma 3.2. *Let the operators L and K be invertible, the Jordan set of the operator-function $A - \lambda B$ $(B - \mu A)$ be three-canonical and Condition I be fulfilled. Then the Condition I for adjoint operator-function $A^* - \lambda B^*$ $(B^* - \mu A^*)$ is also fulfilled and $\mathfrak{B} = \mathfrak{A}$ $(\mathcal{B} = \mathcal{A})$.*

Indeed, in this situation the vectors $\{\zeta_i\}_i^m$ $(\{z_j\}_1^n)$ are transformed by the matrix $\mathfrak{A}'$ $(\mathcal{A}')$, i.e. their spans E_{2,k_A} (E_{2,k_B}) are K-invariant subspaces of E_2.

Corollary 3.2. *In the Lemmas 3.1 and 3.2, the action of the operator K on the subspace $E_{2,k_A} = span\{\zeta_1^{(1)}, \ldots, \zeta_1^{(q_1)}, \ldots, \zeta_{i_1}^{(1)}, \ldots, \zeta_{i_1}^{(q_1)}, \ldots\}$ is expressed* *by the following block-diagonal matrix transformation on* $E_1^{k_A}$

$$
(3.7) \qquad \widehat{\mathfrak{A}} = \begin{pmatrix}
a_{11} & \dots & 0 & \dots & a_{i_1 1} & \dots & 0 & 0 & \dots & 0 & \dots \\
\vdots & \ddots & \vdots & \dots & \vdots & \ddots & \vdots & 0 & \dots & 0 & \dots \\
0 & \dots & a_{11} & \dots & 0 & \dots & a_{i_1 1} & 0 & \dots & 0 & \dots \\
\dots & \dots & \dots & \ddots & \dots & \dots & \dots & \dots & \dots & \dots & \dots \\
a_{1 i_1} & \dots & 0 & \dots & a_{i_1 i_1} & \dots & 0 & 0 & \dots & 0 & \dots \\
\vdots & \ddots & \vdots & \dots & \vdots & \ddots & \vdots & 0 & \dots & 0 & \dots \\
0 & \dots & a_{1 i_1} & \dots & 0 & \dots & a_{i_1 i_1} & 0 & \dots & 0 & \dots \\
0 & \dots & 0 & \dots & 0 & \dots & 0 & a_{i_1+1 i_1+1} & \dots & 0 & \dots \\
\vdots & \ddots & \vdots & \dots & \vdots & \ddots & \vdots & 0 & \dots & 0 & \dots \\
0 & \dots & 0 & \dots & 0 & \dots & 0 & 0 & \dots & a_{i_1+1 i_1+1} & \dots \\
\dots & \dots & \dots & \dots & \dots & \dots & \dots & \dots & \dots & \dots & \dots
\end{pmatrix}
$$

where the matrix $\mathfrak{A} = \|a_{ij}\|_{i,j=1}^{m}$ *is defined in ((3.2)).*

An analogous assertion takes place for $E_{2,k_B} = \operatorname{span}\{z_1^{(1)}, \dots, z_1^{(p_1)}, \dots\}$. The proof is based on the relation $Kz_i^{(j)} = KB\phi_i^{(q_i+1-j)} = BL\phi_i^{(q_i+1-j)}$ and transformation formulae for $\phi_i^{(s)}$ under the action of the operator L (Lemma 3.1 and the Corollary 3.1 ((3.6))).

Lemma 3.3. *The sets* $\mathbf{L} = \{L \in \mathcal{L}(E_1)\}$ *and* $\mathbf{K} = \{K \in \mathcal{L}(E_2)\}$ *of invertible operators subordinated to equalities ((3.1)) are groups.*

In the rest of the communication we assume that the operators L and K are intertwining the operators A, B, f in the equations (1.1) and (2.5) and also that the operator L is invertible by satisfying Condition I.

Lemma 3.4. *We have the following relation*

$$
(3.8) \qquad K\mathbf{q} = \mathbf{q}K
$$

In fact, the operator K is invertible on $\mathcal{R}(B)$ and the projector $\mathbf{q}$ has as its range the subspace $E_{2,k_A} = \operatorname{span}\{\zeta_i^{(j)} = B\phi_i^{(q_i+1-j)}\}$. Therefore, for brevity, $K\mathbf{q}y = \sum_{i=1}^{m}\sum_{j=1}^{q_i} < y, \widehat{\phi}_i^{(j)} > K\zeta_i^{(j)} = \sum_{i=1}^{m}\sum_{j=1}^{q_i} < y, K^*\widehat{\psi}_i^{(j)} > \zeta_i^{(j)} = \mathbf{q}Ky.$
A detailed proof can be obtained by the application of the transformation matrix ((3.7)).

Theorem 3.1. *Let the operators* A, B *and* f *in (1.1) be intertwined by the operators* L *and* K, *let* L *be invertible and Condition I be satisfied. Then in the case of the equation (2.1) unique solvability, the resolving system (2.2) (or (2.3),(2.4)) inherits the intertwining property of the operators* L *and* K, *i.e. they intertwine the matrices* $diag\{A_1,\ldots,A_m\}$ *and* $diag\{B_1,\ldots,B_m\}$ *and the vector-function* $\mathbf{q}f$*. In other words the RS (2.3), (2.4) are intertwined by the matrix* $\widehat{\mathfrak{A}}$ *((3.7)).*

Proof: By virtue of the intertwining properties of operators the equation (1.1) has the solutions $x = w+v = w+\xi\cdot\phi$ and $Lx = Lw+\xi\cdot L\phi$. From the assumption of the unique solvability of (2.1), it follows that $Lw = w(Lv,t) = w(\mathfrak{A}\xi,t)$. Then according to the Lemma 3.4 and formulae ((3.6)) one obtains the desired assertion. The intertwining property of the RS $\mathcal{F}(\xi,t) = 0$ (2.3), (2.4) is expressed by the formula

$$\widetilde{\mathcal{F}}(\xi,t) = \widehat{\mathfrak{A}}\mathcal{F}(\xi,t) = \mathcal{F}(\widehat{\mathfrak{A}}\xi,t), \quad \widetilde{\xi} = \widehat{\mathfrak{A}}\xi,$$

where the matrix $\widehat{\mathfrak{A}}$ is defined in ((3.7)). □

Remark 3.2. *When the sets* $\mathbf{L}$ *and* $\mathbf{K}$ *form the continuous transformation groups this theorem allows the reduction of the RS (2.3), (2.4) (see [6,7]).*

Now we consider the stationary equation (2.5) with the L, K-intertwining property. Together with the solution $x = u+v$, $u \in E_1^{\infty-k_A}$, $v = \xi\cdot\varphi \in E_1^{k_A}$ it has also the solution $Lx = Lu + Lv$. Writing the system ((2.6)-(2.7)) for the solution Lx by using the unique solvability of (2.6) relative to $u \in E_1^{\infty-k_B}$ ($Lu \in E_1^{\infty-k_B}$), the isomorphism $u = u(v,\varepsilon)$ with the property $Lu = u(Lv,\varepsilon)$ is obtained. Analogously to ((3.8)) the following statements can be proved:

Lemma 3.5.

$$K\mathbf{Q} = \mathbf{Q}K \tag{3.9}$$

Theorem 3.2. *Let the operators* A, B *and* R *in the stationary equation (2.5) be intertwined by the pair* L, K*, and let the operator* L *be invertible and Condition I be satisfied. Then the Lyapounov's BEqR (2.8) inherits the intertwining property relative to the matrix* $\widehat{\mathcal{A}}$.

Proof: In fact, in (2.5) operators B, A and R are (L,K)–commuting and according to Lemma 3.5, $\mathbf{Q}K = K\mathbf{Q}$. Consequently by virtue of the Corollary of Lemma 3.2 the BEqR (2.8) is intertwined by the matrix $\widehat{\mathcal{A}}$:

$$\widetilde{\tau}(\xi,\varepsilon) = (\widehat{\mathcal{A}}\tau)(\xi,\varepsilon) = \tau(\widehat{\mathcal{A}}\xi,\varepsilon) = \tau(\widetilde{\xi},\varepsilon), \tag{3.10}$$

where $\widehat{\mathcal{A}}$ has the form ((3.7)) with components α_{ij}.

Remark 3.3.

- *For the case of non-three-canonicity of A-JS the corresponding BEqR is intertwined by the matrices* $\widehat{\mathcal{A}}$ *and* $\widehat{\mathcal{B}}$.

- *Theorem 3. 2 remains true for the more general equation*

$$Bx = A(\varepsilon)x + R(x,\varepsilon), \quad R(0,\varepsilon) = 0, \ D(A(\varepsilon)) = D(A),$$

$$\|R(x,\varepsilon)\| = o(\|x\|), \ \|x\| \to 0$$

under the conditions of existence of three–canonical A-JS for the operator-function $B - A(\varepsilon)$. Here only the equality $\mathbf{Q}A(\varepsilon)u = 0$, $u \in E_1^{\infty - k_A} \cup D_A$ is breaking (see [9]).

Theorem 3.3. *Under the conditions of Theorem 3.2 for equation (2.5) Schmidt's BEqRs (2.14); (2.15) and (2.16);(2.17) inherit the intertwining property ((3.10)) relative to the matrix $\widehat{\mathcal{A}}$.*

In this situation the operators L and K on subspaces $E_1^{k_B}$ and E_{2,k_B} are represented by the same matrix $\widehat{\mathcal{A}}$. Since $Lw(v,\varepsilon) = w(Lv,\varepsilon)$ by virtue of the operator K invertibility, the proof is making on the scheme of the Theorem 3.2.

Remark 3.4. *For the stationary equation (2.5) the case when BEqR inherits the group intertwining property, i. e. the intertwining operators L and K generate the continuous transformations groups is interesting. In a such situation the reduction by unknowns number is possible [6, 7]. However the reduction by the equations number is not always possible in general, because of the nonlinearity $R(x,\varepsilon)$. But here one has the reduction at once by Jordan chains. It is stipulated by the structure of BEqR ((2.9),(2.15),(2.16)) and ((2.17), 2).*

4 Linear problems

4.1 Cauchy problem for linear differential equation.

After the transformation of the unknown function $x = u + x_0$ the Cauchy problem

$$A\frac{dx}{dt} = Bx + f(t), \quad x(0) = x_0 \tag{4.1}$$

is reduced to the problem with homogeneous initial condition

$$A\frac{du}{dt} = Bu + f_1(t), \quad u(0) = 0. \tag{4.2}$$

Rewriting ((4.1)) by using the projections $u(t) = \xi(t)\cdot\phi + w(t)$, one obtains the Cauchy problem in $w(t)$

$$\overset{\sqcap}{A}\frac{dw}{dt} = \overset{\sqcap}{B} w + (I - \mathbf{q})f_1(t), \ w(0) = 0$$

and the system:

$$0 = \xi_{iq_i} + \left\langle f_1(t), \widehat{\psi}_i^{(1)} \right\rangle, \tag{4.3}$$

$$\begin{cases} \dot{\xi}_{iq_i} = \xi_{iq_i-1} + \left\langle f_1(t), \widehat{\psi}_i^{(2)} \right\rangle, \\ \dots\dots\dots\dots\dots\dots\dots\dots \\ \dot{\xi}_{i2} = \xi_{i1} + \left\langle f_1(t), \widehat{\psi}_i^{(q_i)} \right\rangle \end{cases} \tag{4.4}$$

with initial data $\xi_{is}(0) = 0$, $s = 1, \dots, q_i$, $i = 1, \dots, m$. From the solvability conditions it follows $\left\langle f_1(0), \widehat{\psi}_i^{(1)} \right\rangle = 0$.

Further $f(t)$ is assumed to be a sufficiently smooth (or analytic) function in a neighborhood of $t = 0$, i.e. $f_1(t) = a_0 + a_1 t + a_2 t^2 + \dots$, $a_0 = \alpha_0 + Bx_0$, $\alpha_0 = f(0)$ and therefore $\left\langle a_0, \widehat{\psi}_i^{(1)} \right\rangle = 0$. Then ((4.4)) gives successively,

(4.5)
$$\begin{cases} \xi_{i,q_i-1} = -\frac{d}{dt} \left\langle f_1(t), \widehat{\psi}_i^{(1)} \right\rangle - \left\langle f_1(t), \widehat{\psi}_i^{(2)} \right\rangle, \\ \dots\dots\dots\dots\dots\dots\dots\dots \\ \xi_{i,q_i-\sigma} = -\frac{d^\sigma}{dt^\sigma} \left\langle f_1(t), \widehat{\psi}_i^{(1)} \right\rangle - \frac{d^{\sigma-1}}{dt^{\sigma-1}} \left\langle f_1(t), \widehat{\psi}_i^{(2)} \right\rangle - \dots - \frac{d}{dt} \left\langle f_1(t), \widehat{\psi}_i^{(\sigma)} \right\rangle - \\ - \left\langle f_1(t), \widehat{\psi}_i^{(\sigma+1)} \right\rangle, \\ \dots\dots\dots\dots\dots\dots\dots\dots \\ \xi_{i,1} = -\frac{d^{q_i-1}}{dt^{q_i-1}} \left\langle f_1(t), \widehat{\psi}_i^{(1)} \right\rangle - \frac{d^{q_i-2}}{dt^{q_i-2}} \left\langle f_1(t), \widehat{\psi}_i^{(2)} \right\rangle - \dots - \frac{d}{dt} \left\langle f_1(t), \widehat{\psi}_i^{(q_i-1)} \right\rangle - \\ - \left\langle f_1(t), \widehat{\psi}_i^{(q_i)} \right\rangle, \end{cases}$$

and ξ_{ij} are also turned out to be sufficiently smooth in t. The coefficients in the expansion of $\xi_{ij}(t)$ are determined by the following system

(4.6)
$$\begin{cases} \xi_{iq_i} = \xi_{iq_i}^{(0)} + \xi_{iq_i}^{(1)} t + \xi_{iq_i}^{(2)} t^2 + \dots, \xi_{iq_i}^{(s)} = - \left\langle a_s, \widehat{\psi}_i^{(1)} \right\rangle, \quad s = 0, 1, 2, \dots \\ \xi_{iq_i-1}^{(s)} = -(s+1) \left\langle a_{s+1}, \widehat{\psi}_i^{(1)} \right\rangle - \left\langle a_s, \widehat{\psi}_i^{(2)} \right\rangle, \\ \dots\dots\dots\dots\dots\dots\dots\dots \\ \xi_{iq_i-\sigma}^{(s)} = -(s+\sigma)(s+\sigma-1)\dots(s+1) \left\langle a_{s+\sigma-1}, \widehat{\psi}_i^{(1)} \right\rangle - \dots - \\ - (s+1) \left\langle a_{s+1}, \widehat{\psi}_i^{(\sigma)} \right\rangle - \left\langle a_s, \widehat{\psi}_i^{(\sigma+1)} \right\rangle, \\ \dots\dots\dots\dots\dots\dots\dots\dots \\ \xi_{i1}^{(s)} = -(s+q_i-1)\dots(s+1) \left\langle a_{s+1}, \widehat{\psi}_i^{(1)} \right\rangle - \dots - (s+1) \left\langle a_{s+1}, \widehat{\psi}_i^{(q_i-1)} \right\rangle - \\ - \left\langle a_s, \widehat{\psi}_i^{(q_i)} \right\rangle. \end{cases}$$

The successive determination of $\xi_{ij}^{(s)}$, $j = 1, \dots, q_i$, $s = 0, 1, \dots$ from the system ((4.6)) leads to the solvability conditions for the Cauchy problem ((4.2)):

$$
\begin{aligned}
&\left\langle a_0, \widehat{\psi}_i^{(1)} \right\rangle = 0, \quad a_0 = \alpha_0 + Bx_0,\\
&1! \left\langle a_1, \widehat{\psi}_i^{(1)} \right\rangle + \left\langle a_0, \widehat{\psi}_i^{(2)} \right\rangle = 0,\\
&2! \left\langle a_2, \widehat{\psi}_i^{(1)} \right\rangle + 1! \left\langle a_1, \widehat{\psi}_i^{(2)} \right\rangle + \left\langle a_0, \widehat{\psi}_i^{(3)} \right\rangle = 0,\\
&\dots\dots\dots\dots\dots\dots\dots\dots\dots\dots\dots\dots\dots\dots\\
&\sigma! \left\langle a_\sigma, \widehat{\psi}_i^{(1)} \right\rangle + (\sigma - 1)! \left\langle a_{\sigma-1}, \widehat{\psi}_i^{(2)} \right\rangle + \dots + 2! \left\langle a_2, \widehat{\psi}_i^{(\sigma-1)} \right\rangle +\\
&+ 1! \left\langle a_1, \widehat{\psi}_i^{(\sigma)} \right\rangle + \left\langle a_0, \widehat{\psi}_i^{(\sigma+1)} \right\rangle = 0,\\
&\dots\dots\dots\dots\dots\dots\dots\dots\dots\dots\dots\dots\dots\dots\\
&(q_i - 1)! \left\langle a_{q_i-1}, \widehat{\psi}_i^{(1)} \right\rangle + (q_i - 2)! \left\langle a_{q_i-2}, \widehat{\psi}_i^{(2)} \right\rangle + \dots + 2! \left\langle a_2, \widehat{\psi}_i^{(q_i-2)} \right\rangle +\\
&+ 1! \left\langle a_1, \widehat{\psi}_i^{(q_i-1)} \right\rangle + \left\langle a_0, \widehat{\psi}_i^{(q_i)} \right\rangle = 0, \quad i = 1, \dots, n
\end{aligned}
$$

which are consistent with the results in [17]. Equation (2.1) for problem ((4.2)) at $u(t) = \xi(t) \cdot \varphi + \widetilde{u}(t)$ takes the form

$$
\frac{d\widetilde{u}}{dt} = \overset{\sqcap}{A}{}^{-1} \overset{\sqcap}{B}\, \widetilde{u} + \overset{\sqcap}{A}\, (I - \mathbf{q}) f_1(t)
$$

which under the assumption of the sectoriality of the operator $\overset{\sqcap}{A}{}^{-1} \overset{\sqcap}{B}$ has the following solution

$$
\widetilde{u} = \int_0^t \exp\left[(\overset{\sqcap}{A}{}^{-1} \overset{\sqcap}{B})(t - s) \right] \left[\overset{\sqcap}{A}\, (I - \mathbf{q}) f_1(s) \right] ds
$$

Under weaker assumptions about the differentiability of $f(t)$ up to the order $\max(q_i - 1)$, $f(t)$ can be represented as the direct sum $\mathbf{q} f(t) = \eta(t)\zeta$ and $(I - \mathbf{q}) f(t) = \widetilde{f}(t)$. Now the solution of the system ((4.5)) may be written in the following form

$$
\begin{aligned}
&\xi_{iq_i}(t) = -\eta_{i1}(t) - c_{iq_i}^0,\\
&\xi_{iq_i-1}(t) = -\eta_{i1}^{(1)}(t) - \eta_{i2}(t) - c_{iq_i-1}^0,\\
&\dots\dots\dots\dots\dots\dots\dots\dots\dots\dots\dots\dots\\
&\xi_{iq_i-\sigma}(t) = -\eta_{i1}^{(\sigma)}(t) - \eta_{i2}^{(\sigma-1)}(t) - \dots - \eta_{i\sigma+1}(t) - c_{iq_i-\sigma}^0,\\
&\dots\dots\dots\dots\dots\dots\dots\dots\dots\dots\dots\dots\\
&\xi_{i1}(t) = -\eta_{i1}^{(q_i-1)}(t) - \eta_{i2}^{(q_i-2)}(t) - \dots - \eta_{iq_i}(t) - c_{i1}^0.
\end{aligned}
$$

Here c_{ij}'s are the coefficients of the expansion of the initial value into the direct sum $x_0 = \widetilde{x}_0 + \sum\limits_{i,j} c_{ij}^0 \phi_i^{(j)}$. In fact, it is not difficult to verify that $\left\langle Bx_0, \widehat{\psi}_i^{(\sigma)} \right\rangle = c_{i\, q_i - \sigma + 1}$. In this case the solvability conditions for the Cauchy problem ((4.1)) are expressed easier

$$\begin{array}{l}
\eta_{i1}(0) + c_{1q_i}^0 = 0,\\
\eta_{i1}^{(1)}(0) + \eta_{i2}(0) + c_{iq_i-1}^0 = 0,\\
\dots\dots\dots\dots\dots\dots\dots\dots\dots\dots\dots\\
\eta_{i1}^{(\sigma)}(0) + \eta_{i2}^{(\sigma-1)}(0) + \dots + \eta_{i\sigma+1}(0) + c_{iq_i-\sigma}^0 = 0,\\
\dots\dots\dots\dots\dots\dots\dots\dots\dots\dots\dots\\
\eta_{i1}^{(q_i-1)}(0) + \eta_{i2}^{(q_i-2)}(0) + \dots + \eta_{iq_i}(0) + c_{i1}^{(0)} = 0.
\end{array}$$

4.2 The problem about the perturbation of the linear equation by a small linear term.

This problem is expressed by the equation [19]

$$Bx = h + \varepsilon Ax. \tag{4.7}$$

Introducing the A-Jordan chains, equation ((4.7)) can be expanded into direct sum $x = u + v$, $v \in E_1^{k_B}$, $u \in E_1^{\infty - k_B}$, $\widetilde{h} = (I - \mathbf{q})h$
(4.8)

$$\overset{\sqcap}{B} u = \varepsilon \overset{\sqcap}{A} u + \widetilde{h}, \quad B(\xi \cdot \varphi) = \sum_{i=1}^{n} \sum_{j=1}^{p_i} c_{ij} z_i^{(j)} + \varepsilon A(\xi \cdot \varphi), \quad c_{ij} = \left\langle h, \psi_i^{(j)} \right\rangle.$$

The system (2.9) takes the form

$$0 = c_{i1} + \varepsilon \xi_{ip_i} \tag{4.9},$$

$$\left\{ \begin{array}{l}
\xi_{i2} = c_{ip_i} + \varepsilon \xi_{i1},\\
\xi_{i3} = c_{ip_i-1} + \varepsilon \xi_{i2},\\
\dots\dots\dots\dots\dots\dots\\
\xi_{is} = c_{ip_i-s+2} + \varepsilon \xi_{is-1},\\
\dots\dots\dots\dots\dots\dots\\
\xi_{ip_i} = c_{i2} + \varepsilon \xi_{ip_i-1}.
\end{array} \right. \tag{4.10}$$

In fact we obtain

$$B(\xi \cdot \varphi) \overset{(1.2)}{=} \sum_{i=1}^{n} \left[\xi_{i2} A\varphi_1^{(1)} + \cdots + \xi_{ip_i-1} A\varphi_i^{(p_i-2)} + \xi_{ip_i} A\varphi_i^{(p_i-1)}\right] =$$

$$\overset{(1.4)}{=} \sum_{i=1}^{n} \left[\xi_{i2} z_i^{(p_i)} + + \cdots + \xi_{ip_i-1} z_i^{(3)} + \xi_{ip_i} z_i^{(2)}\right] =$$

$$\overset{((4.8))}{=} \sum_{i=1}^{n}\sum_{j=1}^{p_i} c_{ij} z_i^{(j)} + \varepsilon \sum_{i=1}^{n} \left(\xi_{i1} z_i^{(p_i)} + \xi_{i2} z_i^{(p_i-1)} + \cdots + \xi_{ip_i-1} z_i^{(2)} + \xi_{ip_i} z_i^{(1)}\right).$$

Now equations ((4.9)-(4.10)) give

$$\xi_{ip_i} = -\frac{c_{i1}}{\varepsilon},\ \xi_{ip_i-1} = -\frac{c_{i1}}{\varepsilon^2} - \frac{c_{i2}}{\varepsilon}, \ldots,\ \xi_{ip_i-s} = -\frac{c_{i1}}{\varepsilon^{s+1}} - \frac{c_{i2}}{\varepsilon^{s}} - \cdots - \frac{c_{is+1}}{\varepsilon},$$

$$\xi_{i1} = -\frac{c_{i1}}{\varepsilon^{p_i}} - \frac{c_{i2}}{\varepsilon^{p_i-1}} - \cdots - \frac{c_{ip_i}}{\varepsilon}.$$

The first equation of the system ((4.8)) is uniquely solvable and its solution u_0 belongs to $E_1^{\infty - k_B}$. Then the equation ((4.7)) has the following solution

$$x = u_0 - \sum_{i=1}^{n} \left[\left(\frac{c_{i1}}{\varepsilon^{p_i}} + \frac{c_{i2}}{\varepsilon^{p_i-1}} + \cdots + \frac{c_{ip_i}}{\varepsilon}\right)\varphi_i^{(1)} + \cdots +\right.$$

$$\left. + \left(\frac{c_{i1}}{\varepsilon^{s+1}} + \frac{c_{i2}}{\varepsilon^{s}} + \cdots + \frac{c_{is+1}}{\varepsilon}\right)\varphi_i^{(p_i-s)} + \cdots + \left(\frac{c_{i1}}{\varepsilon^2} + \frac{c_{i2}}{\varepsilon}\right)\varphi_i^{(p_i-1)} + \frac{c_{i1}}{\varepsilon}\varphi_i^{(p_i)}\right].$$

As a result we obtain the following theorem

Theorem 4.1. *For the case of one A-JCh of the length p the solution of equation ((4.7)) has a pole of order p if* $c_1 = \langle h, \psi^{(1)} \rangle \neq 0$; *a pole of order* $p-1$ *if* $c_1 = \langle h, \psi^{(1)} \rangle = 0$, $c_2 = \langle h, \psi^{(2)} \rangle \neq 0, \ldots$; *a pole of order* $p - s$, *if* $c_j = \langle h, \psi^{(j)} \rangle = 0$, $j = 1, \ldots, p-s$, $c_{p-s+1} = \langle h, \psi^{(p-s+1)} \rangle \neq 0$; ..., *a pole of the first order if* $c_j = \langle h, \psi^{(j)} \rangle = 0$, $j = 1, \ldots, p-1$, $c_p = \langle h, \psi^{(p)} \rangle \neq 0$; *and it is analytic, when* $c_j = \langle h, \psi^{(j)} \rangle = 0$, $j = 1, \ldots, p$.

In the general case the solution of ((4.7)) has a pole of order $p_0 = \max p_i$ *if there exists an index* $1 \leq i_0 \leq n$ *for which* $c_{i_0 1} \neq 0$. *If* $\left\langle h, \psi_i^{(s+1)} \right\rangle = 0$ *for* $j \leq s$ *and all* i, *but* $\left\langle h, \psi_i^{(s+1)} \right\rangle \neq 0$ *even if one index* i_0 *exists, then the solution has a pole of order* $p - s$. *When* $\left\langle h, \psi_i^{(j)} \right\rangle = 0$ *for all* i *and* $1 \leq j \leq p_i$, *then it is analytic at* ε.

Remark 4.1. *At the presence of symmetry one can decompose the right-hand sides* $f_1(t)$ *and the initial value in the problem (42) or* h *in (47) into* *irreducible invariant subspaces. Then the solutions of the problems ((4.2)) or ((4.7)) will be represented as sums of partial solutions in invariant subspaces. The corresponding resolving systems split into subsystems of smaller orders. Thus the solution of problems ((4.2)), ((4.7)) is simplified.*

5 On Grobman-Hartman theorem for equations with a degenerate operator at the derivative

Let E_1 and E_2 be Banach spaces, and $A : E_1 \supset D_A \to E_2$, $B : E_1 \supset D_B \to E_2$ be densely defined closed linear Fredholm operators, where $D_B \subset D_A$ and A is subordinated to B (i.e. $\|Ax\| \leq \|Bx\| + \|x\|$ on D_B) or $D_A \subset D_B$ and B is subordinated to A (i.e. $\|Bx\| \leq \|Ax\| + \|x\|$ on D_A). We consider the differential equation

$$(5.1) \qquad A\frac{dx}{dt} = Bx - R(x), \quad R(0) = 0, \quad R_x(0) = 0$$

The aim of this part of the communication is to prove the Grobman-Hartman theorem [2,4] for the equation ((5.1)).

It is assumed that for the A-spectrum $\sigma_A(B)$ of the operator B , Re $\sigma_A(B) \neq 0$ and the spectral sets $\sigma_A^-(B) = \{\mu \in \sigma_A(B)|\ Re\mu < 0\}$ and $\sigma_A^+(B) = \{\mu \in \sigma_A(B)|\ \text{Re}\ \mu > 0\}$ are distant from the imaginary axis on some distance $d > 0$. All solutions of the linear Cauchy problem corresponding to ((5.1))

$$(5.2) \qquad A\frac{dx}{dt} = Bx, \qquad x(0) = x_0$$

belong to $E_1^{\infty - k_A}$ and ((5.2)) is solvable iff $x_0 \in E_1^{\infty-k_A}$. In fact, if one sets $x = v + w$, $v(t) = \sum\limits_{i=1}^{m}\sum\limits_{s=1}^{q_i} \xi_{is}(t)\phi_i^{(s)} \in E_1^{k_A}$, $w(t) \in E_1^{\infty-k_A}$, then ((5.2)) is split into the system

$$(5.3) \qquad \frac{d\xi_{is}(t)}{dt} = \xi_{i,s-1},\ s = 2,\dots,q_i,\ i = 1,\dots,m,\ \ \xi_{iq_i} = 0;\ \overset{\sqcap}{A}\frac{dw}{dt} = \overset{\sqcap}{B} w.$$

. Consequently $\xi_{is}(t) = 0$, and the solution of ((5.2)) takes the form

$$(5.4) \qquad x(t) = \exp\left(\overset{\sqcap}{A}{}^{-1}\overset{\sqcap}{B}\,t\right)x_0, \quad x_0 \in E_1^{\infty-k_A}$$

and $\sigma_A(B) = \sigma\left(\overset{\sqcap}{A}{}^{-1}\overset{\sqcap}{B}\right)$. Here the function $\exp\left(\overset{\sqcap}{A}{}^{-1}\overset{\sqcap}{B}\,t\right)$ has the form of the contour integral $\frac{1}{2\pi i}\int\limits_{\gamma}\left(\mu I - \overset{\sqcap}{A}{}^{-1}\overset{\sqcap}{B}\right)^{-1} e^{\mu t}\,dt$ under the assumption of the sectorial property [3] of the operator $\overset{\sqcap}{A}^{-1}\overset{\sqcap}{B}$ (or under the assumption of the A-sectorial property of the operator B [18]) with some special contour γ belonging to sector $S_{\alpha,\theta}(B)$ in the A-resolvent set of the operator B [18].

The more so, this is true when the operator $\overset{\sqcap}{A}^{-1}\overset{\sqcap}{B}$ is bounded. For the generalization of the Grobman-Hartman theorem we will follow [2]. Let us define the spaces D_k, $k = 1, 2$, with graph norms:

1. $D_1 = D_B \subset D_A$ with the norm $\|x\|_1 = \|x\|_{E_1} + \|Bx\|_{E_2}$, $x \in D_1$, if A is subordinated to B,
2. $D_2 = D_A \subset D_B$ with the norm $\|x\|_2 = \|x\|_{E_1} + \|Ax\|_{E_2}$, $x \in D_2$, if B is subordinated to A,

and introduce the spaces $X_{k0}, X_{k1}, X_{k2}, Y_{k0}, Y_{k1}, Y_{k2}$ consisting of uniformly bounded continuous functions $f(t)$ on $[0, \infty)$ with their values correspondingly in D_k, $D_k \cap E_1^{\infty-k_A}$, $E_1^{k_A}$, E_2, $E_{2,\infty-k_A}$, E_{2,k_A} with supremum norms in the relevant spaces, and the spaces

$$X^1_{ks} = \left\{ f(t) \in X_{ks} | \dot{f}(t) \in X_{ks} \right\}, \|f(t)\|_{X^1_{ks}} = \max\left\{ \|f(t)\|_{X_{ks}}, \|\dot{f}(t)\|_{X_{ks}} \right\}.$$

Let the operator $\overset{\sqcap}{A}^{-1}\overset{\sqcap}{B}$ be bounded in X_{k1} (for the case $k = 1$ it is evident), then the operator

$$\mathbf{A}x = A\dot{x} - Bx \tag{5.5}$$

acting from X^1_{k0} to Y_{k0} is linear and continuous with $X_{k2} \subset \mathcal{N}(\mathbf{A})$.

Let be $D_k \supset S_k$ – the set of initial values of the solutions of the equation ((5.2)), which are defined and remain in a small neighborhood of zero in D_k for $t \in [0, +\infty)$ and let U_k be the set of initial values of solutions of ((5.2)), which are defined and remain in a small neighborhood of zero in D_k for $t \in (-\infty, 0]$. It follows from ((5.5)) that $S_k \dot{+} U_k = E_1^{\infty-k_A} \cap D_k$. Then the equality $\sigma_A(B) = \sigma\left(\overset{\sqcap}{A}^{-1}\overset{\sqcap}{B}\right)$ allows us to define the projectors $P^- u = \frac{1}{2\pi i}\int\limits_{\gamma_-}\left(\overset{\sqcap}{A}^{-1}\overset{\sqcap}{B} - \mu I_{E_1^{\infty-k_A}}\right)^{-1} u d\mu$ (γ_- is the contour in $\rho_A(B)$ containing inside the points $\mu \in \sigma_A(B)$ with Re $\mu < 0$), and $P^+ = I_{E_1^{\infty-k_A}} - P^-$. Hence $D_k = D_k^- \dot{+} D_k^0 \dot{+} D_k^+$, $D_k^0 = E_1^{k_A}$, $D_k^\pm = P^\pm D_k$, the operator $\mathbf{A}$ is Noetherian [19] with $R(\mathbf{A}) = Y_{k1}$ and

$$\begin{aligned}
\mathcal{N}(\mathbf{A}) &= \left\{ f(t) \in X^1_{k0} | f(t) = \exp\left(\overset{\sqcap}{A}^{-1}\overset{\sqcap}{B} t\right) P^- f(0) \in D_k^- \right\} \dot{+} \{ f(t) \in D_k^0 \} \\
&= \mathcal{N}_1(\mathbf{A}) \dot{+} \mathcal{N}_2(\mathbf{A}) \quad \text{for} \quad t \geq 0 \\
(\mathcal{N}(\mathbf{A}) &= \left\{ f(t) \in X^1_{k0} | f(t) = \exp\left(\overset{\sqcap}{A}^{-1}\overset{\sqcap}{B} t\right) P^+ f(0) \in D_k^+ \right\} \dot{+} \{ f(t) \in D_k^0 \} \text{ for } t \leq 0)
\end{aligned}$$

Now setting $x = y + z + v$, $z \in D_k^+$, $v \in D_k^0 = E_1^{k_A}$, $y \in D_k^-$ one can write equation ((5.1)) in the form ($w = y + z$ in ((5.3)))

$$\text{(5.6)} \qquad \mathbf{A}z = R(z + y + v) \quad (\mathbf{A}y = R(y + z + v))$$

and apply the implicit operator theorem to ((5.6)) regarding y, v (z, v) as functional parameters (see the relevant theorems 22.1 and 22.2 in [19] for the continuous and analytic operator R respectively). It follows that ((5.6)) has a sufficiently smooth or analytic (according to the properties of the operator R) solution in some neighborhood of the parameters y, v (z, v) with zero values

$$\text{(5.7)} \quad z = z(y + v), \quad z(0) = 0 = Dz(0) \quad (y = y(z + v),\ y(0) = 0 = Dy(0))$$

Consequently the following generalization of Grobman-Hartman Theorem [2,4] is true.

Theorem 5.1. *There exists a neighborhood $\omega^-(\omega^+)$ of zero in $D_k^0 \dot{+} D_k^-$ (in $D_k^0 \dot{+} D_k^+$) and sufficiently smooth mapping $z_R = z_R(\xi, \eta) = z_R(\xi \cdot \phi + \eta) : \omega^- \to D_k^+$, $\eta \in D_k^-$ $(y_R = y_R(\xi, \zeta) = y_R(\xi \cdot \phi + \zeta) : \omega^+ \to D_k^-$, $\zeta \in D_k^+)$, such that*

a) $z_R(0,0) = 0$, $D_\xi z_R(0,0) = 0$, $D_\eta z_R(0,0) = 0$ $(y_R(0,0) = 0$, $D_\xi y_R(0,0) = 0$, $D_\zeta y_R(0,0) = 0)$,

b) for any solution $x(t)$ of ((5.1)) with initial data $x(0) = \xi \cdot \phi + \eta + z_R(\xi \cdot \phi + \eta)$ $(x(0) = \xi \cdot \phi + y_R(\xi \cdot \phi + \zeta) + \zeta)$ one obtains $z(t) = z_R(\xi(t) \cdot \phi + y(t)) \in D_k^+$ for $t \geq 0$ $(y(t) = y_R(\xi(t) \cdot \phi + z(t)) \in D_k^-$ for $t \leq 0)$,

c) any solution $x(t)$ of ((5.1)) with initial data from b) takes the form $x(t) = \xi(t) \cdot \phi + y(t) + z_R(\xi(t) \cdot \phi + y(t))$ $(x(t) = \xi(t) \cdot \phi + y_R(\xi(t) \cdot \phi + z(t)) + z(t))$

and tends to zero when $t \to +\infty (t \to -\infty)$, and belongs, consequently, to the local stable manifold $S_k(R)$ (or to the local unstable manifold $U_k(R)$).

Proof: We give here the proof for the function z_R and the local stable manifold $S_k(R)$, the proof of the second part is analogous. Define the projector $\widetilde{P}^-$ of X_{k1}^1 onto $\mathcal{N}_1(\mathbf{A})$ by the equality $\left(\widetilde{P}^- f\right)(t) = \exp\left(\overset{\sqcap}{A}{}^{-1} \overset{\sqcap}{B}\, t\right) P^- f(0)$, $t \geq 0$. If one sets $x(t) = v(t) + y(t) + z(t)$, $v(t) = \mathbf{p}x(t)$, $v(0) = \xi \cdot \phi = \sum_{i=1}^{m} \sum_{s=1}^{q_i} \xi_{is} \phi_i^{(s)}$, $y(t) = \widetilde{P}^- x(t) = \exp\left(\overset{\sqcap}{A}{}^{-1} \overset{\sqcap}{B}\, t\right) \eta$, $\eta = y(0)$, $z(t) = \left(I_{X_{k1}^1} - \widetilde{P}^-\right) x(t)$, then the Lyapounov-Schmidt method (theorem 27.1 in [19] for Noetherian operators with d-characteristic $(n, 0)$ and the theorems 22.1, 22.2 in [19]) imply that there is a unique solution of ((5.6)) $z =$

$z_R(\xi(t)\cdot\phi + y(t)) \in X^1_{k1}$, such that $x(0) = \xi\cdot\phi + \eta + z_R(\xi\cdot\phi+\eta)$, i.e. the unique solution of (51) $x(t) = v(t) + y(t) + z_R(\xi(t)\cdot\phi + y(t))$, $v(t) = \xi(t)\cdot\phi$, in a sufficiently small semi-neighborhood of $t = 0$, where the function $z_R(\xi,\eta) = z_R(\xi\cdot\phi+\eta)$ is sufficiently smooth by ξ,η, and $z_R(0,0) = 0$, $D_\xi z_R(0,0) = 0$, $D_\eta z_R(0,0) = 0.\Box$

Rewriting equation ((5.1)) in **p**, **q**-projections by using Theorem 5.1, one can get a system for the determination of $\xi_{is}(t)$ (so-called resolving system (RS) for the equation ((5.1)) [8, 11,12,13]. Here $x(t) = \xi(t)\cdot\phi + w(t)$, where $w(t) = y(t) + z_R(\xi(t)\cdot\phi + y(t))$ for $t \geq 0$ and $w(t) = y_R(\xi(t)\cdot\phi + z(t)) + z(t)$ for $t \leq 0$

$$\overset{\sqcap}{A}\frac{dw}{dt} = \overset{\sqcap}{B} w - (I_{D_k} - \mathbf{q})R(\xi\cdot\phi + w) \tag{5.8}$$

$$\begin{aligned}
&0 = \xi_{iq_i}(t) - \left\langle R(\xi(t)\cdot\phi + w), \widehat{\psi}_i^{(1)}\right\rangle,\\
&\dot{\xi}_{iq_i}(t) = \xi_{i,q_i-1}(t) - \left\langle R(\xi(t)\cdot\phi + w), \widehat{\psi}_i^{(2)}\right\rangle,\\
&\dots\dots\dots\dots\dots\dots\dots\dots\dots\dots\dots\dots\\
&\dot{\xi}_{i2}(t) = \xi_{i1}(t) - \left\langle R(\xi(t)\cdot\phi + w), \widehat{\psi}_i^{(q_i)}\right\rangle,\\
&\xi_{is}(0) = \xi_{is}, \quad s = 1,\dots,q_i, \quad i = 1,\dots,m.
\end{aligned} \tag{5.9}$$

Consequently, the manifold $S_k(R)$ containing the initial values of solutions of equation ((5.1)), which are defined and remain in a small neighborhood of $0 \in D_k$ for $t \in [0,+\infty)$ and the manifold $U_k(R)$ containing the initial values of solutions ((5.1)), which are defined and remain in a small neighborhood of $0 \in D_k$ for $t \in (-\infty, 0])$ have the the form $x(0) = \xi\cdot\phi + \eta + z_R(\xi\cdot\phi + \eta)$ ($x(0) = \xi\cdot\phi + y_R(\xi\cdot\phi + \zeta) + \zeta$), where $\eta \in D_k^-$ ($\zeta \in D_k^+$) and ξ are small.

Remark 5.1. *The invariant manifold* $\mathfrak{M}$ *which is determined by the function* $\xi\cdot\phi+\eta+z_R(\xi\cdot\phi+\eta)$ *for* $t \geq 0$ ($\xi\cdot\phi+y_R(\xi\cdot\phi+\zeta)+\zeta$ *for* $t \leq 0$) *can be regarded as the center manifold* ($\xi\cdot\phi \in D_k^0$), *that is nontrivial for the equation ((5.1)) even if* $\{\mu \in \sigma_A(B) | Re\ \mu = 0\} = \emptyset$. *Here* $\{\xi\cdot\phi\}$ *is called the linear center manifold tangent to* $\mathfrak{M}$. *One can say that* $\mathfrak{M}$ *has an hyperbolic structure. Thus the RS ((5.9)) represents the differential-algebraic system on* $\mathfrak{M}$. *Of course, if the operator* A *is invertible,* $\mathfrak{M}$ *and the system ((5.9)) are absent, i.e. in the Grobman-Hartman theorem* $z_R = z_R(\eta)$ *[2.4].*

Theorem 5.2. *Let the operators* A, B *and* R *in ((5.1)) be intertwined by the group* G *representations* L_g *(acting in* E_1*) and* K_g *(acting in* E_2*) and Condition I (direct supplements* $E_1^{\infty-m}$ *to* $\mathcal{N}(A)$ *and* $E_1^{\infty-n}$ *to* $\mathcal{N}(B)$ *are invariant relative to* L_g*) be satisfied. Then the center manifold* $\mathfrak{M}$ *is invariant relative to the operators* L_g.

In fact, according to [12] (see also (38), (39)) the projectors $\mathbf{p}, \mathbf{P}(\mathbf{q}, \mathbf{Q})$ commute with the operators $L_g(K_g)$ and invariant pairs of subspaces reduce the representations $L_g(K_g)$.

It was proved in the articles [7,8] that the stability (instability) of the trivial solution (even for non-autonomous) equation ((5.1)) under sufficiently general conditions is determined by the RS ((5.9)) with corollaries for the investigation of the stability (instability) of bifurcating solutions.

The case when $\sigma_A^+(B) = \emptyset$ is interesting, when $D_k = D_k^- \dot{+} D_k^0$, $x(t) = \xi(t) \cdot \phi + y(t)$ and the center manifold has the form $\xi(t) \cdot \phi + y(\xi(t) \cdot \phi)$. Here equation ((5.8)) becomes

(5.10)

$$\overset{\sqcap}{A}\, y'(\xi(t) \cdot \phi)(\tfrac{d\xi}{dt} \cdot \phi) = \overset{\sqcap}{B}\, y(\xi(t) \cdot \phi) + (I - \mathbf{q})R(\xi(t) \cdot \phi + y(\xi(t) \cdot \phi)),$$
$$y(0) = 0, \quad y'(0) = 0$$

In combination with ((5.9)) this gives a possibility for the determination of the center manifold $w(\xi(t) \cdot \phi) = \xi(t) \cdot \phi + y(\xi(t) \cdot \phi)$ by successive approximations under the conditions of sufficiently smooth operator $y(\xi \cdot \phi)$. However following this approach some essential difficulties arise connected with the fact that the system ((5.9)) is differential-algebraic, i.e. the differential equations for the functions $\xi_{i1}(t)$, $i = 1, \ldots, m$, are absent. One can find $y(\xi \cdot \phi)$ iteratively by differentiating the first equations ((5.9)).

Remark 5.2. *Theorem 5.1 and all its corollaries remain true for the parameter depending equation*

$$A\frac{dx}{dt} = Bx - R(x, \lambda), \quad R(0, \lambda) \equiv 0, \quad R_x(0,0) = 0, \tag{5.11}$$

($\lambda \in \Lambda$, Λ is some Banach space) in a small neighborhood of $\lambda = 0$, when as stated before, Re $\sigma_A(B) \neq 0$, i.e. $\lambda = 0$ is not a bifurcation point. However all functions w, z_R and y_R will depend on a small parameter ε.

We consider now the simplest case when $\sigma_A^+(B) = \emptyset$, but $\sigma_A^0(B) = \{\mu \in \sigma_A(B) | Re\ \mu = 0\} \neq \emptyset$ contains finite numbers $2n = 2n_1 + \cdots + 2n_\ell$, A-eigenvalues $\pm i\alpha_s$ of multiplicities n_s, $s = 1, \ldots, \ell$, $\alpha_s = \kappa_s \alpha$, $\alpha \neq 0$ with coprime $\kappa_s > 0$ or (and) zero-eigenvalue. Without loss of generality it is supposed that equation ((5.1)) is written in the form of the system

$$\begin{matrix} A_1\dot{x} = B_1 x - f(x, y) \\ A_2\dot{y} = B_2 y - R(x, y), \end{matrix} \quad A = \begin{pmatrix} A_1 & 0 \\ 0 & A_2 \end{pmatrix}, \ B = \begin{pmatrix} B_1 & 0 \\ 0 & B_2 \end{pmatrix} \tag{5.12}$$

where the linear operators $A_1, B_1 : E_1^{k_{B_1}} \to E_{2,k_{B_1}}$ ($k_{B_1} = 2n_1p_1 + \cdots + 2n_\ell p_\ell$, p_s are A_1-Jordan chains with lengths for $\pm i\alpha_s, s = 1, \ldots, \ell$) acting on the invariant pair of finite dimensional subspaces $E_1^{k_{B_1}}, E_{2,k_{B_1}}$ and A_2, B_2 act on the invariant pair of subspaces $E_1^{\infty - k_{B_1}}$, $E_{2,\infty - k_{B_1}}$. Thus, $\sigma_{A_1}(B_1) = \sigma_A^0(B)$ and $\sigma_{A_2}^0(B_2) = \emptyset$. Here f and R are C^2-functions vanishing together with their first derivatives at the origin.

The main assumption in the simplest case is

$$\mathcal{N}(A_1) = 0, \quad \mathcal{N}(A_2) = \text{span}\,\{\phi_{(2)1}, \ldots, \phi_{(2)m_2}\cdot\} \tag{5.13}$$

Then there exist a function $y_R(\xi_2(t) \cdot \phi_{(2)}, x)$ vanishing together with its first derivatives at the origin, such that the second equation ((5.12)) reduces to the system

$$\overset{\sqcap}{A}_2 \frac{dy_R}{dt} = \overset{\sqcap}{B}_2\, y_R - (I - \mathbf{q}_{(2)})R(x, \xi_2(t) \cdot \phi_{(2)} + y_R(\xi_2(t) \cdot \phi_{(2)}, x)) \tag{5.14}$$

($\mathbf{q}_{(2)} = \sum_{i=1}^{m_2} \sum_{j=1}^{q_{2,i}} < \cdot, \widehat{\psi}^{(j)}_{(2),i} > \zeta^{(j)}_{(2)} : E_{2,\infty - k_{A_2}} \to \text{span}\{\zeta^{(j)}_{(2)i}\}$, $\overset{\sqcap}{A}_2, \overset{\sqcap}{B}_2$ act in invariant pair of subspaces $E_1^{\infty - k_{B_1} - k_{A_2}}$, $E_{2,\infty - k_{B_1} - k_{A_2}}$)

(5.15)

$$\begin{aligned}
0 &= \xi_{2iq_{2,i}}(t) - \left\langle R(x, \xi_2(t) \cdot \phi_{(2)} + y_R(\xi_2(t) \cdot \phi_{(2)}, x)), \widehat{\psi}^{(1)}_{(2),i} \right\rangle, \\
\dot{\xi}_{2iq_{2,i}}(t) &= \xi_{2i,q_{2,i}-1}(t) - \left\langle R(x, \xi_2(t) \cdot \phi_{(2)} + y_R(\xi_2(t) \cdot \phi_{(2)}, x)), \widehat{\psi}^{(2)}_{(2),i} \right\rangle, \\
\cdots\cdots & \cdots\cdots\cdots\cdots\cdots\cdots\cdots\cdots\cdots\cdots\cdots\cdots \\
\dot{\xi}_{2i2}(t) &= \xi_{2i1}(t) - \left\langle R(x, \xi_2(t) \cdot \phi_{(2)} + y_R(\xi_2(t) \cdot \phi_{(2)}, x)), \widehat{\psi}^{(q_{2,i})}_{(2),i} \right\rangle, \\
\xi_{2i\sigma}(0) &= \xi_{2i\sigma}, \ \sigma = 1, \ldots, q_{2,i}, \ \ i = 1, \ldots, m_2.
\end{aligned}$$

If the system ((5.12)) is equipped with initial values $x(0)$, $y(0)$, then they must satisfy the equality

$$y(0) = \xi_2 \cdot \phi_{(2)} + y_R(\xi_2 \cdot \phi_{(2)}, x(0)). \tag{5.16}$$

Now one has to solve the problem

$$A_1 \dot{x} = B_1 x - f(x, \xi_2(t) \cdot \phi_{(2)} + y_R(\xi_2(t) \cdot \phi_{(2)}, x)) \tag{5.17}$$

with the initial data $x(0)$ satisfying ((5.16)). Thus one has two systems ((5.15)) and ((5.17)) on the center manifold $y = y_R(\xi_2(t) \cdot \phi_{(2)}, x)$.

According to equation ((5.8)) can be written in the form

$$\frac{dw}{dt} = \overset{\sqcap}{A}^{-1} \overset{\sqcap}{B} w - \overset{\sqcap}{A}^{-1} (I_{D_k} - \mathbf{q})R(\xi \cdot \phi + w) \tag{5.18}$$

in the space X^1_{k1}. Then the assumption about the boundedness of the operator $\overset{\sqcap}{A}^{-1} \overset{\sqcap}{B}$ in X_{k1} allows the proof of the Grobman-Hartman theorem for maps [20]. In fact, then for small ξ there exists the resolving operator $U_\xi(t, \cdot) : X_{k1} \to X^1_{k1}, w_0 \mapsto w(t)$ for the problem ((5.18)) with the initial value $w(0) = w_0$ (at $\xi = 0, U_0(t)$ is linear). Thus the following assertion is true:

Theorem 5.3. *At $\sigma_A^0(B) = \emptyset$ and operator $\overset{\sqcap}{A}^{-1}\overset{\sqcap}{B}$ boundedness the assumption for small ξ there exists a resolving operator $U_\xi(t, w_0)$ and a homeomorphism $\Phi_\xi : X_{k1}^1 \to X_{k1}^1, ||\xi|| \ll 1$, such that for $t \in \mathbb{R}$ and $w_0 \in X_{k1}$ the following relation*

$$U_0(t)\Phi_\xi(w_0) = \Phi_\xi(U_\xi(t, w_0)) = \Phi_\xi(w(t)) \tag{5.19}$$

is true, where the function $w(t)$ and the initial values w_0, ξ_0 satisfy the initial value problem for the differential-algebraic systems ((5.9)).

6 Andronov-Hopf bifurcation under cosymmetry conditions

Here the results of the article [21] are generalized to equations with Fredholm operators at the derivative. For the simplicity of presentation, as in [21], a differential equation with real parameter λ in the Hilbert space H is considered

$$A\frac{dx}{dt} = F(x, \lambda). \tag{6.1}$$

It is supposed that the analytical operator $F : H \times \mathbb{R} \to H$ allows a cosymmetry $L : H \to H$, which doesn't depend on λ. It means that the equality

$$(F(x, \lambda), Lx) = 0 \tag{6.2}$$

is satisfied for all $x \in H$ and $\lambda \in \mathbb{R}$. For some $\lambda = \lambda_0$ there exists a non-cosymmetric solution x_0 of (6.1) i.e. $F(x_0, \lambda_0) = 0,\ Lx_0 \neq 0$. The A-spectrum $\sigma_A(B_0)$ of the operator $B_0 = F'_x(x_0, \lambda_0)$ consists of three parts

$$\sigma_A^{\pm}(B_0) = \left\{\lambda \in \sigma_A(B_0) |\ \mathrm{Re}\left\{\begin{array}{c}\lambda > 0\\ \lambda < 0\end{array}\right\}\right\},\ \sigma_A^0(B_0) = \{\lambda \in \sigma_A(B_0) |\ \mathrm{Re}\lambda = 0\}$$

and $\sigma_A^0(B_0)$ contains only three simple A-eigenvalues 0 and $\pm i\omega_0$, where $B_0^* Lx_0 = 0$ [22] in the absence of additional degeneration of B_0. Consequently according to [22] equation (6.1) has a one parameter family of equilibria

$$s \mapsto c(s), \quad s \in (-\eta, \eta); \quad \eta > 0; \quad F(c(s), \lambda_0) = 0, \quad c(0) = x_0, \tag{6.3}$$

and 0 is the simple A-eigenvalue of $B_s = F'_x(c(s), \lambda_0)$ for all s with eigenvector $\gamma = c'(s)$, tangent to the family $c(s)$. Here, without loss of generality, we obtain

$$B_0\gamma_0 = 0, \quad (\gamma_0, Lx_0) = 0. \tag{6.4}$$

Introducing the new time variable $\tau = \omega_0 t$, the linearized equation around the equilibrium x_0 can be written in the form

$$Tu \equiv \omega_0 A\frac{du}{d\tau} - B_0 u = f(\tau) \tag{6.5}$$

where $f = f(\tau)$ is a given 2π-periodic vector-function. The corresponding homogeneous equation $Tu = 0$ has three independent 2π-periodic solutions $\gamma_0, \varphi e^{i\tau}$ and $\bar{\varphi} e^{-i\tau}$, where φ is the A-eigenvector of the operator $B_0 : B_0\varphi = i\omega_0 A\varphi$. The homogeneous conjugate equation $T^*w \equiv -\omega_0 A\frac{dw}{d\tau} - B_0^* w = 0$ also has three independent 2π-periodic solutions: $Lx_0, \Phi e^{i\tau}$ and $\bar{\Phi} e^{-i\tau}$, where Φ is the A^*-eigenvector of the operator B_0^*, i.e. $B_0^*\Phi = -i\omega_0 A^*\Phi$ with the normalization $(A\varphi, \Phi) = 1$ (because of the simplicity of the eigenvalues $\pm i\omega_0$).

Lemma 6.1. *[21] Equation (6.5) with a real continuous 2π-periodic vector-function f has 2π-periodic solutions iff*

$$\left\langle (f, \Phi)e^{-i\tau} \right\rangle = 0, \quad (\langle f \rangle, Lx_0) = 0 \quad \textit{where} \quad \langle f \rangle = \frac{1}{2\pi}\int_0^{2\pi} f(\tau)d\tau.$$

On the basis of this lemma and the Lyapounov-Schmidt procedure for Andronov-Hopf bifurcation for equations with Fredholm operator at the derivative [10] the two theorems of [21] can be proved for equation (6.1). We present here only the first of them, since their formulations and proofs are almost identical as in [21]. The first theorem concerns the stationary solutions in the absence of limit cycles, the second one gives some sufficient conditions for the existence of limit cycles.

Theorem 6.1. *Under the conditions indicated above, equation (6.1) has, for small $\delta = \lambda - \lambda_0$, an equilibrium set $c(s, \delta)$ depending analytically on $(s, \delta) \in (-s_0, s_0) \times (-\delta_0, \delta_0), s_0 > 0, \delta_0 > 0$, i.e. $F(c(s, \delta), \lambda_0 + \delta) = 0$. When the inequality*

$$h_{2000} = (F_0'' \varphi\bar{\varphi}, Lx_0) \neq 0$$

in some neighborhood of x_0 is satisfied, then small limit cycles are absent (here $F_0'' = F''(x_0, \lambda_0)$, i.e. the derivative calculated at the point x_0, λ_0).

Remark 6.1. *The material of this section presents the first step for investigations of bifurcation phenomena for equations with Fredholm operators at the derivative under cosymmetry conditions. We intend to apply group analysis methods [15] to these problems.*

Remark 6.2. *The results of this article remains true for more general operators subordinateness definition (A is subordinated to B if on $D_B \subset D_A$ $\|Ax\| \leq \|Bx\| + \alpha\|x\|$, $\alpha \geq 0$).*

This work was supported by RFBR (Russian Foundation of Fundamental Research), grant N° 0101-00019 and NATO-TÜBITAK PC Program.

References

[1] V. R. Abdullin and N.A. Sidorov, Interlaced equations in branching theory, *Dokl. Akad. Nauk,* **377** 2001, pp. 295–297.

[2] J. Hale, Introduction to dynamic bifurcation, *Bifurcation Theory and Applications, Lecture Notes in Mathematics,* **Vol. 1057**, Springer Verlag, 1984, pp. 106-151,

[3] D. Henry, Geometric Theory of Semilinear Parabolic Equations, *Lecture Notes in Mathematics,* **Vol. 840**, Springer Verlag, 1994

[4] G. Iooss and M. Adelmeyer, Topics in Bifurcation Theory and Applications, *Adv. ser. in Nonl. Dyn., Vol. 3,* World Scientific, 1998

[5] B. Loginov, and V. Trenogin, On the usage of group properties for the determination multiparameter solution families of nonlinear equations, *Matem. Sbornik,* **14**, 1971, pp. 438–452.

[6] B. Loginov, Branching Theory of Solutions of Nonlinear Equations under Group Invariance Condition, *Fan*Tashkent, 1985.

[7] B. Loginov, Branching of solutions of nonlinear equations and group symmetry, *Vestnik of Samara University,* **2**, 1998, pp. 15–75.

[8] B. Loginov and Yu. B. Rousak, Generalized Jordan structure in the problem of the stability of bifurcating solutions, *Nonlinear Analysis, TMA,* **17**, 1991, pp. 219–232.

[9] B. Loginov, Branching equation in the root subspace, *Nonlinear Analysis, TMA,* **32**, 1998, pp. 439-448.

[10] B. Loginov, Determination of the branching equation by its group symmetry-Andronov-Hopf bifurcation, *Nonlinear Analysis, TMA,* **28**, 1997, pp. 2038-2047.

[11] B. Loginov and Ju. B. Rousak, Generalized Jordan structure in branching theory, in *Direct and Inverse Problems for Partial Differential Equations and their Applications*, (edited by M.S.Salakhitdinov and T.D.Dzhuraev), *Fan*, Tashkent, 1978, pp. 113-148.

[12] B. Loginov and I. Konopleva, Symmetry of resolving systems in degenerated functional equations, *Proceedings of International Conference Symmetry and Differential Equations*, edited by V.K.Andreev and V.V.Vasiliev, Krasnoyarsk, 2000, pp. 42-46.

[13] B. Loginov and I. Konopleva, Symmetry of resolving systems for differential equations with Fredholm operator at the derivative, *Proceedings of Internat. MORGAN 2000*, (edited by V.A. Baikov, R.K. Gazizov, N.H. Ibragimov and F.M. Mahomed), USATU, Ufa, 2000, pp. 116-119.

[14] Loginov, B. V., Branching equation in the root subspace: group symmetry and potentiality, *Functional Analysis,* **35**, Ulyanovsk State Pedagogical University, 1994, pp. 16-28/

[15] L. Ovsyannikov, Group Analysis of Differential Equations, *Academic Press*, New York, 1992

[16] Yu. B. Rousak, Some relations between Jordan sets of operator-function and adjoint to it, *Izv.Acad.Nauk Uzbek SSR, Fiz.-Mat. Ser.*, 1978, pp. 15-19.

[17] N. A. Sidorov, General Questions of Regularization in Branching Theory Problems, *Irkutsk University*, 1982

[18] G. Sviridiuk, Phase spaces of semilinear Sobolev-type equations with relatively strong sectorial operator, *Russian Math.,* **38**, 1994 pp. 72-79.

[19] M. Vainberg and V. Trenogin, Branching Theory of Solutions of Nonlinear Equations, *Noordorf Int. Publ.*, Leyden, 1974

[20] L. Volevich and A. Shirikyan, Local dynamics for high-order semilinear hyperbolic equations, *Izvestya RAN: ser. Math.,* **64**, 2000, pp. 3-50; Engl. transl.: *Izvestya Mathematics,* **64**, 2000, pp. 439-485.

[21] V.I. Yudovich, On the bifurcation of the creation of a cycle from a family of equilibria of a dynamical system and its tightening, *J. Appl. Math. Mech.* **62**, 1998, pp. 19-29.

[22] V.I. Yudovich, Cosymmetry, solutions degeneration of operator equations, arising of filtrational convection, *Math. Notes,* **49**, 1991, pp. 540-545.

A Gronwall-like Inequality and Continuous Dependence on Time Scales

P.E. Kloeden

FB Mathematik, Johann Wolfgang Goethe Universität, D-60054 Frankfurt am Main, Germany   e–mail: kloeden@math.uni-frankfurt.de

Abstract. We establish a Gronwall like inequality comparing the solutions of a dynamical equation on different time scales and then use it to prove the continuous dependence of such solutions on time scales. Regressivity is not used explicitly.

AMS Subject Classification. 34C11, 39A13

Key words. Time scales, dynamical equations Gronwall inequality, continuous dependence.

1 Time scales

A *time scale* $\mathbf{T}$ in $\mathbf{R}$ is just a nonempty closed subset of $\mathbf{R}$. The theory of dynamical equations on time scales provides a conceptual framework for the unification of continuous and discrete time dynamical systems generated, respectively, by differential and difference equations [1,4,5].

In numerical dynamics [6,7] one is concerned with the comparison of dynamics on the time scales $\mathbf{R}$ and $h\mathbf{Z} \in \mathcal{T}(\mathbf{R})$, where $h > 0$. Here we wish to compare solutions of a dynamical equation on arbitrary time scales, speficially to prove the continuous of solutions with respect to time scales. To do this we need to establish a Gronwall like inequality with respect to different time scales. We do not explicitly require regressivity of the dynamical equations on the time scales as in [3], although we assume the existence and uniqueness of solutions, which follow from regressivity.

Let $(X, \|\cdot\|_X)$ be a Banach space with norm $\|\cdot\|_X$. Recall that

$$\operatorname{dist}(x, A) = \inf_{a \in A} \|x - a\|_X$$

is the distance of a point $x \in X$ from a nonempty closed set A and that the Hausdorff separation $H^*(A, B)$ of nonempty closed subsets A, B of X is defined as

$$H^*(A, B) := \sup_{a \in A} \operatorname{dist}(a, B) = \sup_{a \in A} \inf_{b \in B} \|a - b\|,$$

while $H(A, B) = \max\{H^*(A, B), H^*(B, A)\}$ is the Hausdorff metric on the space $\mathcal{C}(X)$ of nonempty closed subsets of X (as well as on the subspace $\mathcal{K}(X)$ of nonempty compact subsets of X, in which case the infimum and supremum above are attained and can be replaced by the minimum and maximum, respectively). We will often write H_X^* and H_X when we want to distinguish Hausdorff separation and metric on a particular Banach space X or simply $H_{\mathbf{R}}^*$ and $H_{\mathbf{R}}$ when X is the real space $\mathbf{R}$.

A $\mathbf{T}$ be a time scale in $\mathbf{R}$. For simplicity, here we will consider only time scales which are unbounded in both directions, i.e. with inf $\mathbf{T} = -\infty$ and sup $\mathbf{T} = +\infty$ and denote the set of all such time scales by $\mathcal{T}(\mathbf{R})$ or just $\mathcal{T}$. Thus $\mathcal{T}(\mathbf{R})$ is a closed subspace of the complete metric space $(\mathcal{C}(\mathbf{R}), H_{\mathbf{R}})$, so $(\mathcal{T}(\mathbf{R}), H_{\mathbf{R}})$ is itself a complete metric space.

For example, $H_{\mathbf{R}}(\mathbf{R}, h\mathbf{Z}) = h/2$ for the time scales $\mathbf{R}$, $h\mathbf{Z} \in \mathcal{T}(\mathbf{R})$, where $h > 0$.

2 Continuous functions on time scales

Consider a time scale $\mathbf{T} \in \mathcal{T}(\mathbf{R})$ and denote by $C(\mathbf{T}, X)$ the space of all continuous functions $f : \mathbf{T} \to \mathbf{R}$, where X is a Banach space with norm $|\cdot|_X$. Here continuous means continuous with respect to the relative topology on $\mathbf{T}$ or, equivalently,

$$f(t_n) \to f(t_0) \quad \text{for any} \quad t_n \to t_0 \quad \text{in} \quad \mathbf{T}, \qquad \text{i.e. with} \quad t_n, t_0 \in \mathbf{T}.$$

For any f, $g \in C(\mathbf{T}, \mathbf{R})$ we define

$$\|f - g\|_{\mathbf{T}} = \sup_{t \in \mathbf{T}} \|f(t) - g(t)\|_X,$$

which is a norm on $C(\mathbf{T}, X)$. In particular, the linear space space $C(\mathbf{T}, X)$ is complete. (This is a form of uniform convergence of piecewise continuous functions with the same discontinuity points, but we will not prove it here as we have another aim).

We want to compare such functions on different time scales, i.e. any two functions f, $g \in C(\mathcal{T}, X) := \bigcup\{C(\mathbf{T}, X) : \mathbf{T} \in \mathcal{T}(\mathbf{R})\}$, i.e. we want to define a metric on $C(\mathcal{T}, X)$.

For this purpose, we define the *linear interpolant* $\widehat{f} : \mathbf{R} \to \mathbf{R}$ of a function $f \in C(\mathbf{T}, \mathbf{R})$ by

$$\widehat{f}(t) := \begin{cases} f(t) & : \quad t \in \mathbf{T} \\ f(\tau) + \dfrac{t - \tau}{\sigma_{\mathbf{T}}(\tau) - \tau}\left(f(\sigma_{\mathbf{T}}(\tau)) - f(\tau)\right) & : \quad \begin{array}{c} t \in (\tau, \sigma_{\mathbf{T}}(\tau)) \subset \mathbf{R} \setminus \mathbf{T}, \\ \tau, \sigma_{\mathbf{T}}(\tau) \in \mathbf{T} \end{array} \end{cases}$$

and note that $\widehat{f} \in C(\mathbf{R}, X)$. Moreover, $\widehat{f}$, $\widehat{g} \in C(\mathbf{R}, \mathbf{R})$ for any f, $g \in C(\mathcal{T}, X)$. As usual, we define

$$\|\widehat{f} - \widehat{g}\|_\infty := \sup_{t \in \mathbf{R}} \|\widehat{f}(t) - \widehat{g}(t)\|_X,$$

which is a norm on $C(\mathbf{R}, X)$. However, $\|\cdot\|_\infty$ is only a pseudo-norm on $C(\mathcal{T}, X)$, since for example $\widehat{\chi}_{h\mathbf{Z}} \equiv \widehat{\chi}_{\mathbf{R}}$, so $\|\widehat{\chi}_{\mathbf{R}} - \widehat{\chi}_{h\mathbf{Z}}\|_\infty = 0$, although obviously $\chi_{h\mathbf{Z}} \neq \chi_{\mathbf{R}}$.

Hence, in addition, we consider the *graph* $\mathbf{Gr}(f)$ of a function $f \in C(\mathbf{T}, X)$ defined by

$$\mathbf{Gr}(f) := \{(t, f(t) : \ t \in \mathbf{T}\},$$

which is a closed subset of $\mathbf{R} \times X$ due to the continuity of f on $\mathbf{T}$. However, although the Hausdorff metric $H_{\mathbf{R}\times X}\left(\mathbf{Gr}(f), \mathbf{Gr}(g)\right)$ between the graphs of functions f and g in $C(\mathcal{T}, X)$ provides a metric on $C(\mathcal{T}, X)$, the resulting metric space is not complete because the limit of a Cauchy sequences in the complete metric space $(\mathcal{C}(\mathbf{R} \times X), H_{\mathbf{R}\times X})$ need not be the graph of a function in $C(\mathcal{T}, X)$. For example, consider $f_n \in C(\mathbf{T}_n, \mathbf{R})$, where $\mathbf{T}_n = (-\infty, 0] \cup \left\{1 - \frac{1}{n}\right\} \cup [1, \infty)$, defined by

$$f_n(t) := \begin{cases} 0 & : \ t \in (-\infty, 0] \\ \frac{1}{2} & : \ t = 1 - \frac{1}{n} \\ 1 & : \ t \in [1, \infty) \end{cases}$$

Then $H_{\mathbf{R}^2}\left(\mathbf{Gr}(f_n), \mathbf{Gr}(f_m)\right) \geq \frac{1}{n}$ for all $m > n \geq 1$, so $\{\mathbf{Gr}(f_n)\}$ is a Cauchy sequence in $(\mathcal{C}(\mathbf{R}^2), H_{\mathbf{R}^2})$, which is complete. However, the limit set is not the graph of a function in $C(\mathbf{T}, \mathbf{R})$, where $\mathbf{T} = (-\infty, 0] \cup [1, \infty)$, since it contains the points $(1, \frac{1}{2})$ and $(1, 1)$.

Finally, we combine the above ideas to define a "distance" between two functions f, $g \in C(\mathcal{T}, X)$ by

$$d_\infty(f, g) := \max\left\{\|\widehat{f} - \widehat{g}\|_\infty, H_{\mathbf{R}\times X}\left(\mathbf{Gr}(f), \mathbf{Gr}(g)\right)\right\}, \tag{2.1}$$

which is a metric. In fact,

Theorem 2.1. *$C(\mathcal{T}, X), d_\infty)$ is a complete metric space.*

Moreover, since $H_{\mathbf{R}}(\mathbf{T}, \mathbf{T}') = H_{\mathbf{R}\times X}\left(\mathbf{Gr}(\chi_{\mathbf{R}}), \mathbf{Gr}(\chi_{\mathbf{R}'})\right)$, we also have

Corollary 2.1. *$H_{\mathbf{R}}(\mathbf{T}, \mathbf{T}') \leq H_{\mathbf{R}\times X}\left(\mathbf{Gr}(f), \mathbf{Gr}(g)\right) \leq d_\infty(f, g)$ for all $f \in C(\mathbf{T}, X)$ and $g \in C(\mathbf{T}', X)$.*

3 Dynamical systems on time scales

Let $\mathbf{T}$ be a time scale. We consider a dynamical equation

$$x^\Delta(t) = f(x(t)), \qquad t \in \mathbf{T}, \tag{3.1}$$

on $\mathbf{R}^d$ with respect to a time scale $\mathbf{T}$, i.e. where $x^{\Delta}(t)$ is the derivative $x^{\Delta(\mathbf{T})}(t)$ at time $t \in \mathbf{T}$ with respect to the time scale $\mathbf{T}$ [1,4,5].

We assume that $f : \mathbf{R}^d \to \mathbf{R}^d$ satisfies at least a local Lipschitz condition on $\mathbf{R}^d$ as well as some restricted growth or dissipativity condition to ensure the global existence and uniqueness of a solution $x(t) = \phi(t, t_0, x_0)$ of (3.1) for any initial value problem with $x(t_0) = x_0$. Since the solution depends on the time scale used, we will also write it as $x(t) = \phi(t, t_0, x_0, \mathbf{T})$. Our aim here is to establish the continuity of ϕ as a function of the time scale $\mathbf{T}$.

First, we observe that such a solution satisfies the integral equation

$$x(t) = x(t_0) + \int_{t_0}^{t} f(x(s))\, \Delta_{\mathbf{T}}(s), \tag{3.2}$$

for t, $t_0 \in \mathbf{T}$ with $t \geq t_0$. Here the integral is the Cauchy integral over the time scale $\mathbf{T}$ as defined in [1,4,5].

Then, if $y(\tau) = \phi(\tau, \tau_0, y_0, \mathbf{T}')$ is the solution of (3.1) with respect to another time scale $\mathbf{T}'$, we need to estimate

$$x(t) - y(\tau) = x_0 - y_0 + \int_{t_0}^{t} f(x(s))\, \Delta_{\mathbf{T}}(s) - \int_{\tau_0}^{\tau} f(y(s))\, \Delta_{\mathbf{T}'}(s)$$

for t, $t_0 \in \mathbf{T}$ with $t \geq t_0$ and for τ, $\tau_0 \in \mathbf{T}'$ with $\tau \geq \tau_0$, which requires a Gronwall like inequality. For this we need in turn to compare integrals over different time scales.

4 Comparison of integrals over different time scales

We want to estimate the difference of integrals over different time scales $\mathbf{T}$, $\mathbf{T}' \in \mathcal{T}(\mathbf{R})$, i.e.

$$\Delta(f, g) := \int_{t_0}^{t} f(s)\, \Delta_{\mathbf{T}}(s) - \int_{\tau_0}^{\tau} g(s)\, \Delta_{\mathbf{T}'}(s) \tag{4.1}$$

for t, $t_0 \in \mathbf{T}$ with $t \geq t_0$ and for τ, $\tau_0 \in \mathbf{T}'$ with $\tau \geq \tau_0$, where $f \in C(\mathbf{T}, X)$ and $g \in C(\mathbf{T}', X)$.

First we define the *piecewise constant interpolant* $\bar{f} : \mathbf{R} \to \mathbf{R}$ of a function $f \in C(\mathbf{T}, \mathbf{R})$ by

$$\bar{f}(t) := \begin{cases} f(t) & : \ t \in \mathbf{T} \\ f(\tau) & : \ t \in (\tau, \sigma_{\mathbf{T}}(\tau)) \subset \mathbf{R} \setminus \mathbf{T}, \quad \tau, \sigma_{\mathbf{T}}(\tau) \in \mathbf{T} \end{cases}$$

Then we note from the definition of the time scale integral that

$$\int_{t_0}^{t} \bar{f}(s)\, ds \equiv \int_{t_0}^{t} \bar{f}(s)\, \Delta_{\mathbf{R}}(s) = \int_{t_0}^{t} f(s)\, \Delta_{\mathbf{T}}(s) \tag{4.2}$$

for t, $t_0 \in \mathbf{T}$ with $t \geq t_0$, where the first integral is in fact the Riemann integral of $\bar{f}$ over the interval $[t_0, t]$.

Note that the first equivalence in (4.2) holds for all t, $t_0 \in \mathbf{R}$ with $t \geq t_0$, not just in $\mathbf{T}$. In fact, if for $f \in C(\mathbf{T}, \mathbf{R})$ we define

$$F(t) := \int_{t_0}^{t} f(s)\, \Delta_{\mathbf{T}}(s), \qquad t_0 \leq t \text{ in } \mathbf{T},$$

so $F \in C(\mathbf{T}, \mathbf{R})$, then

$$\widehat{F}(t) = \int_{t_0}^{t} \bar{f}(s)\, ds, \qquad t_0 \leq t \text{ in } \mathbf{R}.$$

Thus our comparison difference (4.1) is equivalent to

$$\Delta(f, g) = \int_{t_0}^{t} \bar{f}(s)\, ds - \int_{\tau_0}^{t} \bar{g}(s)\, ds$$

for $t \geq t_0$, τ_0 in $\mathbf{R}$, where $f \in C(\mathbf{T}, X)$ and $g \in C(\mathbf{T}', X)$.

5 A Gronwall-like inequality

We recall that the dynamical equation (3.1) over a time scale $\mathbf{T}$ can be written equivalently as an integral equation (3.2) over $\mathbf{T}$, namely

$$x(t) = x_0 + \int_{t_0}^{t} f(x(s))\, \Delta_{\mathbf{T}}(s), \qquad t_0 \leq t \text{ in } \mathbf{T}.$$

From our observations above we can rewrite this on $\mathbf{T}$ as

$$\widehat{x}(t) = x_0 + \int_{t_0}^{t} \overline{f \circ x}(s)\, ds, \qquad t_0 \leq t \text{ in } \mathbf{R},$$

which is the same as

$$\widehat{x}(t) = x_0 + \int_{t_0}^{t} f(\bar{x}(s))\, ds, \qquad t_0 \leq t \text{ in } \mathbf{R} \tag{5.1}$$

since $\overline{f \circ x} \equiv f \circ \bar{x}$ for compositions and piecewise constant interpolants.

Now consider the same dynamical equation (5.1) over another time scale $\mathbf{T}'$ with solution y with

$$\widehat{y}(t) = y_0 + \int_{t_0}^{t} f(\bar{y}(s))\, ds, \qquad t_0 \leq t \text{ in } \mathbf{R}$$

(for simplicity we consider the "initial condition" $\widehat{y}(t_0) = y_0$ instead of $y(t_0') = y_0$ for some $t_0' \in \mathbf{T}'$, since t_0 may not be in $\mathbf{T}'$; the general case can be handled with a small correction term of the form $M_\infty|t_0 - t_0'|$,where M_∞ will be defined later).

We want to compare these two solutions, i.e. to estimate

$$\widehat{x}(t) - \widehat{y}(t) = x_0 - y_0 + \int_{t_0}^{t} (f(\bar{x}(s)) - f(\bar{y}(s)))\, ds, \qquad t_0 \leq t \ \text{ in } \ \mathbf{R}$$

We suppose that f is globally Lipschitz with constant L and obtain

$$\begin{aligned} |\widehat{x}(t) - \widehat{y}(t)| &\leq |x_0 - y_0| + \left| \int_{t_0}^{t} (f(\bar{x}(s)) - f(\bar{y}(s)))\, ds \right| \\ &\leq |x_0 - y_0| + \int_{t_0}^{t} |f(\bar{x}(s)) - f(\bar{y}(s))|\, ds \\ &\leq |x_0 - y_0| + L \int_{t_0}^{t} |\bar{x}(s) - \bar{y}(s)|\, ds \end{aligned}$$

for all $t \geq t_0$ in $\mathbf{R}$ (i.e., so long as both solutions exist).

Our Gronwall inequality will involve piecewise linear interpolants like $|\widehat{x}(t) - \widehat{y}(t)|$ on the left hand side, so we need to bound the piecewise constant interpolants $|\bar{x}(s) - \bar{y}(s)|$ in the integral from above by piecewise linear interpolants in the upper bounding integrals below. What happens depends on whether the solutions have a common jump or not during a given subinterval

5.1. A general estimate.

Let B be some big compact ball in $\mathbf{R}^d$ that contains both solutions $x(t)$ and $y(t)$ over the finite time interval $[t_0, T]$, define

$$M_\infty := \max_{x \in B} |f(x)|$$

and let L be the Lipschitz constant for f over this ball B. In addition, define $L_J = L_J(\mathbf{T}, \mathbf{T}', [t_0, T])$ to be the length of the longest overlapping jump interval of the time scales $\mathbf{T}$ and $\mathbf{T}'$ in the finite time interval $[t_0, T]$.

We consider separately the cases of no common jump and a single common jump. These are based on estimates that are proved in Section 7.

5.1.1. Subinterval with no common jump

First, if there is no overlapping jump between t_0 and t_1, then we have

$$|\widehat{x}(t) - \widehat{y}(t)| \leq |x_0 - y_0| + L \int_{t_0}^{t} |\bar{x}(s) - \bar{y}(s)|\, ds$$

$$\leq \ |x_0 - y_0| + L\,\alpha(\mathbf{T}, \mathbf{T}', f)\, H_{\mathbf{R}}(\mathbf{T}, \mathbf{T}') + L \int_{t_0}^{t} |\widehat{x}(s) - \widehat{y}(s)|\, ds$$

for $t \in [t_0, t_1]$, where

$$\alpha(\mathbf{T}, \mathbf{T}', f) := 2\left(H_{\mathbf{R}}(\mathbf{T}, \mathbf{T}') + \max\left\{2H_{\mathbf{R}}(\mathbf{T}, \mathbf{T}'), L_J\right\}\right) M_\infty.$$

We can apply the usual Gronwall inequality to get

$$|\widehat{x}(t) - \widehat{y}(t)| \leq \left(|x_0 - y_0| + L\,\alpha(\mathbf{T}, \mathbf{T}', f)\, H_{\mathbf{R}}(\mathbf{T}, \mathbf{T}')\right)\, e^{L(t-t_0)}$$

This holds until the first $t_1 \in [t_0, T]$ where an overlapping jump interval starts. Hence point we have

$$|\widehat{x}(t) - \widehat{y}(t)| \leq A(t_1, t_0) \tag{5.2}$$

for $t \in [t_0, t_1]$, where

$$A(t_1, t_0) := \left(|\widehat{x}(t_0) - \widehat{y}(t_0)| + L\,\alpha(\mathbf{T}, \mathbf{T}', f)\, H_{\mathbf{R}}(\mathbf{T}, \mathbf{T}')\right)\, e^{L(t_1-t_0)}.$$

5.1.2. Subinterval with a single common jump

Let $t_1 \in (t_0, T]$ be the end of the single common jump interval that started at t_0, but one of the solutions (say y) may have started its jump before t_0, i.e. with $t_0' \leq t_0$ and hence $|t_0 - t_0'| \leq |\sigma_{\mathbf{T}'}(t_0') - t_0'| \leq 2H_{\mathbf{R}}(\mathbf{T}, \mathbf{T}')$. Then for $t \in [t_0, t_1]$ we have

$$\begin{aligned} |\widehat{x}(t) - \widehat{y}(t)| \ &\leq \ (1 + L(t_1 - t_0))\, |\widehat{x}(t_0) - \widehat{y}(t_0)| + 2L(t_1 - t_0) H_{\mathbf{R}}(\mathbf{T}, \mathbf{T}') M_\infty \\ &\leq \ |\widehat{x}(t_0) - \widehat{y}(t_0)|\, e^{L(t_1-t_0)} + 2L(t_1 - t_0) H_{\mathbf{R}}(\mathbf{T}, \mathbf{T}') M_\infty. \end{aligned}$$

Hence

$$|\widehat{x}(t) - \widehat{y}(t)| \leq B(t_1, t_0) \tag{5.3}$$

for $t \in [t_0, t_1]$, where

$$B(t_1, t_0) := |\widehat{x}(t_0) - \widehat{y}(t_0)|\, e^{L(t_1-t_0)} + 2L(t_1 - t_0) H_{\mathbf{R}}(\mathbf{T}, \mathbf{T}') M_\infty$$

.

5.2. Concatenated subintervals

If we start at t_0 and there is no common jump until t_1 we obtain from (5.2) that

$$|\widehat{x}(t) - \widehat{y}(t)| \leq A(t_1, t_0)$$

for $t \in [t_0, t_1]$. We can then start the procedure again at t_1 with a single common jump interval lasting till t_2. From (5.3) we obtain

$$|\widehat{x}(t_2) - \widehat{y}(t_2)| \leq B(t_2, t_0)$$

for $t \in [t_1, t_2]$. We then proceed in one of these previous ways according to whether we have a single common jump or no common jump in the interval. (The situation is more complicated if there is is a convergent sequence of scattered points in a bounded interval, but we delay considering this issue until later, see Section 6.1).

Note that we have a common upper bound for $A(t_1, t_0)$ and $B(t_1, t_0)$, namely

$$C(t_1, t_0) := |\widehat{x}(t_0) - \widehat{y}(t_0)|\, e^{L(t_1 - t_0)} + K(t_1, t_0)\, H_{\mathbf{R}}(\mathbf{T}, \mathbf{T}'),$$

where

$$K(t_1, t_0) := \max\left\{L\, \alpha(\mathbf{T}, \mathbf{T}', f)\, e^{L(t_1 - t_0)}, 2L(t_1 - t_0)\, M_\infty\right\}.$$

Thus for any successive subinterval $[t_n, t_{n+1}]$ of $[t_0, T]$ containing either no common jump or just a single common jump we have the common upper bound

$$|\widehat{x}(t) - \widehat{y}(t)| \leq |\widehat{x}(t_n) - \widehat{y}(t_n)|\, e^{L(t_{n+1} - t_n)} + K(T, t_0)\, H_{\mathbf{R}}(\mathbf{T}, \mathbf{T}') \qquad (5.4)$$

for all $t \in [t_n, t_{n+1}]$. Applying these successively starting at t_0 and ending at $t_N = T$ we obtain

$$\begin{aligned} |\widehat{x}(t) - \widehat{y}(t)| \;\leq\; & |\widehat{x}(t_0) - \widehat{y}(t_0)|\, e^{L(T - t_0)} \qquad (5.5) \\ & + K(T, t_0)\, H_{\mathbf{R}}(\mathbf{T}, \mathbf{T}') \sum_{j=0}^{N-1} e^{L(T - t_{N-j})} \end{aligned}$$

for all $t \in [t_0, T]$. This is our basic Gronwall like inequality.

6 Continuous dependence with respect to time scales

The estimate (5.5) allows one to deduce continuous convergence with respect to time scales of solutions of the dynamical equation (3.1) with a common vector field on different time scales provided that for <u>all</u> of the time scales under consideration we can divide the interval $[t_0, T]$ into a common fixed N subintervals $[t_n, t_{n+1}]$ for n =0, 1, ..., $N - 1$, in each of which a single jump or no jump occurs. This is quite restrictive. Moreover the constant $K(T, t_0)$ depends on both time scales under comparison through $\alpha(\mathbf{T}, \mathbf{T}', f)$ and $L_J = L_J(\mathbf{T}, \mathbf{T}', [t_0, T])$.

Since we wish to consider convergence of time scales $\mathbf{T}'$ to a fixed time scale $\mathbf{T}$, we can restrict attention to time scales $\mathbf{T}'$ for which $H_{\mathbf{R}}(\mathbf{T}', \mathbf{T}) \leq 1$. Then

$$L_J(\mathbf{T}, \mathbf{T}', [t_0, T]) \leq L_J^* < \infty$$

and

$$\alpha(\mathbf{T}, \mathbf{T}', f) \leq \alpha^* := 2\,(1 + \max\{2, L_J^*\})\, M_\infty < \infty$$

for all such time scales $\mathbf{T}'$. We can thus replace $K(T, t_0)$ in (5.5) by $K^*(T, t_0)$ defined with α^*, ie we have

$$\begin{aligned} |\widehat{x}(t) - \widehat{y}(t)| \;&\leq\; |\widehat{x}(t_0) - \widehat{y}(t_0)|\, e^{L(T-t_0)} \\ &\quad + K^*(T, t_0)\, H_{\mathbf{R}}(\mathbf{T}, \mathbf{T}') \sum_{j=0}^{N^*-1} e^{L(T - t_{N^*-j})} \end{aligned} \tag{6.1}$$

and obtain continuous dependence in times scales as $H_{\mathbf{R}}(\mathbf{T}', \mathbf{T}) \to 0$ provided the time scales have a uniformly bounded number N^* of common jump subintervals in the comparison interval $[t_0, T]$.

6.1. Presence of an isolated cluster point of scattered points

Suppose τ is a right–dense point of $\mathbf{T}$ in $[t_0, T)$ which is the limit of a decreasing infinite sequence of right or left–scattered points in $\mathbf{T}$. (Similar arguments can be used when a left-dense point is the limit from below of scattered points). Then for every $\epsilon > 0$ (sufficiently small) there is a point $\tau^\epsilon \in \mathbf{T}$ with $\tau^\epsilon - \tau \leq \epsilon$ such that the subinterval$(\tau, \tau^\epsilon) \cap \mathbf{T}$ of $\mathbf{T}$ contains infinitely many right–scattered points of $\mathbf{T}$ converging to τ. Then $\mathbf{T}_\epsilon := \mathbf{T} \setminus (\tau, \tau^\epsilon)$, which is a time scale with $H_{\mathbf{R}}(\mathbf{T}_\epsilon, \mathbf{T}) \leq \epsilon$.

From Section 7.6 we have

$$|\widehat{x}(t) - \widehat{y}(t)| \leq \left|\widehat{x^\epsilon}(t) - \widehat{y^\epsilon}(t)\right| + 4M_\infty\, \epsilon \tag{6.2}$$

for $t \in [\tau, \tau^\epsilon]$, where $x^\epsilon(t)$ is the solution of the dynamical equation (3.1) with respect to the deleted time scale $\mathbf{T}_\epsilon$ and similarly for $y^\epsilon(t)$. We fix $\epsilon \ll 1$ small enough so that $H_{\mathbf{R}}(\mathbf{T}_\epsilon, \mathbf{T}) \leq 1$ and $H_{\mathbf{R}}(\mathbf{T}'_\epsilon, \mathbf{T}) \leq 1$.

From the estimate (6.2) we have

$$\begin{aligned} |\widehat{x}(t) - \widehat{y}(t)| \;&\leq\; \left|\widehat{x^\epsilon}(\tau^\epsilon) - \widehat{y^\epsilon}(\tau^\epsilon)\right| + 4M_\infty\, \epsilon \\ &\leq\; |\widehat{x}(\tau) - \widehat{y}(\tau)|\, e^{L\epsilon} + K^*(T, t_0)\, H_{\mathbf{R}}(\mathbf{T}, \mathbf{T}') + 4M_\infty\, \epsilon \end{aligned} \tag{6.3}$$

in the interval $[\tau, \tau^\epsilon]$ since the solutions have the same initial conditions at time τ and the ϵ solutions have just a single common jump on this interval.

Then for interval $[\tau^\epsilon, T]$ the times scales $\mathbf{T}$ and $\mathbf{T}_\epsilon$ coincide (as do $\mathbf{T}'$ and $\mathbf{T}'_\epsilon$), so the dynamical equations have the same solutions here for the same initial conditions. For the initial conditions $\widehat{x}(\tau^\epsilon)$ and $\widehat{y}(\tau^\epsilon)$ we can apply the restricted case arguments leading to the estimate (6.1) to obtain

$$|\hat{x}(t) - \hat{y}(t)| \leq |\hat{x}(\tau^\epsilon) - \hat{y}(\tau^\epsilon)|\, e^{L(T-t_0)} + K^*(T,t_0)\, H_{\mathbf{R}}(\mathbf{T},\mathbf{T}') \sum_{j=0}^{N^\epsilon-1} e^{L(T-t_{N^\epsilon-j})} \tag{6.4}$$

for all $t \in [\tau^\epsilon, T)$. (This assumes there are no other cluster points of this kind in $[t_0,T] \setminus (\tau,\tau^\epsilon)$. We can handle finitely many other isolated cluster points by deleting similar subintervals, perhaps with differing lengths ϵ_j). Applying (6.3) for $t = \tau^\epsilon$ we thus have

$$|\hat{x}(t) - \hat{y}(t)| \leq \Big(|\hat{x}(\tau) - \hat{y}(\tau)|\, e^{L\epsilon} + K^*(T,t_0)\, H_{\mathbf{R}}(\mathbf{T},\mathbf{T}') + 4M_\infty\, \epsilon \Big)\, e^{L(T-t_0)} \tag{6.5}$$

we obtain

$$\lim_{\mathbf{T}'\to\mathbf{T}} |\hat{x}(t) - \hat{y}(t)| \leq \begin{cases} \lim_{\mathbf{T}'\to\mathbf{T}} |\hat{x}(\tau) - \hat{y}(\tau)|\, e^{L\epsilon} & : \; t \in [\tau,\tau^\epsilon] \\ \Big(\lim_{\mathbf{T}'\to\mathbf{T}} |\hat{x}(\tau) - \hat{y}(\tau)|\, e^{L\epsilon} + 4M_\infty\, \epsilon\Big)\, e^{L(T-t_0)} & : \; t \in [\tau^\epsilon, T] \end{cases} \tag{6.6}$$

Since the left hand side is independent of ϵ we can take $\epsilon \to 0$ in (6.6) and we obtain

$$\lim_{\mathbf{T}'\to\mathbf{T}} |\hat{x}(t) - \hat{y}(t)| \leq \lim_{\mathbf{T}'\to\mathbf{T}} |\hat{x}(\tau) - \hat{y}(\tau)|\, e^{L(T-t_0)} \qquad \text{for} \quad t \in [\tau, T]. \tag{6.7}$$

Finally, we can apply (6.1) on the subinterval $[t_0, \tau]$, which contains no other cluster points of scattered points, to obtain

$$\lim_{\mathbf{T}'\to\mathbf{T}} |\hat{x}(\tau) - \hat{y}(\tau)| \leq \lim_{\mathbf{T}'\to\mathbf{T}} |\hat{x}(t_0) - \hat{y}(t_0)|\, e^{L(T-t_0)}$$

which we combine with (6.7) to obtain

$$\lim_{\mathbf{T}'\to\mathbf{T}} |\hat{x}(t) - \hat{y}(t)| \leq \lim_{\mathbf{T}'\to\mathbf{T}} |\hat{x}(t_0) - \hat{y}(t_0)|\, e^{2L(T-t_0)} \tag{6.8}$$

for $t \in [\tau, T]$ as well as on $[t_0, \tau]$ in view of (6.1). Now for $x(t_0) = y(t_0)$ we have $\lim_{\mathbf{T}'\to\mathbf{T}} |\hat{x}(t_0) - \hat{y}(t_0)| = 0$ and hence

$$\lim_{\mathbf{T}'\to\mathbf{T}} |\hat{x}(t,t_0,x_0,\mathbf{T}') - \hat{x}(t,t_0,x_0,\mathbf{T})| = 0$$

uniformly on $[t_0, T]$.

This is the desired continuous dependence on time scales results. It applies as long as we restrict attention to time scales for which cluster points of scattered points are isolated. (see the remark at the end of Section 7.6 for the general case).

7 Appendix: Comparison of solutions on subintervals

Our Gronwall like inequality involves piecewise linear interpolants such as $|\widehat{x}(t) - \widehat{y}(t)|$ on the left hand side. Since the integrals on the right hand side involve piecewise constant interpolants $|\bar{x}(s) - \bar{y}(s)|$, we need to estimate these from above by piecewise linear interpolant on subintervals that do not contain a common jump.

We define $x_J^{\Delta}(t)$ to be the constant slope of a linearly interpolating section of a solution $x(t)$ on a jump interval, i.e. for $t \in (\tau, \sigma_{\mathbf{T}}(\tau)) \subset \mathbf{R} \setminus \mathbf{T}$, where τ, $\sigma_{\mathbf{T}}(\tau) \in \mathbf{T}$.

7.1. Both solutions are continuous

Here we suppose that the integration interval $(t_0, t_1) \subset \mathbf{T} \cap \mathbf{T}'$, i.e. neither solution has a jump within the integration interval. Then $\widehat{x}(t) = \bar{x}(t) = x(t)$ for $t \in [t_0, t_1]$, and similarly for the solution y. Thus we have

$$\int_{t_0}^{t_1} |\bar{x}(s) - \bar{y}(s)|\, ds = \int_{t_0}^{t_1} |\widehat{x}(s) - \widehat{y}(s)|\, ds$$

7.2. One solution is continuous, the other jumps

Here we suppose that the integration interval $(t_0, t_1) \subset \mathbf{T}$, but $(t_0, t_1) \subset \mathbf{R} \setminus \mathbf{T}'$ with $t_1 \in (t_0, \sigma_{\mathbf{T}'}(t_0)]$ where t_0, $\sigma_{\mathbf{T}'}(t_0) \in \mathbf{T}'$. This means that the solution x is exists and is continuous on the integration interval $[t_0, t_1]$, whereas the solution y undergoes a jump at t_0 until $\sigma_{\mathbf{T}'}(t_0)$. Hence

$$\widehat{x}(t) = \bar{x}(t) = x(t), \qquad \widehat{y}(t) = \bar{y}(t_0) + (t - t_0)\, y_J^{\Delta}(t_0) = \bar{y}(t_0) + (t - t_0)\, f(\bar{y}(t_0))$$

for $t \in [t_0, t_1]$, since $y(t_0) = \bar{y}(t_0)$ and $y_J^{\Delta}(t_0) = f(\bar{y}(t_0))$ here. Thus we have

$$\begin{aligned}
\int_{t_0}^{t_1} |\bar{x}(s) - \bar{y}(s)|\, ds &= \int_{t_0}^{t_1} \left|\widehat{x}(s) - \widehat{y}(s) - (s - t_0)\, y_J^{\Delta}(t_0)\right| ds \\
&\le \int_{t_0}^{t_1} |\widehat{x}(s) - \widehat{y}(s)|\, ds + \frac{1}{2}(t_1 - t_0)^2 \left|y_J^{\Delta}(t_0)\right| \\
&\le \int_{t_0}^{t_1} |\widehat{x}(s) - \widehat{y}(s)|\, ds + 2H_{\mathbf{R}}(\mathbf{T}, \mathbf{T}')^2 \, |f(\bar{y}(t_0))| \\
&\le \int_{t_0}^{t_1} |\widehat{x}(s) - \widehat{y}(s)|\, ds + 2H_{\mathbf{R}}(\mathbf{T}, \mathbf{T}')^2 \, M_\infty,
\end{aligned}$$

since $|t_1 - t_0| \le |\sigma_{\mathbf{T}'}(t_0) - t_0| \le 2H_{\mathbf{R}}(\mathbf{T}, \mathbf{T}')$.

7.3. One solution is continuous, the other one having started jumping earlier

Suppose that $[t_0, t_1] \subset \mathbf{T}$ whereas $t_0 \notin \mathbf{T}'$ with $t_0' \in \mathbf{T}'$ such that $t_0' < t_0 < t_1 \leq \sigma_{\mathbf{T}'}(t_0')$, i.e. x is continuous on $[t_0, t_1]$, whereas y jumps starting earlier at t_0'. Thus $\widehat{x}(t) = \bar{x}(t) = x(t)$ for $t \in [t_0, t_1]$, whereas $\bar{y}(t) = \bar{y}(t_0') = y(t_0')$ and $\widehat{y}(t) = \bar{y}(t_0') + (t - t_0')\, y_J^{\Delta}(t_0')$ for $t \in [t_0', t_1]$.

In this case we have

$$\begin{aligned}
\int_{t_0}^{t_1} |\bar{x}(s) - \bar{y}(s)|\, ds &= \int_{t_0}^{t_1} \left|\widehat{x}(s) - \widehat{y}(s) - (t - t_0)\, y_J^{\Delta}(t_0') - (t_0 - t_0')\, y_J^{\Delta}(t_0')\right| ds \\
&\leq \int_{t_0}^{t_1} |\widehat{x}(s) - \widehat{y}(s)|\, ds + \frac{1}{2}(t_1 - t_0)^2 \left|y_J^{\Delta}(t_0')\right| \\
&\qquad + (t_1 - t_0)(t_0 - t_0') \left|y_J^{\Delta}(t_0')\right| \\
&\leq \int_{t_0}^{t_1} |\widehat{x}(s) - \widehat{y}(s)|\, ds + 2H_{\mathbf{R}}(\mathbf{T}, \mathbf{T}')^2 \left|f(\bar{y}(t_0'))\right| \\
&\qquad + 2H_{\mathbf{R}}(\mathbf{T}, \mathbf{T}') \max\{2H_{\mathbf{R}}(\mathbf{T}, \mathbf{T}'), L_J\} \left|f(\bar{y}(t_0'))\right| \\
&\leq \int_{t_0}^{t_1} |\widehat{x}(s) - \widehat{y}(s)|\, ds + \alpha(\mathbf{T}, \mathbf{T}', f)\, H_{\mathbf{R}}(\mathbf{T}, \mathbf{T}'),
\end{aligned}$$

where

$$\alpha(\mathbf{T}, \mathbf{T}', f) := 2\left(H_{\mathbf{R}}(\mathbf{T}, \mathbf{T}') + \max\{2H_{\mathbf{R}}(\mathbf{T}, \mathbf{T}'), L_J\}\right) M_\infty,$$

since as above

$$|t_1 - t_0| \leq |\sigma_{\mathbf{T}'}(t_0') - t_0| \leq 2H_{\mathbf{R}}(\mathbf{T}, \mathbf{T}')$$

and

$$|t_0 - t_0'| \leq \max\{2H_{\mathbf{R}}(\mathbf{T}, \mathbf{T}'), L_J\}$$

(we need the "max" here as x may have had a jump ending at t_0).

7.4. Both solutions jump at same starting time

We suppose now that $t_0 \in \mathbf{T} \cap \mathbf{T}'$ and that $t_0 < t_1 \leq \min\{\sigma_{\mathbf{T}}(t_0), \sigma_{\mathbf{T}'}(t_0)\}$. Then

$$\widehat{x}(t) = \bar{x}(t_0) + (t - t_0)\, x_J^{\Delta}(t_0), \qquad \widehat{y}(t) = \bar{y}(t_0) + (t - t_0)\, y_J^{\Delta}(t_0)$$

for $t \in [t_0, t_1]$, since $x(t_0) = \bar{x}(t_0)$ and $y(t_0) = \bar{y}(t_0)$ here. In this case we have

$$|\widehat{x}(t) - \widehat{y}(t)| = \left|\bar{x}(t_0) - \bar{y}(t_0) + (t - t_0)\left(x_J^{\Delta}(t_0) - y_J^{\Delta}(t_0)\right)\right|$$

$$
\begin{aligned}
&\le |\bar{x}(t_0) - \bar{y}(t_0)| + (t_1 - t_0)\, |f(\bar{x}(t_0)) - f(\bar{y}(t_0))| \\
&\le |\bar{x}(t_0) - \bar{y}(t_0)| + (t_1 - t_0)\, L\, |\bar{x}(t_0) - \bar{y}(t_0)| \\
&= (1 + L(t_1 - t_0))\, |\widehat{x}(t_0) - \widehat{y}(t_0)|\, e^{L(t_2 - t_1)}, \\
&\le |\widehat{x}(t_0) - \widehat{y}(t_0)|\, e^{L(t_2 - t_1)}
\end{aligned}
$$

since $\bar{x}(t_0) = \widehat{x}(t_0)$ and $\bar{y}(t_0) = \widehat{x}(t_0)$ here and $|t - t_0| \le |t_1 - t_0| \le L_J$ on an overlapping jump interval.

7.5. Both functions jump with one having started jumping earlier

We suppose now that $t_0 \in \mathbf{T}$ with $t_0 < t_1 \le \sigma_{\mathbf{T}}(t_0)$ whereas $t_0 \notin \mathbf{T}'$ with $t_0' \in \mathbf{T}'$ such that $t_0' < t_0 < t_1 \le \sigma_{\mathbf{T}'}(t_0')$, i.e. x jumps at t_0, whereas y jumps earlier at t_0'. This means that $\widehat{x}(t) = \bar{x}(t_0) + (t - t_0)\, x_J^{\Delta}(t_0) = \widehat{x}(t_0) + (t - t_0)\, x_J^{\Delta}(t_0)$ for $t \in [t_0, t_1]$, whereas $\bar{y}(t) = \bar{y}(t_0') = y(t_0')$ and $\widehat{y}(t) = \bar{y}(t_0') + (t - t_0')\, y_J^{\Delta}(t_0')$ for $t \in [t_0', t_1]$. In particular, $\bar{y}(t_0) = \bar{y}(t_0')$ and

$$
\widehat{y}(t) = \bar{y}(t_0') + (t_0 - t_0')\, y_J^{\Delta}(t_0') + (t - t_0)\, y_J^{\Delta}(t_0') = \widehat{y}(t_0) + (t - t_0)\, y_J^{\Delta}(t_0')
$$

for $t \in [t_0, t_1]$. In this case we thus have

$$
\begin{aligned}
|\widehat{x}(t) - \widehat{y}(t)| &= \left|\widehat{x}(t_0) - \widehat{y}(t_0) + (t - t_0)\left(x_J^{\Delta}(t_0) - y_J^{\Delta}(t_0')\right)\right| \\
&\le |\widehat{x}(t_0) - \widehat{y}(t_0)| + (t - t_0)\, |f(\bar{x}(t_0)) - f(\bar{y}(t_0'))| \\
&\le |\widehat{x}(t_0) - \widehat{y}(t_0)| + L_J L\, |\bar{x}(t_0) - \bar{y}(t_0')|, \\
&\le |\widehat{x}(t_0) - \widehat{y}(t_0)| + L L_J\, |\widehat{x}(t_0) - \widehat{y}(t_0)| + (t_0 - t_0') L L_J \left|y_J^{\Delta}(t_0')\right|
\end{aligned}
$$

$$
\begin{aligned}
&\le (1 + L(t_1 - t_0))\, |\widehat{x}(t_0) - \widehat{y}(t_0)| + 2L(t_1 - t_0) H_{\mathbf{R}}(\mathbf{T}, \mathbf{T}')\, |f(\bar{y}(t_0'))| \\
&\le |\widehat{x}(t_0) - \widehat{y}(t_0)|\, e^{L(t_1 - t_0)} + 2L(t_1 - t_0) H_{\mathbf{R}}(\mathbf{T}, \mathbf{T}')\, M_\infty
\end{aligned}
$$

since $|t - t_0| \le |t_1 - t_0| \le L_J(\mathbf{T}, \mathbf{T}')$ for an overlapping jump interval and $|t_0 - t_0'| \le |\sigma_{\mathbf{T}'}(t_0) - t_0| \le 2H_{\mathbf{R}}(\mathbf{T}, \mathbf{T}')$.

7.6. One or both functions has infinitely many jumps in a bounded subinterval

Consider a right–dense point $\tau \in \mathbf{T}$ which is the limit of a decreasing infinite sequence of right or left–scattered points in $\mathbf{T}$. (Similar arguments can be used when a left-dense point is the limit from below of scattered points).

In particular, for every $\epsilon > 0$ there is a point $\tau^\epsilon \in \mathbf{T}$ with $\tau^\epsilon - \tau \leq \epsilon$ such that the subinterval $(\tau, \tau^\epsilon) \cap \mathbf{T}$ of $\mathbf{T}$ contains infinitely many right–scattered points of $\mathbf{T}$ converging to τ. Then $\mathbf{T}_\epsilon := \mathbf{T} \setminus (\tau, \tau^\epsilon)$, which is a time scale with $H_{\mathbf{R}}(\mathbf{T}_\epsilon, \mathbf{T}) \leq \epsilon$.

Now by the Mean Value Theorem for time scales (Theorem 1.67 and Corollary 1.68 [1]) applied to the solution $x(t)$ of the dynamical equation (3.1) on the time scale $\mathbf{T}$ with initial value $x(\tau) = x_0$ satisfies

$$|x(t) - x_0| \leq M_\infty (t - \tau) \qquad \text{for} \quad t \in [\tau, \tau^\epsilon] \cap \mathbf{T}$$

and hence the linear interpolation of the solution satisfies

$$|\widehat{x}(t) - x_0| \leq M_\infty (t - \tau) \qquad \text{for} \quad t \in [\tau, \tau^\epsilon] \subset \mathbf{R}. \tag{7.1}$$

Similarly, the linear interpolation $\widehat{x^\epsilon}(t)$ of the solution $x^\epsilon(t)$ of the the dynamical equation (3.1) on the time scale $\mathbf{T}_\epsilon$ with the same initial value $x^\epsilon(\tau)$ $= x_0$ satisfies

$$|x^\epsilon(t) - x_0| \leq M_\infty (t - \tau) \qquad \text{for} \quad t \in [\tau, \tau^\epsilon] \cap \mathbf{T}. \tag{7.2}$$

Combining (7.1) and (7.2) gives

$$\left|\widehat{x^\epsilon}(t) - \widehat{x}(t)\right| \leq 2M_\infty (t - \tau) \qquad \text{for} \quad t \in [\tau, \tau^\epsilon]$$

and hence

$$\left|\widehat{x^\epsilon}(t) - \widehat{x}(t)\right| \leq 2M_\infty\, \epsilon \qquad \text{for} \quad t \in [\tau, \tau^\epsilon] \cap \mathbf{T}.$$

We can repeat this with the dynamical equation (3.1) on the time scales $\mathbf{T}'$ and $\mathbf{T}'_\epsilon := \mathbf{T}' \setminus (\tau, \tau^\epsilon)$ (whether or not it contains infinitely many right scattered points of $\mathbf{T}'$) to obtain

$$\left|\widehat{y^\epsilon}(t) - \widehat{y}(t)\right| \leq 2M_\infty\, \epsilon \qquad \text{for} \quad t \in [\tau, \tau^\epsilon].$$

Hence

$$|\widehat{x}(t) - \widehat{y}(t)| \leq \left|\widehat{x^\epsilon}(t) - \widehat{y^\epsilon}(t)\right| + 4M_\infty\, \epsilon \qquad \text{for} \quad t \in [\tau, \tau^\epsilon] \tag{7.3}$$

In this way we can "remove" such a subinterval containing an infinite monotonically decreasing sequence of right-scattered points, i.e. where jumps occur, and just we apply the estimates of the previous subsections for no jumps or just single jumps on a bounded subinterval.

Finally, we note that the above arguments can be refined to handle the situation in which, say, a scattered point is the limit of a sequence of scattered points, which are themselves limits of sequences of scattered points. We leave the details to the interested reader.

References

[1] M. Bohner and A. Peterson, Dynamic Equations on Time Scales, *Birk häuser*, Basel, 2000.

[2] B.M. Garay and S. Hilger, Embeddability of time scales in ode dynamics, *Nonliner Anal.* **47** (2001), 1357–1371.

[3] B.M. Garay, S. Hilger und P.E. Kloeden, Continuous dependence in time scale dynamics, *Proceedings ICDEA 2001, Augsburg* (to appear).

[4] S. Hilger, Ein Maßkettenkälkul mit Anwendungen auf Zentrummannigfaltigkeiten, Dissertation, *Universität Würzburg*, 1988.

[5] S. Keller, Asymptotisches Verhalten invarianter Faserbündel bei Diskret isierung und Mittelwertungbildung im Rahmen der Analysis auf Zeitskalen Dissertation, *Universität Augsburg*, 1999.

[6] P.E. Kloeden and J. Lorenz, Stable attracting sets in dynamical systems and their one–step discretizations, *SIAM J. Numer. Anal.* **23** (1986), 986–995.

[7] A.M. Stuart and A.R. Humphries, Numerical Analysis and Dynamical Systems, *Cambridge University Press*, Cambridge 1996.

On Local Center Unstable Manifolds

Kazuo Kobayasi[1] and Satoru Takagi[2]

[1]Department of Mathematics, School of Education, Waseda University, Tokyo 169-8050 JAPAN e–mail: kzokoba@waseda.jp

[2]Department of Mathematical Sciences, Graduate School of Science and Engineering, Waseda University, Tokyo 169-8555 JAPAN
e–mail: satoru@aoni.waseda.jp

Abstract. We consider the existence of local center unstable manifolds for time dependent evolution equations of parabolic type in Banach spaces.

AMS Subject Classification. 37L10

Key words. Center unstable manifolds, nonlinear diffusion equations, C_0-semigroups.

1 Introduction

We consider the existence of local C^k center unstable manifolds for time dependent evolution equations of parabolic type in Banach spaces. The center unstable manifold theorem is a standard and useful idea in studying the long-time behavior of solutions to a class of partial differential equations in the neighborhood of a stationary point. In its formulation, up to now almost all theorems can apply to only a partial differential equation on a bounded domain. These frameworks are, however, too restrictive for many interesting applications, especially in the application of the equations on unbounded domain. In the case of unbounded domain, the main difficulty arises from the appearance of continuous spectrum of the linearized equation; there is no well-defined spectral gap. Nevertheless, a nonlinear heat equation of the form $u_t = \triangle u + F(u)$ on $\mathbf{R}^d$ does possess finite-dimensional local center unstable manifold, see Wayne [9].

It is thus useful to formulate a local center unstable manifold theorem in order to apply the partial differential equations on unbounded domains. In this paper, we present such an abstract theorem for the evolution equations in Banach spaces as to treat a class of partial differential equations on unbounded domains. Indeed, our result of this paper is used in Kobayasi [5] to construct a local invariant manifold for a nonlinear parabolic equation on the whole space $\mathbf{R}^d$.

Our approach is based on the classical method of Liapunov-Perron and follows closely Chow and Lu [2]. Related results can be found in Miklavčič [7], Galley [3], Mielke [6], Carr [1], Kobayasi [4], etc.

Let X, Y and Z be Banach spaces. The norms of X and Y will be denoted by $\|\cdot\|$ and $|\cdot|$, respectively. Suppose that both X and Y are continuously embedded in Z. Note that X is not necessarily embedded in Y. Let $\{S(t) : t \geq 0\}$ be a C_0-semigroup in Z and $f : \mathbf{R}\times X \to Y$ a nonlinear map of class C^k for some $k \geq 1$. Instead of evolution equations, we would rather consider the integral equation

$$u(t) = S(t)x_0 + \int_0^t S(t-s)f(s,u(s))ds, \qquad t \geq 0. \tag{1.1}$$

We are interested in the asymptotic behavior of the solution of (1.1).

We assume the following conditions on the C_0-semigroup:

(H1) $Z = Z_1 \oplus Z_2$, $\dim Z_1 < \infty$ and $S(t)P_i = P_iS(t)$, $i = 1,2$, where P_i is a continuous projection from Z onto Z_i.

(H2) $Z_1 \subset X\times Y$ and the restriction of $S(t)$ to X also forms a C_0-semigroup on X.

(H3) There exist constants $\alpha, \beta, \gamma, \eta, M, M^*$ such that $\alpha > 0$, $\beta+(k-1)\eta > 0$, $\eta < 0$, $0 \leq \gamma < 1$, $M \geq 1$, $M^* \geq 0$,

$$\begin{aligned}
\|e^{-\eta t}S(t)P_1y\| &\leq Me^{\alpha t}|y|, && \text{for } t \leq 0,\ y \in Y,\\
\|e^{-\eta t}S(t)P_2x\| &\leq Me^{-\beta t}\|x\|, && \text{for } t \geq 0,\ x \in X,\\
\|e^{-\eta t}S(t)P_2y\| &\leq (Mt^{-\gamma} + M^*)e^{-\beta t}|y|, && \text{for } t > 0,\ y \in Y.
\end{aligned}$$

Remark 1.1.

(a) *Condition* (H2) *implies* $X_1 = Y_1 = Z_1$, *where* $X_1 = P_1X$ *and* $Y_1 = P_1Y$, *for* $X_1 \subset Z_1 = P_1Z \subset X_1$. *Therefore, there is a constant* $M_1 \geq 1$ *such that* $M_1^{-1}|y| \leq \|y\| \leq M|y|$ *for* $y \in X_1 = Y_1$.

(b) *The restriction of* P_2 *to* X *becomes a continuous projection from* X *onto* $X_2 = P_2X$, *for* $\|P_2x\| \leq \|x\| + \|P_1x\| \leq \|x\| + C|P_1x|_Z \leq \|x\| + C|x|_Z \leq C\|x\|$ *for* $x \in X$.

(c) *Under the conditions* (H1)-(H3) *there exists* $M_0 \geq 1$ *such that*

$$\|S(t)y\| \leq M_0t^{-\gamma}|y| \qquad \textit{for } t \in (0,1],\ y \in Y. \tag{1.2}$$

We have our primary conclusion.

Theorem 1.2. *Assume that the hypotheses* (H1)-(H3) *above are satisfied. Let the map* $f : \mathbf{R}\times X \to Y$ *satisfy the following conditions:*

(a) *For each* $t \in \mathbf{R}$, $f(t,\cdot)$ *is of class* C^k. *For each* $x \in X$, $f(\cdot,x)$ *is continuous.*

(b) $f(t,0) = 0$ *and* $Df(t,0) = 0$ *for* $t \in \mathbf{R}$, *where* $Df(t,x)$ *is the derivative of* $f(t,x)$ *with respect to* x *evaluated at* (t,x).

(c) $f(t,x)$ *and* $Df(t,x)$ *converge, as* $\|x\| \to 0$ *uniformly in* t, *to* 0 *in* Y *and* $\mathcal{B}(X,Y)$, *respectively.*

Then there exist neighborhoods $U_1 \subset X_1$, $U_2 \subset X_2$ *of zero and a continuous function* $h : \mathbf{R} \times U_1 \to U_2$ *with the following properties:*

(M1) *The set* $\mathcal{M} = \bigcup_{t\in\mathbf{R}} \mathcal{M}(t)$, $\mathcal{M}(t) = \{(t, \xi + h(t,\xi)) : \xi \in U_1\}$, *is a local invariant manifold of* (1.1), *i.e., for each* $x_0 \in \mathcal{M}(0)$ *there exists* T_0, $T_1 \in (0,\infty]$ *such that a solution* u *of* (1.1) *uniquely exists on* $(-T_0, T_1)$ *and* $u(t) \in \mathcal{M}(t)$ *for all* $t \in (-T_0, T_1)$.

(M2) *For each* $t \in \mathbf{R}$, $h(t,\cdot)$ *is of class* C^k, $h(t,0) = 0$ *and* $Dh(t,0) = 0$.

(M3) *For each* $x_0 \in U_1 \times U_2$, (1.1) *has a unique solution on some interval* $[0,T)$. *If in addition* $T = \infty$, *then there exists a unique solution* $\tilde{u}$ *of* (1.1) *on* $\mathcal{M}$ *such that*

$$\sup_{t>0} e^{-\eta t}\|u(t) - \tilde{u}(t)\| < \infty.$$

The proof of Theorem 1.2 is obtained from the global center unstable manifold theorem by using an appropriate cut off function. We thus consider the global theorem in Section 2 and complete the proof of Theorem 1.2 in Section 3.

2 The global center unstable manifold theorem

In this section, as a nonlinear map let us take the continuous map $F : \mathbf{R} \times X \to Y$ satisfying the following conditions:

(H4) For each $t \in \mathbf{R}$, $F(t,\cdot)$ is of class C^k, $F(t,0) = 0$ and $DF(t,0) = 0$.

(H5) There exist constants L, $r > 0$ such that $|DF(t,x)|_{\mathcal{B}(X,Y)} \le L$ and $F(t,\tilde{x}) = 0$ for all $t \in \mathbf{R}$, $x \in B_2(r)$ and $\tilde{x} \in X$ with $\|P_1\tilde{x}\| > r$, where $B_2(r) = \{x \in X; \|P_2 x\| \le r\}$.

Let $J \subset \mathbf{R}$ be an interval. For any $\mu \in \mathbf{R}$ we denote by $C_\mu(J,X)$ the Banach space

$$C_\mu(J,X) = \left\{ v \in C(J,X) : \sup_{t\in J} e^{-\mu t}\|v(t)\| < \infty \right\}$$

with the norm $\|v\|_{C_\mu(J,X)} = \sup_{t\in J} e^{-\mu t}\|v(t)\|$.
Let

$$C(J, B_2(r)) = \{u \in C(J,X) : u(t) \in B_2(r) \quad \text{for all} \quad t \in J\}$$

and

$$C_\eta(r) = C_\eta(\mathbf{R}^-, X) \cap C(\mathbf{R}^-, B_2(r)).$$

Clearly, $C_\eta(r)$ is a closed subset of $C_\eta(\mathbf{R}^-, X)$.
Set

$$\begin{aligned} \mathcal{J}_\tau(\varphi, \xi)(t) &= S(t)\xi + \int_t^0 S(t-s)P_1F(s+\tau, \varphi(s))ds \\ &\quad + \int_{-\infty}^t S(t-s)P_2F(s+\tau, \varphi(s))ds, \qquad t \in \mathbf{R}^- \end{aligned}$$

for $\xi \in X_1$, $\varphi \in C_\eta(r)$ and $\tau \in \mathbf{R}$. Note that by virtue of (H3) - (H5), these integrals exist.

Lemma 2.1. *If $K(\alpha, \beta + (k-1)\eta, \gamma)L < 1/\{2(M+1)\}$, then there exists $\varepsilon_0 \in (0, \alpha)$ such that for each $\varepsilon \in [0, \varepsilon_0]$, $\tau \in \mathbf{R}$ and $\xi \in X_1$, the equation*

$$\varphi(t) = \mathcal{J}_\tau(\varphi, \xi)(t), \qquad t \in \mathbf{R}^- \tag{2.1}$$

has a unique solution $\varphi(\tau, \xi)(\cdot) \in C_{\eta+\varepsilon}(r/(M+1))$ independent of ε, where

$$K(\alpha, \beta, \gamma) = M\left(\alpha^{-1} + \Gamma(1-\gamma)\beta^{\gamma-1}\right) + M^*\beta^{-1}.$$

and Γ is the gamma function.
Moreover, the map $\varphi(\tau, \cdot) : X_1 \to C_{k\eta}(r/(M+1))$ is of class C^k.

Proof. By the continuity of $K(\alpha, \beta, \gamma)$ in α and β, there exists $\varepsilon_0 > 0$ such that $K(\alpha - \varepsilon, \beta + \varepsilon, \gamma)L < 1/\{2(M+1)\}$ for every $\varepsilon \in [0, \varepsilon_0]$. We show that $\mathcal{J}_\tau(\cdot, \xi) : C_{\eta+\varepsilon}(r/(M+1)) \to C_{\eta+\varepsilon}(r/(M+1))$ is a uniform contraction with respect to ξ and τ. We first prove that $\mathcal{J}_\tau(\cdot, \xi)$ maps $C_{\eta+\varepsilon}(r/(M+1))$ into itself. Let $\varphi \in C_{\eta+\varepsilon}(r/(M+1))$. By (H3)-(H5), we have $\mathcal{J}_\tau(\varphi, \xi) \in C_{\eta+\varepsilon}(\mathbf{R}^-, X)$ and

$$\begin{aligned} &\|P_2\mathcal{J}_\tau(\varphi, \xi)(t)\| \\ &\quad = \left\| \int_{-\infty}^t S(t-s)P_2F(s+\tau, \varphi(s))ds \right\| \\ &\quad \le \int_{-\infty}^t (M(t-s)^{-\gamma} + M^*)e^{-(\beta-\eta)(t-s)}|F(s+\tau, \varphi(s))|ds. \end{aligned}$$

Since $F(s+\tau, \varphi(s)) = 0$ if $\|P_1\varphi(s)\| > r$ by (H5), we have

$$\begin{aligned} |F(s+\tau, \varphi(s))| &= |F(s+\tau, \varphi(s)) - F(s+\tau, 0)| \\ &\le L(r + \|P_2\varphi(s)\|) \\ &= \frac{M+2}{M+1}Lr. \end{aligned}$$

Therefore,

$$\|P_2\mathcal{J}_\tau(\varphi,\xi)(t)\| \leq K(\alpha,\beta-\eta,\gamma)\frac{M+2}{M+1}Lr \leq \frac{r}{M+1},$$

and so $\mathcal{J}_\tau(\varphi,\xi) \in C_{\eta+\varepsilon}(r/(M+1))$.

Next we prove that $\mathcal{J}_\tau(\cdot,\xi)$ is a contraction uniformly with respect to ξ and τ. For $\varphi_1,\varphi_2 \in C_{\eta+\varepsilon}(\frac{r}{M+1})$, $\xi \in X_1$ and $\tau \in \mathbf{R}$, from (H3) and (H5) we have (see [2]) 9

$$\begin{aligned}
&\|\mathcal{J}_\tau(\varphi_1,\xi) - \mathcal{J}_\tau(\varphi_2,\xi)\|_{C_{\eta+\varepsilon}(\mathbf{R}^-,X)} \\
&\quad\leq K(\alpha-\varepsilon,\beta+\varepsilon,\gamma)L\|\varphi_1-\varphi_2\|_{C_{\eta+\varepsilon}(\mathbf{R}^-,X)} \\
&\quad\leq \frac{1}{2(M+1)}\|\varphi_1-\varphi_2\|_{C_{\eta+\varepsilon}(\mathbf{R}^-,X)}.
\end{aligned}$$

The strict contraction theorem assures that there exists a unique $\varphi_\varepsilon(\tau,\xi) \in C_{\eta+\varepsilon}(r/(M+1))$ such that $\mathcal{J}_\tau(\varphi_\varepsilon(\tau,\xi),\xi) = \varphi_\varepsilon(\tau,\xi)$. Since $C_{\eta+\varepsilon}(r/(M+1)) \subset C_\eta(r/(M+1))$, by uniqueness we have $\varphi_\varepsilon(\tau,\xi) = \varphi_0(\tau,\xi)$ for every $\varepsilon \in [0,\varepsilon_0]$.

Finally, according to [2, Lemma 3.4], $\varphi(\tau,\cdot)$ is C^k as a mapping from X_1 into $C_{k\eta}(r/(M+1))$. (The proof of [2] works well only by replacing $C_{\eta+\delta}(\mathbf{R}^-,X)$ with $C_{\eta+\delta}(r/(M+1))$). □

Now consider the equation

$$\text{(2.2)} \qquad u(t) = S(t-t_0)u(t_0) + \int_{t_0}^t S(t-s)F(s,u(s))ds, \qquad t \geq t_0.$$

We shall say that $u \in C(J,X)$ is a solution of (2.2) on J if it satisfies (2.2) for all t, $t_0 \in J$ with $t_0 \leq t$. Proceeding as in the proof of [2, Lemma 4.2] we can also obtain that a function $u \in C((-\infty,\tau],B_2(r))$ is a solution of (2.2) on $(-\infty,\tau]$ if and only if the function $\varphi(t)$ defined by $\varphi(t) = u(t+\tau)$ is a solution of (2.1) with $\xi = P_1u(\tau)$.

Lemma 2.2. *Let* $1 < \rho < 1 + 1/M$ *and*

$$K(\alpha,\beta,\gamma)L < \frac{1-(\rho-1)M}{2(M+1)}.$$

Then, for each $x_0 \in B_2(\rho r/(M+1))$ *and* $t_0 \in \mathbf{R}$, *the equation (2.2) has a unique solution* $u \in C([t_0,\infty),B_2(r))$ *such that* $u(t_0) = x_0$.

Proof. We may assume that $t_0 = 0$. Let $x_0 \in B_2(\rho r/(M+1))$. For $w \in C([0,T],B_2(r))$ set

$$(Gw)(t) = S(t)x_0 + \int_0^t S(t-s)F(s,w(s))ds, \qquad t \in [0,T].$$

We first show that G maps $C([0,T],B_2(r))$ into itself. Indeed, by our hypotheses

$$\begin{aligned}
&\|P_2(Gw)(t)\| \\
&\le \frac{M\rho r}{M+1} + \int_0^t (M(t-s)^{-\gamma} + M^*)e^{-\beta(t-s)}L(r + \|P_2 w(s)\|)ds \\
&\le \frac{M\rho r}{M+1} + K(\alpha,\beta,\gamma)L2r \le r.
\end{aligned}$$

It follows from (H5) and (1.2) that for $w, \tilde{w} \in C([0,T],B_2(r))$

$$\|(Gw)(t) - (G\tilde{w})(t)\| \le \frac{LM_0}{1-\gamma} t^{1-\gamma} \|w - \tilde{w}\|_{C([0,T],X)}.$$

By induction on n it follows easily that

$$\|(G^n w)(t) - (G^n \tilde{w})(t)\| \le \frac{(LM_0\Gamma(1-\gamma))^n}{\Gamma(n+1-n\gamma)} t^{n-n\gamma} \|w - \tilde{w}\|_{C([0,T],X)}$$

Since

$$\lim_{n\to\infty} \frac{(LM_0\Gamma(1-\gamma)T)^n}{\Gamma(n+1-n\gamma)} = 0,$$

by the fixed point theorem G has a unique fixed point u_T in $C([0,T],B_2(r))$. We then define $u \in C(\mathbf{R}^+, B_2(r))$ by $u(t) = u_T(t)$ for $t \in [0,T]$, which is well-defined by the uniqueness of fixed points. Clearly, u becomes a solution of (2.2).

The uniqueness of u is a consequence of the following argument.Let $\tilde{u}$ be another solution. Let $t_1 \ge 0$ and $t \in [t_1, t_1+1]$. By (H5) and (1.2) we have

$$\|u(t) - \tilde{u}(t)\| \le K\|u(t_1) - \tilde{u}(t_1)\| + M_0 L \int_{t_1}^t (t-s)^{n-n\gamma-1} \|u(s) - \tilde{u}(s)\| ds,$$

where $K = \max_{0\le t\le 1} \|S(t)\|_{\mathcal{B}(X)}$. Therefore, it follows from Lemma 6.7 of [8] that

$$\begin{aligned}
\|u(t) - \tilde{u}(t)\| \le\ & K\|u(t_1) - \tilde{u}(t_1)\| + \sum_{j=0}^{n-1} \left(\frac{M_0 L (t_1+1)^{1-\gamma}}{1-\gamma} \right)^j \\
& + \frac{(M_0 L\Gamma(1-\gamma))^n}{\Gamma(n-n\gamma)} \int_{t_1}^t (t-s)^{n-n\gamma-1} \|u(s) - \tilde{u}(s)\| ds.
\end{aligned}$$

We now fix n sufficiently large such that $n(1-\gamma) > 1$. Then, we find that there exist positive constants $C_1(t_1)$ and $C_2(t_1)$ such that

$$\|u(t) - \tilde{u}(t)\| \le C_1(t_1)\|u(t_1) - \tilde{u}(t_1)\| + C_2(t_1) \int_{t_1}^t \|u(s) - \tilde{u}(s)\| ds.$$

Using Gronwall's inequality, we obtain

$$\|u(t)-\tilde{u}(t)\| \leq C_1(t_1)e^{C_2(t_1)}\|u(t_1)-\tilde{u}(t_1)\|.$$

Since $t_1 \geq 0$ is arbitrary, this inequality immediately yields the uniqueness of u. □

Proposition 2.3. *Suppose that* (H1)-(H5) *are satisfied. Let* $1<\rho<1+1/M$ *and*

$$K(\alpha,\beta,\gamma)L < \frac{1-(\rho-1)M}{2(M+1)}.$$

For $\tau \in \mathbf{R}$*, define*

$$\mathcal{M}(\tau) = \{u(\tau) \quad ; \quad u \in C((-\infty,\tau], B_2(r/(M+1))) \text{ is a solution of } (2.2) \text{ on } (-\infty,\tau]\}.$$

Then we have that

(P1) *There exists a function* $h \in C(\mathbf{R}\times X_1, B_2(\frac{r}{M+1}))$ *such that* $h(t,\xi)$ *is* C^k *in* ξ *and*

$$\mathcal{M}(\tau) = \{\xi + h(\tau,\xi) : \xi \in X_1\}.$$

(P2) *For a solution* u *of* (2.2) *on* $[\tau,\infty)$*, we have that* $u(\tau)\in\mathcal{M}(\tau)$ *implies* $u(t)\in\mathcal{M}(t)$ *for* $t\geq\tau$*.*

Proof. By Lemma 2.1 we see that $\mathcal{M}(\tau)\neq\emptyset$. Let $x_0\in\mathcal{M}(\tau)$ and $u\in C((-\infty,\tau],B_2(r/(M+1)))$ a solution of (2.2) with $u(\tau)=x_0$. As noted above, $\varphi(\tau,\xi)(\cdot)\equiv u(\cdot+\tau)$ is the unique solution of (2.1) with $\xi_1=P_1x_0$. Then we set

$$h(\tau,\xi) = \int_{-\infty}^{0} S(-s)P_2F(s+\tau,\varphi^\tau(s))ds.$$

It is easy to see that $x_0=\xi+h(\tau,\xi)$ and $h(\tau,\xi)=\varphi(\tau,\xi)(0)-\xi$. Hence, by Lemma 2.1, $h(\tau,\xi)$ is a C^k mapping from X_1 into X_2 with respect to ξ. To see the continuity of $h(t,\xi)$ in t, we write

$$\begin{aligned} &h(t,P_1u(t)) - h(\sigma,P_1u(t)) \\ &\quad = P_2(u(t)-u(\sigma)) + h(\sigma,P_1u(\sigma)) - h(\sigma,P_1u(t)). \end{aligned}$$

Hence, $\|h(t,P_1u(t))-h(\sigma,P_1u(t))\|\leq C\|u(\sigma)-u(t)\|$ for some constant C. Thus (P1) is proved. Next, let $x_0\in\mathcal{M}(\tau)$. Since $x_0\in B_2(r/(M+1))$, by Lemma 2.2 we can extend u to the solution of (2.2) on $\mathbf{R}$ satisfying $u(t)\in B_2(r)$ for all $t\in\mathbf{R}$; in particular, we have for $\tau\prime>\tau$

$$u(t+\tau\prime) = \mathcal{J}_{\tau\prime}(u(\cdot+\tau\prime), P_1u(\tau\prime))(t), \quad t\in\mathbf{R}^-.$$

Hence, for $t \leq \tau\prime$

$$\begin{aligned}\|P_2u(t)\| &= \left\| \int_{-\infty}^{t-\tau\prime} S(t-\tau\prime - s)P_2F(s+\tau\prime, u(s+\tau\prime))ds \right\| \\ &\leq \int_{-\infty}^{t} (M(t-s)^{-\gamma} + M^*)e^{-\beta(t-s)} L(r + \|P_2u(s)\|)ds \\ &\leq K(\alpha,\beta,\gamma)L2r \leq \frac{r}{M+1}.\end{aligned}$$

Therefore, by definition we have $u(\tau\prime) \in \mathcal{M}(\tau\prime)$. This proves (P2). □

Proposition 2.4. *Suppose that* (H1)-(H5) *are satisfied. Let* $1 < \rho < 1 + 1/M$,

$$K(\alpha,\beta,\gamma)L < \frac{1-(\rho-1)M}{2(M+1)}$$

and

$$\frac{MK(\alpha,\beta,\gamma)L}{1-K(\alpha,\beta,\gamma)L} < 1.$$

Then, for each $x_0 \in B_2(\rho r/(M+1))$, *there exists a unique* $x_0^* \in \mathcal{M}(0)$ *such that*

$$\sup_{t\geq 0} e^{-\eta t}\|u(t,x_0) - u(t,x_0^*)\| < \infty,$$

where $u(t,x_0)$ *is the solution of* (2.2) *on* $\mathbf{R}^+$ *with* $u(0,x_0) = x_0$ *and* M_1 *is the constant given in Remarks* 1.1 (a).

Proof. Fix the solution $u(t) = u(t,x_0) \in C(\mathbf{R}^+, B_2(r))$ of (2.2) and put

$$\dot{E}_r = \{w \in C_\eta(\mathbf{R}^+, X) : w(t) + u(t) \in B_2(r) \quad \text{for all } t \geq 0\}.$$

Let $\omega_2 \in B_2(\rho r/(M+1))$. For $w \in \dot{E}_r$ define

$$\begin{aligned}&\mathcal{L}(w)(t) \\ &\quad = S(t)(\omega_2 - x_0) + \int_0^t S(t-s)P_2(F(s,w+u) - F(s,u))ds \\ &\qquad - \int_t^\infty S(t-s)P_1(F(s,w+u) - F(s,u))ds, \qquad \text{for } t \geq 0.\end{aligned}$$

Then, we have for $w,\ \tilde{w} \in \dot{E}_r$ and $t \geq 0$

$$\|P_2(\mathcal{L}(w)(t) + u(t)\| \leq \frac{M\rho r}{M+1} + K(\alpha,\beta,\gamma)L2r \leq r.$$

and

$$\|e^{-\eta t}(\mathcal{L}(w)(t) - \mathcal{L}(\tilde{w})(t))\| \leq K(\alpha,\beta,\gamma)L\|w-\tilde{w}\|_{C_\eta(\mathbf{R}^+,X)}.$$

Thus, $\mathcal{L}$ is a strict contraction from $\dot{E}_r$ into itself and hence there exists a unique $\hat{w}(\omega_2)(\cdot) \in \dot{E}_r$ such that $\mathcal{L}(\hat{w}(\omega_2)) = \hat{w}(\omega_2)$. Define

$$g(\omega_2) = P_1\hat{w}(\omega_2)(0) \qquad \text{for } \omega_2 \in B_2\left(\frac{\rho r}{M+1}\right).$$

Since we have

$$\begin{aligned} &\|g(\omega_2) - g(\tilde{\omega}_2)\| \\ &\leq \frac{MK(\alpha,\beta,\gamma)L}{1-K(\alpha,\beta,\gamma)L}\|\omega_2 - \tilde{\omega}_2\|, \quad \omega_2,\ \tilde{\omega}_2 \in B_2\left(\frac{\rho r}{M+1}\right) \end{aligned}$$

and

$$\begin{aligned} &\|h(0,\xi) - h(0,\tilde{\xi})\| \\ &\leq \frac{MM_1K(\alpha,\beta,\gamma)L}{1-K(\alpha,\beta,\gamma)L}|\xi - \tilde{\xi}|, \quad \xi,\ \tilde{\xi} \in X_1, \end{aligned}$$

there exists a unique $\omega_1^* \in X_1$ such that $\omega_1^* = g(h(0, \omega_1^* + P_1x_0))$. Then we set $\omega_2^* = h(0, \omega_1^* + P_1x_0)$ and $\hat{w}^*(t) = \hat{w}(\omega_2^*)(t)$. Obviously, we get $\omega_1^* = g(\omega_2^*) = P_1\hat{w}^*(0)$ and $\hat{w}^*$ satisfies the equation

$$\hat{w}^*(t) + u(t) = S(t)(\hat{w}^*(0) + x_0) + \int_0^t S(t-s)F(s, \hat{w}^* + u)ds.$$

Now set $x_0^* = \hat{w}^*(0)x_0$. Since $x_0^* = \omega_1^* + P_1x_0 + h(0, \omega_1^* + P_1x_0) \in \mathcal{M}(0)$, by Lemma 2.2 we must have $\hat{w}^*(t) + u(t) = u(t, x_0^*)$ the unique solution of (2.2) on $\mathbf{R}^+$ with $u(0, x_0^*) = x_0^*$. Hence, $u(t, x_0^*) - u(t) = \hat{w}^*(t) \in C_\eta(\mathbf{R}^+, X)$. □

3 Proof of Theorem 1.2.

Let $\rho : X_1 \to \mathbf{R}$ be a smooth function such that $0 \leq \rho(\xi) \leq 1$ for $\xi \in X_1$, $\rho(\xi) = 1$ for $\|\xi\| \leq 1/2$ and $\rho(\xi) = 0$ for $\|\xi\| \geq 1$. Since X_1 is finite dimensional, the existence of such a function is obvious. For $r > 0$ set

$$F_r(t,x) = f(t,x)\rho\left(\frac{P_1x}{r}\right), \qquad x \in X,$$

and denote by $L(t,r)$ the maximum of the Lipschitz constant of F_r with respect to x over $\{x \in X : \|P_2x\| \leq r\}$. Then $\lim_{r\to+0} L(t,r) = 0$ uniformly in t by assumption (c) of Theorem 1.2. Hence, applying Propositions 2.3 and 2.4 to the nonlinear map $F_r(t,u)$ with sufficiently small $r > 0$, we obtain the conclusion of Theorem 1.2 with $U_1 = \{x_1 \in X_1 : \|x_1\| \leq r\}$ and $U_2 = \{x_2 \in X_2 : \|x_2\| \leq r\}$.

References

[1] J. Carr, The center manifold theorem and its applications, *Springer-Verlag*, New York, 1983.

[2] S.-N. Chow and K. Lu, Invariant manifolds for flows in Banach spaces, *J. Differential Equations,* **74** (1988), 285–317.

[3] Th. Gallay, A center-stable manifold theorem for differential equations in Banach spaces, *Comm. Math. Phys.*, **152** (1993), 249–268.

[4] K. Kobayasi, C^1 approximations of inertial manifolds via finite differences, *Proc. Amer. Math. Soc.*, **127** (1999), 1143–1150.

[5] K. Kobayasi, An L^p theory of invariant manifolds for parabolic partial differential equations on $\mathbf{R}^d$, *J. Differential Equations,* **179** (2002), 195–212.

[6] A. Mielke, Locally invariant manifolds for quasilinear parabolic equations, *Rocky Mountain J. Math.*, **21** (1991), 707–714.

[7] M. Miklavčič, A sharp condition for existence of an inertial manifold, *J. Dyn. Differential Equations,* **3** (1991), 437–456.

[8] A. Pazy, Semigroups of linear operators and applications to partial differential equations, *Springer-Verlag*, New York, 1983.

[9] C. E. Wayne, Invariant manifolds for parabolic partial differential equations on unbounded domains, *Arch. Rat. Mech. Anal.*, **138** (1997), 279–306.

Monotonicity Properties and Inequalities of the Zeros of q-Associated Polynomials

C. G. Kokologiannaki[1], P. D. Siafarikas[2] and J. D. Stabolas[2,3]

[1]Department of Mathematics, University of Patras, 26500 Patras, Greece
e–mail: chrykok@math.upatras.gr

[2]Department of Mathematics, University of Patras, 26500 Patras, Greece
e–mail: panos@math.upatras.gr, stabol@math.upatras.gr

Abstract. Using a functional analytic method, based on the three terms recurrence relations that the q-associated polynomials satisfy , we present some monotonicity results and inequalities of the zeros of the orthogonal polynomials under consideration. The obtained results unify, generalize and improve previously known results.

AMS Subject Classification. 33C47, 33D45

Key words. q-associated polynomials, zeros, monotonicity pro-perties, inequalities..

1 Introduction

The theory of q-analogues or q-extensions of classical formulas and functions is based on the observation that $\lim_{q\to 1} \frac{1-q^a}{1-q} = a$. Therefore the number $\frac{1-q^a}{1-q}$ is called the basic number $[a]$. A q-analogue of the Pochhammer symbol $(a)_k$, $(a)_0 = 1$ and $(a)_k = a(a+1)(a+2)\dots(a+k-1)$, $k = 1,2,3,\dots$ is $(a;q)_0 = 1$ and $(a;q)_k = (1-a)(1-aq)(1-aq^2)\dots(1-aq^{k-1})$, $k = 1,2,3\dots$.

By use of the q-Pochhammer symbol we can define the q-hypergeometric series (or basic hypergeometric series) in the following way

$$
{}_r\phi_s\left(\begin{array}{c} a_1,\dots,a_r \\ b_1,\dots,b_s \end{array}\middle| q; z\right) = \\
= \sum_{k=0}^{\infty} \frac{(a_1;q)_k \cdots (a_r;q)_k}{(b_1;q)_k \cdots (b_s;q)_k} (-1)^{(1+s-r)k} q^{(1+s-r)k\binom{k}{2}} \frac{z^k}{(q;q)_k}.
$$

The q-hypergeometric series is a q-analogue of the hypergeometric series since

[3]Partially supported by the Greek State Scholarship's Foundation (I.K.Y.)

$$\lim_{q\to 1^-} {}_r\phi_s\left(\begin{matrix} q^{a_1},\dots,q^{a_r} \\ q^{b_1},\dots,q^{b_s}\end{matrix}\;\middle|\; q;(q-1)^{(1+s-r)}z\right) = {}_rF_s\left(\begin{matrix} a_1,\dots,a_r \\ b_1,\dots,b_s\end{matrix}\;\middle|\; z\right).$$

Using the above limit one can obtain the orthogonal polynomials which can be defined in terms of a hypergeometric function from their q-analogues, the so called q-orthogonal polynomials.

In 1884, Markov introduced a family of q-polynomials. Later, in 1949, Hahn [10] introduced the q-classical orthogonal polynomials, the q-derivatives of which are also orthogonal. This definition is analogous to the definition of the classical orthogonal polynomials, the derivatives of which are also orthogonal. In [2], Andrews and Askey continued the work of Hahn. A collection of formulas of the hypergeometric orthogonal polynomials which appear in the so-called Askey-scheme, as well as of their q-analogues, can be found in [23].

Lately, there has been an increasing interest in the q-orthogonal polynomials. This is due to their numerous applications in several areas of mathematics (e.g. continued fractions, Eulerian series, theta functions, elliptic functions, etc) and physics (e.g. angular momentum and its q-analogue, the q-Schrödinger equation and q-harmonic oscillators). There is also a connection between the representation theory of quantum algebras (Glebsch-Gordan coefficients, 3j and 6j symbols), which play an important role in physical applications, and the q-orthogonal polynomials, (see [27, and the references there in]).

The investigation of the behavior of the zeros of the classical orthogonal polynomials as functions of a parameter, say $\rho_{n,k}(a)$, $1 \le k \le n$, is very important due to their relevance in several physical contexts. It is also of great interest to find a suitable positive and differentiable function (even the best if this is possible) $f(a)$, which forces the products $f(a)\rho_{n,k}(a)$, $1 \le k \le n$ to increase or decrease.

In [24], Laforgia discussed the above problem for the zeros $x_{n,k}(\lambda)$, $n \ge 2$, $1 \le k \le [\frac{n}{2}]$ of the Ultraspherical (or Gegenbauer) polynomials. More precisely he proved in [24] that the function $\lambda x_{n,k}(\lambda)$ increases as λ increases for $\lambda \in (0,1)$ and in [25] conjectured that this holds for all $\lambda > 0$. Later, Ahmed et al. in [1], found that the function $[\lambda+(2n^2+1)/(4n+2)]^{1/2}x_{n,k}(\lambda)$ increases with respect to λ, for $-1/2 < \lambda \le 3/2$. Then in [17] and [19], Ismail-Letessier-Askey formulated the conjecture (ILAC) that the function $(\lambda+1)^{1/2}x_{n,k}(\lambda)$ increases with respect to λ, for $\lambda > -1/2$. This conjecture was proved by Ifantis and Siafarikas in [13] for the largest positive zero and by Dimitrov in [5] for all the positive zeros $x_{n,k}(\lambda)$ for $-1/2 < \lambda \le 9/2$ or $-1/2 < \lambda < 3/2 + \nu$ and $n > 1 + (\nu^2 + 3\nu + 3/2)^{1/2}$ where $\nu \in \mathbf{N}$. Finally in [8], Elbert and Siafarikas proved that the function $[\lambda + (2n^2 + 1)/(4n+2)]^{1/2}x_{n,k}(\lambda)$ increases with respect to λ for $-1/2 < \lambda$. This result extends the result of Ahmed et al. and implies the ILAC. The sharpness of the above result was established in [7]. In [26], Natalini and Palumbo gave some results in this direction for the zeros $x_{n,k}(a)$ of the generalized Laguerre polynomials.

To the best of our knowledge, there are only very few results concerning the zeros of q-orthogonal polynomials. In this paper, we continue the investigation of the above problem for the zeros of some q-associated orthogonal polynomials obtained from the q-orthogonal polynomials when we replace n by $n + c$ in the coefficients of the recurrence relation.

It is known that the classical orthogonal polynomials are given as the polynomial solutions of a second order Sturm-Liouville type differential equation. The associated polynomials do not satisfy a differential equation of such type, but a fourth order differential equation. Therefore, methods based on the properties of Sturm-Liouville type equations cannot be applied. Finally, we mention that the classical q-orthogonal polynomials of the Hahn tableau are the polynomial solutions of the q-difference equation

$$\sigma(x)D_qD_{1/q}y(x) + \tau(x)D_qy(x) + \lambda_{q,n}y(x) = 0,$$

where $\sigma(x)$ is a polynomial of at most second degree, $\tau(x)$ is a polynomial of at most first degree and

$$D_qf(x) = \begin{cases} \dfrac{f(qx) - f(x)}{(q-1)x}, & q \neq 1, \ x \neq 0, \\ f'(0), & x = 0 \end{cases}$$

denotes the q-difference operator.

The method we use for the study of the zeros of the q-associated polynomials, is a functional analytic method. It is based on the three terms recurrence relation of the q-associated polynomials and it was introduced by Ifantis and the second named author in [13] for classical orthogonal polynomials and in [14], [29] for the corresponding associated orthogonal polynomials. The method is briefly described in §2.

Our main results are presented in §3 whereas their proofs are given in §4. More precisely in §3.1, we give monotonicity properties and inequalities for the largest zero of the q-associated Pollaczek polynomials which is a generalization of the q-Pollaczek polynomials introduced in [16] and [4]. Special cases of the q-associated Pollaczek polynomials are the q-associated continuous Ultraspherical polynomials and the q-associated Hermite polynomials for which we can immediately obtain the corresponding results.

In §3.2 - §3.5, we study the monotonicity and we give inequalities for the zeros of the q-associated continuous Jacobi polynomials, the q-associated Laguerre polynomials, the associated Al-Salam-Carlitz II polynomials and the q-Meixner polynomials, special case of which are the q-associated Charlier polynomials. The above polynomials are generalizations of the corresponding polynomials that appear in the q-analogue of the Askey scheme.

Finally in §3.6, we study the monotonicity and we give inequalities for the largest zero of the q-Lommel polynomials introduced in [15] in association with the Jackson q-Bessel function $J_\nu^{(2)}(x;q)$ [22]. Corresponding results for the first zero of $J_\nu^{(2)}(x;q)$ are also given.

The above results unify, generalize and improve previously known results.

2 The method

A sequence of q-polynomials, $\{P_n(x;q)\}$, is orthogonal with respect to a positive measure if and only if it satisfies the three term recurrence relation

$$xP_n(x;q) = \alpha_n(q)P_{n+1}(x;q)+\beta_n(q)P_n(x;q)+c_n(q)P_{n-1}(x;q), \quad n=0,1,\ldots$$

with $P_{-1}(x;q) = 0$, $P_0(x;q) = 1$ and $\alpha_{n-1}(q)c_n(q) > 0$ (this theorem is usually attributed to Favard [9] but it is older, see e.g. the comments in [2]. Therefore, throughout the rest of the paper we will investigate only the cases that the above condition holds. We will also always assume that $0 < q < 1$.

Setting $Q_n(x;q) = U_nP_n(x;q)$, where $U_n = \sqrt{\dfrac{\alpha_{n-1}(q)}{c_n(q)}}U_{n-1}$, $U_{-1} = 0$, $U_0 = 1$, we obtain the corresponding orthonormal polynomials $Q_n(x;q)$, which have the same zeros with the polynomials $P_n(x;q)$ and satisfy the recurrence relation

$$\begin{gathered} a_n(q)Q_{n+1}(x;q) + a_{n-1}(q)Q_{n-1}(x;q) + b_n(q)Q_n(x;q) = xQ_n(x;q), \\ Q_{-1}(x;q) = 0, \quad Q_0(x;q) = 1, \end{gathered} \tag{2.1}$$

where $a_n(q) = \sqrt{\alpha_n(q)c_{n+1}(q)}$ and $b_n(q) = \beta_n(q)$, $n = 0,\ldots$ are real sequences.

The normalized q-associated orthonormal polynomials $Q_n(x,c;q)$ are obtained when we replace n by $n+c$, for arbitrary real $c \geq 0$ or $c > -1$, in the coefficients a_n and b_n of (2.1), i.e.

$$\begin{gathered} a_{n+c}(q)Q_{n+1}(x,c;q) + a_{n+c-1}(q)Q_{n-1}(x,c;q)+ \\ +b_{n+c}(q)Q_n(x,c;q) = xQ_n(x,c;q), \\ Q_{-1}(x,c;q) = 0, \quad Q_0(x,c;q) = 1. \end{gathered} \tag{2.2}$$

In the following we briefly present the method we'll use.

Let e_j, $j = 0,\ldots,n-1$ be an orthonormal base in a finite dimensional Hilbert space H_n. The truncated shift operator V and its adjoint V^* are defined by

$$Ve_j = e_{j+1}, \quad j=0,\ldots,n-2, \quad Ve_{n-1} = 0,$$

$$V^*e_j = e_{j-1}, \quad j=1\ldots,n-1, \quad V^*e_0 = 0.$$

Let also, $A(c;q)$ and $B(c;q)$ be the diagonal operators:

$$A(c;q)e_j = a_{j+c}(q)e_j \text{ and } B(c;q)e_j = b_{j+c}(q)e_j, \quad j=0\ldots,n-1.$$

Then we can see that the zeros of the polynomials $Q_n(x,c;q)$ of degree n defined by (2.2) are the eigenvalues of the operator

$$T(c;q) = A(c;q)V^* + VA(c;q) + B(c;q)$$

i.e.

$$(A(c;q)V^* + VA(c;q) + B(c;q))x_k(c;q) = \lambda_k(c;q)x_k(c;q)$$

or

$$\lambda_k(c;q) = \langle (A(c;q)V^* + VA(c;q) + B(c;q))x_k(c;q), x_k(c;q)\rangle, \tag{2.3}$$

where $\|x_k(c;q)\| = 1$, $0 \le k \le n-1$ and vice versa.

Remark 2.1. The identification of the zeros of orthogonal polynomials as eigenvalues of a tridiagonal matrix is a very old result. What is new in the above approach is the separation of this tridiagonal matrix as a sum of products of simple matrices in the operator form $T = AV^* + VA + B$. Due to this separation, some results have been found very easily.

When the operators $A(c;q)$ and $B(c;q)$ of the operator $T(c;q)$ depend on a parameter ν and $A(c,\nu;q)$, $B(c,\nu;q)$ are uniformly bounded for ν in some interval and differentiable in the operator norm, the eigenvectors of $T(c,\nu;q)$ are strongly differentiable and the derivative of the corresponding eigenvalues $\lambda_k(c,\nu;q)$ is given by

$$\begin{aligned}\frac{d\lambda_k(c,\nu;q)}{d\nu} &= \langle (A'(c,\nu;q)V^* + VA'(c,\nu;q) + \\ &+B'(c,\nu;q))x_k(c,\nu;q), x_k(c,\nu;q)\rangle, \quad \|x_k(c,\nu;q)\| = 1,\end{aligned} \tag{2.4}$$

where $A'(c,\nu;q)e_j = a'_{j+c}(\nu;q)e_j$, $B'(\nu)e_j = b'_{j+c}(\nu;q)e_j$, $j = 0\ldots,n-1$ and the prime means differentiation with respect to ν (for details see [11]).

As far as the largest eigenvalue $\lambda_{\max}(c,\nu;q)$ of $T(c,\nu;q)$ is concerned, it is known that the components of the corresponding eigenvector $x_{\max}(c,\nu;q)$ are strictly positive numbers [12]. From this result and (2.4) it follows that if $a'_{j+c}(\nu;q) > 0$ and $b'_{j+c}(\nu;q) > 0$ ($a'_{j+c}(\nu;q) < 0$ and $b'_{j+c}(\nu;q) < 0$), $j = 0,\ldots,n-1$, the largest eigenvalue of $T(c,\nu;q)$, i.e. the largest zero of $Q_n(x,c;q)$, is a strictly increasing (decreasing) function of ν.

A positive function $f(\nu)$ that forces the product $f(\nu)\lambda_{\max}(c,\nu;q)$ to decrease (increase) can be found in the following way:

Let $a_{j+c}(\nu;q) > 0$ and $b_{j+c}(\nu;q) > 0$, $j = 0,\ldots,n-1$. Let also $\frac{d}{d\nu}a_{j+c}(\nu;q) == \gamma_j(c,\nu;q)a_{j+c}(\nu;q)$ and $\frac{d}{d\nu}b_{j+c}(\nu;q) = \delta_j(c,\nu;q)b_{j+c}(\nu;q)$. From (2.4) and the fact that the components of $x_{\max}(c,\nu;q)$ are strictly positive numbers it follows that, if there exists a function $h(\nu) \neq 0$ such that $\gamma_j(c,\nu;q) < h(\nu)$ and $\delta_j(c,\nu;q) < h(\nu)$ ($\gamma_j(\nu,c;q) > h(\nu)$ and $\delta_j(\nu,c;q) > h(\nu)$), $j = 0,\ldots,n-1$ the following differential inequality holds

$$\frac{d}{d\nu}\lambda_{\max}(c,\nu;q) < h(\nu)\lambda_{\max}(c,\nu;q), \quad (\frac{d}{d\nu}\lambda_{\max}(c,\nu;q) > h(\nu)\lambda_{\max}(c,\nu;q)),$$

from which it follows that the product $e^{-\int h(\nu)d\nu}\lambda_{\max}(c,\nu;q)$ is a decreasing (increasing) function of ν.

3 Main results

3.1. q-Associated Pollaczek Polynomials

The orthonormal q-associated Pollaczek polynomials satisfy the recurrence relation (2.2) with

$$\alpha_n = \frac{1}{2}\sqrt{\frac{(1-q^{n+c+1})(1-\lambda^2 q^{n+c})}{(1-a\lambda q^{n+c})(1-a\lambda q^{n+c+1})}} > 0, \quad n \geq 0, \quad c > -1, \tag{3.1.1}$$

and

$$\beta_n = -\frac{q^{n+c}b}{(1-a\lambda q^{n+c})}. \tag{3.1.2}$$

and they are orthonormal for $-q^{-c/2} \leq \lambda \leq q^{-c/2}$ and $a\lambda < q^{-c}$.

Theorem 3.1.1 *(Monotonicity with respect to λ)*

(i) The largest zero $x_{n,1}(c,\lambda,a,b;q)$ with $c > -1$, of the q-associated Pollaczek polynomials decreases with respect to λ, for:

$0 < a \leq q^{-c/2}, \ \lambda'' < \lambda < q^{-c/2}, \ b \geq 0,$

or

$-q^{-c/2} \leq a < 0, \ \lambda' < \lambda < q^{-c/2}, \ b \leq 0,$

and increases with respect to λ, for

$-q^{-c/2} \leq a < 0, \ -q^{-c/2} < \lambda < \lambda'', \ b \geq 0,$

or

$0 < a \leq q^{-c/2}, \ -q^{-c/2} < \lambda < \lambda', \ b \leq 0.$

(ii) For $a\lambda q^c < 1$, $c > -1$, $b \leq 0$, the function

$$(1-a\lambda q^c)x_{n,1}(c,\lambda,a,b;q),$$

increases with respect to λ, for

$a < 0, \quad -q^{-c/2} < \lambda < 0,$

and decreases with respect to λ, for

$0 < a, \quad 0 < \lambda < q^{-c/2}.$

(iii) For $b \leq 0$, the function

$$\frac{1-a\lambda q^c}{\sqrt{1-\lambda^2 q^c}}\, x_{n,1}(c,\lambda,a,b;q)$$

increases with respect to λ, for

$a < 0, \quad 0 \le \lambda < q^{-c/2}$

and decreases with respect to λ, *for*

$0 < a, \quad -q^{-c/2} < \lambda \le 0.$

(iv) For $b = 0$, $c > -1$ *the function*

$$\left(\frac{1 - a\lambda q^{c+1}}{|\lambda|}\right)^{1/2} x_{n,1}(c, \lambda, a, 0; q),$$

increases with respect to λ, *for*

$-q^{-c/2} \le a < 0, \quad -q^{-c/2} < \lambda < 0$

and decreases with respect to λ, *for*

$0 < a \le q^{-c/2}, \quad 0 < \lambda < q^{-c/2}$, *where*

$$\lambda' = \frac{1 + a^2 q^{n+c} - \sqrt{(1 - a^2 q^{n+c-1})(1 - a^2 q^{n+c+1})}}{a q^{n+c-1}(1+q)}$$

and

$$\lambda'' = \frac{1 + a^2 q^{c+1} + \sqrt{(1 - a^2 q^{c})(1 - a^2 q^{c+2})}}{a q^{c}(1+q)}.$$

From the previous monotonicity results we can obtain the following inequalities:

Corollary 3.1.2. (i) The largest zero $x_{n,1}(c, \lambda, a, b; q)$ with $a\lambda q^c < 1$, $b \ge 0$, $c > -1$ of the q-associated Pollaczek polynomials satisfies the inequalities:

$$\frac{1 + a q^{c/2}}{1 - a\lambda q^c} x_{n,1}(c, -q^{-c/2}, a, b; q) < x_{n,1}(c, \lambda, a, b; q) < \tag{3.1.3}$$
$$< (1 - a\lambda q^c)^{-1} x_{n,1}(c, 0, a, b; q),$$

for $a < 0$, $-q^{-c/2} < \lambda < 0$,

$$\frac{1 - a q^{c/2}}{1 - a\lambda q^c} x_{n,1}(c, q^{-c/2}, a, b; q) < x_{n,1}(c, \lambda, a, b; q) < \tag{3.1.4}$$
$$< (1 - a\lambda q^c)^{-1} x_{n,1}(c, 0, a, b; q),$$

for $0 < a$, $0 < \lambda < q^{-c/2}$.

(ii) For $b = 0$, $c > -1$, the largest zero $x_{n,1}(c, \lambda, a, 0; q)$, of the q-associated Pollaczek polynomials satisfies the inequalities:

$$x_{n,1}(c, \lambda, a, 0; q) > \sqrt{\frac{-q^{c/2}\lambda(1 + a q^{c/2+1})}{(1 - a\lambda q^{c+1})}} x_{n,1}(c, -q^{-c/2}, a, 0; q), \tag{3.1.5}$$

for $-q^{-}c/2 \leq a < 0$, $-q^{-c/2} < \lambda < 0$, and

$$(3.1.6) \qquad x_{n,1}(c,\lambda,a,0;q) > \sqrt{\frac{q^{c/2}\lambda(1-aq^{c/2+1})}{(1-a\lambda q^{c+1})}}\, x_{n,1}(c,-q^{-c/2},a,0;q),$$

for $0 < a \leq q^{-c/2}$, $0 < \lambda < q^{-c/2}$.

The associated Pollaczek polynomials are obtained from the q-associated Pollaczek polynomials by replacing λ by q^{λ}, a by q^{α}, b by $1-q^{b}$, and taking the limit $q \to 1^{-}$. In this case, we obtain corresponding results for the largest zero $x_{n,1}(c,\lambda,\alpha,b)$ of the associated Pollaczek polynomials.

Corollary 3.1.3.(i) The largest zero $x_{n,1}(c,\lambda,\alpha,b)$ with $c > -1$ of the associated Pollaczek polynomials increases with respect to λ, for

$-\frac{c}{2} < \lambda < \lambda''$, $\alpha \geq -\frac{c}{2}$, $b \geq 0$

and decreases with respect to λ, for

$\lambda > \lambda'$, $\alpha \geq -\frac{c}{2}$, $b \leq 0$.

(ii) For $b \leq 0$, $c > -1$ the function

$$(c+\lambda+\alpha)x_{n,1}(c,\lambda,\alpha,b),$$

increases with respect to λ, for $\lambda > -\frac{c}{2}$, $\alpha+\lambda+c > 0$.

(iii) For $b = 0$, $c > -1$ the function

$$(c+\lambda+\alpha+1)^{1/2}x_{n,1}(c,\lambda,\alpha,0),$$

increases with respect to λ, for $\lambda > -\frac{c}{2}$, $\alpha \geq -\frac{c}{2}$,

where

$$(3.1.7) \qquad \lambda' = \frac{1-c-n+\sqrt{(n+c+2\alpha-1)(n+c+2\alpha+1)}}{2},$$

and

$$(3.1.8) \qquad \lambda'' = \frac{-c+\sqrt{(c+2\alpha)(c+2\alpha+2)}}{2}.$$

From the above monotonicity properties, we obtain the following inequalities:

Corollary 3.1.4. The largest zero $x_{n,1}(c,\lambda,\alpha,b)$ with $c > -1$ of the associated Pollaczek polynomials satisfies the inequalities:

$$(3.1.9) \qquad x_{n,1}(c,\lambda,\alpha,b) > \frac{c/2+\alpha}{c+\lambda+\alpha}x_{n,1}(c,-c/2,\alpha,b), \text{ with } b \leq 0,$$

for $\lambda > -c/2$, $\alpha + \lambda + c > 0$,
and

$$(3.1.10) \quad x_{n,1}(c, \lambda, \alpha, 0) > \sqrt{\frac{c/2 + \alpha + 1}{c + \lambda + \alpha + 1}} x_{n,1}(c, -c/2, \alpha, 0), \text{ with } b = 0$$

for $\alpha \geq -c/2$ and $\lambda > -c/2$.

Remark 3.1.5. For $c > -1$, $b = 0$, $\alpha > -c/2$ and $\lambda > -c/2$ the inequality (3.1.10) is better than the inequality (3.1.9).

Remark 3.1.6. For $c = 0$, we obtain the corresponding results for the largest zero of the q-Pollaczek and Pollaczek polynomials.

Theorem 3.1.7.*(Monotonicity with respect to a)*

(i) The largest zero $x_{n,1}(c, \lambda, a, b; q)$ with $b \leq 0$, $c > -1$ and $a\lambda q^c < 1$ of the q-associated Pollaczek polynomials increases with respect to a, for

$0 < \lambda < q^{-c/2}$

and decreases with respect to a, for

$-q^{-c/2} < \lambda < 0.$

(ii) For $b \leq 0$, $c > -1$ and $a\lambda q^c < 1$ the function

$$(1 - a\lambda q^c) x_{n,1}(c, \lambda, a, b; q),$$

decreases with respect to a, for

$0 < \lambda < q^{-c/2}$

and increases with respect to a, for

$-q^{-c/2} < \lambda < 0.$

Since, for $a < q^{-c/2}$, $0 < \lambda < q^{-c/2}$ and for $-q^{-c/2} < a$, $-q^{-c/2} < \lambda < 0$ we obtain $a < q^{-c}\lambda^{-1}$ and $q^{-c}\lambda^{-1} < a$ respectively. Then from the previous monotonicity properties we obtain the following:

Corollary 3.1.8. The largest zero of the q-associated Pollaczek polynomials for $b \leq 0$ satisfy the inequalities

$$x_{n,1}(c, \lambda, a, b; q) > \frac{1 + \lambda q^{c/2}}{1 - a\lambda q^c} x_{n,1}(c, \lambda, -q^{-c/2}, b; q)$$

for $-q^{-c/2} < a$, $-q^{-c/2} < \lambda < 0$ and

$$x_{n,1}(c, \lambda, a, b; q) > \frac{1 - \lambda q^{c/2}}{1 - a\lambda q^c} x_{n,1}(c, \lambda, q^{-c/2}, b; q)$$

for $a < q^{-c/2}$, $0 < \lambda < q^{-c/2}$.

By replacing, λ by q^λ, a by q^α, b by $1-q^b$, and taking the limit $q \to 1^-$, we obtain ana-logous results for the largest zero $x_{n,1}(c, \lambda, \alpha, b)$ of the associated Pollaczek polynomials. More precisely we have the following:

Corollary 3.1.9.

(i) The largest zero $x_{n,1}(c,\lambda,\alpha,b)$ of the associated Pollaczek polynomials decreases with respect to α, for $\lambda > -c/2$, $\lambda+\alpha+c>0$, $b\leq 0$.

(ii) The function $(c+\lambda+\alpha)x_{n,1}(c,\lambda,\alpha,b)$ increases with respect to α, for $\lambda > -c/2$, $\lambda+\alpha+c>0$, $b\leq 0$.

(iii) The largest zero $x_{n,1}(c,\lambda,\alpha,b)$ of the associated Pollaczek polynomials satisfy the inequality

$$x_{n,1}(c,\lambda,\alpha,b) > \frac{c/2+\lambda}{c+\lambda+\alpha}x_{n,1}(c,\lambda,-c/2,b)$$

for $-c/2<\alpha$, $-c/2<\lambda$, $b\leq 0$.

Remark 3.1.10. For $c=0$ we obtain the corresponding results for the largest zero of the q-Pollaczek and Pollaczek polynomials.

Theorem 3.1.11.

(i) (Monotonicity with respect to b)
All the zeros $x_{n,k}(c,\lambda,a,b;q)$, $k=1,2,\ldots,n$ of the q-associated Pollaczek polynomials decreases with respect to b.

(ii)(Differential inequalities with respect to b)
The zeros $x_{n,k}(c,\lambda,a,b;q)$, $k=1,2,\ldots,n$ of the q-associated Pollaczek polynomials satisfy the differential inequalities:

$$-\frac{q^c}{1-a\lambda q^c} < \frac{\partial}{\partial b}x_{n,k}(c,\lambda,a,b;q) < -\frac{q^{n+c-1}}{1-a\lambda q^{n+c-1}} < 0, \quad k=1,2,\ldots,n \tag{3.1.11}$$

Substituting λ by q^λ, a by q^α, b by $1-q^b$ and taking the limit $q\to 1^-$, we obtain from (3.1.11) the corresponding results for the zeros $x_{n,k}(c,\lambda,\alpha,b)$, $k=1,2,\ldots,n$ of the associated Pollaczek polynomials.

Corollary 3.1.12. (i) All the zeros $x_{n,k}(c,\lambda,a,b)$, $k=1,2,\ldots,n$ of the associated Pollaczek polynomials decrease with respect to b.

(ii) The zeros $x_{n,k}(c,\lambda,a,b)$, $k=1,2,\ldots,n$ of the associated Pollaczek polynomials satisfy the differential inequalities:

$$\frac{-1}{c+\lambda+a} < \frac{\partial}{\partial b}x_{n,k}(c,\lambda,a,b) < \frac{-1}{n+c+\lambda+a-1} \tag{3.1.12}$$

Remark 3.1.13. For $c=0$, we obtain the analogous results for the zeros $x_{n,k}(\lambda,\alpha,b)$, $k=1,2,\ldots,n$ of the Pollaczek polynomials, which have also been found in [20].

Theorem 3.1.14.*(Monotonicity with respect to c)*

(i) The largest zero $x_{n,1}(c,\lambda,a,b;q)$ of the q-associated Pollaczek polynomials increases with respect to c, for $c>-1$, $b\geq 0$ in the following cases:

a) $a\lambda\leq\min\{\lambda^2,1\}$
b) $-q^{-c/2}\leq\lambda\leq a$, $-1\leq a\leq 0$,
c) $-q^{-c}<a\lambda<q^{-c}$, $-q^{-c/2}\leq\lambda\leq aq$, $0\leq a\leq 1$
d) $0<\lambda<q^{-c/2}$, $a\leq\min\{\frac{q}{\lambda},\frac{\lambda}{q}\}$
e) $-q^{-c/2}<\lambda<0$, $a\geq\max\{\frac{q}{\lambda},\frac{\lambda}{q}\}$

(ii) For $b=0$ and $c>-1$ the function

$$\left(\frac{1}{(1-q^{c+1})(1-\lambda^2q^c)}\right)^{1/2} x_{n,1}(c,\lambda,a,0;q)$$

decreases with respect to c for $-q^{-c/2}<\lambda<q^{-c/2}$, $0<a\lambda<q^{-c}$.

(iii) For $b=0$ and $c>-1$, the function

$$\left(\frac{1}{1-q^{c+1}}\right)^{1/2} x_{n,1}(c,\lambda,a,0;q)$$

decreases with respect to c, for $0<\lambda<q^{-c/2}$, $0<\lambda<a$.

Using the previous monotonicity properties we obtain the following:

Corollary 3.1.15. The largest zero $x_{n,1}(c,\lambda,a,b;q)$ of the q-associated Pollaczek polynomials for $b=0$ and $c\geq 0$ satisfies the inequalities

$$x_{n,1}(c,\lambda,a,0;q)\leq\sqrt{\frac{(1-q^{c+1})(1-\lambda^2q^c)}{(1-q)(1-\lambda^2)}}x_{n,1}(\lambda,a,0;q),$$

for $-q^{-c/2}<\lambda<q^{-c/2}$, $0<a\lambda<q^{-c}$ and

$$x_{n,1}(c,\lambda,a,0;q)\leq\sqrt{\frac{1-q^{c+1}}{1-q}}x_{n,1}(\lambda,a,0;q)$$

for $0<\lambda<q^{-c/2}$, $0<\lambda<a$, where $x_{n,1}(\lambda,a,b;q)$ is the largest zero of the q-Pollaczek polynomials.

Substituting λ by q^λ, a by q^α, b by $1-q^b$ and taking the limit $q\to 1^-$, we obtain from theorem 3.1.14 the analogous results for the largest zero $x_{n,1}(c,\lambda,\alpha,b)$ of the associated Pollaczek polynomials:

Corollary 3.1.16.(i) The largest zero $x_{n,1}(c,\lambda,\alpha,b)$ of the associated Pollaczek polynomials increases with respect to c, for $b\geq 0$, $\alpha\geq 0$ and λ in one of the intervals $[-\alpha,\alpha]$, $[1-\alpha,\infty)$.

(ii)For $b = 0$, the function

$$\left(\frac{1}{(c+1)(c+2\lambda)}\right)^{1/2} x_{n,1}(c,\lambda,\alpha,0)$$

decreases with respect to c, for $\alpha+\lambda+c>0$ and $\lambda > -c/2$.

(iii) For $b = 0$, the function

$$\left(\frac{1}{c+1}\right)^{1/2} x_{n,1}(c,\lambda,\alpha,0)$$

decreases with respect to c, for $\alpha\lambda + c > 0$ and $\lambda \geq \alpha$.

(iv) For $b = 0$ and $c \geq 0$, the largest zero $x_{n,1}(c,\lambda,\alpha,b)$ of the associated Pollaczek polynomials satisfies the inequalities

$$x_{n,1}(c,\lambda,\alpha,0) < \sqrt{\frac{(c+1)(c+2\lambda)}{2\lambda}}\, x_{n,1}(0,\lambda,\alpha,0),$$

for $\alpha+\lambda+c>0$, $\lambda > -c/2$ and

$$x_{n,1}(c,\lambda,\alpha,0) < \sqrt{c+1}\, x_{n,1}(0,\lambda,\alpha,0),$$

for $\alpha+\lambda+c>0$, $\lambda \geq \alpha$.

The q-associated continuous Ultraspherical polynomials are obtained from the q-associated Pollaczek polynomials for $b = 0$ and $a = 1$. Thus, the analogous results for the largest zero of the q-associated continuous Ultraspherical polynomials follow immediately from the corresponding results for the largest zero of the q-associated Pollaczek polynomials that hold for $b = 0$ and $a = 1$. Also, substituting, as in the case of the q-associated Pollaczek polynomials, λ by q^λ and taking the limit $q \to 1^-$ we obtain the corresponding results for the largest zero of the associated Ultraspherical polynomials.

Remark 3.1.17. (i) In [3], Askey and Wilson proved that the zeros of the q-Ultraspherical polynomials increase in absolute value with respect to λ, for $0 < q, \lambda < 1$.

(ii) From case (iv) of theorem 3.1.1. for $c \geq 0$, $a = 1$, substituting λ by q^λ and taking the limit $q \to 1^-$ we obtain that the function $(c+\lambda+1)^{1/2} x_{n,1}(c,\lambda)$ increases with respect to λ for $-c/2 < \lambda$. Also, since $-1 < c \Rightarrow -c/2 < 1/2$, the above function increases also for $\frac{1}{2} \leq \lambda$, as it was found in [29].

(iii) From case (ii) of theorem 3.1.1 for $c \geq 0$, $a = 1$, substituting λ by q^λ and taking the limit $q \to 1^-$ we obtain that the function $(c+\lambda)x_{n,1}(c,\lambda)$ increases with respect to λ for $-c/2 < \lambda$. For $c = 0$, which is the case of Ultraspherical polynomials we obtain that the functions $(\lambda+1)^{1/2} x_{n,1}(\lambda)$ and $\lambda x_{n,1}(\lambda)$ increase with respect to λ from where easily follows that the functions $(\lambda+1)^{3/2}x_{n,1}(\lambda)$ and $\lambda^{3/2}x_{n,1}(\lambda)$ also increase with respect to λ. Dimitrov in [6] proved that this functions are convex functions of λ for $\lambda \geq 0$.

Moreover for the largest zero of the q-associated Ultraspherical polynomials we can prove the following differential inequalities:

Theorem 3.1.18. *The largest zero $x_{n,1}(c,\lambda;q)$ of the associated continuous q-Ultra-spherical polynomials satisfies the following inequalities:*

$$
\begin{aligned}
& e^{\left[\frac{(1-\lambda)(q-\lambda)}{2\lambda^3 q^2}\left(\frac{2\lambda^2 q^n+\lambda q-q}{2(q-\lambda^2 q^n)^2}-\frac{q(1+\lambda)}{2(q-\lambda^2 q^{n+c})^2}+\frac{1}{q-\lambda^2 q^{n+c}}\right)\right]} x_{n,1}(\lambda;q) < \\
& < x_{n,1}(c,\lambda;q) < \\
& < e^{\left[\frac{(q-\lambda)(1-\lambda)(1-q^c)(2-q+\lambda q-q^{c+1}+\lambda q^{c+1}-2\lambda q^{c+2})}{4(1-q)^2(1-q^{c+1})^2}\right]} x_{n,1}(\lambda;q), \\
& \qquad 0<\lambda<q, \quad c>-1,
\end{aligned} \tag{3.1.13}
$$

and

$$
\begin{aligned}
& e^{\frac{(q-\lambda)(1-\lambda)(1-q^c)(2-\lambda+\lambda q-\lambda q^c+\lambda q^{c+1}-2\lambda^2 q^{c+1})}{4(1-\lambda)^2(1-\lambda q^c)^2}} x_{n,1}(\lambda;q) < \\
& < x_{n,1}(c,\lambda;q) < \\
& < e^{\frac{q^n(q-\lambda)(1-\lambda)(1-q^c)(2q+\lambda q^n-\lambda q^{n+1}+\lambda q^{n+c}-\lambda q^{n+c+1}-2\lambda^2 q^{2n+c})}{4q^2(1-\lambda q^n)^2(1-\lambda q^{n+c})^2}} x_{n,1}(\lambda;q), \\
& \qquad q<\lambda<1, \quad c\geq 0.
\end{aligned} \tag{3.1.14}
$$

Remark 3.1.19. For $\lambda = q^\lambda$ and $q \to 1$ we obtain the corresponding results for the largest zero $x_{n,1}(c,\lambda)$ of the associated Ultraspherical polynomials, which are the same with the inequalities which were given in [29].

The q-associated continuous Hermite polynomials are obtained from the q-associated continuous Ultraspherical polynomials for $\lambda = 0$. Thus, using the results that we just obtained for the q-associated continues Ultraspherical polynomials, for $\lambda = 0$, we obtain the analogous results for the largest zero of the q-associated continuous Hermite polynomials.

Remark 3.1.20. Taking the limit $q \to 1^-$ we obtain the analogous results for the zeros of the associated Hermite polynomials, which have also been proved in [14].

3.2. q-Associated continuous Jacobi polynomials

The continuous q-Jacobi polynomials $P_n^{(a,b)}(x;q)$ satisfy [23 p. 83] the recurrence relation

$$2xP_n^{(a,b)}(x;q) = A_n P_{n+1}^{(a,b)}(x;q)+$$

$$+[q^{a/2+1/4} + q^{-a/2-1/4} - (A_n + C_n)]P_n^{(a,b)}(x;q) + C_n P_{n-1}^{(a,b)}(x;q)$$

where

$$A_n = \frac{(1-q^{n+a+1}))(1-q^{n+a+b+1})(1+q^{n+(a+b+1)/2})(1+q^{n+(a+b+2)/2})}{q^{a/2+1/4}(1-q^{2n+a+b+1})(1-q^{2n+a+b+2})}$$

$$C_n = \frac{q^{a/2+1/4}(1-q^n))(1-q^{n+b})(1+q^{n+(a+b)/2})(1+q^{n+(a+b+1)/2})}{(1-q^{2n+a+b})(1-q^{2n+a+b+1})},$$

with $a > -1$, $b > -1$.

If instead of q we set q^2 in the recurrence relation, we obtain another continuous q-analogue of Jacobi polynomials $R_n^{(a,b)}(x;q)$, given by Rahman in [28], which is not really different, since they are connected by the quadratic transformation:

$$R_n^{(a,b)}(x;q) = P_n^{(a,b)}(x;q^2) = \frac{(-q;q)_n}{(-q^{a+b+1};q)_n} q^{na} P_n^{(a,b)}(x;q).$$

The orthonormal q-associated continuous Jacobi polynomials $Q_n^{(a,b)}(x;c,q)$ satisfy the recurrence relation (2.2), with

$$\alpha_n(c,a,b;q) = \frac{1}{2}\sqrt{\frac{(1-q^{2n+2c+2})(1-q^{2n+2c+2b+2})}{(1-q^{2n+2c+a+b+2})(1-q^{2n+2c+a+b+3})}} \times$$

$$\sqrt{\frac{(1-q^{2n+2c+2a+2})(1-q^{2n+2c+2a+2b+2})}{(1-q^{2n+2c+a+b+1})(1-q^{2n+2c+a+b+2})}} \tag{3.2.1}$$

and

$$\beta_n(c,a,b;q) = -\frac{(1+q)(q^b-q^a)(1-q^{b+a})q^{2n+2c+1/2}}{2(1-q^{2n+2c+a+b})(1-q^{2n+2c+a+b+2})}. \tag{3.2.2}$$

The coefficients (3.2.1) and ((3.2.2)) are both positive for $a > -1$, $b > |a|$, $n \geq 0$ and $c \geq 0$.

Remark 3.2.1. For $a = b = \lambda - \frac{1}{2}$ and setting q instead of q^2 in (3.2.1) and ((3.2.2)), we obtain the three term recurrence relation, which is satisfied by the q-associated continues Ultraspherical polynomials.

Remark 3.2.2. In [3] Askey and Wilson proved that the zeros of the q-Jacobi polynomials decrease with respect to a and increase with respect to b just as the zeros of the classical Jacobi polynomials do.

In this section we will only find a suitable positive function that changes the monotonicity of the largest zero $x_{n,1}(c,a,b;q)$ of the q-Jacobi polynomials.

Theorem 3.2.3. *(Monotonicity with respect to a)*

(i) The function

$$\frac{(1-q^{2c+a+b})(1-q^{2c+a+b+2})}{q^{2a}} x_{n,1}(c,a,b;q)$$

increases with respect to a, for $b>0,\ -b<a<b-\dfrac{ln(\frac{1+q^{2b}}{2})}{lnq},\quad c\geq 0.$

(ii) The function

$$\frac{1-q^{2c+a+b}}{q^a-q^b} x_{n,1}(c,a,b;q)$$

increases with respect to a, for $b>0,\ b-\dfrac{ln(\frac{1+q^{2b}}{2})}{lnq}<a<b,\ c\geq 0.$

Taking the limit $q\to 1^-$, we obtain analogous results for the zeros $x_{n,1}(c,a,b)$ of the associated Jacobi polynomials. More precisely we obtain the following:

Corollary 3.2.4. (i) The function

$$(2c+a+b)(2c+a+b+2)x_{n,1}(c,a,b)$$

increases with respect to a, for $-b<a<0$ and $c\geq 0$.

(ii) The function

$$\frac{2c+a+b+2}{b-a} x_{n,1}(c,a,b)$$

increases with respect to a, for $0<a<b$ and $c\geq 0$.

From the above monotonicity properties we obtain:

Corollary 3.2.5. (i) The largest zero $x_{n,1}(c,a,b;q)$ with $c\geq 0$, $b>0$ of the q-associated continuous Jacobi polynomials satisfies the inequalities

$$\frac{(1-q^{2c+b})(1-q^{2c+b+2})q^{2a}}{(1-q^{2c+a+b})(1-q^{2c+a+b+2})} x_{n,1}(c,0,b;q) < x_{n,1}(c,a,b;q) \tag{3.2.3}$$

for $0<a<b-\dfrac{ln(\frac{1+q^{2b}}{2})}{lnq}$, and

$$x_{n,1}(c,a,b;q) < \frac{(1-q^{2c+b})(1-q^{2c+b+2})q^{2a}}{(1-q^{2c+a+b})(1-q^{2c+a+b+2})} x_{n,1}(c,0,b;q) \tag{3.2.4}$$

for $-b < a < 0$.

(ii) The largest zero $x_{n,1}(c,a,b)$ with $c \geq 0$ of the associated Jacobi polynomials satisfies the inequalities

$$x_{n,1}(c,a,b) < \frac{(2c+b)(2c+b+2)}{(2c+a+b)(2c+a+b+2)} x_{n,1}(c,0,b) \tag{3.2.5}$$

for $-b < a < 0$ and

$$x_{n,1}(c,a,b) > \frac{(2c+b+2)(b-a)}{b(2c+a+b+2)} x_{n,1}(c,0,b) \tag{3.2.6}$$

for $0 < a < b$.

Theorem 3.2.6. *(Monotonicity with respect to b)*

(i) The function

$$\frac{q^{2a}(1-q^{2n+2c+a+b})(1-q^{2n+2c+a+b+1})}{(q^a-q^b)^2} x_{n,1}(c,a,b;q)$$

decreases with respect to b, *for* $0 < a < b$, $c \geq 0$.

(ii) The function

$$\frac{(1-q^{2n+2c+a+b})(1-q^{2n+2c+a+b+1})}{(1-q^{a+b})^2} x_{n,1}(c,a,b;q)$$

decreases with respect to b, *for* $-b < a < 0$, $c \geq 0$.

Taking the limit $q \to 1^-$, we obtain analogous results for the zeros $x_{n,1}(c,a,b)$ of the associated Jacobi polynomials. Moreover for $c = 0$ we obtain analogous results for the zeros $x_{n,1}(a,b)$ of the Jacobi polynomials. More precisely we have

Corollary 3.2.7. (i) The function

$$\frac{(2n+2c+a+b)(2n+2c+a+b+1)}{(b-a)^2} x_{n,1}(c,a,b)$$

decreases with respect to b, for $0 < a < b$ and $c \geq 0$.

(ii) The function

$$\frac{(2n+2c+a+b)(2n+2c+a+b+1)}{(b+a)^2} x_{n,1}(c,a,b)$$

decreases with respect to b, for $-b < a < 0$ and $c \geq 0$.

Remark 3.2.8. For $c = 0$ we obtain the analogous results for the largest zero of the q-Jacobi and Jacobi polynomials.

3.3. q-Associated Laguerre polynomials

The q-associated Laguerre polynomials $L_n^{(a)}(x;q)$ satisfy [23 p.108] the recurrence relation

$$(1-q^{n+c+1})L_{n+1}^{(a)}(x;q) + q(1-q^{n+c+a})L_{n-1}^{(a)}(x;q) -$$

$$-[1-q^{n+c+1} + q(1-q^{n+c+a})]L_n^{(a)}(x;q) = -xq^{2n+2c+a+1}L_n^{(a)}(x;q).$$

The orthonormal q-associated Laguerre polynomials $L_n(x;a,q)$ satisfy the recurrence relation (2.2), with

$$a_n = a_n(c,a;q) = \sqrt{\frac{(1-q^{n+c+1})(1-q^{n+c+a+1})}{q^{4n+4c+2a+3}}} \tag{3.3.1}$$

and

$$b_n = b_n(c,a;q) = \frac{1-q^{n+c+1}+q-q^{n+c+a+1}}{q^{2n+2c+a+1}}, \tag{3.3.2}$$

where $n \geq 0$, $a > -1$ and $c \geq 0$.

Theorem 3.3.1. *The zeros $x_{n,k}(c,a;q)$ $k = 1,2,\ldots$ of the q-associated Laguerre polynomials increase with respect to a, for $a \geq 0$, $c > -1$.*

Theorem 3.3.2. *The largest zero $x_{n,1}(c,a;q)$ of the q-associated Laguerre polynomials increases with respect to a, while the function*

$$\left(\frac{q^a}{1-q^{a+c+1}}\right) x_{n,1}(c,a;q)$$

decreases with respect to a, for $c \geq 0$, $a > -c-1$.

Taking the limit $q \to 1^-$, we obtain the following results for the largest zero $x_{n,1}(c,a)$ of the associated Laguerre polynomials:

Corollary 3.3.3. (i) All the zeros of the associated Laguerre polynomials increase with respect to a, for $a \geq 0$ and $c > -1$.

(ii) The function

$$\frac{1}{a+c+1} x_{n,1}(c,a)$$

decreases with respect to a, for $c \geq 0$, $a > -c-1$.

Remark 3.3.4. (i) Case (i) of the above corollary was also proved in [20].

Also, in the special case $c = 0$, it has been proved in [30 p. 122] that the zeros $x_{nk}(a)$ increase with respect to a, for $a > -1$.

(ii) For $c = 0$ we obtain analogous results for the largest zero $x_{n,1}(a)$ of the Laguerre polynomials. These results were given in [14].

(iii) Other results concerning the zeros of Laguere polynomials were obtained recently in [26].

From the above monotonicity properties we obtain

Corollary 3.3.5.(i) The largest zero $x_{n,1}(c,a;q)$ of the q-associated Laguerre polynomials satisfies the inequality

$$x_{n,1}(c,a;q) < \frac{1-q^{a+c+1}}{(1-q^{c+1})q^a} x_{n,1}(c,0;q) \tag{3.3.3}$$

for $c \geq 0$, $a > 0$.

(ii) The largest zero $x_{n,1}(c,a)$ of the associated Laguerre polynomials satisfies the inequality

$$x_{n,1}(c,a) < \frac{a+c+1}{c+1} x_{n,1}(c,0) \tag{3.3.4}$$

for $c \geq 0$, $a > 0$.

Theorem 3.3.6.(*Monotonicity with respect to c*)

(i) The largest zero $x_{n,1}(c,a;q)$ of the q-associated Laguerre polynomials $L_n(x;a,c,q)$ increases with respect to c, for $c > -1$ and $a + c > 0$.

(ii) The function

$$\left(\frac{q^{c(2+q)}}{(1-q^{c+1})^{1+q}}\right) x_{n,1}(c,a;q)$$

decreases with respect to c, for $c \geq 0$ and $a > 0$.

Using the above monotonicity properties we obtain:

Corollary 3.3.7. The largest zero $x_{n,1}(c,a;q)$ of the q-associated Laguerre polynomials $L_n(x;a,c,q)$ satisfies the inequality

$$x_{n,1}(c,a;q) \leq \frac{(1-q^{c+1})^{1+q}}{q^{c(2+q)}(1-q)^{1+q}} x_{n,1}(a,q), \tag{3.3.5}$$

for $c \geq 0$, $a > 0$, where $x_{n,1}(a,q)$ is the largest zero of the q-Laguerre polynomials.

Taking the limit $q \to 1^-$, we obtain analogous results for the largest zero $x_{n,1}(c,a)$ of the associated Laguerre polynomials:

Corollary 3.3.8.(i) The largest zero $x_{n,1}(c,a)$ of the associated Laguerre polynomials $L_n(x;a,c)$ increases with respect to c, for $c \geq 0$ and $a > 0$.

(ii) The function

$$\frac{1}{(c+1)^2} x_{n,1}(c,a)$$

decreases with respect to c, for $c \geq 0$ and $a > 0$.

(iii) Using the above monotonicity, we obtain the inequality

$$x_{n,1}(c,a) \leq (c+1)^2 x_{n,1}(a), \tag{3.3.6}$$

for $c \geq 0$, $a > 0$, where $x_{n,1}(a)$ is the largest zero of Laguerre polynomials.

Remark 3.3.9. The inequalities (3.3.5) and (3.3.6) are sharp for small values of c, since for $c = 0$ they become equalities.

3.4. Associated Al-Salam-Carlitz II polynomials

The orthonormal associated Al-Salam-Carlitz II polynomials, are obtained from the Al-Salam-Carlitz II Polynomials [23 p.114] and satisfy the recurrence relation (2.2), with

$$\alpha_n = \sqrt{\frac{a(1-q^{n+c+1})}{q^{2n+2c+1}}} \tag{3.4.1}$$

and

$$\beta_n = \frac{a+1}{q^{n+c}} \tag{3.4.2}$$

Theorem 3.4.1. *(Monotonicity with respect to a)*

(i) All the zeros $x_{n,k}(c,a;q)$, $k = 1, 2, \ldots, n$ of the associated Al-Salam and Carlitz II polynomials increase with respect to a, for $a > 1$, $c > -1$.

(ii) The largest zero $x_{n,1}(c,a;q)$ of the associated Al-Salam and Carlitz II polynomials increases with respect to a, for $a > 0$ and $c > -1$.

(iii) The function

$$\left(\frac{x_{n,1}(c,a;q)}{a}\right)$$

decreases with respect to a, for $a > 0$, $c > -1$.

Theorem 3.4.2. *(Monotonicity with respect to c)*

(i) The largest zero $x_{n,1}(c,a;q)$ of the associated Al-Salam and Carlitz II polynomials increases with respect to c, for $a > 0$ and $c > -1$.

(ii) The function

$$\frac{q^c}{\sqrt{1-q^{c+1}}} x_{n,1}(c,a;q)$$

decreases with respect to c, for $a > 0$, $c > -1$.

From the above monotonicity properties we obtain the following:

Corollary 3.4.3. The largest zero $x_{n,1}(c, a; q)$ of the associated Al-Salam and Carlitz II polynomials satisfy the inequality

$$x_{n,1}(c, a; q) \leq \sqrt{\frac{1-q^{c+1}}{(1-q)q^{2c}}} x_{n,1}(0, a; q) \tag{3.4.3}$$

for $c \geq 0$, $a > 0$.

3.5. q-Associated Meixner polynomials

The orthonormal q-associated Meixner polynomials, which are obtained from the q-Meixner polynomials [23 p. 95], satisfy the recurrence relation (2.2) with $x = q^{-x} - 1$,

$$\alpha_n = \sqrt{\frac{r(1-q^{n+c+1})(r+q^{n+c+1})(1-bq^{n+c+1})}{q^{4n+4c+3}}} > 0 \tag{3.5.1}$$

and

$$\beta_n = q^{-n-c} + r\left[1 + q - (1+b)q^{n+c+1}\right] q^{-2n-2c-1} - 1, \tag{3.5.2}$$

for $0 < q < 1$, $c > -1$, $r > 0$ and $b < q^{-c-1}$.

Theorem 3.5.1. *(Monotonicity with respect to b)*

(i) The largest zero $x_{n,1}(c, r, b; q)$ of the q-associated Meixner polynomials decreases with respect to b, for $c > -1$, $r > 0$ and $b < q^{-c-1}$.

(ii) The function

$$\frac{1}{1-bq^{c+1}}\left(q^{-x_{n,1}(c,r,b;q)} - 1\right)$$

increases with respect to b, for $b < q^{-c-1}$, $c \geq 0$ and $r > 0$.

The associated Meixner polynomials are obtained from the q-associated Meixner polynomials [23 p.141] if we replace b by $q^{\beta-1}$, $\beta > -c$, r by $(1-d)^{-1}d$, $0 < d < 1$ and let $q \to 1^-$.

Corollary 3.5.2. (i) The largest zero $x_{n,1}(c, d, \beta)$ of the associated Meixner polynomials increases with respect to β, for $c \geq 0$, $0 < d < 1$ and $\beta > -c$.

(ii) The function $\dfrac{x_{n,1}(c, d, \beta)}{c+\beta}$ decreases with respect to β, for $\beta > -c$, $c \geq 0$, $0 < d < 1$.

Theorem 3.5.3. *(Monotonicity with respect to r)*

(i) The largest zero $x_{n,1}(c,r,b;q)$ of the q-associated Meixner polynomials increases with respect to r, for $c > 0$, $r > 0$ and $b < q^{-c-1}$.

(ii) The function

$$\frac{1}{r}\left(q^{-x_{n,1}(c,r,b;q)} - 1\right)$$

decreases with respect to r, for $c > 0$, $r > 0$ and $b < q^{-c-1}$.

Substituting b by $q^{\beta-1}$, $\beta > -c$, r by $(1-d)^{-1}d$, $0 < d < 1$ and taking $q \to 1^-$ we obtain the following:

Corollary 3.5.4. (i) The largest zero $x_{n,1}(c,d,\beta)$ of the associated Meixner polynomials increases with respect to d, for $0 < d < 1$, $c \geq 0$, and $\beta > -c$.

(ii) The function

$$\frac{1-d}{d}x_{n,1}(c,d,\beta)$$

decreases with respect to d, for $0 < d < 1$, $c > 0$ $\beta > -c$

Remark 3.5.5. The first result of the above corollary extends the interval $\frac{n-1}{\beta+n-1} < d < 1$ of monotonicity for the largest zero of Meixner polynomials which was given in [20] for all the zeros of Meixner polynomials.

Theorem 3.5.6. *(Monotonicity with respect to c)*

(i) The largest zero of the q-associated Meixner polynomials increases with respect to c for $c > 0$, $r > 0$ and $b < q^{-c-1}$.

(ii) The function

$$\frac{q^{2c}}{1-q^c}(q^{-x_{n,1}(c,r,b;q)} - 1)$$

decreases with respect to c, for $c > 0$, $b < 1$.

Substituting b by $q^{\beta-1}$, $\beta > -c$, r by $(1-d)^{-1}d$, $0 < d < 1$ and taking $q \to 1^-$ we obtain the following:

Corollary 3.5.7. (i) The largest zero $x_{n,1}(c,d,\beta)$ of the associated Meixner polynomials increases with respect to c for $c > 0$, $\beta > 1$, $0 < d < 1$.

(ii) The function

$$\frac{x_{n,1}(c,d,\beta)}{c}$$

decrease with respect to c for $c > 0$.

Remark 3.5.8. The above results hold also for $b = 0$ and $r = a$, which correspond to the q-associated Charlier polynomials.

3.6. q-Lommel polynomials

The orthonormal q-Lommel polynomials [15], satisfy the recurrence relation (2.2) with

$$a_n(\nu;q) = \frac{1}{2}\sqrt{\frac{q^{\nu+n}}{(1-q^{\nu+n+1})(1-q^{\nu+n})}}, \quad b_n(\nu;q) = 0, \tag{3.6.1}$$

for $0 < q < 1$ and $\nu \geq 0$.

Theorem 3.6.1. *(Monotonicity with respect to ν)*

(i) The largest zero $x_{n,1}(\nu, q)$ of the q-Lommel polynomials decreases with respect to ν.

(ii) The function

$$\sqrt{\frac{(1-q^\nu)(1-q^{\nu+1})}{q^\nu}}\; x_{n,1}(\nu;q)$$

increases with respect to ν, for $\nu \geq 0$.

Using the above monotonicity property we obtain the following:

Corollary 3.6.2. The largest zero $x_{n,1}(c, \nu, q)$ of the q-Lommel polynomials satisfy the inequality

$$x_{n,1}(\nu;q) > \sqrt{\frac{(1-q^{1/2})(1-q^{3/2})q^{\nu-1/2}}{(1-q^\nu)(1-q^{\nu+1})}}\; x_{n,1}(1/2;q), \qquad \text{for } \nu \geq 1/2.$$

Since, $\lim_{n\to\infty} x_{n,k}(\nu;q) = \frac{1}{j_{\nu-1,k}}$, $\quad k = 1, 2, \ldots$, where $j_{\nu,k}(q)$ are the zeros of Jackson's q-Bessel function

$$J_\nu^{(2)}(x;q) = \frac{(q^{\nu+1};q)_\infty}{(q;q)_\infty} \sum_{n=0}^{\infty} \frac{(-1)^n (x/2)^{\nu+2n}}{(q;q)_n (q^{\nu+1};q)_n} q^{n(\nu+n)}, \quad 0 < q < 1,$$

we obtain the following:

Corollary 3.6.3.

(i) The first zero $j_{\nu,1}(q)$ of the q-Bessel function $J_\nu^{(2)}(x;q)$ increases with respect to ν for $\nu > -1$.

(ii) The function

$$\sqrt{\frac{q^{\nu+1}}{(1-q^{\nu+1})(1-q^{\nu+2})}}\; j_{\nu,1}(q), \tag{3.6.2}$$

decreases with respect to ν for $\nu > -1$.

(iii) The first zero $j_{\nu,1}(q)$ of $J_\nu^{(2)}(x;q)$ satisfies the inequality

$$j_{\nu,1}(q) \leq \sqrt{\frac{(1-q^{\nu+1})(1-q^{\nu+2})}{q^{\nu+1/2}(1-q^{1/2})(1-q^{3/2})}}\, j_{-1/2,1}(q) \tag{3.6.3}$$

for $\nu \geq -1/2$.

Remark 3.6.4. Since, [23 p. 20] $\lim_{q\to 1} J_\nu^{(2)}((1-q)x;q) = J_\nu(x)$, where $J_\nu(x)$ is the Bessel function, we obtain from (3.6.3) the following upper bound for the first zero of $J_\nu(x)$:

$$j_{\nu,1} \leq \frac{2}{\sqrt{3}}\sqrt{(\nu+1)(\nu+2)}\, j_{-1/2,\,1} = \frac{\pi}{\sqrt{3}}\sqrt{(\nu+1)(\nu+2)}$$

for $\nu \geq -1/2$.

Remark 3.6.5. (i) In [21] it was proved that all the positive zeros $j_{n,k}(q)$ of $J_\nu^{(2)}(x;q)$ increase with respect to q for $\nu > -1$.

(ii) The result (3.6.2) is stronger, for $\nu > 0$, than the corresponding result that the function $\dfrac{j_{\nu,1}(q)}{(q^{-\nu-1}-q^{\nu-1})}$ decreases with respect to ν for $\nu > -1$, which was given in [21], while the latter is stronger for $-1 < \nu < 0$.

Theorem 3.6.6. *(Monotonicity with respect to q)*

(i) The largest zero $x_{n,1}(\nu,q)$ of the q-Lommel polynomials increases with respect to q for $0 < q < 1$, $0 \leq \nu$.

(ii) The function $\sqrt{\dfrac{(1-q^{n+\nu})(1-q^{n+\nu-1})}{q^{n+\nu-1}}}\, x_{n,1}(\nu;q)$ decreases with respect to q.

Corollary 3.6.7. The first zero of $J_\nu^{(2)}(x;q)$ decreases with respect to q for $0 < q < 1$, $-1 < \nu$.

Remark 3.6.8. The above corollary extends the interval $e^{-4} < q < 1$ of monotonicity, for the first zero of $J_\nu^{(2)}(x;q)$, which was given in [21] for all the zeros of $J_\nu^{(2)}(x;q)$.

4 Proofs

4.1. q-Associated Pollaczek polynomials

Proof of theorem 3.1.1.

(i) Differentiating α_j and β_j, given in (3.1.1) and (3.1.2), $j = 0, \ldots, n-1$, with respect to λ we obtain:

(4.1.1)
$$\frac{\partial}{\partial\lambda}\alpha_j = \frac{1}{2}\alpha_j\gamma_j,$$

(4.1.2)
$$\frac{\partial}{\partial\lambda}\beta_j = \beta_j\delta_j,$$

where

$$\gamma_j = \frac{aq^{j+c}}{1-a\lambda q^{j+c}} + \frac{aq^{j+c+1}}{1-a\lambda q^{j+c+1}} - 2\frac{\lambda q^{j+c}}{1-\lambda^2 q^{j+c}} =$$

(4.1.3)
$$= \frac{1}{\lambda}\left[\frac{1}{1-a\lambda q^{j+c}} + \frac{1}{1-a\lambda q^{j+c+1}} - \frac{2}{1-\lambda^2 q^{j+c}}\right]$$

$$= q^{j+c}\frac{-2\lambda + a + qa + a\lambda^2 q^{j+c} + a\lambda^2 q^{j+c+1} - 2a^2\lambda q^{j+c+1}}{(1-\lambda^2 q^{j+c})(1-a\lambda q^{j+c})(1-a\lambda q^{j+c+1})},$$

and

(4.1.4)
$$\delta_j = \frac{aq^{j+c}}{1-a\lambda q^{j+c}}.$$

For $-q^{-c/2} \le a \le q^{-c/2}$, the equation

$$\sigma_j = -2\lambda + a + qa + a\lambda^2 q^{j+c} + a\lambda^2 q^{j+c+1} - 2a^2\lambda q^{j+c+1} = 0$$

has two real zeros:

$$\lambda_1 = \frac{1 + a^2 q^{j+c+1} + \sqrt{(1-a^2 q^{j+c})(1-a^2 q^{j+c+2})}}{aq^{j+c}(1+q)}$$

and

$$\lambda_2 = \frac{1 + a^2 q^{j+c+1} - \sqrt{(1-a^2 q^{j+c})(1-a^2 q^{j+c+2})}}{aq^{j+c}(1+q)}.$$

For $0 < a \le q^{-c/2}$ we have $0 < \lambda_2 \le \lambda_1$ and:

$$\lambda_1 - q^{-c/2} \ge \frac{1 - aq^{j+c/2} - aq^{j+c/2+1} + a^2 q^{j+c+1} + 1 - a^2 q^{j+c}}{aq^{j+c}(1+q)} \ge$$

$$\ge \frac{2(1-aq^{j+c/2})}{aq^{j+c}(1+q)} \ge 0.$$

Hence, $\lambda_1 \geq q^{-c/2}$.
Also, taking in account that

$$\frac{\partial}{\partial j}\lambda_2 = \frac{\ln q}{q^{j+c}} \frac{\left(\sqrt{1-a^2q^{j+c}} - \sqrt{1-a^2q^{j+c+2}}\right)^2}{2a(1+q)\sqrt{(1-a^2q^{j+c})(1-a^2q^{j+c+2})}}, \tag{4.1.5}$$

which is negative for $0 < a \leq q^{-c/2}$ and $a^2q^{j+c} \neq 1$, we obtain

(4.1.6)

$$\min_{j=0,\ldots,n-1}\{\lambda_2\} = \lambda' = \frac{1+a^2q^{n+c} - \sqrt{(1-a^2q^{n+c-1})(1-a^2q^{n+c+1})}}{aq^{n+c-1}(1+q)} >$$

$$> \lim_{j\to\infty}\{\lambda_2\} = a\frac{q+1}{2} > aq^{1/2} > 0,$$

and

$$\max_{j=0,\ldots,n-1}\{\lambda_2\} = \lambda'' = \frac{1+a^2q^{c+1} - \sqrt{(1-a^2q^{c})(1-a^2q^{c+2})}}{aq^{c}(1+q)} \leq \tag{4.1.7}$$

$$\leq a \leq q^{-c/2}.$$

Also, in the special case $a = q^{-c/2}$, $j = 0$ it holds $\lambda_2 = \lambda'' > \lambda'$.
Since, for $\lambda = 0$, $a > 0$ we have that $\sigma_j = a(1+q) > 0$, we obtain:

$$\sigma_j > 0, \quad j = 0, 1, \ldots, n-1, \quad -q^{-c/2} < \lambda < \lambda', \tag{4.1.8}$$

$$\sigma_j < 0, \quad j = 0, 1, \ldots, n-1, \quad \lambda'' < \lambda < q^{-c/2}. \tag{4.1.9}$$

Similarly, for $-q^{-c/2} \leq a < 0$ we have $\lambda_1 \leq \lambda_2 < 0$ and:

$$\lambda_1 + q^{-c/2} \leq \frac{1 + aq^{j+c/2} + aq^{j+c/2+1} + a^2q^{j+c+1} + 1 - a^2q^{j+c}}{aq^{j+c}(1+q)} \leq$$

$$\leq \frac{2(1+aq^{j+c/2})}{aq^{j+c}(1+q)} \leq 0.$$

Hence, $\lambda_1 \leq -q^{-c/2}$.

Also, from (4.1.5) we obtain that $\frac{\partial}{\partial j}\lambda_2 > 0$ for $-q^{-c/2} \le a < 0$ and $a^2 q^{j+c} \ne 1$. Hence,

$$\min_{j=0,\dots,n-1}\{\lambda_2\} = \lambda'' = \frac{1 + a^2 q^{c+1} - \sqrt{(1 - a^2 q^c)(1 - a^2 q^{c+2})}}{a q^c (1+q)} \ge \tag{4.1.10}$$

$$\ge a \ge -q^{-c/2},$$

$$\max_{j=0,\dots,n-1}\{\lambda_2\} = \lambda' = \frac{1+a^2q^{n+c} - \sqrt{(1-a^2q^{n+c-1})(1-a^2q^{n+c+1})}}{aq^{n+c-1}(1+q)} \tag{4.1.11}$$

$$< \lim_{j\to\infty}\{\lambda_2\} = a\frac{q+1}{2} < aq^{1/2} < 0.$$

Also, in the special case $a = -q^{-c/2}$, $j = 0$ it holds $\lambda_2 = \lambda'' < \lambda'$. Since, now, for $\lambda = 0$, $a < 0$ we have that $\sigma_j = a(1+q) < 0$, we obtain:

$$\sigma_j > 0, \quad j = 0, 1, \dots, n-1, \quad -q^{-c/2} < \lambda < \lambda'', \tag{4.1.12}$$

$$\sigma_j < 0, \quad j = 0, 1, \dots, n-1, \quad \lambda' < \lambda < q^{-c/2}. \tag{4.1.13}$$

From the above we conclude that:

$$\gamma_j < 0, \quad 0 < a \le q^{-c/2}, \quad \lambda'' < \lambda < q^{-c/2}, \quad j = 0, \dots, n-1, \tag{4.1.14}$$

$$\gamma_j < 0, \quad -q^{-c/2} \le a < 0, \quad \lambda' < \lambda < q^{-c/2}, \quad j = 0, \dots, n-1, \tag{4.1.15}$$

and that

$$\gamma_j > 0, \quad 0 < a \le q^{-c/2}, \quad -q^{-c/2} < \lambda < \lambda', \quad j = 0, \dots, n-1, \tag{4.1.16}$$

$$\gamma_j > 0, \quad -q^{-c/2} \le a < 0, \quad -q^{-c/2} < \lambda < \lambda'', \quad j = 0, \dots, n-1. \tag{4.1.17}$$

For the coefficients β_j, the following inequalities hold:

$$\frac{\partial}{\partial \lambda}\beta_j \ge 0, \quad ab \le 0, \quad a\lambda q^c < 1, \quad j = 0, \dots, n-1, \tag{4.1.18}$$

$$\frac{\partial}{\partial \lambda}\beta_j \le 0, \quad ab \ge 0, \quad a\lambda q^c < 1, \quad j = 0, \dots, n-1. \tag{4.1.19}$$

Inequalities (4.1.14) - (4.1.19) complete the proof of case (i).

(ii) For the derivatives of α_j and β_j with respect to λ, the following inequalities also hold:

$$\frac{\partial}{\partial \lambda}\beta_j < \beta_j \frac{aq^c}{1 - a\lambda q^c}, \qquad a \ge 0, \quad b \le 0, \quad a\lambda q^c < 1, \quad j = 0, \dots, n-1, \tag{4.1.20}$$

$$\frac{\partial}{\partial \lambda}\beta_j > \beta_j \frac{aq^c}{1 - a\lambda q^c}, \qquad a \le 0, \quad b \le 0, \quad a\lambda q^c < 1, \quad j = 0, \dots, n-1 \tag{4.1.21}$$

and

$$\gamma_j < \left[\frac{aq^{j+c}}{1 - a\lambda q^{j+c}} + \frac{aq^{j+c+1}}{1 - a\lambda q^{j+c+1}}\right] < 2\frac{aq^c}{1 - a\lambda q^c}, \qquad 0 < a, \ a\lambda q^c < 1, \ 0 < \lambda < q^{-c/2},$$

$$\gamma_j > \left[\frac{aq^{j+c}}{1 - a\lambda q^{j+c}} + \frac{aq^{j+c+1}}{1 - a\lambda q^{j+c+1}}\right] > 2\frac{aq^c}{1 - a\lambda q^c}, \qquad a < 0, \ a\lambda q^c < 1 \ -q^{-c/2} < \lambda < 0.$$

Hence,

$$\frac{\partial}{\partial \lambda}\alpha_j < \alpha_j \frac{aq^c}{1 - a\lambda q^c}, \quad 0 < a, \ a\lambda q^c < 1, \ 0 < \lambda < q^{-c/2}, \tag{4.1.22}$$

and

$$\frac{\partial}{\partial \lambda}\alpha_j > \alpha_j \frac{aq^c}{1 - a\lambda q^c}, \quad a < 0, \ a\lambda q^c < 1, \ -q^{-c/2} < \lambda < 0. \tag{4.1.23}$$

From (4.1.20) - (4.1.23) we obtain the inequalities

$$\frac{aq^c}{1-a\lambda q^c}x_{n,1}(c,\lambda,a,b;q) < \frac{\partial}{\partial\lambda}x_{n,1}(c,\lambda,a,b;q) < 0, \tag{4.1.24}$$

for $a < 0$, $a\lambda q^c < 1$, $-q^{-c/2} < \lambda < 0$, $b \leq 0$ and

$$0 < \frac{\partial}{\partial\lambda}x_{n,1}(c,\lambda,a,b;q) < \frac{aq^c}{1-a\lambda q^c}x_{n,1}(c,\lambda,a,b;q), \tag{4.1.25}$$

for $0 < a$, $a\lambda q^c < 1$, $0 < \lambda < q^{-c/2}$, $b \leq 0$, which prove that the function $(1-a\lambda q^c)x_{n,1}(c,\lambda,a,b;q)$ increases with respect to λ, for $a < 0$, $a\lambda q^c < 1$, $-q^{-c/2} < \lambda < 0$, $b \leq 0$, $c > -1$ and decreases with respect to λ, for $0 < a$, $a\lambda q^c < 1$, $0 < \lambda < q^{-c/2}$, $b \leq 0$, $c > -1$.

(iii) For $a < 0$, $0 \leq \lambda < q^{-c/2}$ we obtain

$$\gamma_j > 2\left(\frac{aq^c}{1-a\lambda q^c} - \frac{\lambda q^c}{1-\lambda^2 q^c}\right)$$

and

$$\delta_j > \frac{aq^c}{1-a\lambda q^c} - \frac{\lambda q^c}{1-\lambda^2 q^c},$$

from which, for $b \leq 0$, follows that

$$\frac{\partial}{\partial\lambda}x_{n,1}(c,a,\lambda,b;q) > \left(\frac{aq^c}{1-a\lambda q^c} - \frac{\lambda q^c}{1-\lambda^2 q^c}\right)x_{n,1}(c,a,\lambda,b;q).$$

Also, for $0 < a$, $-q^{-c/2} < \lambda \leq 0$, we obtain

$$\gamma_j < 2\left(\frac{aq^c}{1-a\lambda^c} - \frac{\lambda q^c}{1-\lambda^2 q^c}\right)$$

and

$$\delta_j < \frac{aq^c}{1-a\lambda q^c} - \frac{\lambda q^c}{1-\lambda^2 q^c},$$

from which, for $b \leq 0$, follows that

$$\frac{\partial}{\partial\lambda}x_{n,1}(c,a,\lambda,b;q) < \left(\frac{aq^c}{1-a\lambda q^c} - \frac{\lambda q^c}{1-\lambda^2 q^c}\right)x_{n,1}(c,a,\lambda,b;q).$$

The above complete the proof of the case (iii) of the theorem.

(iv) For the proof of the case (iv) we have to prove the inequalities

$$0 < \frac{\partial}{\partial \lambda} x_{n,1}(c, \lambda, a, 0; q) < \frac{1}{2\lambda}\left(\frac{1}{1 - a\lambda q^{c+1}}\right) x_{n,1}(c, \lambda, a, 0; q), \tag{4.1.26}$$

$$0 < \lambda < q^{-c/2}, \quad 0 < a \leq q^{-c/2}$$

and

$$\frac{1}{2\lambda}\left(\frac{1}{1 - a\lambda q^{c+1}}\right) x_{n,1}(c, \lambda, a, b; q) < \frac{\partial}{\partial \lambda} x_{n,1}(c, \lambda, a, b; q) < 0, \tag{4.1.27}$$

$$-q^{-c/2} < \lambda < 0, \ -q^{-c/2} \leq a < 0.$$

To do this, since $b = 0 \Rightarrow \beta_j = 0$, we have to prove that

$$\gamma_j < \frac{1}{\lambda}\left[\frac{1}{1 - a\lambda q^{c+1}}\right], \quad 0 < \lambda < q^{-c/2}, \quad 0 < a \leq q^{-c/2}$$

and

$$\gamma_j > \frac{1}{\lambda}\left[\frac{1}{1 - a\lambda q^{c+1}}\right], \quad -q^{-c/2} < \lambda < 0, \quad -q^{-c/2} \leq a < 0.$$

We have that:

$$\frac{1}{1 - a\lambda q^{j+c}} - \frac{2}{1 - \lambda^2 q^{j+c}} = \frac{-1 - \lambda^2 q^{j+c} + 2a\lambda q^{j+c}}{(1 - a\lambda q^{j+c})(1 - \lambda^2 q^{j+c})}$$

and

$$\frac{\partial}{\partial \lambda}(-1 - \lambda^2 q^{j+c} + 2a\lambda q^{j+c}) = 2q^{j+c}(a - \lambda),$$

Hence, $-1 - \lambda^2 q^{j+c} + 2a\lambda q^{j+c}$ has a maximum at $\lambda = a$, which is:

$$-1 + a^2 q^{j+c} < 0, \quad -q^{-c/2} < a < q^{-c/2}.$$

Thus,

$$\frac{1}{1 - a\lambda q^{j+c}} - \frac{2}{1 - \lambda^2 q^{j+c}} < 0, \quad -q^{-c/2} \leq a \leq q^{-c/2}, \tag{4.1.28}$$

from where it follows that

$$\gamma_j < \frac{1}{\lambda}\left[\frac{1}{1-a\lambda q^{j+c+1}}\right] < \frac{1}{\lambda}\left[\frac{1}{1-a\lambda q^{c+1}}\right], \tag{4.1.29}$$

$$0<\lambda<q^{-c/2}, \quad 0<a\le q^{-c/2}$$

and

$$\gamma_j > \frac{1}{\lambda}\left[\frac{1}{1-a\lambda q^{j+c+1}}\right] > \frac{1}{\lambda}\left[\frac{1}{1-a\lambda q^{c+1}}\right], \tag{4.1.30}$$

$$-q^{-c/2}\le\lambda<0, \quad -q^{-c/2}\le a<0.$$

Proof of theorem 3.1.7.

(i) Differentiating α_j and β_j with respect to a we obtain:

$$\frac{\partial}{\partial a}\alpha_j = \frac{1}{2}\alpha_j\left[\frac{\lambda q^{j+c}}{1-a\lambda q^{j+c}} + \frac{\lambda q^{j+c+1}}{1-a\lambda q^{j+c+1}}\right] \tag{4.1.31}$$

and

$$\frac{\partial}{\partial a}\beta_j = \beta_j\frac{\lambda q^{j+c}}{1-a\lambda q^{j+c}}. \tag{4.1.32}$$

For $b\le 0$ it is $\beta_j\ge 0$ and hence,

$$\frac{\partial}{\partial a}\beta_j\ge 0, \quad 0<\lambda<q^{-c/2}, \quad a\lambda q^c<1,$$

$$\frac{\partial}{\partial a}\beta_j\le 0, \quad -q^{-c/2}<\lambda<0, \quad a\lambda q^c<1.$$

Also, since $a_j>0$, from (4.1.31) we obtain

$$\frac{\partial}{\partial a}\alpha_j>0, \quad 0<\lambda<q^{-c/2}, \quad a\lambda q^c<1,$$

$$\frac{\partial}{\partial a}\alpha_j<0, \quad -q^{-c/2}<\lambda<0, \quad a\lambda q^c<1.$$

Hence,

$$\frac{\partial}{\partial a}x_{n,1}(c,\lambda,a,b;q)>0, \quad 0<\lambda<q^{-c/2}, \quad a\lambda q^c<1, \quad b\le 0 \tag{4.1.33}$$

and

$$\frac{\partial}{\partial a} x_{n,1}(c, \lambda, a, b; q) < 0, \quad -q^{-c/2} < \lambda < 0, \quad a\lambda q^c < 1, \quad b \leq 0. \tag{4.1.34}$$

Also,

$$\frac{\partial}{\partial a}\alpha_j < \alpha_j \frac{\lambda q^{j+c}}{1 - a\lambda q^{j+c}} < \alpha_j \frac{\lambda q^c}{1 - a\lambda q^c}, \quad 0 < \lambda < q^{-c/2}, \quad a\lambda q^c < 1 \tag{4.1.35}$$

and

$$\frac{\partial}{\partial a}\alpha_j > \alpha_j \frac{\lambda q^{j+c}}{1 - a\lambda q^{j+c}} > \alpha_j \frac{\lambda q^c}{1 - a\lambda q^c}, \quad -q^{-c/2} < \lambda < 0, \quad a\lambda q^c < 1. \tag{4.1.36}$$

From (4.1.32) and (4.1.35) the inequality

$$\frac{\partial}{\partial a} x_{n,1}(c, \lambda, a, b; q) < \frac{\lambda q^c}{1 - a\lambda q^c} x_{n,1}(c, \lambda, a, b; q), \tag{4.1.37}$$

follows, which means that the function $(1 - a\lambda q^c)x_{n,1}(c, \lambda, a, b; q)$ decreases with respect to a, for $a\lambda q^c < 1$, $0 < \lambda < q^{-c/2}$, $b \leq 0$.

(ii) Also, from (4.1.32) and (4.1.36) the inequality

$$\frac{\partial}{\partial a} x_{n,1}(c, \lambda, a, b; q) > \frac{\lambda q^c}{1 - a\lambda q^c} x_{n,1}(c, \lambda, a, b; q), \tag{4.1.38}$$

follows, which means that the function $(1 - a\lambda q^c)x_{n,1}(c, \lambda, a, b; q)$ increases with respect to a, for $a\lambda q^c < 1$, $-q^{-c/2} < \lambda < 0$, $b \leq 0$.

Proof of theorem 3.1.11.

(i) The differential equation for the zeros $x_{n,k}(c, \lambda, a, b; q)$ with respect to b is

$$\begin{aligned} \frac{\partial}{\partial b} x_{n,k}(c, \lambda, a, b; q) &= \langle \frac{\partial}{\partial b}(AV^* + VA + B) f_{n,k}, f_{n,k} \rangle = \\ &= \langle \frac{\partial}{\partial b} B f_{n,k}, f_{n,k} \rangle = \sum_{j=0}^{n-1} \frac{\partial}{\partial b}\beta_j \, \|\langle f_{n,k}, e_j \rangle\|^2 = \\ &= -\sum_{j=0}^{n-1} \frac{q^{j+c}}{1 - a\lambda q^{j+c}} \, \|\langle f_{n,k}, e_j \rangle\|^2, \end{aligned} \tag{4.1.39}$$

from where it follows immediately that all the zeros $x_{n,k}(c,\lambda,a,b;q)$, $k=1,2,\ldots,n$ of the q-associated Pollaczek polynomials decrease with respect to b.

(ii) It holds

$$-\frac{q^c}{1-a\lambda q^c}<-\frac{q^{j+c}}{1-a\lambda q^{j+c}}<-\frac{q^{n+c-1}}{1-a\lambda q^{n+c-1}},\quad j=1,\ldots,n-2$$

and since,

$$\sum_{j=0}^{n-1}\|< f_{n,k},e_j\|^2=1,$$

it follows that,

$$-\frac{q^c}{1-a\lambda q^c}<-\sum_{j=0}^{n-1}\frac{q^{j+c}}{1-a\lambda q^{j+c}}\|\langle f_{n,k},e_j\rangle\|^2<$$

$$<-\frac{q^{n+c-1}}{1-a\lambda q^{n+c-1}},\quad k=1,2,\ldots,n$$

or

$$-\frac{q^c}{1-a\lambda q^c}<\frac{\partial}{\partial b}x_{n,k}(c,\lambda,a,b;q)<-\frac{q^{n+c-1}}{1-a\lambda q^{n+c-1}},\ k=1,2,\ldots,n.$$

Proof of theorem 3.1.14.

(i) Differentiating α_j and β_j with respect to c we obtain:

$$\frac{\partial}{\partial c}\alpha_j=\frac{1}{2}\alpha_j\gamma_j \tag{4.1.40}$$

where

$$\gamma_j=\ln q\left[\frac{1}{1-a\lambda q^{j+c}}+\frac{1}{1-a\lambda q^{j+c+1}}-\frac{1}{1-q^{j+c+1}}-\frac{1}{1-\lambda^2q^{j+c}}\right]$$

and

$$\frac{\partial}{\partial c}\beta_j=\beta_j\ln q\left[\frac{1}{1-a\lambda q^{j+c}}\right]. \tag{4.1.41}$$

a) For $a\lambda < \min\{\lambda^2, 1\}$ we obtain

$$\frac{1}{1-a\lambda q^{j+c+1}} - \frac{1}{1-q^{j+c+1}} < 0 \text{ and } \frac{1}{1-a\lambda q^{j+c}} - \frac{1}{1-\lambda^2 q^{j+c}} < 0.$$

Hence,

$$\gamma_j > 0 \Rightarrow \frac{\partial}{\partial c}\alpha_j > 0, \quad a\lambda < \min\{\lambda^2, 1\}. \tag{4.1.42}$$

b), e) Differentiating γ_j with respect to λ we obtain:

$$\frac{\partial}{\partial \lambda}\gamma_j = \ln q \left[\frac{aq^{j+c}}{(1-a\lambda q^{j+c})^2} + \frac{aq^{j+c+1}}{(1-a\lambda q^{j+c+1})^2} - 2\frac{\lambda q^{j+c}}{(1-\lambda^2 q^{j+c})^2}\right].$$

Taking in account that

$$\frac{\partial}{\partial j}\left[\frac{aq^{j+c}}{(1-a\lambda q^{j+c})^2}\right] = \frac{aq^{j+c}(1-a^2\lambda^2 q^{2(j+c)})\ln q}{(1-a\lambda q^{j+c})^4}$$

and

$$\frac{\partial}{\partial a}\left[\frac{aq^{j+c}}{(1-a\lambda q^{j+c})^2}\right] = \frac{q^{j+c}(1-a^2\lambda^2 q^{2(j+c)})}{(1-a\lambda q^{j+c})^4},$$

we obtain for $-q^{-c} < a\lambda < q^{-c}$,

$$\frac{aq^{j+c+1}}{(1-a\lambda q^{j+c+1})^2} \le \frac{aq^{j+c}}{(1-a\lambda q^{j+c})^2}, \quad a \ge 0 \tag{4.1.43}$$

$$\frac{\lambda q^{j+c}}{(1-\lambda^2 q^{j+c})^2} \le \frac{aq^{j+c+1}}{(1-a\lambda q^{j+c+1})^2}, \quad -q^{-c/2} < \lambda \le aq. \tag{4.1.44}$$

From inequalities (4.1.43) and (4.1.44) follows that

$$\frac{\partial}{\partial \lambda}\gamma_j \le 0, \quad a \ge 0, \quad -q^{-c/2} < \lambda \le aq. \tag{4.1.45}$$

Also, for $-q^{-c} < a\lambda < q^{-c}$ holds

$$\frac{aq^{j+c}}{(1-a\lambda q^{j+c})^2} \le \frac{aq^{j+c+1}}{(1-a\lambda q^{j+c+1})^2}, \quad a \le 0 \tag{4.1.46}$$

(4.1.47) $$\frac{\lambda q^{j+c}}{(1-\lambda^2 q^{j+c})^2} \leq \frac{a q^{j+c}}{(1-a\lambda q^{j+c})^2}, \quad -q^{-c/2} < \lambda \leq a.$$

From inequalities (4.1.46) and (4.1.47) follows that

(4.1.48) $$\frac{\partial}{\partial \lambda}\gamma_j \leq 0, \quad a \leq 0, \quad -q^{-c/2} < \lambda \leq a.$$

For $\lambda = aq$ or $\lambda = a$ we have:

(4.1.49) $$\gamma_j = \ln q \left[\frac{1}{1-a^2 q^{j+c+1}} - \frac{1}{1-q^{j+c+1}}\right] \geq 0, \quad -1 \leq a \leq 1.$$

From (4.1.45), (4.1.48) and (4.1.49) we obtain

(4.1.50) $$\gamma_j > 0 \Rightarrow \frac{\partial}{\partial c}\alpha_j > 0,$$

for
$-q^{-c} < a\lambda < q^{-c}$, $-1 \leq a \leq 0$, $-q^{-c/2} < \lambda \leq a$
or
$-q^{-c} < a\lambda < q^{-c}$, $0 \leq a \leq 1$, $-q^{-c/2} < \lambda \leq aq$.

d), e) Writing γ_j as follows

$$\gamma_j = q^{j+c} \ln q \left[\frac{a\lambda - q}{(1-a\lambda q^{j+c})(1-q^{j+c+1})} + \frac{\lambda(aq-\lambda)}{(1-a\lambda q^{j+c+1})(1-\lambda^2 q^{j+c})}\right]$$

we easily obtain that $\gamma_j > 0 \Rightarrow \frac{\partial}{\partial c}\alpha_j > 0$, for
$0 < \lambda < q^{-c/2}$, $a\lambda \leq q$, $aq \leq \lambda$
or
$-q^{-c/2} < \lambda < 0$, $a\lambda \leq q$, $aq \geq \lambda$.

Also,

(4.1.51) $$\frac{\partial}{\partial c}\beta_j \geq 0, \quad b \geq 0.$$

This completes the proof of case (i) of the theorem.

(ii) For $a\lambda > 0$, from the inequality

$$0 < \frac{a\lambda q^{j+c+1}}{1-a\lambda q^{j+c+1}} + \frac{a\lambda q^{j+c}}{1-a\lambda q^{j+c}}$$

we find

$$\gamma_j = \ln q\left[\frac{a\lambda q^{j+c+1}}{1-a\lambda q^{j+c+1}} + \frac{a\lambda q^{j+c}}{1-a\lambda q^{j+c}} - \frac{q^{j+c+1}}{1-q^{j+c+1}} - \frac{\lambda^2 q^{j+c}}{1-\lambda^2 q^{j+c}}\right] <$$

$$(4.1.52) \qquad < \ln q\left[-\frac{q^{j+c+1}}{1-q^{j+c+1}} - \frac{\lambda^2 q^{j+c}}{1-\lambda^2 q^{j+c}}\right] < \ln q\left[-\frac{q^{c+1}}{1-q^{c+1}} - \frac{\lambda^2 q^{c}}{1-\lambda^2 q^{c}}\right]$$

or

$$(4.1.53) \qquad \frac{1}{2}\alpha_j\gamma_j < \frac{1}{2}\alpha_j \ln q\left[-\frac{q^{c+1}}{1-q^{c+1}} - \frac{\lambda^2 q^{c}}{1-\lambda^2 q^{c}}\right].$$

Also for $b=0$ we have $\beta_j = 0$ and thus

$$(4.1.54) \qquad \frac{\partial}{\partial c}\beta_j = 0, \quad b = 0.$$

So, for $b=0$ from (4.1.53) and (4.1.54), we obtain

$$\frac{\partial}{\partial c}x_{n,1}(c,\lambda,a,0;q) < -\frac{1}{2}\ln q\left[\frac{q^{c+1}}{1-q^{c+1}} + \frac{\lambda^2 q^{c}}{1-\lambda^2 q^{c}}\right] x_{n,1}(c,\lambda,a,0;q),$$

which means that the function $\left(\frac{1}{(1-q^{c+1})(1-\lambda^2 q^c)}\right)^{1/2} x_{n,1}(c,\lambda,a,0;q)$ decreases with respect to c for $a\lambda > 0$. Combining this result with the intervals of monotonicity of $x_{n,1}(c,\lambda,a,b;q)$ given in case (i), the proof of case (ii) of the theorem follows.

(iii) From the inequality

$$\frac{a\lambda q^{j+c}}{1-a\lambda q^{j+c}} + \frac{a\lambda q^{j+c+1}}{1-a\lambda q^{j+c+1}} - \frac{\lambda^2 q^{j+c}}{1-\lambda^2 q^{j+c}} > 0, \quad 0<\lambda<a$$

we find in a similar way as before that

$$\gamma_j < -\ln q\left[\frac{q^{j+c+1}}{1-q^{j+c+1}}\right] < -\ln q\left[\frac{q^{c+1}}{1-q^{c+1}}\right], \quad 0<\lambda<a$$

or

$$\frac{1}{2}\alpha_j\gamma_j < -\frac{1}{2}\alpha_j \ln q \left[\frac{q^{c+1}}{1-q^{c+1}}\right], \quad 0<\lambda<a. \tag{4.1.55}$$

From (4.1.54) and (4.1.55) and cases (i,d) and (i,c) of the theorem it follows

$$\frac{\partial}{\partial c}x_{n,1}(c,\lambda,a,0;q) < -\frac{1}{2}\ln q\frac{q^{c+1}}{1-q^{c+1}}x_{n,1}(c,\lambda,a,0;q), \tag{4.1.56}$$
$$0<\lambda<a, \quad \lambda<q^{-c/2}, \quad a\lambda q^c<1,$$

which means that the function $\left(\frac{1}{1-q^{c+1}}\right)^{1/2} x_{n,1}(c,\lambda,a,0;q)$ decreases with respect to c, for $0<\lambda<a$, $\lambda<q^{-c/2}$, $a\lambda q^c<1$.

Proof of theorem 3.1.18.

Differentiating the coefficients a_j, $j=0,\ldots,n-1$ of the recurrence relation of the q-associated continuous Ultraspherical polynomials, with respect to c we obtain:

$$\frac{\partial}{\partial c}\alpha_j = \frac{1}{2}\alpha_j\gamma_j, \quad j=0,1,\ldots,n-1, \tag{4.1.57}$$

where

$$\gamma_j = -q^{j+c}\ln q\frac{(1-\lambda)(q-\lambda)(1-\lambda^2q^{2j+2c+1})}{(1-q^{j+c+1})(1-\lambda^2q^{j+c})(1-\lambda q^{j+c+1})(1-\lambda q^{j+c})}.$$

Obviously $\gamma_j>0$ for $0<\lambda<q$.
Also,

$$1-\lambda^2q^{j+c} < 1-\lambda^2q^{2j+2c+1} < 1-\lambda^2q^{2(j+c+1)}, \quad j=0,1,\ldots,n-1 \tag{4.1.58}$$

and

$$\frac{1}{1-q^{j+c+1}} > \frac{1}{1-\lambda q^{j+c}} > \frac{1}{1-\lambda q^{j+c+1}} > \frac{1}{1-\lambda^2q^{j+c}}, \tag{4.1.59}$$
$$0<\lambda<q, \quad j=0,1,\ldots,n-1.$$

From (4.1.58) and (4.1.59) it follows that

$$\gamma_j < -q^{j+c}\ln q\ \frac{(1-\lambda)(q-\lambda)(1-\lambda^2 q^{2(j+c+1)})}{(1-q^{j+c+1})^3(1-\lambda q^{j+c+1})} <$$

$$(4.1.60) \qquad < -q^c \ln q \frac{(1-\lambda)(q-\lambda)(1+\lambda q^{(c+1)})}{(1-q^{c+1})^3},$$

$$0<\lambda<q,\ j=0,1,\dots,n-1$$

and

$$\gamma_j > -q^{j+c}\ln q\ \frac{(1-\lambda)(q-\lambda)(1-\lambda^2 q^{2(j+c)})}{(1-\lambda^2 q^{j+c})^3(1-\lambda q^{j+c})} >$$

$$(4.1.61) \qquad > \ -q^{n+c-1}\ln q \frac{(1-\lambda)(q-\lambda)(1+\lambda q^{(n+c-1)})}{(1-\lambda^2 q^{n+c-1})^3},$$

$$0<\lambda<q,\ j=0,1,\dots,n-1.$$

From (4.1.57), (4.1.60) and (4.1.61) we obtain

$$-\frac{1}{2}q^{n+c-1}\ln q \frac{(1-\lambda)(q-\lambda)(1+\lambda q^{(n+c-1)})}{(1-\lambda^2 q^{n+c-1})^3}\ \alpha_j < \frac{\partial}{\partial c}\alpha_j <$$

$$(4.1.62) \qquad < -\frac{1}{2}q^c \ln q \frac{(1-\lambda)(q-\lambda)(1+\lambda q^{(c+1)})}{(1-q^{c+1})^3}\ \alpha_j,$$

$$0<\lambda<q,\ j=0,1,\dots,n-1,$$

from which the inequality

$$0 < -q^{n+c}\ln q \frac{(1-\lambda)(q-\lambda)(1+\lambda q^{(n+c-1)})}{2(1-\lambda^2 q^{n+c-1})^3} x_{n,1}(c,\lambda;q) <$$

$$(4.1.63) \qquad < \frac{\partial}{\partial c} x_{n,1}(c,\lambda;q) <$$

$$< -q^c \ln q \frac{(1-\lambda)(q-\lambda)(1+\lambda q^{(c+1)})}{2(1-q^{c+1})^3} x_{n,1}(c,\lambda;q),$$

$$\lambda < q, \quad c > -1$$

follows. Integrating (4.1.63) with respect to c , from 0 to c, we obtain the inequality (3.1.13).

For $q < \lambda < 1$ and $c \geq 0$ we have $\gamma_j < 0, \quad j = 0, 1 \ldots, n-1$.

Also,

$$1 - \lambda^2 q^{2(j+c)} < 1 - \lambda^2 q^{2j+2c+1} < 1 - \lambda^2 q^{2(j+c+1)}, \qquad q < \lambda < 1, \; j = 0, 1, \ldots, n-1 \tag{4.1.64}$$

and

$$\frac{1}{(1-\lambda q^{j+c+1})^3 (1-\lambda q^{j+c})} < \frac{1}{(1-q^{j+c+1})(1-\lambda^2 q^{j+c})(1-\lambda q^{j+c+1})(1-\lambda q^{j+c})} < \frac{1}{(1-\lambda q^{j+c})^3(1-\lambda q^{j+c+1})}, \qquad q < \lambda < 1, \; j = 0, 1, \ldots, n-1. \tag{4.1.65}$$

From (4.1.64) and (4.1.65) we obtain, since $q - \lambda < 0$,

$$\gamma_j > -q^{j+c} \ln q \frac{(1-\lambda)(q-\lambda)(1-\lambda^2 q^{2(j+c+1)})}{(1-\lambda q^{j+c})^3(1-\lambda q^{j+c+1})} > -q^c \ln q \frac{(1-\lambda)(q-\lambda)(1+\lambda q^{(c+1)})}{(1-\lambda q^c)^3}, \qquad q < \lambda < 1, \; j = 0, 1, \ldots, n-1 \tag{4.1.66}$$

and

$$\gamma_j < -q^{j+c} \ln q \frac{(1-\lambda)(q-\lambda)(1-\lambda^2 q^{2(j+c)})}{(1-\lambda q^{j+c+1})^3(1-\lambda q^{j+c})} < -q^{n+c-1} \ln q \frac{(1-\lambda)(q-\lambda)(1+\lambda q^{(n+c-1)})}{(1-\lambda q^{n+c})^3}, \qquad q < \lambda < 1, \; j = 0, 1, \ldots, n-1. \tag{4.1.67}$$

Relation (4.1.57) together with inequalities (4.1.66) and (4.1.67) gives

$$(4.1.68)\quad \begin{aligned} &-\frac{1}{2}q^c \ln q \frac{(1-\lambda)(q-\lambda)(1+\lambda q^{(c+1)})}{(1-\lambda q^c)^3}\alpha_j < \frac{\partial}{\partial c}\alpha_j < \\ &< -\frac{1}{2}q^{n+c-1}\ln q \frac{(1-\lambda)(q-\lambda)(1+\lambda q^{(n+c-1)})}{(1-\lambda q^{n+c})^3}\alpha_j, \\ &q<\lambda<1,\ \ j=0,1,\ldots,n-1, \end{aligned}$$

from which the differential inequality

$$(4.1.69)\quad \begin{aligned} &-q^c \ln q \frac{(1-\lambda)(q-\lambda)(1+\lambda q^{(c+1)})}{2(1-\lambda q^c)^3} x_{n,1}(c,\lambda;q) < \\ &< \frac{\partial}{\partial c} x_{n,1}(c,\lambda;q) < \\ &< -q^{n+c-1}\ln q \frac{(1-\lambda)(q-\lambda)(1+\lambda q^{(n+c-1)})}{2(1-\lambda q^{n+c})^3} x_{n,1}(c,\lambda;q) < 0, \\ &q<\lambda<1, \quad c \geq 0 \end{aligned}$$

follows. Integrating (4.1.69) with respect to c, from 0 to c, we obtain the inequalities (3.1.14).

4.2. q-Associated continuous Jacobi polynomials

Proof of theorem 3.2.3.

The partial derivatives with respect to a of $\alpha_j(c,a,b;q)$ and $\beta_j(c,a,b;q)$, given in (3.2.1) and (3.2.2), are

$$(4.2.1)\quad \frac{\partial \alpha_j(c,a,b;q)}{\partial a} = \alpha_j(c,a,b;q)\gamma_j(c,a,b;q)$$

$$(4.2.2)\quad \frac{\partial \beta_j(c,a,b;q)}{\partial a} = \beta_j(c,a,b;q)\delta_j(c,a,b;q)$$

where

$$(4.2.3)\quad \begin{aligned} \gamma_j(c,a,b;q) = \frac{1}{2}\ln q \Big[&-\frac{2}{1-q^{2j+2c+2a+2}} - \frac{2}{1-q^{2j+2c+2a+2b+2}} + \\ &+ \frac{1}{1-q^{2j+2c+a+b+1}} + \frac{2}{1-q^{2j+2c+a+b+2}} + \frac{1}{1-q^{2j+2c+a+b+3}} \Big] \end{aligned}$$

and

$$\delta_j\,(c,a,b;q) = \ln q\left[-1 + \frac{1}{1-q^{b-a}} - \frac{1}{1-q^{a+b}} + \right.$$

$$\left. + \frac{1}{1-q^{2j+2c+a+b}} + \frac{1}{1-q^{2j+2c+a+b+2}}\right] =$$

(4.2.4)

$$= \ln q\left[\frac{q^a - 2q^b + q^{a+2b}}{(q^b - q^a)(1-q^{a+b})} + \frac{1}{1-q^{2j+2c+a+b}} + \right.$$

$$\left. + \frac{1}{1-q^{2j+2c+a+b+2}}\right].$$

Since, for $c \geq 0$

$$-\frac{2}{1-q^{2j+2c+2a+2b+2}} + \frac{1}{1-q^{2j+2c+a+b+1}} <$$

(4.2.5)

$$< -\frac{(1-q^{j+c+a+b+1})^2}{(1-q^{2j+2c+2a+2b+2})(1-q^{2j+2c+a+b+1})} < 0,$$

we obtain from (4.2.3):

$$\gamma_j(c,a,b;q) > \frac{1}{2}\ln q\left[\frac{2}{1-q^{2j+2c+a+b+2}} + \frac{1}{1-q^{2j+2c+a+b+3}}\right] =$$

$$\ln q\left[\frac{1}{1-q^{2j+2c+a+b+2}} + \frac{1}{2(1-q^{2j+2c+a+b+3})}\right] >$$

$$\ln q\left[\frac{1}{1-q^{2j+2c+a+b+2}} + \frac{1}{1-q^{2j+2c+a+b+3}}\right] >$$

(4.2.6)

$$\ln q\left[\frac{1}{1-q^{2c+a+b+2}} + \frac{1}{1-q^{2c+a+b}}\right].$$

The function $q^a - 2q^b + q^{a+2b}$ decreases with respect to a, since

$$\frac{\partial(q^a - 2q^b + q^{a+2b})}{\partial a} = \ln q(q^a + q^{a+2b}) < 0$$

and becomes equal to 0 for $a = b - \dfrac{\ln(\frac{1+q^{2b}}{2})}{\ln q}$.

(i) For $|a| < b$, $a < b - \frac{\ln(\frac{1+q^{2b}}{2})}{\ln q}$ we have that $\frac{q^a - 2q^b + q^{a+2b}}{(q^b - q^a)(1 - q^{a+b})} < 0$ and from (4.2.4) we obtain

$$\delta_j(c,a,b;q) > \ln q \left(\frac{1}{1 - q^{2j+2c+a+b}} + \frac{1}{1 - q^{2j+2c+a+b+2}} \right) >$$

$$lnq \left(\frac{1}{1 - q^{2c+a+b}} + \frac{1}{1 - q^{2c+a+b+2}} \right). \tag{4.2.7}$$

Thus, from (4.2.6) and (4.2.7) we obtain the inequality

$$\ln q(\frac{1}{1 - q^{2c+a+b}} + \frac{1}{1 - q^{2c+a+b+2}})x_{n,1}(c,a,b;q) < \frac{\partial x_{n,1}(c,a,b;q)}{\partial a}, \tag{4.2.8}$$

which means that the function $\frac{(1 - q^{2c+a+b})(1 - q^{2c+a+b+2})}{q^{2a}} x_{n,1}(c,a,b;q)$ increases with respect to a, for $|a| < b$, $a < b - \frac{\ln(\frac{1+q^{2b}}{2})}{\ln q}$, $c \geq 0$, $0 < q < 1$.

(ii) In the case where $0 < b - \frac{\ln(\frac{1+q^{2b}}{2})}{\ln q} < a < b$, it holds

$$-1 + \frac{1}{1 - q^{b-a}} - \frac{1}{1 - q^{a+b}} = \frac{q^a - 2q^b + q^{a+2b}}{(q^b - q^a)(1 - q^{a+b})} > 0$$

and thus

$$-1 + \frac{1}{1 - q^{b-a}} > \frac{1}{1 - q^{a+b}}. \tag{4.2.9}$$

Using (4.2.9) we obtain from (4.2.6)

$$\begin{aligned} \gamma_j(c,a,b;q) &> \ln q \left[\frac{1}{1 - q^{2c+a+b+2}} + \frac{1}{1 - q^{a+b}} \right] > \\ &> \ln q \left[-1 + \frac{1}{1 - q^{b-a}} + \frac{1}{1 - q^{2c+a+b+2}} \right]. \end{aligned} \tag{4.2.10}$$

Also, from (4.2.4), we have

$$\delta_j(c,a,b;q) > \ln q \left[-1 + \frac{1}{1 - q^{b-a}} + \frac{1}{1 - q^{2c+a+b+2}} \right]. \tag{4.2.11}$$

Thus from (4.2.10) and (4.2.11) we obtain the inequality

(4.2.12)
$$\ln q(-1+\frac{1}{1-q^{b-a}}+\frac{1}{1-q^{2c+a+b+2}})x_{n,1}(c,a,b;q)<$$
$$<\frac{\partial x_{n,1}(c,a,b;q)}{\partial a},$$

which means that the function $\frac{1-q^{2c+a+b}}{q^a-q^b}x_{n,1}(c,a,b;q)$ increases with respect to a, for $0<b-\frac{ln(\frac{1+q^{2b}}{2})}{lnq}<a<b$, $c\geq 0$ and $0<q<1$.

Proof of theorem 3.2.6.

(i) The partial derivatives of $\alpha_j(c,a,b;q)$ and $\beta_j(c,a,b;q)$ with respect to b are

(4.2.13)
$$\frac{\partial\alpha_j(c,a,b;q)}{\partial b}=\alpha_j(c,a,b;q)\gamma_j(c,a,b;q)$$

(4.2.14)
$$\frac{\partial\beta_j(c,a,b;q)}{\partial b}=\beta_j(c,a,b;q)\delta_j(c,a,b;q)$$

where

(4.2.15)
$$\gamma_j(c,a,b;q)=\frac{1}{2}\ln q\left[-\frac{2}{1-q^{2j+2c+2b+2}}-\frac{2}{1-q^{2j+2c+2a+2b+2}}+\right.$$
$$\left.+\frac{1}{1-q^{2j+2c+a+b+1}}+\frac{2}{1-q^{2j+2c+a+b+2}}+\frac{1}{1-q^{2j+2c+a+b+3}}\right]$$

and

(4.2.16)
$$\delta_j(c,a,b;q)=\ln q\left[-\frac{1}{1-q^{b-a}}-\frac{1}{1-q^{a+b}}+\right.$$
$$\left.\frac{1}{1-q^{2j+2c+a+b}}+\frac{1}{1-q^{2j+2c+a+b+2}}\right].$$

Since for $0<a<b$ the inequalities

$$\frac{1}{1-q^{b-a}}>\frac{1}{1-q^{2j+2c+2b+2}}>\frac{1}{1-q^{2j+2c+2a+2b+2}},$$

$$\frac{1}{1-q^{2j+2c+a+b+1}} > \frac{1}{1-q^{2j+2c+a+b+3}},$$

$$\frac{1}{1-q^{b-a}} > \frac{1}{1-q^{b+a}}$$

hold, we obtain
from (4.2.15) and (4.2.16):

$$\gamma_j(c,a,b;q) < \ln q\left[\frac{1}{1-q^{2j+2c+a+b+2}}+ \right.$$

(4.2.17)

$$\left.+\frac{1}{1-q^{2j+2c+a+b+3}} - \frac{2}{1-q^{b-a}}\right], \quad j=0,\dots,n-1$$

and

$$\delta_j(c,a,b;q) < \ln q\left[\frac{1}{1-q^{2j+2c+a+b+2}}+ \right.$$

(4.2.18)

$$\left.+\frac{1}{1-q^{2j+2c+a+b+3}} - \frac{2}{1-q^{b-a}}\right], \quad j=0,\dots,n-1.$$

From (4.2.17) and (4.2.18) we obtain the differential inequality

$$\ln q\left(\frac{1}{1-q^{2n+2c+a+b}} + \frac{1}{1-q^{2n+2c+a+b+1}} - \frac{2}{1-q^{b-a}}\right)\times$$

(4.2.19)

$$x_{n,1}(c,a,b;q) > \frac{\partial x_{n,1}(c,a,b;q)}{\partial b},$$

which means that the function

$$\frac{q^{2a}(1-q^{2n+2c+a+b})(1-q^{2n+2c+a+b+1})}{(q^a-q^b)^2} x_{n,1}(c,a,b;q)$$

decreases with respect to b, for $0<a<b$, $c\geq 0$ and $0<q<1$.

(ii) Similarly, since for $-b<a<0$, $j=0,\dots,n-1$ the following inequalities

$$\frac{1}{1-q^{b+a}} > \frac{1}{1-q^{b-a}},$$

$$\frac{1}{1-q^{b+a}} > \frac{1}{1-q^{2j+2c+2b+2}} > \frac{1}{1-q^{2j+2c+2a+2b+2}},$$

$$\frac{1}{1-q^{2j+2c+a+b+1}} > \frac{1}{1-q^{2j+2c+2a+b+2}},$$

$$\frac{1}{1-q^{2j+2c+a+b}} > \frac{1}{1-q^{2j+2c+a+b+3}}$$

hold, we obtain from (4.2.15)and (4.2.16):

(4.2.20)

$$\gamma_n(c,a,b;q) < lnq\left[\frac{1}{1-q^{2n+2c+a+b}} + \frac{1}{1-q^{2n+2c+a+b+1}} - \frac{2}{1-q^{b+a}}\right]$$

and

(4.2.21).

$$\delta_n(c,a,b;q) < lnq\left[\frac{1}{1-q^{2n+2c+a+b}} + \frac{1}{1-q^{2n+2c+a+b+1}} - \frac{2}{1-q^{b+a}}\right]$$

From (4.2.20) and (4.2.21) we obtain the differential inequality

(4.2.22)

$$\ln q\left(\frac{1}{1-q^{2n+2c+a+b}} + \frac{1}{1-q^{2n+2c+a+b+1}} - \frac{2}{1-q^{b+a}}\right) \times x_{n,1}(c,a,b;q) > \frac{\partial x_{n,1}(c,a,b;q)}{\partial b},$$

which means that the function

$$\frac{(1-q^{2n+2c+a+b})(1-q^{2n+2c+a+b+1})}{(1-q^{a+b})^2} x_{n,1}(c,a,b;q)$$

decreases with respect to b, for $0 < -a < b$, $c \geq 0$ and $0 < q < 1$.

4.3. q-Associated Laguerre polynomials

Proof of theorem 3.3.1.

For the proof of the theorem it suffice [20] to prove that

(4.3.1)

$$\frac{\left(\frac{\partial a_j}{\partial a}\right)^2}{\frac{\partial b_j}{\partial a}\frac{\partial b_{j+1}}{\partial a}} < \frac{1}{4}.$$

Differentiating (3.3.1) and (3.3.2) with respect to a, we obtain

(4.3.2)

$$\frac{\partial a_j}{\partial a} = \frac{a_j(-\ln q)(2-q^{j+c+a+1})}{2(1-q^{j+c+a+1})}$$

and

$$\frac{\partial b_j}{\partial a} = \frac{b_j(-\ln q)(1 - q^{j+c+1} + q)}{(1 - q^{j+c+1} + q - q^{j+c+a+1})}. \tag{4.3.3}$$

Thus, (4.3.1) becomes

$$\frac{q(1 - q^{j+c+1})(2 - q^{j+c+a+1})^2}{4(1 + q - q^{j+c+2})(1 - q^{j+c+a+1})(1 + q - q^{j+c+1})} < \frac{1}{4}$$

or equivalently

$$q(1-q^{j+c+1})(2-q^{j+c+a+1})^2-(1+q-q^{j+c+2})(1-q^{j+c+a+1})(1+q-q^{j+c+1}) < 0.$$

But, for $a \geq 0$, it is

$$\begin{aligned} & q(1 - q^{j+c+1})(2 - q^{j+c+a+1})^2 - \\ & \quad -(1 + q - q^{j+c+2})(1 - q^{j+c+a+1})(1 + q - q^{j+c+1}) \leq \\ & \quad \leq (1 - q^2)(-1 + q^{j+c+1} + q^{j+c+a+1} - q^{2j+2c+a+2}) \end{aligned}$$

which is negative for $c > -1$.

Proof of theorem 3.3.2.

From (4.3.2) and (4.3.3) it follows that $\frac{\partial \alpha_j}{\partial a}, \frac{\partial \beta_j}{\partial a} > 0$, for $c \geq 0$, $a + c > -1$. Also, we can easily obtain the following inequalities

$$\frac{\partial a_j}{\partial a} < a_j(-\ln q)\left(1 + \frac{q^{a+c+1}}{1 - q^{a+c+1}}\right) \tag{4.3.4}$$

and

$$\frac{\partial b_j}{\partial a} < b_j(-\ln q)\left(1 + \frac{q^{a+c+1}}{1 - q^{a+c+1}}\right) \tag{4.3.5}$$

Thus, using (4.3.4)and (4.3.5) we obtain the inequality

$$\frac{\partial x_{n,1}(c, a; q)}{\partial a} < -\ln q\left(1 + \frac{q^{a+c+1}}{1 - q^{a+c+1}}\right) x_{n,1}(c, a; q), \tag{4.3.6}$$

which means that the function $\left(\dfrac{q^a}{1-q^{a+c+1}}\right) x_{n,1}(c,a;q)$ decreases with respect to a, for $c \geq 0$, $a > -c-1$.

Proof of theorem 3.3.6.

(i) Differentiating (3.3.1) and (3.3.2) with respect to c we obtain

$$\frac{\partial a_j}{\partial c} = \frac{a_j}{2}(-\ln q)\left(\frac{1}{1-q^{j+c+1}} + \frac{1}{1-q^{j+c+a+1}} + 2\right) \tag{4.3.7}$$

and

$$\frac{\partial b_j}{\partial c} = b_j(-\ln q)\left(\frac{1+q}{1+q-q^{j+c+1}-q^{j+c+a+1}} + 1\right). \tag{4.3.8}$$

Since $\dfrac{\partial a_j}{\partial c} > 0$ and $\dfrac{\partial b_j}{\partial c} > 0$, for $c > -1$ and $a + c > 0$, it follows immediately that the largest zero $x_{n,1}(c,a;q)$ increases with respect to c, for $c > -1$ and $a + c > 0$.

(ii) For $a > 0$ we have

$$\frac{\partial a_j}{\partial c} \leq a_j(-\ln q)\left(1 + \frac{1+q}{1-q^{j+c+1}}\right) \leq a_j(-\ln q)\left(1 + \frac{1+q}{1-q^{c+1}}\right). \tag{4.3.9}$$

Also, the inequality

$$\frac{1+q}{1+q-q^{j+c+1}-q^{j+c+a+1}} < \frac{1+q}{1-q^{j+c+1}} \tag{4.3.10}$$

holds, because $0 < q(1-q^{j+c+a})$ holds, for $a > 0$ and $c \geq 0$. Thus, from (4.3.8) we have

$$\frac{\partial b_j}{\partial c} < b_j(-\ln q)\left(1 + \frac{1+q}{1-q^{j+c+1}}\right) < b_j(-\ln q)\left(1 + \frac{1+q}{1-q^{c+1}}\right). \tag{4.3.11}$$

Using (4.3.9) and (4.3.11) we obtain the inequality

$$\frac{\partial x_{n,1}(c,a;q)}{\partial c} < -\ln q\left(1 + \frac{1+q}{1-q^{c+1}}\right) x_{n,1}(c,a;q), \tag{4.3.12}$$

which means that the function $\left(\dfrac{q^{c(2+q)}}{(1-q^{c+1})^{1+q}}\right) x_{n,1}(c,a;q)$ decreases with respect to c, for $c \geq 0$ and $a > 0$.

4.4. Associated Al-Salam-Carlitz II polynomials

Proof of theorem 3.4.1.

(i) Differentiation of α_j and β_j, $j = 0, \ldots, n-1$, given in (3.4.1) and (3.4.2), with respect to a gives:

$$\frac{\partial}{\partial a}\alpha_j = \frac{1}{2}\sqrt{\frac{1-q^{j+c+1}}{aq^{2j+2c+1}}} \tag{4.4.1}$$

and

$$\frac{\partial}{\partial a}\beta_j = \frac{1}{q^{j+c}}. \tag{4.4.2}$$

It holds that

$$\frac{\left(\frac{\partial}{\partial a}\alpha_j\right)^2}{\left(\frac{\partial}{\partial a}\beta_j\right)\left(\frac{\partial}{\partial a}\beta_{j+1}\right)} = \frac{1-q^{j+c+1}}{4a} < \frac{1}{4}, \quad a > 1.$$

Hence [20] the operator $\frac{\partial}{\partial a}(AV^* + VA + B)$ is positive definite for $a > 1$ and the first part of the theorem has been proved.

(ii)-(iii) From (4.4.1) and (4.4.2) it follows that

$$\frac{\partial}{\partial a}\alpha_j > 0, \quad a > 0 \tag{4.4.3}$$

and

$$\frac{\partial}{\partial a}\beta_j > 0, \quad a > 0. \tag{4.4.4}$$

Also,

$$\frac{\partial}{\partial a}\alpha_j = \frac{1}{2}\alpha_j\frac{1}{a} < \alpha_j\frac{1}{a}, \quad a > 0 \tag{4.4.5}$$

and

$$\frac{\partial}{\partial a}\beta_j = \beta_j\frac{1}{a+1} < \beta_j\frac{1}{a}, \quad a > 0. \tag{4.4.6}$$

From the inequalities (4.4.3) - (4.4.6) we obtain the inequality

$$0 < \frac{\partial}{\partial a} x_{n,1}(c,a;q) < \frac{1}{a} x_{n,1}(c,a;q), \quad a > 0, \quad c > -1, \tag{4.4.7}$$

from which we obtain the second and third part of the theorem.

Proof of theorem 3.4.2.

Differentiating α_j and β_j with respect to c we obtain:

$$\frac{\partial}{\partial c}\alpha_j = -\frac{1}{2}\alpha_j \ln q \left(1 + \frac{1}{1-q^{j+c+1}}\right) < -\frac{1}{2}\alpha_j \ln q \left(1 + \frac{1}{1-q^{c+1}}\right) \tag{4.4.8}$$

and

$$\frac{\partial}{\partial c}\beta_j = -\beta_j \ln q < -\frac{1}{2}\beta_n \ln q \left(1 + \frac{1}{1-q^{c+1}}\right), \quad c > -1. \tag{4.4.9}$$

From the inequalities (4.4.8) and (4.4.9)we obtain the inequality

(4.4.10)

$$0 < \frac{\partial}{\partial c} x_{n,1}(c,a;q) < -\ln q \frac{1}{2}\left(1 + \frac{1}{1-q^{c+1}}\right) x_{n,1}(c,a;q), \quad a > 0, \quad c > -1,$$

from which it follows immediately the proof of the theorem.

4.5. q-Associated Meixner polynomials

Proof of theorem 3.5.1.

Differentiating α_j and β_j, $j = 0, \ldots, n-1$, given in (3.5.1) and (3.5.2), with respect to b we obtain:

$$\frac{\partial}{\partial b}\alpha_j = -\frac{1}{2}\alpha_j \frac{q^{j+c+1}}{1 - bq^{j+c+1}} < 0 \tag{4.5.1}$$

and

(4.5.2)

$$\frac{\partial}{\partial b}\beta_j = -\beta_j \frac{q^{j+c+1}r}{r + rq + q^{j+c+1} - q^{2j+2c+1} - rq^{j+c+1} - brq^{j+c+1}} = -\frac{r}{q^{j+c}} < 0.$$

Since $\beta_j > 0$, $r > 0$, $b < q^{-c-1}$ from (4.5.2) it follows that

$$\frac{q^{j+c+1}r}{r + rq + q^{j+c+1} - q^{2j+2c+1} - rq^{j+c+1} - brq^{j+c+1}} > 0.$$

But,

$$rq - rq^{j+c+1} + q^{j+c+1} - q^{2j+2c+1} > 0, \quad c > 0.$$

Hence,

(4.5.3)
$$\frac{q^{j+c+1}r}{r+rq+q^{j+c+1}-q^{2j+2c+1}-rq^{j+c+1}-brq^{j+c+1}} < \frac{q^{j+c+1}}{1-bq^{j+c+1}} \Rightarrow$$
$$\Rightarrow \frac{\partial}{\partial b}\beta_j > -\beta_j\frac{q^{j+c+1}}{1-bq^{j+c+1}} > -\beta_j\frac{q^{c+1}}{1-bq^{c+1}}, \quad c > 0, \quad j = 0, \ldots.$$

Also,

(4.5.4)
$$\frac{q^{j+c+1}}{2(1-bq^{j+c+1})} < \frac{q^{j+c+1}}{1-bq^{j+c+1}} < \frac{q^{c+1}}{1-bq^{c+1}} \Rightarrow$$
$$\Rightarrow \frac{\partial}{\partial b}\alpha_j > -\alpha_j\frac{q^{c+1}}{1-bq^{c+1}}, \quad j = 0, 1, \ldots.$$

From the inequalities (4.5.3) and (4.5.4) follows the inequality

(4.5.5)
$$-\frac{q^{c+1}}{1-bq^{c+1}}(q^{-x_{n,1}(c,r,b;q)} - 1) < \frac{\partial}{\partial b}(q^{-x_{n,1}(c,r,b;q)} - 1) < 0,$$

which means that the function $q^{-x_{n,1}(c,r,b;q)} - 1$ decreases with respect to b, or the largest zero $x_{n,1}(c, r, b; q)$ decreases with respect to b and that the function $\dfrac{\left(q^{-x_{n,1}(c,r,b;q)} - 1\right)}{1 - bq^{c+1}}$ increases with respect to b.

Proof of theorem 3.5.3.

Differentiating α_j and β_j with respect to r we obtain:

(4.5.6)
$$\frac{\partial}{\partial r}\alpha_j = \frac{1}{2}\alpha_j\left[\frac{1}{r} + \frac{1}{r+q^{n+c+1}}\right] > 0$$

and

(4.5.7)
$$\frac{\partial}{\partial r}\beta_j = \beta_j\frac{1+q-q^{j+c+1}-bq^{j+c+1}}{r+rq+q^{j+c+1}-q^{2j+2c+1}-rq^{j+c+1}-brq^{j+c+1}} > 0.$$

Also, since $\dfrac{1}{r} > \dfrac{1}{r+q^{n+c+1}}$ and $q^{j+c+1} - q^{2j+2c+1} > 0$, $c > 0$, it follows from (4.5.6) and (4.5.7) that

(4.5.8)
$$\frac{\partial}{\partial r}\alpha_j < \alpha_j\frac{1}{r}, \quad j = 0, 1, \ldots, \quad c > 0$$

and

$$\frac{\partial}{\partial r}\beta_j < \beta_j\frac{1}{r}, \quad j = 0, 1, \ldots. \tag{4.5.9}$$

From (4.5.6) - (4.5.9) it follows the inequality

$$\frac{1}{r}(q^{-x_{n,1}(c,r,b;q)} - 1) > \frac{\partial}{\partial r}(q^{-x_{n,1}(c,r,b;q)} - 1) > 0, \tag{4.5.10}$$

from which we obtain that the function $q^{-x_{n,1}(c,r,b;q)} - 1$ or the largest zero $x_{n,1}(c,r,b;q)$ of the q-associated Meixner polynomials increases with respect to r and that the function $\frac{1}{r}\left(q^{-x_{n,1}(c,r,b;q)} - 1\right)$ decreases with respect to r.

Proof of theorem 3.5.6.

Writing α_j and β_j as

$$\alpha_j = \sqrt{A_j C_{j+1}} \quad \text{and} \quad \beta_j = A_j + C_j,$$

where

$$A_j = \frac{r(1 - bq^{j+c+1})}{q^{2j+2c+1}} \quad \text{and} \quad C_j = \frac{q(1 - q^{j+c})(r + q^{j+c})}{q^{2j+2c+1}},$$

we obtain

$$\frac{\partial \alpha_j}{\partial c} = \frac{1}{2}\alpha_j(\Gamma_j + \Delta_{j+1}) \tag{4.5.11}$$

and

$$\frac{\partial \beta_j}{\partial c} = A_j\Gamma_j + C_j\Delta_j, \tag{4.5.12}$$

where

$$\Gamma_j = \frac{1}{A_j}\frac{\partial A_j}{\partial c} = -\ln q\left(1 + \frac{1}{1 - bq^{j+c+1}}\right)$$

and

$$\Delta_j = \frac{1}{C_j}\frac{\partial C_j}{\partial c} = -\ln q\left(\frac{1}{1 - q^{j+c}} + \frac{r}{r + q^{j+c}}\right).$$

Since, A_j, Γ_j, C_j and Δ_j are positives for $c > 0$, the quantity $(q^{-x_{n,1}(c,r,b;q)} - 1)$ increases with respect to c from which it follows that the largest zero $x_{n,1}(c,r,b;q)$ of the q-associated Meixner polynomials also increases with respect to c for $c > 0$. Also,

$$\Gamma_j < -\ln q\left(1+\frac{1}{1-q^{j+c+1}}\right) < -\ln q\left(1+\frac{1}{1-q^c}\right), \tag{4.5.13}$$
$$b<1, \quad c>0$$

and

$$\Delta_j < -\ln q\left(1+\frac{1}{1-q^{j+c}}\right) < -\ln q\left(1+\frac{1}{1-q^c}\right), \quad c>0. \tag{4.5.14}$$

Using (4.5.13) and (4.5.14) we obtain from (4.5.11) and (4.5.12)

$$\frac{\partial \alpha_j}{\partial c} < \alpha_j(-\ln q)\left(1+\frac{1}{1-q^c}\right), \quad b<1, \quad c>0$$

and

$$\frac{\partial \beta_j}{\partial c} < \beta_j(-\ln q)\left(1+\frac{1}{1-q^c}\right), \quad c>0,$$

from which it follows that

$$\frac{\partial}{\partial c}(q^{-x_{n,1}(c,r,b;q)}-1) < -\ln q\left(1+\frac{1}{1-q^c}\right)(q^{-x_{n,1}(c,r,b;q)}-1), \quad b<1, \quad c>0$$

which means that the function $\frac{q^{2c}}{1-q^c}(q^{-x_{n,1}(c,r,b;q)} - 1)$ decreases with respect to c for $c > 0$, $b < 1$.

4.6. q-Lommel polynomials

Proof of theorem 3.6.1.

Differentiating $a_j(\nu, c, q)$ with respect to ν we obtain:

$$\frac{\partial a_j(\nu;q)}{\partial \nu} = \frac{1}{2}a_j(\nu;q)\ln q\left\{1+\frac{q^{\nu+j+1}}{1-q^{\nu+j+1}}+\frac{q^{\nu+j}}{1-q^{\nu+j}}\right\}. \tag{4.6.1}$$

Since $\ln q < 0$, we obtain from (4.6.1) that the largest zero $x_{n,1}(c,\nu,q)$ of the q-Lommel polynomials decreases with respect to ν, for $\nu \geq 0$. Also, the following inequality holds

$$\frac{\partial \alpha_j}{\partial \nu} > \frac{1}{2}\alpha_j \ln q \left(1 + \frac{q^{\nu+1}}{1-q^{\nu+1}} + \frac{q^{\nu}}{1-q^{\nu}}\right),$$

from which it follows that the function

$$\sqrt{\frac{(1-q^{\nu})(1-q^{\nu+1})}{q^{\nu}}}\, x_n(\nu;q),$$

increases with respect to ν for $\nu > 0$.

Proof of theorem 3.6.6.

Differentiating $a_j(\nu;q)$ with respect to q we obtain

$$\frac{\partial}{\partial q} a_j(\nu;q) = \frac{1}{2} a_j(\nu;q)\left(\frac{j+\nu}{q} + \frac{(j+\nu+1)q^{j+\nu}}{1-q^{j+\nu+1}} + \frac{(j+\nu)q^{j+\nu-1}}{1-q^{j+\nu}}\right)$$

which is positive for $j \geq 0$, $\nu > 0$. Thus the largest zero $x_{n,1}(\nu;q)$ of the q-Lommel polynomials increases with respect to q.

Setting $j+\nu = k$ we obtain

$$\frac{\partial}{\partial q} a_j(\nu;q) = \frac{1}{2} a_j(\nu;q)\left(\frac{k}{q} + \frac{(k+1)q^{k}}{1-q^{k+1}} + \frac{kq^{k-1}}{1-q^{k}}\right). \tag{4.6.2}$$

It holds that

$$\frac{\partial^2}{\partial k^2}\left(\frac{(k+1)q^k}{1-q^{k+1}}\right) = \frac{q^k \ln q\left[2 - 2q^{k+1} + (1+k)(1+q^{k+1})\ln q\right]}{(1-q^{k+1})^3}. \tag{4.6.3}$$

But,

$$\frac{\partial}{\partial q}\left(2 - 2q^{k+1} + (1+k)(1+q^{k+1})\ln q\right) = \frac{(k+1)}{q}\left(1 - q^{k+1} + q^{k+1}\ln q + kq^{k+1}\ln q\right). \tag{4.6.4}$$

Also,

$$\frac{\partial}{\partial q}\left(1 - q^{k+1} + q^{k+1}\ln q + kq^{k+1}\ln q\right) = (1+k)^2 q^k \ln q < 0, \quad k \geq 0. \tag{4.6.5}$$

For $q = 1$ we have $1 - q^{k+1} + q^{k+1}\ln q + kq^{k+1}\ln q = 0$, hence from (4.6.5) we obtain

$$1 - q^{k+1} + q^{k+1}\ln q + kq^{k+1}\ln q > 0, \quad k > 0, \quad 0 < q < 1. \tag{4.6.6}$$

Since for $q = 1$ we have $2 - 2q^{k+1} + (1+k)(1+q^{k+1})\ln q = 0$, we obtain from (4.6.3), (4.6.4) and (4.6.6)

$$\frac{\partial^2}{\partial k^2}\left(\frac{(k+1)q^k}{1-q^{k+1}}\right) > 0, \quad k > 0,$$

from which it follows that

$$\frac{\partial}{\partial k}\left(\frac{(k+1)q^k}{1-q^{k+1}}\right) > \frac{\partial}{\partial k}\left(\frac{kq^{k-1}}{1-q^k}\right)$$

or

$$\frac{\partial}{\partial k}\left(\frac{k}{q} + \frac{(k+1)q^k}{1-q^{k+1}} + \frac{kq^{k-1}}{1-q^k}\right) > \frac{\partial}{\partial k}\left(\frac{k}{q} + 2\frac{kq^{k-1}}{1-q^k}\right). \tag{4.6.7}$$

But,

$$\frac{\partial}{\partial k}\left(\frac{k}{q} + 2\frac{kq^{k-1}}{1-q^k}\right) = \frac{1 - q^{2k} + 2kq^k\ln q}{q(1-q^k)^2} \tag{4.6.8}$$

and

$$\frac{\partial}{\partial k}\left(1 - q^{2k} + 2kq^k\ln q\right) = -2q^k\ln q(-1 + q^k - k\ln q). \tag{4.6.9}$$

Also,

$$\frac{\partial}{\partial k}(-1 + q^k - k\ln q) = -\ln q(1-q^k) > 0, \quad k > 0. \tag{4.6.10}$$

Since for $k = 0$ we have $-1 + q^k - k\ln q = 0$ and $1 - q^{2k} + 2kq^k\ln q = 0$, we obtain from (4.6.7), (4.6.8), (4.6.9) and (4.6.10)

$$\frac{\partial}{\partial k}\left(\frac{k}{q} + \frac{(k+1)q^k}{1-q^{k+1}} + \frac{kq^{k-1}}{1-q^k}\right) > 0,$$

from which it follows that

$$\frac{j+\nu}{q}+\frac{(j+\nu+1)q^{j+\nu}}{1-q^{j+\nu+1}}+\frac{(j+\nu)q^{j+\nu-1}}{1-q^{j+\nu}}<$$

$$<\frac{n+\nu-1}{q}+\frac{(n+\nu)q^{n+\nu-1}}{1-q^{n+\nu}}+\frac{(n+\nu-1)q^{n+\nu-2}}{1-q^{n+\nu-1}},$$

$$j=0,\dots,n-1,\ \nu>0,\ n\geq 2.$$

Hence,

$$\frac{\partial}{\partial q}x_{n,1}(\nu;q)<x_{n,1}(\nu;q)\frac{1}{2}\left(\frac{n+\nu-1}{q}+\frac{(n+\nu)q^{n+\nu-1}}{1-q^{n+\nu}}+\right.$$

$$\left.+\frac{(n+\nu-1)q^{n+\nu-2}}{1-q^{n+\nu-1}}\right),\nu>0,\ n\geq 2,$$

from which it follows that

$$\frac{\partial}{\partial q}\left[\sqrt{\frac{(1-q^{n+\nu})(1-q^{n+\nu-1})}{q^{n+\nu-1}}}\;x_{n,1}(\nu;q)\right]<0.$$

References

[1] S. Ahmed, M. E. Muldoon and Spigler, Inequalities and numerical bounds for zeros of ultraspherical polynomials, *SIAM J. Math. Anal.* **17** (1986), 1000-1007.

[2] G.E. Andrews and R. Askey, Classical orthogonal polynomials, *in Polynômes Orthogonaux et Applications, C. Brezinski et al., eds., Lecture Notes in Mathematics 1171, Springer-Verlag. Berlin*, 1985, 33-62.

[3] R. Askey and J. Wilson, t Some basic hypergeometric polynomials that generalize Jacobi polynomials, *Mem. Amer. Math. Soc.*, **54 (319)** (1985), iv+55 pp.

[4] J. Charris and M. E. Ismail, On sieved orthogonal polynomials V: Sieved Pollaczek polynomials, *SIAM J. Math. Anal.*, **18** *(4)* (1987), 1177-1218.

[5] D. Dimitrov, On a conjecture concerning monotonicity of zeros of ultraspherical polynomials, *J. Approx. Theory* **85** *(1)* (1996), 88-97.

[6] D. Dimitrov, Convecity of the Extreme Zeros of Gegnbauer and Laguerre Polynomials, *J. Comput. Appl. Math* to appear.

[7] D. K. Dimitrov and R. O. Rodrigues, On the behaviour of zeros of Jacobi polynomials, *J. Approx. Theory* **116** *(2)* (2002), 224-239.

[8] Á. Elbert and P D. Siafarikas, Monotonicity properties of the zeros of ultraspherical polynomials, *J. Approx. Theory* **97** (1999), 31-39.

[9] J. Favard, Sur les polynomes de Tchebicheff, *C. R. Acad. Sci. Paris Sér I Math.*, **200** (1935), 2052-2053.

[10] W. Hahn, Über Orthogonalpolynome, die q-Differenzengleichungen genügen, *Math. Nachr.* **2** (1949), 4-34.

[11] E. K. Ifantis, A theorem concerning differentiability of eigenvectors and eigenvalues with some applications, *Appl. Anal.* **28** (1988), 257-283.

[12] E. K. Ifantis, Perron-Frobenius-type theorems for tridiagonal operators, *Appl. Anal.*, **44** (1992), 159-169.

[13] E. K. Ifantis and P. D. Siafarikas, Differential inequalities for the largest zero of Laguerre and ultraspherical polynomials, *in "Actas del VI Symposium on Polinomios Orthogonales y Aplicationes"*, Gijon, Spain, 1989, 187-197.

[14] E. K. Ifantis and P. D Siafarikas, Differential inequalities and monotonicity properties of the zeros of associated Laquerre and Hermite polynomials, *Ann. Numer. Math.*, **2** (1995), 79-91.

[15] M. E. Ismail, The zeros of basic Bessel functions, the functions $J_{\nu+ax}(x)$, and associated orthogonal polynomials, *J. Math. Anal. Appl.*, **86** (1982), 1-19.

[16] M. E. Ismail, On sieved orthogonal polynomials. I. Symmetric Pollaczek analogues, *SIAM J. Math. Anal.*, **16** (1985), 1093-1113.

[17] M. E. Ismail, Monotonicity of zeros of orthogonal polynomials, *in "q-Series and Partitions"* (D. Stanton, Ed.), 177-190, Springer-Verlag, New York, 1989.

[18] M. E. Ismail, The variation of zeros of certain orthogonal polynomials, *Adv. in Appl. Math.*, **8** (1987), 111-119.

[19] M. E. Ismail and J. Letessier, Monotonicity of zeros of ultraspherical polynomials, sl in "Orthogonal Polynomials and their Applications, Proceedings, International Symposium, Segovia, Spain, Sept. 22-27, 1986" Problem Section, Lecture Notes in Mathematics, Vol 1329, 329-333, Springer-Verlag, Berlin/Heidelberg/New York, 1988.

[20] M. E. Ismail and M. E. Muldoon, A discrete approach to monotonicity of zeros of orthogonal polynomials, *Trans. Amer. Math. Soc.*, **323** (1) (1991), 65-78.

[21] M.E.H. Ismail and M.E. Muldoon, On the variation with respect to a parameter of zeros of Bessel and q-Bessel functions, *J. Math. Anal. Appl.*, **135** (1988), 187-207.

[22] F. H. Jackson, The basic gamma function and elliptic functions, *Proc. Roy. Soc. Edinburgh Sect. A* **76**, (1905), 127-144.

[23] R. Koekoek and R.F. Swarttouw, The Askey-scheme of hypergeometric orthogonal polynomials and its q-analogue, *Delft University of Technology, Faculty of Information Technology and Systems, Dep. of Technical Mathematics and Informatics, Report no. 98-17* (1998).

[24] A. Laforgia, A monotonic property for the zeros of ultraspherical polynomials, *Proc. Amer. Math. Soc.*, **83** (1981), 757-758.

[25] A. Laforgia, Monotonicity properties for the zeros of orthogonal polynomials and Bessel functions, *in "Polynomes Orthogonaux et Applicationes, Proceedings of the Laguerre Symposium, Bar-le-Duk, Spain, 1984", Lecture Notes in Mathematics*, **1171**, 267-277, Springer-Verlag, Berlin/ Heidelberg/ New York, 1985.

[26] P. Natalini and B. Palumbo, Some monotonicity results on the zeros of the generalized Laguerre polynomials, *J. Comput. Appl. Math. (Proceedings of the 6th international Symposium on Orthogonal Polynomials, Special Functions and their Apllications,* Rome, June 18-22, 2001) to appear.

[27] R. Alvarez-Nodarse, E. Buendia and J.S.Dehesa, The distribution of zeros of general q-polynomials, *J.Phys. A* **30** (1997), 6743-6768.

[28] M. Rahman, The linearization of the product of continuous q-Jacobi polynomials, *Canad. J. Math.* **33** (1981), 961-987.

[29] P. D. Siafarikas, Inequalities for the zeros of the associated ultraspherical polynomials, *Math. Inequal. Appl.*, **2** *(2)* (1999), 233-241.

[30] G. Szegö, Orthogonal polynomials. Fourth eidtion. American Mathematical Society, Colloquium Publications, Vol. XXIII. American Mathematical Society, Providence, R.I., 1975. xiii+432 pp.

Seminorms Related to Weak Compactness Under Real and Complex Interpolation for Finite Families of Banach Spaces

Andrzej Kryczka

Institute of Mathematics, Maria Curie-Skłodowska University, 20031 Lublin, Poland e–mail: andrzej.kryczka@umcs.lublin.pl

Current address: Departamento de Análisis Matemático, Facultad de CC. Matemáticas, Universidad Complutense de Madrid, 28040 Madrid, Spain e–mail: kryczka@mat.ucm.es

Abstract. We consider certain seminorms on the space of bounded linear operators between Banach spaces which measure the deviation of operators from weak compactness. We prove logarithmically convex-type inequalities for the seminorms of operators under real and complex interpolation with respect to finite families of Banach spaces. These estimates give a quantitative extension to weakly noncompact operators of the results on interpolation of weakly compact operators with respect to finite families.

AMS Subject Classification. 46B70, 46M35, 47B07

Key words. Real interpolation, complex interpolation, measure of weak noncompactness.

1 Introduction

The classical interpolation theorems provide estimates on the norms of interpolated operators in the form of logarithmically convex-type inequalities. In this article we establish analogous estimates for certain seminorms which measure the deviation of bounded linear operators between Banach spaces from weak compactness.

We deal with two interpolation methods: the real method of Yoshikawa [34] (equivalent to the K-method of Sparr [32]) and the complex method of Coifman, Cwikel, Rochberg, Sagher and Weiss [14] for finite families of Banach spaces. The seminorms we consider are related to the measures of weak noncompactness defined in [24] and [25] for subsets of Banach spaces. These seminorms do not coincide with those based on the outer and inner measures related to operator ideals, and which, in the case of the ideal of weakly compact operators, are strongly related to De Blasi's [16] measure of weak noncompactness (see [20]). As shown in [11] and [13], in the general case, the estimates of the above-mentioned type do not exist for De Blasi's measure. In contrast, the seminorms used in this paper 'behave well' and no additional restrictions on interpolated families or operators are needed. In the proofs of our main results we develop techniques applied in [24] and [25] for the real interpolation of Lions and Peetre [27] and the complex interpolation of Calderón [7] for pairs of Banach spaces.

Various quantitative approaches to weak compactness (or other properties of operators) in the case of pairs of Banach spaces are treated in [2, 11, 12, 13, 24, 25]. The qualitative results on interpolation of weakly compact operators for several interpolation methods with respect to finite families are given in [8] and [10] (see also results for pairs in [5, 6, 7, 21, 28, 29, 30, 31]). Some estimates for the measures related to operator ideals under interpolation methods associated with polygons are obtained in [9].

The open unit ball of a Banach space X will be denoted by B_X and its closure by $\overline{\mathsf{B}}_X$. We write $\mathcal{M}_X$ for the family of all nonempty and bounded subsets of X. By $\operatorname{conv} A$ we mean the convex hull of a set $A \subset X$. The space of all bounded linear operators between Banach spaces X and Y is denoted by $\mathcal{L}(X,Y)$. The subspace of $\mathcal{L}(X,Y)$ consisting of all weakly compact operators will be denoted by $\mathcal{W}(X,Y)$. We use the same symbols to designate elements or subsets of X and their respective canonical images in the second dual X^{**}. The abbreviations with the beginning w^* refer to the weak-star topology. Unless otherwise stated we consider Banach spaces over the real field. However, the presented results are valid for complex Banach spaces as well.

2 Preliminaries

Measures of weak noncompactness and related seminorms

Our basic tools are the *measures of weak noncompactness* $\overline{\gamma}$ and γ introduced in [24] and [25] respectively. Let X be a Banach space and let (x_n) be a sequence in X. We say that (y_n) is a sequence of *successive convex combinations* (scc for short) for (x_n) if there exists a sequence of integers $0 = r_1 < r_2 < \cdots$ such that $y_n \in \operatorname{conv}\{x_i\}_{i=r_n+1}^{r_{n+1}}$ for each n. Vectors u_1, u_2 are said to be a pair of scc for (x_n) if $u_1 \in \operatorname{conv}\{x_i\}_{i=1}^{r}$ and $u_2 \in \operatorname{conv}\{x_i\}_{i=r+1}^{\infty}$ for some integer $r \geq 1$. For each $A \in \mathcal{M}_X$, by definition,

$$\overline{\gamma}(A) = \sup \operatorname{dist}(x^{**}, X),$$

where the supremum is taken over all w^*-cluster points $x^{**} \in X^{**}$ of sequences $(x_n) \subset \operatorname{conv} A$, and

$$\gamma(A) = \sup\{\operatorname{csep}(x_n) \colon (x_n) \subset \operatorname{conv} A\}$$

with

$$\operatorname{csep}(x_n) = \inf\{\|u_1 - u_2\| : u_1, u_2 \text{ is a pair of scc for } (x_n)\}.$$

It is proved [25] that for every $A \in \mathcal{M}_X$

$$\begin{aligned} \gamma(A) = \sup\{ &\lim_n \lim_k F_n(x_k) - \lim_k \lim_n F_n(x_k) \colon \\ &(x_k) \subset \operatorname{conv} A, (F_n) \subset \overline{\mathsf{B}}_{X^*} \text{ and the limits exist} \} \end{aligned} \tag{2.1}$$

and

$$\gamma(A) = \sup \operatorname{dist}(x^{**}, \operatorname{conv}\{x_n\}), \tag{2.2}$$

where the supremum in (2.2) is taken as for $\overline{\gamma}(A)$. The measures $\overline{\gamma}$ and γ are equivalent, namely $\overline{\gamma}(A) \le \gamma(A) \le 2\overline{\gamma}(A)$ for every $A \in \mathcal{M}_X$ (see [24]), but in general, neither of these is equivalent to De Blasi's [16] measure of weak noncompactness ω (see [3] and [25]), where for each $A \in \mathcal{M}_X$

$$\omega(A) = \inf\{\, t > 0 \colon A \subset C + t\mathsf{B}_X,\ C \subset X \text{ is weakly compact}\,\}.$$

For more properties of measures of weak noncompactness we refer to [4].

The following theorem will be useful in the sequel.

Theorem 2.1. **([25])** *Let (x_n) be a bounded sequence in a Banach space X. For every $\varepsilon > 0$ there exists a sequence (y_n) of scc for (x_n) such that if u_1, u_2 and v_1, v_2 are any pairs of scc for (y_n), then $|\|u_1 - u_2\| - \|v_1 - v_2\|| \le \varepsilon$.*

Let X and Y be Banach spaces and $T \in \mathcal{L}(X,Y)$. Using the fact that a measure of weak noncompactness of a set A is equal zero if and only if A is relatively weakly compact, one can define various seminorms in $\mathcal{L}(X,Y)$ which vanish on $\mathcal{W}(X,Y)$. We investigate the behaviour under interpolation of the following seminorms (measures of weak noncompactness of operators):

$$\Gamma(T) = \gamma(T(\mathsf{B}_X)) \quad \text{and} \quad \overline{\Gamma}(T) = \overline{\gamma}(T(\mathsf{B}_X)).$$

We also consider another approach which appears in the literature mainly in the context of measures of noncompactness (see [1]). Let μ denote one of the measures of weak noncompactness: ω, γ or $\overline{\gamma}$, and let μ be given in X and Y. An operator $T \in \mathcal{L}(X,Y)$ is called a *weak k-contraction* if $\mu(T(A)) \le k\mu(A)$ for every $A \in \mathcal{M}_X$. Each $T \in \mathcal{L}(X,Y)$ is a weak k-contraction for $\mu = \omega$ with $k = \omega(T(\mathsf{B}_X))$, and for $\mu = \gamma, \overline{\gamma}$ with $k = \|T\|$. Clearly, the formula

$$|T|_\mu = \inf\{\, k \colon T \text{ is a weak } k\text{-contraction}\,\}$$

defines a seminorm in $\mathcal{L}(X,Y)$. It is evident that

$$|T|_\mu = \sup\{\, \mu(T(A)) \colon \mu(A) \le 1\,\}.$$

It is also easy to see that $|T|_\omega = \omega(T(\mathsf{B}_X))$ for every $T \in \mathcal{L}(X,Y)$. The equivalence of the above seminorms will be discussed in the last section.

A crucial fact in our considerations is the behaviour of the seminorms on some vector-valued sequence or function spaces. For Γ this behaviour is described by the next two theorems. Let us fix $n \in N$, $p \in (1,\infty)$ and $\xi \in R^n$. For $z \in Z^n$ let $\xi \cdot z$ denote the standard scalar product of ξ and z. Let $l_p(\xi, X)$ denote the Banach space of all families $u = (u(z))_{z \in Z^n}$ such that $u(z) \in X$ for every $z \in Z^n$ and

$$\|u\|_{l_p(\xi,X)} = \left(\sum_{z \in Z^n} \left(2^{-\xi \cdot z} \|u(z)\| \right)^p \right)^{1/p} < \infty.$$

Since the spaces $l_p(\xi, X)$ and $l_p(X)$, where the latter is just $l_p(0, X)$ with families indexed by N instead of Z^n, are isometrically isomorphic and γ is invariant under linear isometries, we can rephrase Theorem 3.6 from [25] as follows.

Theorem 2.2. *Let X and Y be Banach spaces and $T \in \mathcal{L}(X, Y)$. Let the operator $\widetilde{T} \in \mathcal{L}(l_p(\xi, X), l_p(\xi, Y))$ be given by $\widetilde{T}u = (Tu(z))_{z\in Z^n}$ for every $u = (u(z))_{z\in Z^n} \in l_p(\xi, X)$. Then $\Gamma(\widetilde{T}) = \Gamma(T)$.*

Let (Ω, Σ, ν) be a probability space. Let $L_\infty(\nu, X)$ denote the Banach space of all (equivalence classes of) ν-measurable vector-valued functions $f : \Omega \to X$ such that $\|f\| = \operatorname{ess\,sup}_{\tau\in\Omega} \|f(\tau)\| < \infty$. The Banach space of all (equivalence classes of) ν-measurable vector-valued functions $g : \Omega \to Y$ such that $\|g\| = \int_\Omega \|g(\tau)\| \, d\nu(\tau) < \infty$ will be denoted by $L_1(\nu, Y)$. In the case of scalar-valued functions, this space will be denoted by $L_1(\nu)$. For $L_\infty(\nu, X)$ and $L_1(\nu, Y)$ we can state the analogue of Theorem 3.3 from [24] and adapt the proof with some slight changes. For the convenience of the reader we present that proof.

Theorem 2.3. *Let X and Y be Banach spaces and $T \in \mathcal{L}(X, Y)$. Let the operator $\widehat{T} \in \mathcal{L}(L_\infty(\nu, X), L_1(\nu, Y))$ be given by $(\widehat{T}f)(\tau) = T(f(\tau))$, ν-a.e. for every $f \in L_\infty(\nu, X)$. Then $\Gamma(\widehat{T}) = \Gamma(T)$.*

Proof. Considering the subspace of all constant functions in $L_\infty(\nu, X)$, we see at once that $\Gamma(\widehat{T}) \geq \Gamma(T)$. To prove that $\Gamma(\widehat{T}) \leq \Gamma(T)$ we use formula (2.1). We fix $\varepsilon > 0$ and choose sequences $(y_n) \subset T(\mathsf{B}_{L_\infty(\nu,X)})$ and $(F_n) \subset (L_1(\nu, Y))^*$ with $\|F_n\| \leq 1$ for every n such that $\Gamma(\widehat{T}) - \varepsilon \leq \vartheta_1 - \vartheta_2$ where $\vartheta_1 = \lim_n \lim_k F_n(y_k)$ and $\vartheta_2 = \lim_k \lim_n F_n(y_k)$. Let $\Lambda(Y^*)$ denote the space of all functions $f : \Omega \to Y^*$ such that the function $\tau \mapsto \langle y, f(\tau)\rangle$ is ν-integrable for every $y \in Y$. We can assume that $L_1(\nu)$ is separable, for if not, we consider a countably generated σ-algebra $\Sigma_0 \subset \Sigma$ such that $(y_n) \subset L_1(\nu|\Sigma_0, Y)$. Then for each n there exists $f_n \in \Lambda(Y^*)$ such that $\|f_n(\tau)\| \leq 1$ and

$$F_n(y) = \int_\Omega \langle y(\tau), f_n(\tau)\rangle \, d\nu(\tau)$$

for every $y \in L_1(\nu, Y)$ (see [19, p. 588]). Let $\langle y_k, f_n\rangle$ denote the function $\tau \mapsto \langle y_k(\tau), f_n(\tau)\rangle$. Obviously, the functions $\langle y_k, f_n\rangle$, $k, n \geq 1$ are equiintegrable, so they form a relatively weakly compact set in $L_1(\nu)$ (see [17, p. 93]). By the Eberlein-Šmulian and Mazur theorems (see [17, pp. 11, 18]), for each sequence in a relatively weakly compact set of a Banach space there exists a convergent sequence of its scc. Next, each convergent sequence in $L_1(\nu)$ has a subsequence which converges ν-a.e. to the same limit (see [18, p. 150]). Therefore we can find a sequence (y_k^1) of scc for (y_k) such that $\langle y_k^1, f_1\rangle$ tends ν-a.e. to some $g_1 \in L_1(\nu)$. We now proceed by induction. For $n \geq 2$ let us choose a sequence (y_k^n) of scc for (y_k^{n-1}) such that $\langle y_k^n, f_n\rangle$ tends ν-a.e. to $g_n \in L_1(\nu)$ as $k \to \infty$. Write $y'_k = y_k^k$ for every k. Then $(y'_k)_{k\geq m}$ is a sequence of scc for (y_k^m) and $\langle y'_k, f_n\rangle$ also tends ν-a.e. to g_n as $k \to \infty$ for every n. We next choose a sequence (g'_n) of scc for (g_n) convergent ν-a.e. to some $g \in L_1(\nu)$. Then $g'_n = \sum_{i=m_n+1}^{m_{n+1}} \lambda_i^{(n)} g_i$, where (m_n) is an increasing sequence of positive integers, $\lambda_i^{(n)} \geq 0$ and $\sum_{i=m_n+1}^{m_{n+1}} \lambda_i^{(n)} = 1$. We set

$$f'_n = \sum_{i=m_n+1}^{m_{n+1}} \lambda_i^{(n)} f_i \quad \text{and} \quad F'_n = \sum_{i=m_n+1}^{m_{n+1}} \lambda_i^{(n)} F_i.$$

In this way we have obtained the sequences $(y'_k) \subset T(\mathsf{B}_{L_\infty(\nu,X)})$, $(f'_n) \subset \Lambda(Y^*)$ and $(F'_n) \subset (L_1(\nu,Y))^*$ with the following properties: $\|F'_n\| \leq 1$, $\lim_n \lim_k F'_n(y'_k) = \vartheta_1$, $\lim_k \lim_n F'_n(y'_k) = \vartheta_2$,

$$F'_n(y) = \int_\Omega \langle y(\tau), f'_n(\tau)\rangle \, d\nu(\tau)$$

for all $y \in L_1(\nu, Y)$, $n \geq 1$, and $\lim_n \lim_k \langle y'_k, f'_n\rangle = g$, ν-a.e.

Similarly, we can choose a sequence (f''_n) of scc for (f'_n) such that $\langle y'_k, f''_n\rangle$ tends ν-a.e. to some $h_k \in L_1(\nu)$ for each k. Next, we take a sequence (y''_k) of scc for (y'_k) and the corresponding sequence (F''_n) of scc for (F'_n) such that the following limit exists

$$\lim_k \lim_n \langle y''_k, f''_n\rangle = h, \quad \nu\text{-a.e.}$$

Then $\lim_n \lim_k \langle y''_k, f''_n\rangle = g$, ν-a.e.

The Lebesgue dominated convergence theorem yields

$$\begin{aligned}
&\int_\Omega (g(\tau) - h(\tau)) d\nu(\tau) \\
&\quad = \lim_n \lim_k \int_\Omega \langle y''_k(\tau), f''_n(\tau)\rangle \, d\nu(\tau) - \lim_k \lim_n \int_\Omega \langle y''_k(\tau), f''_n(\tau)\rangle \, d\nu(\tau) \\
&\quad = \lim_n \lim_k F''_n(y''_k) - \lim_k \lim_n F''_n(y''_k) = \vartheta_1 - \vartheta_2 \geq \Gamma(\widehat{T}) - \varepsilon.
\end{aligned}$$

Since $\nu(\Omega) = 1$, there exists $\tau_0 \in \Omega$ such that

$$\lim_n \lim_k \langle y''_k(\tau_0), f''_n(\tau_0)\rangle - \lim_k \lim_n \langle y''_k(\tau_0), f''_n(\tau_0)\rangle \geq \Gamma(\widehat{T}) - \varepsilon,$$

which gives $\Gamma(T) \geq \Gamma(\widehat{T}) - \varepsilon$. By an arbitrary choice of $\varepsilon > 0$ we conclude that $\Gamma(T) \geq \Gamma(\widehat{T})$. □

The real and complex interpolation methods

We say that a family $(A_0, \ldots, A_n)$ of Banach spaces forms a *Banach* $(n+1)$*-tuple*, if there exists a Hausdorff topological vector space E such that

A_j for $j = 0, \ldots, n$ is linearly and continuously embedded in E. For a given Banach $(n+1)$-tuple $\vec{A} = (A_0, \ldots, A_n)$ the spaces

$$\Delta(\vec{A}) = A_0 \cap \cdots \cap A_n \quad \text{and} \quad \Sigma(\vec{A}) = A_0 + \cdots + A_n$$

are Banach spaces with the norms

$$\|a\|_{\Delta(\vec{A})} = \max_{0 \le j \le n} \|a\|_{A_j} \quad \text{and} \quad \|a\|_{\Sigma(\vec{A})} = \inf \sum_{j=0}^{n} \|a_j\|_{A_j},$$

where the infimum is taken over all decompositions $a = \sum_{j=0}^{n} a_j$, $a_j \in A_j$, $j = 0, \ldots, n$. A Banach space A is said to be an *intermediate space* with respect to $\vec{A}$, if $\Delta(\vec{A}) \subset A \subset \Sigma(\vec{A})$ with both continuous inclusions. For Banach $(n+1)$-tuples $\vec{A} = (A_0, \ldots, A_n)$ and $\vec{B} = (B_0, \ldots, B_n)$, we write $T : \vec{A} \to \vec{B}$, if $T : \Sigma(\vec{A}) \to \Sigma(\vec{B})$ is a linear operator and each restriction $T|A_j$, $j = 0, \ldots, n$ is a bounded operator into B_j.

The interpolation methods with which we deal assign to each Banach $(n+1)$-tuple $\vec{A}$ a parameterized family of *interpolation spaces* A_α, so that each space A_α is intermediate with respect to $\vec{A}$, and the following *interpolation property* is satisfied for all Banach $(n+1)$-tuples $\vec{A}$ and $\vec{B}$: if $T : \vec{A} \to \vec{B}$, then $T|A_\alpha$ is a bounded operator into B_α.

We first recall a discrete real interpolation method of Yoshikawa [34] for Banach $(n+1)$-tuples, which is equivalent to Sparr's K-method (see [32, Remarks 4.5 and 4.6]).

Let $H_+^{n+1} = \{(t_0, \ldots, t_n) \in R^{n+1} \colon \sum_{j=0}^{n} t_j = 1,\ t_j > 0,\ j = 0, \ldots, n\}$. Let us fix $\theta = (\theta_0, \ldots, \theta_n) \in H_+^{n+1}$ and let us define $\xi_j = (\xi_1^j, \ldots, \xi_n^j) \in R^n$ for $j = 0, \ldots, n$ by the formula

$$\xi_i^j = \begin{cases} \theta_i - 1, & \text{if } j > 0 \text{ and } i = j \\ \theta_i, & \text{elsewhere} \end{cases},$$

where $i = 1, \ldots, n$. Let $\vec{A} = (A_0, \ldots, A_n)$ be a Banach $(n+1)$-tuple and $p \in (1, \infty)$. For $a \in \Sigma(\vec{A})$ let

$$\|a\|_{\theta,p} = \inf \max_{0 \le j \le n} \|a_j\|_{l_p(\xi_j, A_j)}, \tag{2.3}$$

where the infimum is taken over all decompositions

$$a = \sum_{j=0}^{n} a_j(z) \quad \text{for every } z \in Z^n, \tag{2.4}$$
$$a_j = (a_j(z))_{z \in Z^n} \in l_p(\xi_j, A_j) \quad \text{for } j = 0, \ldots, n.$$

By $A_{\theta,p}$ we denote the interpolation space with respect to $\vec{A}$ with the norm (2.3), that is

$$A_{\theta,p} = \{\, a \in \Sigma(\vec{A}) \colon \|a\|_{\theta,p} < \infty \,\}.$$

From now on, let c_θ be given by the equation

$$\log_2 c_\theta = \sum_{j=1}^{n} \theta_j (1 - \theta_j). \tag{2.5}$$

By a change of the variable $z \in Z^n$ in (2.4) one can easily prove the following lemma (compare [34, Proposition 2.6] and [26, p. 223]). The proof is left to the reader.

Lemma 2.4. *Let $a \in A_{\theta,p}$. Then*

$$\|a\|_{\theta,p} \leq c_\theta \inf \prod_{j=0}^{n} \|a_j\|_{l_p(\xi_j, A_j)}^{\theta_j}, \tag{2.6}$$

where the infimum is taken over all decompositions (2.4).

We now recall the complex interpolation method of Coifman, Cwikel, Rochberg, Sagher and Weiss [14] reduced to the case of a finite family of Banach spaces (see also [15, 22, 23]).

Let D be a simply connected domain in the complex plain, whose boundary ∂D is a rectifiable simple closed curve. The harmonic measure on ∂D with respect to $\zeta \in D$ will be denoted by P_ζ. For $j = 0, \ldots, n$ let $\Omega_j \subset \partial D$ be P_ζ-measurable disjoint subsets with positive measures, and whose union is ∂D. For such a partition of the boundary we write $\partial D = (\Omega_0, \ldots, \Omega_n)$. Let $\vec{A} = (A_0, \ldots, A_n)$ be a $(n+1)$-tuple of complex Banach spaces. To each $\tau \in \partial D$ we assign a complex Banach space A_τ so that if $\tau \in \Omega_j$, $j = 0, \ldots, n$ then $A_\tau = A_j$. Let $\mathcal{G}(\vec{A})$ denote the space of all functions, which are finite sums of all functions of the form $\varphi(\cdot)x$, where $x \in \Delta(\vec{A})$ and φ is a scalar-valued bounded and analytic function on D. Then for each $g \in \mathcal{G}(\vec{A})$ there exists the nontangential limit $\lim_{\eta \rhd \tau} g(\eta) = g(\tau)$ for P_ζ-a.e. $\tau \in \partial D$. Let $\mathcal{F}(\vec{A})$ denote the completion of $\mathcal{G}(\vec{A})$ with respect to the norm $\|g\|_{\mathcal{F}(\vec{A})} = \operatorname{ess\,sup}_{\tau \in \partial D} \|g(\tau)\|_{A_\tau}$. Thus $\mathcal{F}(\vec{A})$ consists of $\Sigma(\vec{A})$-valued analytic functions on D, whose boundary values are in A_j for P_ζ-a.e. $\tau \in \Omega_j$. The (St. Louis) interpolation space $A_{[\zeta]}$ with respect to $\vec{A}$ is defined for each $\zeta \in D$ by

$$A_{[\zeta]} = \{\, a \in \Sigma(\vec{A}) \colon a = f(\zeta),\ f \in \mathcal{F}(\vec{A}) \,\}$$

with the norm

$$\|a\|_{[\zeta]} = \inf\{\, \|f\|_{\mathcal{F}(\vec{A})} \colon a = f(\zeta),\ f \in \mathcal{F}(\vec{A}) \,\}.$$

If $f \in \mathcal{F}(\vec{A})$ and $f(\zeta) = a$, then, by [14, formula (2.4a)],

$$\|a\|_{[\zeta]} \le \exp \int_{\partial D} \log \|f(\tau)\|_{A_\tau} \, dP_\zeta(\tau). \tag{2.7}$$

For $j = 0, \dots, n$ let $\zeta_j = P_\zeta(\Omega_j)$. Then $\sum_{j=0}^n \zeta_j = 1$ and, by assumption, $\zeta_j > 0$ for $j = 0, \dots, n$. Therefore, the formula $\nu_j(K) = P_\zeta(K)/\zeta_j$ for every P_ζ-measurable set $K \subset \Omega_j$ defines a probabilistic measure on Ω_j. For $f \in \mathcal{F}(\vec{A})$ and $j = 0, \dots, n$ let $f_j = f|\Omega_j$. Then (2.7) can be rewritten in the form

$$\|a\|_{[\zeta]} \le \prod_{j=0}^{n} \|f_j\|_{L_1(\nu_j, A_j)}^{\zeta_j} \tag{2.8}$$

for all $a = f(\zeta)$, $f \in \mathcal{F}(\vec{A})$.

For relations between various interpolation methods for finite families of Banach spaces we refer to [15, p. 247].

3 Main results

Theorem 3.1. *Let $\vec{A} = (A_0, \dots, A_n)$ and $\vec{B} = (B_0, \dots, B_n)$ be Banach $(n+1)$-tuples, $\theta = (\theta_0, \dots, \theta_n) \in H_+^{n+1}$ and $p \in (1, \infty)$. If $T : \vec{A} \to \vec{B}$, then*

$$\Gamma(T_{\theta,p}) \le c_\theta \prod_{j=0}^{n} \Gamma(T_j)^{\theta_j},$$

where c_θ is given by (2.5), $T_{\theta,p} = T|A_{\theta,p}$ and $T_j = T|A_j$ for $j = 0, \dots, n$.

Proof. Fix $\varepsilon > 0$ and a sequence $(a_m) \subset \mathsf{B}_{A_{\theta,p}}$. For each a_m and $j = 0, \dots, n$ there exist families $(a_{j,m}(z))_{z \in Z^n} \in \mathsf{B}_{l_p(\xi_j, A_j)}$ such that $a_m = \sum_{j=0}^n a_{j,m}(z)$ for every $z \in Z^n$. Let $y_{j,m} = (T_j a_{j,m}(z))_{z \in Z^n}$ and $b_m = T a_m$ for every $m \in N$ and $j = 0, \dots, n$.

In the first step we take a sequence $(y''_{0,m})$ of scc for $(y_{0,m})$ satisfying the assertion of Theorem 2.1. Then

$$y''_{0,m} = \sum_{i=r_{0,m}+1}^{r_{0,m+1}} \lambda_{0,i} y_{0,i}$$

for some sequence of integers $0 = r_{0,1} < r_{0,2} < \cdots$ and nonnegative numbers $\lambda_{0,r_{0,m}+1}, \dots, \lambda_{0,r_{0,m+1}}$ such that $\sum_{i=r_{0,m}+1}^{r_{0,m+1}} \lambda_{0,i} = 1$.

In the next n steps we proceed as follows. For $j = 1, \dots, n$ set

$$y'_{j,m} = \sum_{i=r_{j-1,m}+1}^{r_{j-1,m+1}} \lambda_{j-1,i} y_{j,i}.$$

We choose a sequence $(y''_{j,m})$ of scc for $(y'_{j,m})$ satisfying the assertion of Theorem 2.1. There exists a sequence of integers $0 = r_{j,1} < r_{j,2} < \cdots$ such that

$$y''_{j,m} = \sum_{i=r_{j,m}+1}^{r_{j,m+1}} \lambda_{j,i} y_{j,i},$$

where $\lambda_{j,r_{j,m}+1}, \ldots, \lambda_{j,r_{j,m+1}}$ are nonnegative and $\sum_{i=r_{j,m}+1}^{r_{j,m+1}} \lambda_{j,i} = 1$.

The sequence $0 = r_{n,1} < r_{n,2} < \cdots$ and numbers $\lambda_{n,r_{n,m}+1}, \ldots, \lambda_{n,r_{n,m+1}}$ obtained in the last step for $(y''_{n,m})$ of scc for $(y_{n,m})$ will now be used in order to get a sequence $(y'''_{j,m})$ of scc for $(y_{j,m})$ with common coefficients of convex combinations for $j = 0, \ldots, n$, namely

$$y'''_{j,m} = \sum_{i=r_{n,m}+1}^{r_{n,m+1}} \lambda_{n,i} y_{j,i}.$$

Of course, each $(y'''_{j,m})$ satisfies the assertion of Theorem 2.1. Set

$$v_m = \sum_{i=r_{n,m}+1}^{r_{n,m+1}} \lambda_{n,i} b_i.$$

Then v_1, v_2 form a pair of scc for (b_m). Thus

$$\operatorname{csep}(b_m) \leq \|v_1 - v_2\|_{\theta,p} \leq c_\theta \prod_{j=0}^{n} \left\| y'''_{j,1} - y'''_{j,2} \right\|_{l_p(\xi_j, B_j)}^{\theta_j},$$

where the last inequality is a consequence of (2.6). But

$$\left\| y'''_{j,1} - y'''_{j,2} \right\|_{l_p(\xi_j, B_j)} \leq \operatorname{csep}(y'''_{j,m}) + \varepsilon$$

for every $j = 0, \ldots, n$. Moreover, $y'''_{j,m} \in \widetilde{T}_j(\mathsf{B}_{l_p(\xi_j, A_j)})$ for every $m \in N$, where $\widetilde{T}_j : l_p(\xi_j, A_j) \to l_p(\xi_j, B_j)$, $j = 0, \ldots, n$ is defined as in Theorem 2.2. Therefore

$$\operatorname{csep}(b_m) \leq c_\theta \prod_{j=0}^{n} (\Gamma(\widetilde{T}_j) + \varepsilon)^{\theta_j}.$$

By Theorem 2.2 it follows that $\Gamma(\widetilde{T}_j) = \Gamma(T_j)$ for $j = 0, \ldots, n$. An arbitrary choice of ε and (a_m) gives the desired conclusion. □

Theorem 3.2. *Let $\vec{A} = (A_0, \ldots, A_n)$ and $\vec{B} = (B_0, \ldots, B_n)$ be Banach $(n+1)$-tuples, $\zeta \in D$ and $\partial D = (\Omega_0, \ldots, \Omega_n)$. If $T : \vec{A} \to \vec{B}$, then*

$$\Gamma(T_{[\zeta]}) \leq \prod_{j=0}^{n} \Gamma(T_j)^{\zeta_j},$$

where $\zeta_j = P_\zeta(\Omega_j)$, $T_{[\zeta]} = T|A_{[\zeta]}$ *and* $T_j = T|A_j$ *for* $j = 0, \ldots, n$.

Proof. Fix $\varepsilon > 0$ and a sequence $(a_m) \subset \mathsf{B}_{A_{[\zeta]}}$, and set $b_m = Ta_m$. For each a_m there exists $f_m \in \mathcal{F}(\vec{A})$ such that $\|f_m\|_{\mathcal{F}(\vec{A})} < 1$ and $f_m(\zeta) = a_m$. Then $T \circ f_m = g_m \in \mathcal{F}(\vec{B})$ and $g_m(\zeta) = b_m$. For $j = 0, \ldots, n$ let $g_{j,m} = g_m|\Omega_j$. Of course, $g_{j,m} \in L_1(\nu_j, B_j)$ with $\nu_j = P_\zeta/\zeta_j$ on Ω_j. By applying to $g_{j,m}$ the same procedure as to $y_{j,m}$ in the proof of Theorem 3.1, we can assume that we have the sequences $(g'_{j,m})$, $j = 0, \ldots, n$ satisfying for every k, l

$$\left\|g'_{j,k} - g'_{j,l}\right\|_{L_1(\nu_j, B_j)} \le \operatorname{csep}(g'_{j,m}) + \varepsilon,$$

where $g'_{j,m} \in \widehat{T}_j(\mathsf{B}_{L_\infty(\nu_j, A_j)})$ and $\widehat{T}_j : L_\infty(\nu_j, A_j) \to L_1(\nu_j, B_j)$, $j = 0, \ldots, n$ is defined as in Theorem 2.3. Let $g'_1(\zeta), g'_2(\zeta)$ be a pair of scc for (b_m) obtained just as the pair v_1, v_2 of scc for (b_m) in the proof of Theorem 3.1. Then

$$\begin{aligned}\operatorname{csep}(b_m) &\le \|g'_1(\zeta) - g'_2(\zeta)\|_{[\zeta]} \le \prod_{j=0}^{n} \left\|g'_{j,1} - g'_{j,2}\right\|^{\zeta_j}_{L_1(\nu_j, B_j)} \\ &\le \prod_{j=0}^{n} (\operatorname{csep}(g'_{j,m}) + \varepsilon)^{\zeta_j} \le \prod_{j=0}^{n} (\Gamma(\widehat{T}_j) + \varepsilon)^{\zeta_j},\end{aligned}$$

where the second inequality follows from (2.8). By Theorem 2.3 and an arbitrary choice of ε and (a_m) the proof is completed. □

Corollary 3.3. *Under the hypotheses of Theorems 3.1 and 3.2 respectively, if* $T_j : A_j \to B_j$ *is weakly compact for at least one* $j = 0, \ldots, n$, *then so are* $T_{\theta,p} : A_{\theta,p} \to B_{\theta,p}$ *and* $T_{[\zeta]} : A_{[\zeta]} \to B_{[\zeta]}$. *In particular, the reflexivity of any* A_j *implies that of* $A_{\theta,p}$ *and* $A_{[\zeta]}$.

Another proof of the result for real interpolation in the above corollary is given in [8, Corollary 6.2] (see also [23, Theorem 2]).

4 Other estimates

The deviation of operators from weak compactness can be evaluated with no use of measures of weak noncompactness for sets. Each $T \in \mathcal{L}(X, Y)$ generates the operator $R(T) : X^{**}/X \to Y^{**}/Y$ by $R(T)(x^{**} + X) = T^{**}x^{**} + Y$ for every $x^{**} \in X^{**}$ with $\|R(T)\| = \sup\{\operatorname{dist}(T^{**}x^{**}, Y) : \operatorname{dist}(x^{**}, X) \le 1\}$. Then $\|R(T)\| = 0$ if and only if $T \in \mathcal{W}(X, Y)$ (see [18, p. 482]). The norm $\|T\|_{\mathcal{W}} = \operatorname{dist}(T, \mathcal{W}(X, Y))$ in the quotient space $\mathcal{L}(X, Y)/\mathcal{W}(X, Y)$, viewed as a seminorm in $\mathcal{L}(X, Y)$, has the same property. An essential difference between these quantities is connected with Gantmacher's duality theorem, which states that $T \in \mathcal{W}(X, Y)$ if and only if $T^* \in \mathcal{W}(Y^*, X^*)$. A quantitative version of this theorem for $\|R(\cdot)\|$ was obtained in [20, Proposition 1.3].

In contrast, a result of that sort holds neither for $\|\cdot\|_{\mathcal{W}}$ (see [33]) nor for $|\cdot|_{\omega}$ (see [20]), which in addition are not equivalent (see [3, Theorem 1]). The seminorms Γ, $\overline{\Gamma}$ and $\|R(\cdot)\|$ are equivalent (see [24, Theorem 3.2]) and certainly none of these is equivalent to $|\cdot|_{\omega}$ or $\|\cdot\|_{\mathcal{W}}$.

Theorem 4.1. *If $T \in \mathcal{L}(X,Y)$, then $|T|_{\gamma} \leq \Gamma(T) \leq \gamma(\mathsf{B}_X)\,|T|_{\gamma}$.*

Proof. Fix $\varepsilon > 0$ and $A \in \mathcal{M}_X$ such that $\gamma(A) \leq 1$. Let us choose $(x_n) \subset \operatorname{conv} A$ such that $\gamma(T(A)) - \varepsilon < \operatorname{csep}(Tx_n)$. If (y_n) is a sequence of scc for (x_n), then $\operatorname{csep}(x_n) \leq \operatorname{csep}(y_n)$ and $\operatorname{csep}(Tx_n) \leq \operatorname{csep}(Ty_n)$. By Theorem 2.1, we can assume that $|\|x_k - x_l\| - \|x_m - x_n\|| \leq \varepsilon$ for $k \neq l$ and $m \neq n$. Then $\|x_n - x_1\| \leq \operatorname{csep}(x_n) + \varepsilon \leq 1 + \varepsilon$ for $n \in N$. Moreover, $\operatorname{csep}(T(x_n - x_1)) = \operatorname{csep}(Tx_n)$. Hence

$$\gamma(T(A)) - \varepsilon < (1+\varepsilon)\operatorname{csep}\left(T\left(\frac{x_n - x_1}{1+\varepsilon}\right)\right) \leq (1+\varepsilon)\Gamma(T),$$

and we conclude that $|T|_{\gamma} \leq \Gamma(T)$. On the other hand, for a nonreflexive space X (the reflexive case is obvious) we have $\gamma\left(T\left(\mathsf{B}_X/\gamma(\mathsf{B}_X)\right)\right) \leq |T|_{\gamma}$, which gives $\Gamma(T) \leq \gamma(\mathsf{B}_X)\,|T|_{\gamma}$. □

The estimates in the above theorem cannot be improved. Indeed, let us consider the space $V = c_0 \times c$ with the norm $\|(x,y)\| = \max\{\|x\|, \|y\|\}$ and the projections $P_1 : V \to V$ and $P_2 : V \to V$, where $P_1(x,y) = (x,0)$ and $P_2(x,y) = (0,y)$. Then $\Gamma(P_1) = \gamma(\mathsf{B}_{c_0}) = 1 = |P_1|_{\gamma}$ (see [25, Theorem 2.8]) and $\Gamma(P_2) = \gamma(\mathsf{B}_c) = 2 = 2\,|P_2|_{\gamma}$.

Theorem 4.2. *If $T \in \mathcal{L}(X,Y)$ and the weak-star topology in X^{**} is metrizable for bounded subsets, then $|T|_{\overline{\gamma}} = \overline{\Gamma}(T)$.*

Proof. Fix $\varepsilon > 0$ and $A \in \mathcal{M}_X$ such that $\overline{\gamma}(A) \leq 1$. Let us choose $(x_n) \subset \operatorname{conv} A$ and a w^*-cluster point y^{**} of (Tx_n) such that $\overline{\gamma}(T(A)) - \varepsilon \leq \operatorname{dist}(y^{**}, Y)$. Then $y^{**} = w^*\text{-}\lim_{\mathcal{U}} Tx_n$ for some free ultrafilter $\mathcal{U}$ in N. Let $x^{**} = w^*\text{-}\lim_{\mathcal{U}} x_n$. Since T^{**} is w^*-continuous, we have $y^{**} = T^{**}x^{**}$. Let us take $x \in X$ such that $\|x^{**} - x\| \leq \operatorname{dist}(x^{**}, X) + \varepsilon$. By the hypothesis we can assume that $x^{**} = w^*\text{-}\lim_{k\to\infty} x_{n_k}$ for some subsequence (x_{n_k}). Let $\delta = \sup_k \operatorname{dist}(x, \operatorname{conv}\{x_{n_i}\}_{i=k}^{\infty})$. We show that $\delta \leq 1 + \varepsilon$. Let us assume that $\delta > 1 + \varepsilon$. Then there exist m and $1 + \varepsilon < \delta_0 < \delta$ such that $\delta_0 \leq \operatorname{dist}(x, \operatorname{conv}\{x_{n_i}\}_{i=m}^{\infty})$. By a separation theorem there exists $u^* \in \mathsf{B}_{X^*}$ such that $\delta_0 \leq u^*(x - u)$ for every $u \in \operatorname{conv}\{x_{n_i}\}_{i=m}^{\infty}$. Hence $\delta_0 \leq u^*(x - x_{n_i})$ for $i \geq m$, and therefore $\delta_0 \leq u^*(x) - x^{**}(u^*) \leq \|x - x^{**}\| \leq 1 + \varepsilon$, which leads to a contradiction. It follows that $\operatorname{dist}(x, \operatorname{conv}\{x_{n_i}\}_{i=k}^{\infty}) \leq 1 + \varepsilon$ for every k. Hence there exists a sequence (z_n) of scc for (x_{n_i}) such that $\|x - z_n\| \leq 1+2\varepsilon$. Of course, $w^*\text{-}\lim_{n\to\infty} z_n = x^{**}$. Then

$$\begin{aligned}\overline{\gamma}(T(A)) - \varepsilon &\leq \operatorname{dist}(T^{**}x^{**}, Y) = \operatorname{dist}(T^{**}(x^{**} - x), Y)\\ &\leq (1+2\varepsilon)\left\|T^{**}\left(\frac{x^{**} - x}{1+2\varepsilon}\right) - \frac{y}{1+2\varepsilon}\right\|\end{aligned}$$

for every $y \in Y$. But $T^{**}\left(\frac{x^{**}-x}{1+2\varepsilon}\right)$ is a w^*-limit of $\left(T\left(\frac{z_n-x}{1+2\varepsilon}\right)\right) \subset T(\overline{\mathsf{B}}_X)$. It follows that $\overline{\gamma}(T(A)) - \varepsilon \leq (1+2\varepsilon)\overline{\Gamma}(T)$, and consequently $|T|_{\overline{\gamma}} \leq \overline{\Gamma}(T)$. Since $\overline{\gamma}(\mathsf{B}_X) \leq 1$, the inequality $|T|_{\overline{\gamma}} \geq \overline{\Gamma}(T)$ is obvious. □

Corollary 4.3. *Under the hypotheses of Theorem 3.1,*

$$\overline{\Gamma}(T_{\theta,p}) \leq 2c_\theta \prod_{j=0}^{n} \overline{\Gamma}(T_j)^{\theta_j}, \quad |T_{\theta,p}|_\gamma \leq c_\theta \prod_{j=0}^{n} \left(\gamma(\mathsf{B}_{A_j})\, |T_j|_\gamma\right)^{\theta_j}.$$

Under the hypotheses of Theorem 3.2,

$$\overline{\Gamma}(T_{[\zeta]}) \leq 2 \prod_{j=0}^{n} \overline{\Gamma}(T_j)^{\zeta_j}, \quad \left|T_{[\zeta]}\right|_\gamma \leq \prod_{j=0}^{n} \left(\gamma(\mathsf{B}_{A_j})\, |T_j|_\gamma\right)^{\zeta_j}.$$

Most results of this work come from the author's doctoral thesis written under the supervision of Stanisław Prus at the Maria Curie-Skłodowska University.

References

[1] R.R. Akhmerov, M.I. Kamenskii, A.S. Potapov, A.E. Rodkina, B.N. Sadovskii, Measures of noncompactness and condensing operators, *Birkhäu-ser Verlag*, Basel Boston Berlin 1992.

[2] A.G. Aksoy, L. Maligranda, Real interpolation and measure of weak noncompactness, *Math. Nachr.* **175** (1995), 5–12.

[3] K. Astala, H.-O. Tylli, Seminorms related to weak compactness and to Tauberian operators, *Math. Proc. Camb. Phil. Soc.* **107** (1990), 367–375.

[4] J. Banaś, J. Rivero, On measures of weak noncompactness, *Ann. Mat. Pura Appl.* **151** (1988), 213–224.

[5] B. Beauzamy, Espaces d'interpolation réels: topologie et géométrie, *Springer-Verlag*, Berlin Heidelberg New York 1978.

[6] Yu.A. Brudnyi, N.Ya. Krugljak, Interpolation functors and interpolation spaces, Vol. I, *North-Holland*, Amsterdam New York Oxford Tokyo 1991.

[7] A.P. Calderón, Intermediate spaces and interpolation, the complex method, *Studia Math.* **24** (1964), 113–190.

[8] M.J. Carro, L.I. Nikolova, Interpolation of limited and weakly compact operators on families of Banach spaces, *Acta Appl. Math.* **49** (1997), 151–177.

[9] F. Cobos, J.M. Cordeiro, A. Martínez, Quantitative estimates for interpolated operators by multidimensional methods, *Revista Math. Complutense* **12** (1999), 85–103.

[10] F. Cobos, P. Fernández-Martínez, A. Martínez, On reiteration and the behaviour of weak compactness under certain interpolation methods, *Collect. Math.* **50** (1999), 53–72.

[11] F. Cobos, A. Manzano, A. Martínez, Interpolation theory and measures related to operator ideals, *Quart. J. Math. Oxford Ser.* (2) **50** (1999), 401–416.

[12] F. Cobos, A. Martínez, Extreme estimates for interpolated operators by the real method, *J. London Math. Soc.* (2) **60** (1999), 860–870.

[13] F. Cobos, A. Martínez, Remarks on interpolation properties of the measure of weak non-compactness and ideal variations, *Math. Nachr.* **208** (1999), 93–100.

[14] R.R. Coifman, M. Cwikel, R. Rochberg, Y. Sagher, G. Weiss, A theory of complex interpolation for families of Banach spaces, *Adv. Math.* **43** (1982), 203–229.

[15] M. Cwikel, S. Janson, Real and complex interpolation methods for finite and infinite families of Banach spaces, *Adv. Math.* **66** (1987), 234–290.

[16] F.S. De Blasi, On a property of the unit sphere in a Banach space, *Bull. Math. Soc. Sci. Math. R.S. Roumanie* **21**(69) (1977), 259–262.

[17] J. Diestel, Sequences and series in Banach spaces, *Springer-Verlag*, New York Berlin Heidelberg Tokyo, 1984.

[18] N. Dunford, J.T. Schwartz, Linear operators. Part I: General theory, *Interscience Publishers, Inc.*, New York, 1958.

[19] R. Edwards, Functional analysis. Theory and applications, *Holt, Rinehart and Winston*, New York Chicago San Francisco Toronto London, 1965.

[20] M. González, E. Saksman, H.-O. Tylli, Representing non-weakly compact operators, *Studia Math.* **113** (1995), 265–282.

[21] S. Heinrich, Closed operator ideals and interpolation, *J. Funct. Anal.* **35** (1980), 397–411.

[22] S.G. Krein, L.I. Nikolova, A complex interpolation method for a family of Banach spaces, *Ukrain. Mat. Zh.* **34** (1982), 31–42 [Russian], *Ukrainian Math. J.* **34** (1982), 26–36 [English translation].

[23] S.G. Krein, L.I. Nikolova, Holomorphic functions in a family of Banach spaces and interpolation, *Dokl. Akad. Nauk SSSR* **250** (1980), 547–550 [Russian], *Soviet Math. Dokl.* **21** (1980), 131–134 [English translation].

[24] A. Kryczka, S. Prus, Measure of weak noncompactness under complex interpolation, *Studia Math.* **147** (2001), 89–102.

[25] A. Kryczka, S. Prus, M. Szczepanik, Measure of weak noncompactness and real interpolation of operators, *Bull. Austral. Math Soc.* **62** (2000), 389–401.

[26] [26]. D. Kutzarova, L.I. Nikolova, S. Prus, Infinite dimensional geometric properties of real interpolation spaces, *Math. Nachr.* **191** (1998), 215–228.

[27] J.-L. Lions, J. Peetre, Sur une classe d'espaces d'interpolation, *Inst. Hautes Études Sci. Publ. Math.* **19** (1964), 5–68.

[28] L. Maligranda, Interpolation between sum and intersection of Banach spaces, *J. Approx. Theory* **47** (1986), 42–53.

[29] L. Maligranda, Weakly compact operators and interpolation, *Acta Appl. Math.* **27** (1992), 79–89.

[30] L. Maligranda, A. Quevedo, Interpolation of weakly compact operators, *Arch. Math.* **55** (1990), 280–284.

[31] M. Mastyło, On interpolation of weakly compact operators, *Hokkaido Math. Jour.* **22** (1993), 105–114.

[32] G. Sparr, Interpolation of several Banach spaces, *Ann. Mat. Pura Appl.* **99** (1974), 247–316.

[33] H.-O. Tylli, Duality of the weak essential norm, *Proc. Amer. Math. Soc.* **129** (2001), 1437–1443.

[34] A. Yoshikawa, Sur la théorie d'espaces d'interpolation – les espaces de moyenne de plusieurs espaces de Banach, *J. Fac. Sci. Univ. Tokyo,* **16** (1970), 407–468.

A Differentiability Result for the First Eigenvalue of the p-Laplacian upon Domain Perturbation

Pier Domenico Lamberti

Dipartimento di Matematica Pura ed Applicata, Università degli Studi di Padova, via Belzoni 7, 35131 Padova, Italy e–mail: lamberti@math.unipd.it

Abstract. We study the dependence of the first eigenvalue of the p-Laplacian upon the variation of the domain: we prove a differentiability result and we find a Hadamard-type formula. Finally, we indicate how this formula yields a class of overdetermined nonlinear eigenvalue problems.

AMS Subject Classification. 35P30, 47H30

Key words. p-Laplacian, eigenvalues, domain perturbation.

1 Introduction

This paper is devoted to the study of the dependence of the first eigenvalue of a nonlinear differential operator upon variation of the domain.
Namely, for a given bounded connected regular open subset Ω of $\mathbb{R}^n$ and $p \in]1, \infty[$, we consider the nonlinear eigenvalue problem

$$-\Delta_p u = \lambda |u|^{p-2} u, \tag{1.1}$$

for $\lambda \in \mathbb{R}$ and u in the Sobolev space $W_0^{1,p}(\Omega)$. Here $\Delta_p u \equiv \operatorname{div}\left(|Du|^{p-2} Du\right)$ is the so-called p-Laplacian. See Definition 2.1. We note that for $p = 2$ we obtain the usual Laplace operator.
It is well-known that the set of real numbers λ such that (1.1) has a nontrivial solution, admits a minimum $\lambda_p > 0$ which is called the first eigenvalue of $-\Delta_p$. In order to study the dependence of λ_p on Ω, we focus our attention on a suitable class of domains which are diffeomorphic to Ω. Thus, if ϕ is a sufficiently regular diffeomorphism of $\mathrm{cl}\Omega$ into $\mathrm{cl}\phi(\Omega)$, we consider the first eigenvalue of $-\Delta_p$ on $\phi(\Omega)$, which we denote by $\lambda_p[\phi]$, and we study the map which takes ϕ to $\lambda_p[\phi]$.
Our goal is to prove a Frechét differentiability result for the dependence of $\lambda_p[\phi]$ on ϕ and to find a Hadamard-type formula for the derivatives of the map $\lambda_p[\cdot]$.
The case $p = 2$ has been largely investigated and analyticity results are known (see *e.g.* Henry [9], Prodi [16], and the extensive monograph [18] by Sokolowski and Zolesio. See also Cox [5], Rousselet and Chenais [17] and [10]). Nevertheless, little seems to be known in the case $p \neq 2$.

This paper is organized as follows: in Section 2, we introduce some notation and some preliminary results. In Section 3, we prove a continuity result for the dependence of the eigenvectors upon the variation of ϕ (see Theorem 3.1). In Section 4, we prove our main differentiability result (see Theorem 4.1) and we find a natural generalisation of the well-known Hadamard formula (see Theorem 4.2), which yields a class of overdetermined nonlinear eigenvalue problems (see Remark 4.1).
Finally, we mention that the key idea of the proof of Theorem 3.1 has been inspired by a careful reading of Lindqvist [14, Theorem 6.3, p. 214], which is concerned with the dependence of λ_p on p.
We would also like to thank Prof. Massimo Lanza de Cristoforis and Prof. Peter Lindqvist for some useful discussions and references.

2 Preliminaries and notation

We denote by $\mathbb{N}$ the set of natural numbers including 0. Let $(\mathcal{X}, \|\cdot\|_{\mathcal{X}})$, $(\mathcal{Y}, \|\cdot\|_{\mathcal{Y}})$, be normed spaces. We endow the product space $\mathcal{X} \times \mathcal{Y}$ with the norm $\|\cdot\|_{\mathcal{X}\times\mathcal{Y}} = \|\cdot\|_{\mathcal{X}} + \|\cdot\|_{\mathcal{Y}}$, while we use the Euclidean norm $|\cdot|$ for $\mathbb{R}^n$. For all $x \in \mathbb{R}^n$ and $r > 0$, $B(x, r)$ denotes the open ball $\{y \in \mathbb{R}^n : |y - x| < r\}$. A dot '$\cdot$' denotes the scalar product in $\mathbb{R}^n$ or the product between matrices. The elements of $\mathbb{R}^n$ are thought as row vectors. The inverse and the transpose of a matrix A are denoted by A^{-1} and A^t respectively, while the inverse of an invertible function f is denoted by $f^{(-1)}$.
Let Ω be a bounded open subset of $\mathbb{R}^n$. Then $\mathrm{cl}\Omega$ and $\partial\Omega$ denote the closure and the boundary of Ω respectively. By $C(\mathrm{cl}\Omega)$ we denote the space of the continuous functions of $\mathrm{cl}\Omega$ to $\mathbb{R}$ endowed with the sup-norm $\|\cdot\|_0$ of the uniform convergence on $\mathrm{cl}\Omega$. The space of those elements of $C(\mathrm{cl}\Omega)$ which are continuously differentiable in Ω and whose derivatives can be extended with continuity to $\mathrm{cl}\Omega$, is denoted by $C^1(\mathrm{cl}\Omega)$. $C^1(\mathrm{cl}\Omega)$ is equipped with the norm $\|f\|_1 \equiv \|f\|_0 + \sum_{i=1}^n \|\frac{\partial f}{\partial x_i}\|_0$. By $C_0^1(\mathrm{cl}\Omega)$ we denote the closed subspace of $C^1(\mathrm{cl}\Omega)$ of those functions which vanish at the boundary.
Let $\alpha \in]0, 1]$. We recall that a function f of $\mathrm{cl}\Omega$ to $\mathbb{R}$ is Hölder continuous of exponent α, if its Hölder quotient $|f|_\alpha \equiv sup\left\{\frac{|f(x)-f(y)|}{|x-y|^\alpha} : x, y \in \mathrm{cl}\Omega,\ x \neq y\right\}$ is finite. The subspace of $C^1(\mathrm{cl}\Omega)$ of those functions whose first order derivatives are Hölder continuous of exponent α, is denoted by $C^{1,\alpha}(\mathrm{cl}\Omega)$, and is equipped with the norm $\|f\|_{1,\alpha} \equiv \|f\|_1 + \sum_{i=1}^n |\frac{\partial f}{\partial x_i}|_\alpha$. By $C^{1,\alpha}(\mathrm{cl}\Omega, \mathbb{R}^m)$ we denote the space of those functions $f = (f_1, \ldots, f_m)$ of $\mathrm{cl}\Omega$ to $\mathbb{R}^m$ whose entries f_i belong to $C^{1,\alpha}(\mathrm{cl}\Omega)$, for all $i \in \{1, \ldots, m\}$. Thus $C^{1,\alpha}(\mathrm{cl}\Omega, \mathbb{R}^m)$ is identified with $\left(C^{1,\alpha}(\mathrm{cl}\Omega)\right)^m$. We denote by Df the Jacobian matrix $\left(\frac{\partial f_i}{\partial x_j}\right)_{i,j}$ of a function $f = (f_1, \ldots, f_m) \in \left(C^1(\mathrm{cl}\Omega)\right)^m$. Let $p \in [1, \infty[$ and $k \in \mathbb{N}$. By $W^{k,p}(\Omega)$, we denote the space of the (equivalence classes of) real-valued functions in $L^p(\Omega)$ which have all distributional derivatives up to order k in $L^p(\Omega)$. By $C_c^\infty(\Omega)$ we denote the space of the $C^\infty(\Omega)$ functions with compact support in Ω. The closure of $C_c^\infty(\Omega)$ in $W^{1,p}(\Omega)$ with respect

to the norm $\|f\|_{1,p} \equiv \left(\int_\Omega |f|^p \mathrm{dx} + \sum_{i=1}^n \int_\Omega |\frac{\partial f}{\partial x_i}|^p \mathrm{dx}\right)^{\frac{1}{p}}$, is the well-known Sobolev Space $W_0^{1,p}(\Omega)$.

As we have announced in the introduction, we will consider problem (1.1) in a class of domains which are parametrized by means of diffeomorphisms defined on a fixed domain Ω. Namely, let $\alpha \in]0,1[$ and Ω be a bounded connected open subset of $\mathbb{R}^n$ of class $C^{1,\alpha}$. We set

$$\Phi_{1,\alpha}(\Omega) \equiv \{\phi \in C^{1,\alpha}(\mathrm{cl}\Omega, \mathbb{R}^n) : \phi \text{ is injective and } \det D\phi(x) \neq 0, \text{ for all } x \in \mathrm{cl}\Omega\}.$$

It is known that $\Phi_{1,\alpha}(\Omega)$ is an open subset of $C^{1,\alpha}(\mathrm{cl}\Omega, \mathbb{R}^n)$ (cf. Lanza de Cristoforis [11, Lemma 5.2, p. 475]) and that $\phi(\Omega)$ is again a bounded connected open subset of $\mathbb{R}^n$ of class $C^{1,\alpha}$ for all $\phi \in \Phi_{1,\alpha}(\Omega)$ (cf. *e.g.* [10, §2]). Furthermore, $\partial\phi(\Omega) = \phi(\partial\Omega)$ for all $\phi \in \Phi_{1,\alpha}(\Omega)$.
Thus, it makes sense to consider problem (1.1) in the domain $\phi(\Omega)$ for all $\phi \in \Phi_{1,\alpha}(\Omega)$. Of course equation (1.1) must be interpreted in a weak sense. We do so by means of the following.

Definition 2.1. *Let $\alpha \in]0,1[$, $p \in]1,\infty[$. Let Ω be a bounded connected open subset of $\mathbb{R}^n$ of class $C^{1,\alpha}$. Let $\phi \in \Phi_{1,\alpha}(\Omega)$. A real number λ is an eigenvalue of $-\Delta_p$ on $\phi(\Omega)$ if there exists $u \in W_0^{1,p}(\phi(\Omega)) \setminus \{0\}$ such that*

$$\int_{\phi(\Omega)} |Du|^{p-2} Du \cdot D\zeta^t \mathrm{d}y = \lambda \int_{\phi(\Omega)} |u|^{p-2} u\zeta \mathrm{d}y, \tag{2.1}$$

for all $\zeta \in W_0^{1,p}(\phi(\Omega))$. We also say that u is an eigenvector of λ.

It is easy to check that each eigenvalue is positive and it is well-known that there exists a minimum eigenvalue of $-\Delta_p$ on $\phi(\Omega)$ which we denote by $\lambda_p[\phi]$, for all $\phi \in \Phi_{1,\alpha}(\Omega)$. As a matter of fact, it turns out that

$$\lambda_p[\phi] = \min_{u \in W_0^{1,p}(\phi(\Omega)) \setminus \{0\}} \frac{\int_{\phi(\Omega)} |Du|^p \mathrm{d}y}{\int_{\phi(\Omega)} |u|^p \mathrm{d}y}, \tag{2.2}$$

for all $\phi \in \Phi_{1,\alpha}(\Omega)$. Furthermore, the eigenvectors of $\lambda_p[\phi]$ are exactly the minimizing elements in (2.2).
We need to recall that the set of eigenvectors of $\lambda_p[\phi]$ is a linear subspace of dimension 1 (cf. Lindqvist [13]), and that there exists $\beta \in]0,1[$ depending only on $\phi(\Omega)$, α, n, p such that the eigenvectors corresponding to $\lambda_p[\phi]$ belong to $C^{1,\beta}(\mathrm{cl}\phi(\Omega))$ (cf. Lieberman [12], see also DiBenedetto [7], Tolksdorf [19]). Furthermore, the eigenvectors of $\lambda_p[\phi]$ do not change sign on $\phi(\Omega)$ (cf. Lindqvist [15]).
Since the open set $\phi(\Omega)$ has a regular boundary, the eigenvectors attain the boundary value zero in the classical sense.

It follows that there exists a uniquely determined eigenvector of $\lambda_p[\phi]$, that we denote by $u_p[\phi]$, such that

$$\begin{cases} \int_{\phi(\Omega)} |u_p[\phi](y)|^p dy = 1 \\ u_p[\phi](y) > 0, & \textit{for all } y \in \phi(\Omega), \end{cases} \tag{2.3}$$

for all $\phi \in \Phi_{1,\alpha}(\Omega)$. In particular, we have that

$$\lambda_p[\phi] = \min_{u \in C_0^1(cl\phi(\Omega)) \setminus \{0\}} \frac{\int_{\phi(\Omega)} |Du|^p dy}{\int_{\phi(\Omega)} |u|^p dy} = \int_{\phi(\Omega)} |Du_p[\phi]|^p dy. \tag{2.4}$$

Our goal is to prove that the function of $\Phi_{1,\alpha}(\Omega)$ to $\mathbb{R}$ which takes $\phi \in \Phi_{1,\alpha}(\Omega)$ to $\lambda_p[\phi]$ is continuously Frechét differentiable.
We end this section with a technical statement concerning the local boundedness of the functions $\lambda_p[\cdot]$ and $\|u_p[\cdot]\|_0$.

Proposition 2.1. *Let $\alpha \in]0,1[$, $p \in]1,\infty[$. Let Ω be a bounded connected open subset of $\mathbb{R}^n$ of class $C^{1,\alpha}$. Let $\tilde{\phi} \in \Phi_{1,\alpha}(\Omega)$. Then there exist an open neighborhood $\mathcal{U}$ of $\tilde{\phi}$ in $\Phi_{1,\alpha}(\Omega)$ and $\tilde{M} > 0$ such that*

$$\tilde{M}^{-1} \le \lambda_p[\phi] \le \tilde{M},$$
$$\|u_p[\phi]\|_0 \le \tilde{M},$$

for all $\phi \in \mathcal{U}$.

Proof. Let $\tilde{y} \in \tilde{\phi}(\Omega)$ be fixed. Let $\tilde{r}_1, \tilde{r}_2 \in \mathbb{R}$ be such that

$$0 < \tilde{r}_1 < \min_{z \in \partial\tilde{\phi}(\Omega)} |z - \tilde{y}| \le \max_{z \in \partial\tilde{\phi}(\Omega)} |z - \tilde{y}| < \tilde{r}_2.$$

It's easy to see that there exists an open neighborhood $\mathcal{U}$ of $\tilde{\phi}$ in $\Phi_{1,\alpha}(\Omega)$ such that $B(\tilde{y}, \tilde{r}_1) \subseteq \phi(\Omega) \subseteq B(\tilde{y}, \tilde{r}_2)$, for all $\phi \in \mathcal{U}$. Then, we have that

$$\min_{u \in W_0^{1,p}(B(\tilde{y},\tilde{r}_2)) \setminus \{0\}} \frac{\int_{\phi(\Omega)} |Du|^p dy}{\int_{\phi(\Omega)} |u|^p dy} \le \lambda_p[\phi] \le \min_{u \in W_0^{1,p}(B(\tilde{y},\tilde{r}_1)) \setminus \{0\}} \frac{\int_{\phi(\Omega)} |Du|^p dy}{\int_{\phi(\Omega)} |u|^p dy},$$

and thus there exists $\tilde{M}_1 > 0$ such that

$$\tilde{M}_1^{-1} \le \lambda_p[\phi] \le \tilde{M}_1, \tag{2.5}$$

for all $\phi \in \mathcal{U}$.
We now recall that

$$\|u_p[\phi]\|_0 \le 4^n \lambda_p[\phi]^{\frac{n}{p}} \int_{\phi(\Omega)} |u_p[\phi](y)| dy, \tag{2.6}$$

for all $\phi \in \Phi_{1,\alpha}(\Omega)$ (cf. Lindqvist [14, Lemma 4.1, p. 208]). By (2.3), (2.5), (2.6) and by the *Hölder* inequality, we obtain

$$\begin{aligned}\|u_p[\phi]\|_0 &\leq 4^n \lambda_p[\phi]^{\frac{n}{p}} \left(\int_{\phi(\Omega)} |u_p[\phi](y)|^p \mathrm{d}y\right)^{\frac{1}{p}} \left(\int_\Omega |\mathrm{det} D\phi(x)| \mathrm{d}x\right)^{1-\frac{1}{p}} \\ &\leq 4^n \tilde{M}_1^{\frac{n}{p}} \left(\int_\Omega |\mathrm{det} D\phi(x)| \mathrm{d}x\right)^{1-\frac{1}{p}},\end{aligned}$$

for all $\phi \in \mathcal{U}$. Then, by possibly shrinking $\mathcal{U}$, we deduce that there exists $\tilde{M}_2 > 0$ such that $\|u_p[\phi]\|_0 \leq \tilde{M}_2$, for all $\phi \in \mathcal{U}$. Thus we can choose $\tilde{M} = \max\left\{\tilde{M}_1, \tilde{M}_2\right\}$. □

3 Dependence of the first eigenvector of $-\Delta_p$ upon the domain

In this section we study the dependence of the eigenvectors corresponding to $\lambda_p[\phi]$ upon ϕ. Actually, we shall not consider the function $u_p[\phi]$ itself. Indeed, the domain of $u_p[\phi]$ changes when ϕ is perturbed and thus $u_p[\phi]$ is difficult to handle. To circumvent this difficulty, we will consider the function $v_p[\phi] \equiv u_p[\phi] \circ \phi$ which is defined in the fixed domain Ω for $\phi \in \Phi_{1,\alpha}(\Omega)$ and we will prove that the (nonlinear) operator of $\Phi_{1,\alpha}(\Omega)$ to $C^1(cl\Omega)$ which takes $\phi \in \Phi_{1,\alpha}(\Omega)$ to $v_p[\phi]$ is continuous. To do so, we need to transform our eigenvalue problem (2.1) on $\phi(\Omega)$ into an eigenvalue problem on Ω, which is solved by $v_p[\phi]$. We do so by means of the following Proposition, whose validity can be easily verified by a simple change of variables in (2.1) and (2.4).

Proposition 3.1. *Let $\alpha \in]0,1[$, $p \in]1,\infty[$. Let Ω be a bounded connected open subset of $\mathbb{R}^n$ of class $C^{1,\alpha}$. Let $\phi \in \Phi_{1,\alpha}(\Omega)$, $\lambda \in]0,\infty[$. Then $u \in W_0^{1,p}(\phi(\Omega))$ satisfies the eigenvalue problem* (2.1) *if and only if the function $v \equiv u \circ \phi$ belongs to $W_0^{1,p}(\Omega)$ and*

$$(3.1)\quad \int_\Omega |Dv\cdot(D\phi)^{-1}|^{p-2} Dv\cdot\Gamma_\phi\cdot D\vartheta^t |\mathrm{det} D\phi| \mathrm{d}x = \lambda \int_\Omega |v|^{p-2} v\vartheta |\mathrm{det} D\phi| \mathrm{d}x,$$

for all $\vartheta \in W_0^{1,p}(\Omega)$, where $\Gamma_\phi \equiv (D\phi)^{-1} \cdot \left[(D\phi)^{-1}\right]^t$, for all $\phi \in \Phi_{1,\alpha}(\Omega)$. Furthermore,

$$(3.2)\quad \begin{aligned}\lambda_p[\phi] &= \min_{C_0^1(cl\Omega)\setminus\{0\}} \frac{\int_\Omega |Dv\cdot(D\phi)^{-1}|^p |\mathrm{det} D\phi| \mathrm{d}x}{\int_\Omega |v|^p |\mathrm{det} D\phi| \mathrm{d}x} \\ &= \int_\Omega |Dv_p[\phi]\cdot(D\phi)^{-1}|^p |\mathrm{det} D\phi| \mathrm{d}x,\end{aligned}$$

where $v_p[\phi] \equiv u_p[\phi] \circ \phi$, *for all* $\phi \in \Phi_{1,\alpha}(\Omega)$.

In order to prove the continuity of the function $v_p[\cdot]$, we will exploit a well-known boundary regularity result for equation (3.1) which provides us with an estimate of the Hölder exponents and quotients of the solutions $v_p[\phi]$ (cf. Lieberman [12, Theorem 1, p. 1203]). To do so we need the following technical Lemma which allows us to control the ellipticity constants of equation (3.1).

Lemma 3.1. *Let the same assumptions of Proposition 3.1 hold. Let* $M_0 > 0$ *and*

$$\begin{aligned} A(\phi, x, q) &\equiv |\det D\phi(x)||q \cdot (D\phi(x))^{-1}|^{p-2} q \cdot \Gamma_\phi(x), \\ B(\phi, x, z) &\equiv \lambda_p[\phi]\,|\det D\phi(x)||z|^{p-2} z, \end{aligned}$$

for all $x \in cl\Omega$, $q \in \mathbb{R}^n$, $z \in]M_0, M_0[$, $\phi \in \Phi_{1,\alpha}(\Omega)$.
Let

$$a_{ij}(\phi, x, q) \equiv \frac{\partial A_i(\phi, x, q)}{\partial q_j},$$

for all $x \in cl\Omega$, $q \in \mathbb{R}^n \setminus \{0\}$, $\phi \in \Phi_{1,\alpha}(\Omega)$, *where* $A_i(\phi, x, q)$ *is the* $i-th$ *component of* $A(\phi, x, q)$. *Let* $\tilde{\phi} \in \Phi_{1,\alpha}(\Omega)$ *be fixed. Then there exist an open neighborhood* $\mathcal{V}$ *of* $\tilde{\phi}$ *in* $\Phi_{1,\alpha}(\Omega)$ *and* $c_1, c_2 \in]0, \infty[$ *such that*

$$\sum_{i,j=1}^{n} a_{ij}(\phi, x, q)\,\xi_i \xi_j \geq c_1 |q|^{p-2} |\xi|^2, \tag{3.3}$$

$$|a_{ij}(\phi, x, q)| \leq c_2 |q|^{p-2}, \tag{3.4}$$

$$|A(\phi, x, q) - A(\phi, y, q)| \leq c_2 (1 + |q|)^{p-1} |x - y|^\alpha, \tag{3.5}$$

$$|B(\phi, x, z)| \leq c_2 (1 + |q|)^p, \tag{3.6}$$

for all $\phi \in \mathcal{V}$, $x, y \in cl\Omega$, $q \in \mathbb{R}^n \setminus \{0\}$, $z \in]M_0, M_0[$.

Proof. We first compute explicitly $a_{ij}(\phi, x, q)$. We have

$$\begin{aligned} \frac{\partial A_i(\phi, x, q)}{\partial q_j} &= |\det D\phi(x)||q \cdot (D\phi(x))^{-1}|^{p-2} (\Gamma_\phi)_{ij} + \\ &+ (p-2)\,|\det D\phi(x)||q \cdot (D\phi(x))^{-1}|^{p-4} \textstyle\sum_{r,s} (\Gamma_\phi)_{ri} (\Gamma_\phi)_{sj}\, q_r q_s\,, \end{aligned} \tag{3.7}$$

where $(\Gamma_\phi)_{ij}$ is the $(i,j)-$*component* of the matrix Γ_ϕ. We now prove (3.3). By equality (3.7), we have

$$\begin{aligned} \sum_{i,j=1}^{n} a_{ij}(\phi, x, q)\,\xi_i \xi_j &= |\det D\phi(x)||q \cdot (D\phi(x))^{-1}|^{p-2} |\xi \cdot (D\phi(x))^{-1}|^2 + \\ & (p-2)\,|\det D\phi(x)||q \cdot (D\phi(x))^{-1}|^{p-4} |q \cdot \Gamma_\phi \cdot \xi^t|^2, \end{aligned}$$

for all $x \in cl\Omega$, $q \in \mathbb{R}^n \setminus \{0\}$, $\xi = (\xi_i)_{1 \leq i \leq n} \in \mathbb{R}^n$, $\phi \in \Phi_{1,\alpha}(\Omega)$. By the previous equality, we deduce that

$$\begin{aligned}(3.8) \qquad &\sum_{i,j=1}^{n} a_{ij}(\phi, x, q)\, \xi_i \xi_j \geq \\ &\min\{1, p-1\} \, |detD\phi(x)| \, |q \cdot (D\phi(x))^{-1}|^{p-2} |\xi \cdot (D\phi(x))^{-1}|^2,\end{aligned}$$

for all $x \in cl\Omega$, $q \in \mathbb{R}^n \setminus \{0\}$, $\xi \in \mathbb{R}^n$, $\phi \in \Phi_{1,\alpha}(\Omega)$. Indeed, inequality (3.8) is trivial when $p \geq 2$; in case $1 < p < 2$, inequality (3.8) can be easily verified by noting that $\left(q \cdot \Gamma_\phi \cdot \xi^t\right)^2 \leq |q \cdot (D\phi(x))^{-1}|^2 |\xi \cdot (D\phi(x))^{-1}|^2$. By (3.8) and a continuity argument, it can be easily proved that there exists a bounded open neighborhood $\mathcal{V}$ of $\tilde{\phi}$ in $\Phi_{1,\alpha}(\Omega)$ and $c_1 \in]0, +\infty[$ such that

$$|detD\phi(x)| \left[\frac{|q \cdot (D\phi(x))^{-1}|}{|q|}\right]^{p-2} \left[\frac{|\xi \cdot (D\phi(x))^{-1}|}{|\xi|}\right]^2 \geq c_1,$$

for all $x \in cl\Omega$, $q, \xi \in \mathbb{R}^n \setminus \{0\}$, $\phi \in \mathcal{V}$. Thus, the proof of (3.3) is complete. We now prove (3.5). By the triangular inequality and by a continuity argument, it can be shown that, by possibly shrinking $\mathcal{V}$, there exists $c_2 \in]0, \infty[$ such that

$$\begin{aligned}&|A(\phi, x, q) - A(\phi, y, q)| \\ &\leq c_2 \left[|q|^{p-1} \big||detD\phi(x)| - |detD\phi(y)|\big| + |q|^{p-1} |\Gamma_\phi(x) - \Gamma_\phi(y)| + \right. \\ &\left. |q| \big||q \cdot (D\phi(x))^{-1}|^{p-2} - |q \cdot (D\phi(y))^{-1}|^{p-2}\big|\right],\end{aligned}$$

for all $x, y \in cl\Omega$, $q \in \mathbb{R}^n$, $\phi \in \mathcal{V}$. Then, by the boundedness of the set $\{\|\phi\|_{1,\alpha} : \phi \in \mathcal{V}\}$, and by simple considerations about the sum, the product, and the reciprocal of *Hölder* continuous functions, it follows that, by possibly replacing c_2 with a bigger constant, $|A(\phi, x, q) - A(\phi, y, q)| \leq c_2 |q|^{p-1} |x - y|^\alpha$. Thus, inequality (3.5) follows.
By possibly shrinking $\mathcal{V}$ and by Proposition 2.1, it is easy to see that $|B(\phi, x, z)|$ is uniformly bounded with respect to $x \in cl\Omega$, $z \in]M_0, M_0[$, $\phi \in \mathcal{V}$. Thus, by possibliy choosing a larger c_2, we obtain (3.6). Inequality (3.4) can be obtained by similar considerations. □

We are now ready to prove the announced continuity result for the dependence of $v_p[\phi]$ on $\phi \in \Phi_{1,\alpha}(\Omega)$.

Theorem 3.1. *Let the same assumptions of Proposition 3.1 hold. Then the function of* $\Phi_{1,\alpha}(\Omega)$ *to* $C^1(\mathrm{cl}\Omega)$ *which takes* $\phi \in \Phi_{1,\alpha}(\Omega)$ *to* $v_p[\phi] \equiv u_p[\phi] \circ \phi \in C^1(\mathrm{cl}\Omega)$ *is continuous.*

Proof. It suffices to prove that if a sequence $\{\phi_h\}_{h \in \mathbb{N}}$ converges to $\tilde{\phi}$ in $\Phi_{1,\alpha}(\Omega)$, then there exists a subsequence $\{\phi_{h_k}\}_{k \in \mathbb{N}}$ of the sequence $\{\phi_h\}_{h \in \mathbb{N}}$

such that $v_p[\phi_{h_k}]$ converges to $v_p\left[\tilde{\phi}\right]$ in $C^1(cl\Omega)$. By Proposition 2.1, we know that

$$M \equiv \sup_{h\in\mathbb{N}} \|v_p[\phi_h]\|_0 < +\infty.$$

By Lemma 3.1 , it follows that there exist $\bar{h} \in \mathbb{N}$ and $c_1, c_2 \in]0, \infty[$ such that inequalities (3.3), (3.4), (3.5), (3.6) hold for $\phi = \phi_h$, for all $h \geq \bar{h}$, $x, y \in cl\Omega$, $q \in \mathbb{R}^n \setminus \{0\}$, $\xi \in \mathbb{R}^n$, $z \in]-M, M[$. By Proposition 3.1, we have that

$$divA(\phi_h, x, Dv_p[\phi_h]) + B(\phi_h, x, v_p[\phi_h]) = 0, \tag{3.9}$$

(in the weak sense, as in (3.1)), for all $h \in \mathbb{N}$. Thus, by a well-known boundary regularity result (cf. Lieberman [12, Theorem 1, p. 1203]), by Lemma 3.1 and the previous considerations, by equation (3.9), it follows that there exist a constant $\beta \in]0, 1[$ depending only on α, c_1, c_2, p, n and a constant $C > 0$ depending only on α, c_1, c_2, p, n, M, Ω, such that

$$v_p[\phi_h] \in C^{1,\beta}(cl\Omega), \qquad \|v_p[\phi_h]\|_{1,\beta} \leq C,$$

for all $h \geq \bar{h}$. Since $C^{1,\beta}(cl\Omega)$ is compactly imbedded in $C^1(cl\Omega)$, there exists a subsequence $\{\phi_{h_k}\}_{k\in\mathbb{N}}$ of the sequence $\{\phi_h\}_{h\in\mathbb{N}}$ and $\tilde{v} \in C^1(cl\Omega)$ such that

$$\lim_{k\to\infty} v_p[\phi_{h_k}] = \tilde{v}, \qquad in\ C^1(cl\Omega) \tag{3.10}$$

By (3.10) it follows that $\tilde{v}(x) = 0$, for all $x \in \partial\Omega$. Furthermore, we have that $\int_\Omega |\tilde{v}|^p |detD\tilde{\phi}| dx = 1$ and thus $\tilde{v} \neq 0$. By equality (3.2), we have that (3.11)

$$\frac{\int_\Omega |Dv_p[\phi_{h_k}] \cdot (D\phi_{h_k})^{-1}|^p |detD\phi_{h_k}| dx}{\int_\Omega |v_p[\phi_{h_k}]|^p |detD\phi_{h_k}| dx} \leq \frac{\int_\Omega |Dv \cdot (D\phi_{h_k})^{-1}|^p |detD\phi_{h_k}| dx}{\int_\Omega |v|^p |detD\phi_{h_k}| dx},$$

for all $k \in \mathbb{N}$ and $v \in C_0^1(cl\Omega) \setminus \{0\}$. By taking the limit in (3.11) as k tends to infinity and by (3.10), we deduce that $\tilde{v}$ is a minimizing element in (3.2) for $\phi = \tilde{\phi}$. Thus, since $\tilde{v} \geq 0$, we have that $\tilde{v} = v_p\left[\tilde{\phi}\right]$, and the proof is complete. □

4 Differentiability of the first eigenvalue of $-\Delta_p$

This section is devoted to the main result of this paper. Namely, the continuous Frechét differentiability of the function $\lambda_p[\cdot]$. The proof will exploit Proposition 3.1, Theorem 3.1 and the following technical Lemma.

Lemma 4.1. *Let Ω be a bounded open subset of $\mathbb{R}^n$, and $p \in]1, +\infty[$. Let F be the map of $(C(cl\Omega))^n \times C(cl\Omega)$ to $\mathbb{R}$ defined by*

$$F(u, w) \equiv \int_\Omega |u|^p w dx,$$

for all $(u,w) \in (C(cl\Omega))^n \times C(cl\Omega)$. *Then* F *is of class* C^1 *and*

$$\mathrm{d}F(u,w)[\dot{u},\dot{w}] = \int_\Omega p\,|u|^{p-2}\,u\cdot\dot{u}w\mathrm{d}x + \int_\Omega |u|^p\,\dot{w}\mathrm{d}x,$$

for all (u,w), $(\dot{u},\dot{w}) \in (C(cl\Omega))^n \times C(cl\Omega)$.

We are now ready to prove the announced differentiabilty result. As for linear eigenvalue problems, we shall exploit the variational representation of the first eigenvalue (cf. *e.g.* Sokolowski and Zolesio [18, §3.6]). We have the following.

Theorem 4.1. *Let* $\alpha \in]0,1[$, $p \in]1,\infty[$. *Let* Ω *be a bounded open connected subset of* $\mathbb{R}^n$ *of class* $C^{1,\alpha}$. *Then the function of* $\Phi_{1,\alpha}(\Omega)$ *to* $\mathbb{R}$ *which takes* $\phi \in \Phi_{1,\alpha}(\Omega)$ *to* $\lambda_p[\phi]$ *is of class* C^1. *Furthermore,*

$$\begin{aligned}\mathrm{d}\lambda_p[\phi][\psi] &= \int_{\phi(\Omega)} [|Du_p[\phi]|^p - \lambda_p[\phi]\,|u_p[\phi]|^p]\,div\left(\psi\circ\phi^{(-1)}\right)\mathrm{d}y + \\ &\quad -p\int_{\phi(\Omega)} |Du_p[\phi]|^{p-2}Du_p[\phi]\cdot D\left(\psi\circ\phi^{(-1)}\right)\cdot(Du_p[\phi])^t\,\mathrm{d}y,\end{aligned} \tag{4.1}$$

for all $\phi \in \Phi_{1,\alpha}(\Omega)$, $\psi \in C^{1,\alpha}(cl\Omega,\mathbb{R}^n)$.

Proof. Ler R be the function of $\left(C_0^1(cl\Omega)\setminus\{0\}\right)\times\Phi_{1,\alpha}(\Omega)$ to $\mathbb{R}$ which takes the pair $(v,\phi) \in \left(C_0^1(cl\Omega)\setminus\{0\}\right)\times\Phi_{1,\alpha}(\Omega)$ to

$$R(v,\phi) \equiv \frac{\int_\Omega |Dv\cdot(D\phi)^{-1}|^p|detD\phi|\mathrm{d}x}{\int_\Omega |v|^p|detD\phi|\mathrm{d}x},$$

for all $(v,\phi) \in \left(C_0^1(cl\Omega)\setminus\{0\}\right)\times\Phi_{1,\alpha}(\Omega)$. By the previous Lemma and by noting that the map of $\Phi_{1,\alpha}(\Omega)$ to $C(cl\Omega,\mathbb{R}^{n\times n})$ which takes $\phi \in \Phi_{1,\alpha}(\Omega)$ to $(D\phi)^{-1} \in C(cl\Omega,\mathbb{R}^{n\times n})$ is real-analytic, it follows that the function R is continuously differentiable. Let $\tilde{\phi}$ be fixed. By Proposition 3.1 and by Taylor's formula for real-valued functions, there exists a convex open neighborhood U of $\tilde{\phi}$ in $\Phi_{1,\alpha}(\Omega)$ such that

$$\begin{aligned}\lambda_p[\phi]-\lambda_p\left[\tilde{\phi}\right] &= \min_{v\in C_0^1(cl\Omega)\setminus\{0\}} R(v,\phi) - \min_{v\in C_0^1(cl\Omega)\setminus\{0\}} R\left(v,\tilde{\phi}\right) \\ &\le R\left(v_p\left[\tilde{\phi}\right],\phi\right) - R\left(v_p\left[\tilde{\phi}\right],\tilde{\phi}\right) \\ &= \partial_\phi R\left(v_p\left[\tilde{\phi}\right],\tilde{\phi}+t_\phi\left(\phi-\tilde{\phi}\right)\right)\left[\phi-\tilde{\phi}\right],\end{aligned}$$

for some $t_\phi \in [0,1]$ depending on ϕ, for all $\phi \in U$. By the previous inequality, it follows that

$$\begin{aligned}&\lambda_p[\phi]-\lambda_p\left[\tilde{\phi}\right]-\partial_\phi R\left(v_p\left[\tilde{\phi}\right],\tilde{\phi}\right)\left[\phi-\tilde{\phi}\right] \\ &\le \partial_\phi R\left(v_p\left[\tilde{\phi}\right],\tilde{\phi}+t_\phi\left(\phi-\tilde{\phi}\right)\right)\left[\phi-\tilde{\phi}\right] - \partial_\phi R\left(v_p\left[\tilde{\phi}\right],\tilde{\phi}\right)\left[\phi-\tilde{\phi}\right],\end{aligned} \tag{4.2}$$

and thus, by the continuity of the function $\partial_\phi R\left(v_p\left[\tilde{\phi}\right],\cdot\right)$, we have that

$$\limsup_{\phi\to\tilde{\phi}} \frac{\lambda_p\left[\phi\right]-\lambda_p\left[\tilde{\phi}\right]-\partial_\phi R\left(v_p\left[\tilde{\phi}\right],\tilde{\phi}\right)\left[\phi-\tilde{\phi}\right]}{\|\phi-\tilde{\phi}\|_{1,\alpha}}\leq 0. \tag{4.3}$$

By arguing as in the proof of (4.2), we obtain

$$\begin{aligned}&\lambda_p\left[\phi\right]-\lambda_p\left[\tilde{\phi}\right]-\partial_\phi R\left(v_p\left[\tilde{\phi}\right],\tilde{\phi}\right)\left[\phi-\tilde{\phi}\right]\geq\\ &\partial_\phi R\left(v_p\left[\phi\right],\tilde{\phi}+\hat{t}_\phi\left(\phi-\tilde{\phi}\right)\right)\left[\phi-\tilde{\phi}\right]-\partial_\phi R\left(v_p\left[\tilde{\phi}\right],\tilde{\phi}\right)\left[\phi-\tilde{\phi}\right],\end{aligned} \tag{4.4}$$

for some $\hat{t}_\phi\in[0,1]$ depending on ϕ. Thus by (4.4) and by the continuity of the functions $\partial_\phi R\left(\cdot,\cdot\right)$ and $v_p\left[\cdot\right]$ (see Theorem 3.1), we have that

$$\liminf_{\phi\to\tilde{\phi}} \frac{\lambda_p\left[\phi\right]-\lambda_p\left[\tilde{\phi}\right]-\partial_\phi R\left(v_p\left[\tilde{\phi}\right],\tilde{\phi}\right)\left[\phi-\tilde{\phi}\right]}{\|\phi-\tilde{\phi}\|_{1,\alpha}}\geq 0. \tag{4.5}$$

Finally, by inequalities (4.3) and (4.5), we deduce that

$$\lim_{\phi\to\tilde{\phi}} \frac{\left|\lambda_p\left[\phi\right]-\lambda_p\left[\tilde{\phi}\right]-\partial_\phi R\left(v_p\left[\tilde{\phi}\right],\tilde{\phi}\right)\left[\phi-\tilde{\phi}\right]\right|}{\|\phi-\tilde{\phi}\|_{1,\alpha}}=0. \tag{4.6}$$

By equality (4.6), by Theorem 3.1 and by Lemma 4.1, we obtain that $\lambda_p\left[\cdot\right]$ is continuously Frechét differentiable and that

$$\mathrm{d}\lambda_p\left[\tilde{\phi}\right]\left[\psi\right]=\partial_\phi R\left(v_p\left[\tilde{\phi}\right],\tilde{\phi}\right)\left[\psi\right], \tag{4.7}$$

for all $\psi\in C^{1,\alpha}\left(cl\Omega,\mathbb{R}^n\right)$.

We now compute explicitly $\partial_\phi R\left(v_p\left[\tilde{\phi}\right],\tilde{\phi}\right)\left[\psi\right]$. Let $\psi\in C^{1,\alpha}\left(cl\Omega,\mathbb{R}^n\right)$ be fixed. Let $\delta>0$ be such that $\phi_t\equiv\tilde{\phi}+t\psi\in\Phi_{1,\alpha}\left(\Omega\right)$, for all $t\in]-\delta,\delta[$. To shorten our notation, we set $\tilde{v}\equiv v_p\left[\tilde{\phi}\right]$, $\tilde{u}\equiv u_p\left[\tilde{\phi}\right]$, $\chi\equiv\psi\circ\tilde{\phi}^{(-1)}$. By standard calculus, and a change of variables, we have

$$\begin{aligned}&\frac{\mathrm{d}\left(\int_\Omega|D\tilde{v}\cdot\left(D\phi_t\right)^{-1}|^p|detD\phi_t|\mathrm{d}x\right)}{\mathrm{d}t}_{|t=0}\\ &=-p\int_{\tilde{\phi}(\Omega)}|D\tilde{u}|^{p-2}D\tilde{u}\cdot D\chi\cdot\left(D\tilde{u}\right)^t\mathrm{d}y+\int_{\tilde{\phi}(\Omega)}|D\tilde{u}|^p div\chi \mathrm{d}y.\end{aligned} \tag{4.8}$$

By the same argument, we obtain

$$\frac{\mathrm{d}\left(\int_\Omega|\tilde{v}|^p|detD\phi_t|\mathrm{d}x\right)}{\mathrm{d}t}_{|t=0}=\int_{\tilde{\phi}(\Omega)}|\tilde{u}|^p div\chi\mathrm{d}y. \tag{4.9}$$

Thus, by combining (4.7), (4.8), (4.9), and by equalities $\int_{\tilde{\phi}(\Omega)} |\tilde{u}|^p \mathrm{d}y = 1$, and $\lambda_p\left[\tilde{\phi}\right] = \int_{\tilde{\phi}(\Omega)} |D\tilde{u}|^p \mathrm{d}y$, we obtain (4.1). □

Our goal is now to simplify formula (4.1) in order to obtain a nicer formula for $\mathrm{d}\lambda_p[\phi]$, which we expect to be the natural generalisation to the p-Laplacian of the well-known Hadamard formula. To do so, we prove the following Theorem.

Theorem 4.2. *Let the same assumptions of Theorem 4.1 hold. Let $\tilde{\phi} \in \Phi_{1,\alpha}(\Omega)$ be fixed. Let $\tilde{u} \equiv u_p\left[\tilde{\phi}\right] \in W^{2,1}\left(\tilde{\phi}(\Omega)\right)$. Then*

$$\mathrm{d}\lambda_p\left[\tilde{\phi}\right][\psi] = (1-p)\int_{\partial\tilde{\phi}(\Omega)} \left|\frac{\partial \tilde{u}}{\partial \nu}\right|^p \left(\psi \circ \tilde{\phi}^{(-1)}\right) \cdot \nu \mathrm{d}\sigma, \tag{4.10}$$

for all $\psi \in C^{1,\alpha}(\mathrm{cl}\Omega, \mathbb{R}^n)$, where ν denotes the exterior unit normal to $\partial\tilde{\phi}(\Omega)$.

Proof. To shorten our notation, we set $\chi \equiv \psi \circ \tilde{\phi}^{(-1)}$. Let $i, j \in \{1, \ldots, n\}$ and $p \in]1, +\infty[$, be fixed. We notice that the function $f_{p,i,j}$ of $\mathbb{R}^n$ to $\mathbb{R}$ defined by $f_{p,i,j}(t) \equiv |t|^{p-2} t_i t_j$ for all $t = (t_1, \ldots, t_n) \in \mathbb{R}^n \setminus \{0\}$, $f_{p,i,j}(0) = 0$, is of class C^1. The function $|D\tilde{u}|^{p-2} \frac{\partial \tilde{u}}{\partial y_i} \frac{\partial \tilde{u}}{\partial y_j}$ (which is understood to be equal to zero where $|D\tilde{u}|$ vanishes) equals $f_{p,i,j} \circ D\tilde{u}$. Since $f_{p,i,j}$ is of class C^1, we have $f_{p,i,j} \circ D\tilde{u} \in W^{1,1}\left(\tilde{\phi}(\Omega)\right)$. Since $\tilde{u} = 0$ on $\partial\tilde{\phi}(\Omega)$ and $\tilde{u} \in C^1\left(cl\tilde{\phi}(\Omega)\right)$, we have that $D\tilde{u}_{|\partial\tilde{\phi}(\Omega)} = \frac{\partial \tilde{u}}{\partial \nu}\nu$. Thus, by the divergence Theorem, we conclude that

(4.11)

$$\begin{aligned}
\int_{\tilde{\phi}(\Omega)} |D\tilde{u}|^{p-2} D\tilde{u} \cdot D\chi \cdot (D\tilde{u})^t \, \mathrm{d}y &= \sum_{i,j=1}^n \int_{\tilde{\phi}(\Omega)} |D\tilde{u}|^{p-2} \frac{\partial \tilde{u}}{\partial y_i} \frac{\partial \tilde{u}}{\partial y_j} \frac{\partial \chi_j}{\partial y_i} \mathrm{d}y \\
= \sum_{i,j=1}^n \Bigg[\int_{\partial\tilde{\phi}(\Omega)} |D\tilde{u}|^{p-2} \frac{\partial \tilde{u}}{\partial y_i} \frac{\partial \tilde{u}}{\partial y_j} \chi_j \nu_i \mathrm{d}\sigma &- \int_{\tilde{\phi}(\Omega)} \frac{\partial |D\tilde{u}|^{p-2} \frac{\partial \tilde{u}}{\partial y_i} \frac{\partial \tilde{u}}{\partial y_j}}{\partial y_i} \chi_j \mathrm{d}y \Bigg] \\
= \int_{\partial\tilde{\phi}(\Omega)} \left|\frac{\partial \tilde{u}}{\partial \nu}\right|^p \chi \cdot \nu \mathrm{d}\sigma &- \sum_{i,j=1}^n \int_{\tilde{\phi}(\Omega)} \frac{\partial}{\partial y_i} \left(|D\tilde{u}|^{p-2} \frac{\partial \tilde{u}}{\partial y_i} \frac{\partial \tilde{u}}{\partial y_j} \right) \chi_j \mathrm{d}y.
\end{aligned}$$

Let $\tilde{\phi}(\Omega)^\sharp = \left\{x \in \tilde{\phi}(\Omega) : |D\tilde{u}(x)| \neq 0\right\}$. It can be easily verified that

$$-div\left(|D\tilde{u}(y)|^{p-2} D\tilde{u}(y)\right) = \lambda_p\left[\tilde{\phi}\right] |\tilde{u}(y)|^{p-2} \tilde{u}(y), \tag{4.12}$$

almost everywhere in $\tilde{\phi}(\Omega)^\sharp$. (Actually, by standard regularity theory, equality (4.12) is verified everywhere in $\tilde{\phi}(\Omega)^\sharp$, in the classical sense.) Thus, by computing the derivatives in the last integral of (4.11), and by exploiting equality (4.12), we obtain that

$$\sum_{i,j=1}^{n} \frac{\partial}{\partial y_i}\left(|D\tilde{u}|^{p-2}\frac{\partial \tilde{u}}{\partial y_i}\frac{\partial \tilde{u}}{\partial y_j}\right)\chi_j$$

$$= \sum_{j=1}^{n} div\left(|D\tilde{u}|^{p-2} D\tilde{u}\right)\frac{\partial \tilde{u}}{\partial y_j}\chi_j + \sum_{i,j=1}^{n}\left(|D\tilde{u}|^{p-2}\frac{\partial \tilde{u}}{\partial y_i}\frac{\partial^2 \tilde{u}}{\partial y_j \partial y_i}\right)\chi_j$$

$$= -\lambda_p\left[\tilde{\phi}\right]\sum_{j=1}^{n}|\tilde{u}|^{p-2}\,\tilde{u}\frac{\partial \tilde{u}}{\partial y_j}\chi_j + \sum_{i,j=1}^{n}\left(|D\tilde{u}|^{p-2}\frac{\partial \tilde{u}}{\partial y_i}\frac{\partial^2 \tilde{u}}{\partial y_j \partial y_i}\right)\chi_j$$

$$= -p^{-1}\lambda_p\left[\tilde{\phi}\right]\sum_{j=1}^{n}\frac{\partial\,|\tilde{u}|^p}{\partial y_j}\chi_j + p^{-1}\sum_{j=1}^{n}\frac{\partial\,|D\tilde{u}|^p}{\partial y_j}\chi_j,$$

almost everywhere in $\tilde{\phi}(\Omega)$. Then, by the previous equality, by (4.11), by the membership of the functions $|\tilde{u}|^p$ and $|D\tilde{u}|^p$ in $W^{1,1}\left(\tilde{\phi}(\Omega)\right)$, and by the divergence Theorem, it follows that

(4.13)

$$\int_{\tilde{\phi}(\Omega)} |D\tilde{u}|^{p-2} D\tilde{u}\cdot D\chi\cdot(D\tilde{u})^t\,\mathrm{d}y = \left(1-p^{-1}\right)\int_{\partial\tilde{\phi}(\Omega)}\left|\frac{\partial \tilde{u}}{\partial \nu}\right|^p \chi\cdot\nu \mathrm{d}\sigma +$$
$$-p^{-1}\lambda_p\left[\tilde{\phi}\right]\int_{\tilde{\phi}(\Omega)}|\tilde{u}|^p\, div\chi \mathrm{d}y + p^{-1}\int_{\tilde{\phi}(\Omega)}|D\tilde{u}|^p\, div\chi \mathrm{d}y.$$

Finally, by (4.1) and (4.13), we obtain (4.10). □

The condition $u_p\left[\tilde{\phi}\right] \in W^{2,1}\left(\tilde{\phi}(\Omega)\right)$ in Theorem 4.2 has been used only for technical reasons. At the moment all the author can say is that such condition can be easily verified when $\tilde{\phi}(\Omega)$ is a ball. (We refer to Bhattacharya [2] for the study of equation (1.1) in the ball.) We also mention that interior estimates for the second order weak derivatives of the eigenvectors of the p-Laplacian are known: for example, it is known that $u_p\left[\tilde{\phi}\right] \in W^{2,p}_{loc}\left(\tilde{\phi}(\Omega)\right)$ when $1 < p \leq 2$ (cf. Tolksdorf [19, Proposition 1, p. 129]).

Remark 4.1. As in the linear case $p = 2$ (cf. e.g. Chatelain [4], Henry [9]), formula (4.10) naturally leads to an overdetermined problem. Indeed, under the assumptions of Theorem 4.2, if we assume that $\tilde{\phi}$ is a critical point of the functional $\phi \mapsto \lambda_p[\phi]$ under the volume constraint $Vol(\phi(\Omega)) = Vol\left(\tilde{\phi}(\Omega)\right)$ and if $\partial\tilde{\phi}(\Omega)$ is connected, then we easily find out that $\tilde{u} \equiv u_p\left[\tilde{\phi}\right]$ and

$\tilde{\lambda} \equiv \lambda_p\left[\tilde{\phi}\right]$ solve the following overdetermined problem

$$
(4.14) \qquad \begin{cases} -\Delta_p \tilde{u} = \tilde{\lambda}\, |\tilde{u}|^{p-2}\, \tilde{u} \\ \frac{\partial \tilde{u}}{\partial \nu} = const. \end{cases}
$$

In particular, this happens when $\tilde{\phi}\,(\Omega)$ is a ball. Indeed, the ball minimizes the first eigenvalue for perturbations of the domain which preserve the volume (cf. Bhattacharya [3, Theorem 1, p. 226]) and the first eigenfunction is known to be radial when the domain is a ball (cf. Bhattacharya [1]). Thus, a natural question arises. Let $\lambda \in]0,\infty[$ and $u \in W_0^{1,p}\left(\tilde{\phi}\,(\Omega)\right) \setminus \{0\}$ solve problem (4.14). Is $\tilde{\phi}\,(\Omega)$ necessarily a ball? Under the assumptions that u does not change sign and $1 < p \leq 2$, the answer is positive: we refer to Henry [8] for the case $p = 2$, and to Damascelli and Pacella [6] for the case $1 < p < 2$.

References

[1] T. Bhattacharya, Radial symmetry of the first eigenfunction for the p-Laplacian in the ball, *Proc. Amer. Math. Soc.*, **104**(1988), 169-174.

[2] T. Bhattacharya, Some results concerning the eigenvalue problem for the p-Laplacian, *Ann. Acad. Sci. Fenn. Ser. AI Math.*, **14**(1989), 325-343.

[3] T. Bhattacharya, A proof of the Faber-Krahn inequality for the first eigenvalue of the p-Laplacian, *Annali di Matematica Pura ed Applicata (IV)*, **177**(1999), 225-240.

[4] T. Chatelain, A new approach to two overdetermined eigenvalue problems of Pompeiu type, Élasticité, viscoélasticité et controle optimal, Lyon 1995, *ESAIM Proc.*, **2**, *Soc. Math. Appl. Indust.*, Paris, 1997, 235-242 (electronic).

[5] S.J. Cox, Extremal eigenvalue problems for the Laplacian, Recent advances in partial differential equations, El Escorial 1992, *RAM Res. Appl. Math.*, **30**, *Masson*, Paris, 1994, 37-53.

[6] L. Damascelli, F. Pacella, Monotonicity and symmetry results for p-Laplace equations and applications, *Advances in Differential Equations*, **5** (2000), 1179-1200.

[7] E. DiBenedetto, $C^{1,\alpha}$ local regularity of weak solutions of degenerate elliptic equations, *Nonlinear Anal. TMA*, **7**(1983), 827-850.

[8] D. Henry, Topics in nonlinear analysis, in *Trabalho de Matematica*, **192**, *Universidade de Brasilia, Departamento de Matematica-IE*, 1982.

[9] D. Henry, Perturbation of boundary value problems of partial differential equations, manuscript, Brasilia, 1994.

[10] P.D. Lamberti, M. Lanza de Cristoforis, An analyticity result for the dependence of multiple eigenvalues and eigenspaces of the Laplace operator upon perturbation of the domain, *Glasgow Math. J.*, **44**(2002), 29-43.

[11] M. Lanza de Cristoforis, Higher order differentiability properties of the composition and of the inversion operator, *Indag. Math. (N.S.)*, **5**(1994), 457-482.

[12] G. Lieberman, Boundary regularity for solutions of degenerate elliptic equations, *Nonlinear Anal. TMA*, **12**(1988), 1203-1219.

[13] P. Lindqvist, On the equation $div\left(|\nabla u|^{p-2}\nabla u\right)+\lambda|u|^{p-2}u=0$, *Proc. Amer. Math. Soc.*, **109**(1990), 157-164.

[14] P. Lindqvist, On nonlinear Rayleigh quotients, *Potential Anal.*, **2**(1993), 199-218.

[15] P. Lindqvist, On a nonlinear eigenvalue problem, Fall School in Analysis, Jyväskylä 1994, *Univ. Jyväskylä*, **68**, Jyväskylä, 1995, 33-54.

[16] G. Prodi, Dipendenza dal dominio degli autovalori dell'operatore di Laplace, *Istit. Lombardo Accad. Sci. Lett. Rend. A*, **128**(1994), 3-18.

[17] B. Rousselet, D. Chenais, Continuité et différentiabilité d'éléments propres: application à l'optimisation de structures, *Appl. Math. Optim.*, **22**(1990), 27-59.

[18] J. Sokolowski, J.P. Zolesio, Introduction to shape optimization. Shape sensitivity analysis, *Springer-Verlag*, Berlin, 1992.

[19] P. Tolksdorf, Regularity for a more general class of quasilinear elliptic equations, *Journal of Differential Equations*, **51**(1984), 126-150.

Fixed Point and Related Properties for the Fourier and Fourier Stieltjes Algebra of a Locally Compact Group

Anthony To-Ming Lau[1]

Department of Mathematical and Statistical Sciences, University of Alberta, Edmonton, Alberta, Canada T6G 2G1, e–mail:tlaumath.ualberta.ca

1 Introduction

Let E be a Banach space and K be a weakly compact convex subset of E. We say that E has the *FPP* (- *fixed point property*) if every nonexpansive mapping $T : K \to K$ (i.e. $\|Tx - Ty\| \leq \|x - y\|$ for every $x, y \in K$) has a fixed point. The space E has the *FPP* if every weakly compact convex set $K \subseteq E$ has the FPP.

It is well known (Shauder's Theorem) that compact convex nonempty subsets of a Banach space have the *FPP*. In particular, any Banach space E having the Shur's property (i.e. weakly compact subsets of E are norm compact) has the *FPP*. It is also well-known (Browder's Theorem [5]) that uniformly convex Banach spaces have that FPP. In [17] Kirk, extending Browder's Theorem showed that a weakly compact convex subset of a Banach space with normal structure has the FPP. In [1], Alspach exhibited a weakly compact convex subset K of the Lesbegue space $L^1[0,1]$ does not have the FPP. On the other hand, in [23], Maurey, using the techniques of ultraproducts, showed that reflexive subspaces of $L^1[0,1]$, as well as the sequence space c_0, have the FPP (see [16] and [17] for more details).

When $G = (\mathbb{Z}, +)$, the integer group, then the Fourier algebra $A(G)$ is isometrically isomorphic to $L^1[0,1]$. In particular, $A(G)$ does not have the FPP. On the other hand, if $G = \mathbb{T}$, the circle group, then $A(G) \cong \ell^1(\mathbb{Z})$ which has the FPP (even for weak*-compact convex subsets [22]). It is the purpose of this chapter of the book to discuss various properties related to the FPP for the Fourier algebra $A(G)$ and Fourier Stieltjes algebra $B(G)$of a locally compact group G. Since this book is intended for researchers and graduate students, some detailed proofs, open problems and historical remarks will be provided. It is our hope that our effort will generate further research in the direction of non-linear analysis and various geometric properties of the Fourier and Fourier Stieltjes algebra of a locally compact group (see also [4], [19] and [28]).

I would like to thank Professor Agarwal for kindly inviting me to contribute to this special volume on Nonlinear Analysis.

[1]This research is supported by an NSERC-Grant.

2 Some preliminaries

Let G be a locally compact group with a fixed left Haar measure λ. For $1 \leq p < \infty$ let $L^p(G)$ denote the equivalence classes of λ-measurable functions $f : G \mapsto \mathbb{C}$ such that $\int |f(x)|^p d\lambda(x) < \infty$ with norm $\|f\|_p = \int |f(x)|^p d\lambda(x)$ as defined in [13]. Then $L^1(G)$ is a Banach algebra with product

$$(f * g)(x) = \int f(y)g(y^{-1}x)d\lambda(y) \tag{2.1}$$

$f, g \in L^1(G)$ (called the *group algebra* of G). We define $C^*(G)$, the *group $C^*(G)$-algebra* of G, to be the completion of $L^1(G)$ with the norm

$$\|f\| = \sup\{\|\pi(f)\|\}$$

where the supremum is taken over all nondegenerate representation π of $L^1(G)$ as an algebra of bounded operator on a Hilbert space. Let $C(G)$ be the Banach space of bounded continuous complex-valued functions on G with the supremum norm. Denote the set of continuous positive definite functions on G by $P(G)$ i.e. all continuous functions ϕ on G such that

$$\sum_{j=1}^{n}\sum_{i=1}^{n} \lambda_i \overline{\lambda}_j \phi(x_i x_j^{-1}) \geq 0$$

for any $\lambda_1, \ldots, \lambda_n \in \mathbb{C}$, and $x_1, \ldots, x_n \in G$.

Define the *Fourier Stieltjes algebra* of G, denoted by $B(G)$, to be the closed linear span of $P(G)$. Then $B(G)$ may be identified with the dual of $C^*(G)$ by

$$\langle \phi, f \rangle = \int f(x)\phi(x)d\lambda(x) \tag{2.2}$$

with $\phi \in B(G)$, $f \in L^1(G)$. Then $B(G)$ with the dual norm and pointwise multiplication is a commutative Banach algebra. Let $C_{00}(G)$ denote all continuous functions on G with compact support, and $A(G)$ (the Fourier algebra of G) be the closure of $C_{00}(G) \cap B(G)$. Then $A(G)$ is an ideal in $B(G)$. When G is abelian, $B(G) \cong M(\widehat{G})$ the measure algebra of the dual group $\widehat{G}$ and $A(G) \cong L^1(\widehat{G})$, the group algebra of $\widehat{G}$.

Let ρ be the *left regular representation* of G, i.e. for each $f \in L^1(G)$, $\rho(f)$ is the operator in $\mathcal{B}(L^2(G))$ defined by $\rho(f)(h) = f * h$, $h \in L^2(G)$ (the convolution of f and h as defined by (1)). Then denote by $VN(G)$ to be the closure of $\{\rho(f);\ f \in L^1(G)\}$ in the weak operator topology in $\mathcal{B}(L^2(G))$. It is known that $A(G)^* = VN(G)$ with $\langle \phi, f \rangle$ as defined by (2) for $\phi \in A(G)$, $f \in L^1(G)$. We refer the reader to [11] for more details on these spaces.

A semigroup S is called *amenable* if there is a left invariant mean m on $\ell^\infty(S)$ (space of bounded complex-valued functions on S) i.e. $m \in \ell^\infty(S)^*$,

$m(1) = \|m\| = 1$ and $\langle m, \ell_a f\rangle = \langle m, r_a f\rangle \ \langle m, f\rangle$ for all $f \in \ell^\infty(S)$, $a \in S$, where $(\ell_a f)(t) = f(at)$, and $(r_a f)(t) = f(ta)$ for all $t \in S$. Amenable semigroups include all abelian semigroups, but the free semigroup (or group) on two generators is not amenable.

If G is a locally compact group, then G is *amenable* if there is an invariant mean on $CB(G)$. Amenable groups include all compact groups and solvable groups (see [24] or [25] for details). In this case

$$C^*(G) \cong \text{ mean closure of } \{\rho(f);\ f \in: L^1(G)\}.$$

3 Fixed point property and Kadec-Klee-type properties

A Banach space E is said to have the *Kadec-Klee property*

(KK): if whenever (x_n) is a sequence in the unit ball of E that converges weakly to x, and sup $\big((x_n)\big) > 0$ where

$$\sup\big((x_n)\big) \equiv \inf\{\|x_n - x_m\| : n \neq m\},$$

then $\|x\| < 1$.

This property, given in different form, is known as property (H) in [9] or the Radon-Riesz property in [2]. The definition as given above is due to Huff [15].

A Banach space E is said to have the *uniform Kadec-Klee property*

(UKK) : if for every $\varepsilon > 0$ there is an $0 < \delta < 1$ such that whenever (x_n) is a sequence in the unit ball of E converging weakly to x and $\text{sep}\big((x_n)\big) > \varepsilon$, then $\|x\| \leq \delta$.

This property was introduced by Huff [15] who showed that UKK is strictly stronger than property KK. van Dulst and Sims showed that a Banach space with UKK has FPP [10].

A dual Banach space E is said to have the property

(UKK^*) : if for every $\varepsilon > 0$ there exists a $0 < \delta < 1$ such that whenever A is a subset of the closed unit ball of E containing a sequence (x_n) with $\text{sep}\big((x_n)\big) > \varepsilon$ then there exists an x in the weak$*$-closure (A) such that $\|x\| \leq \delta_0$,

van Dulst and Sims [10] proved that if a dual Banach space E has property UKK^* then E has the *weak*-fixed point property* (FPP*) i.e. for any weak*-compact convex subset K of E, and any nonexpansive mapping T from K into K, then K contains a fixed point for T.

Let K be a bounded closed convex subset of a Banach space E. A point x in K is called a *diametral point* of K if

$$\sup\{\|x - y\|;\ y \in K\} = \text{ diam}\,(K)$$

where diam (K) denotes the diameter of K. The set K is said to have *normal structure* if every nontrivial convex subset (i.e. containing at least two points) H of K contains a nondiametral point of H. A Banach space has *weak-normal structure* if every non-trivial weakly compact convex subset has normal structure. If the Banach space is also a dual space, then it has *weak*-normal structure* if every nontrivial weak*-compact convex subset has normal structure.

Kirk [17] proved that if E has weak-normal structure, then E has FPP. Subsequently, Lim [22] proved that a dual Banach space has FPP* whenever it has weak*-normal structure.

Theorem 3.1 *Let G be a compact group. Then $B(G)$ has UKK^*. In particular, $B(G)$ has* FPP*.

Proof It suffices to show that

(*) for any $\varepsilon > 0$, there is a $0 < \delta < 1$ such that whenever (ϕ_α) is a net in the unit ball of $B(G)$ that converges to ϕ in the weak*-topology and $\operatorname{sep}((\phi_\alpha)) > \varepsilon$, then $\|\phi\| \leq \delta$.

First we observe that (since G is compact), $B(G) = A(G)$ and so $C^*(G)^{**} = A(G)^* = VN(G)$. Now

$$C^*(G) = \{T \in VN(G) : T \text{ compact}\}$$

(see [7, Proposition 2.2]). Consequently $C^*(G)$ is an ideal in the enveloping von Neumann algebra $C^*(G)^{**}$. Hence $C^*(G)$ has a bounded approximate identity $\{P_i\}_{i\in I}$ consisting of projections of finite rank [27, p. 157].

Suppose (*) fails. Then there exists an $\varepsilon_0 > 0$ such that for every $0 < \delta < 1$, there is a net (ϕ_α) and ϕ in the closed unit ball with $\operatorname{sep}((\phi_\alpha)) > \varepsilon_0$, and converging to ϕ in the weak* topology but $\|\phi\| > \delta$. Now let $\varepsilon_1 < \varepsilon_0$, $0 < \varepsilon_1 < 1$ and pick $0 < \delta_0 < 1$ so that $\delta_0 > (1 - \varepsilon_1^2/12)^{1/2}$. Pick the net (ϕ_α) and ϕ with the properties indicated above corresponding to $\delta = \delta_0$. Let $X_0 \in C^*(G)$, $\|X_0\| = 1$, be a weak* linear functional which separates ϕ and the closed ball of radius δ_0 centred at 0. Denote the slice $\{\phi \in B(G) : \|\phi\| \leq 1, \operatorname{Re}\langle X_0, \phi\rangle \geq \delta_0\}$ by $S[X_0, \delta_0]$. Since (ϕ_α) converges to ϕ in the weak* topology the net (ϕ_α) is eventually in $S[X_0, \delta_0]$. Since this subnet also has separation exceeding ε_0 we may assume the original net is contained in $S[X_0, \delta_0]$.

Now for each $\psi \in S[X_0, \delta_0]$, define, for each τ,

$$\alpha_\tau(\psi) = \|P_\tau \psi Q_\tau\| + \|Q_\tau \psi P_\tau\| + \|Q_\tau \psi Q_\tau\|,$$

with $Q_\tau = I - P_\tau$, where I is the identity in the von Neumann algebra $VN(G)$ of G. We first show that

$$\alpha_\tau(\psi) \leq 3^{1/2}(1 - \|P_\tau \psi P_\tau\|^2)^{1/2}. \tag{3.1}$$

Let $\psi^{\sim} \in \mathcal{T}(L^2(G))$ be an extension of ψ such that $\|\psi^{\sim}\| = \|\psi\|$ [27, p. 156]. Then by Proposition 2.2 in [21], we have, for each τ,

$$\begin{aligned} 1 &\geq \|\psi^{\sim}\|^2 \geq \|P_\tau \psi^{\sim} P_\tau\|^2 + \|P_\tau \psi^{\sim} Q_\tau\|^2 + \|Q_\tau \psi^{\sim} P_\tau\|^2 + \|Q_\tau \psi^{\sim} Q_\tau\|^2 \\ &\geq \|P_\tau \psi P_\tau\|^2 + \|P_\tau \psi Q_\tau\|^2 + \|Q_\tau \psi P_\tau\|^2 + \|Q_\tau \psi Q_\tau\|^2, \end{aligned}$$

and so the inequality (3.1) follows. Now,

$$\begin{aligned} \|P_\tau \psi P_\tau\| &\geq |\psi(P_\tau X_0 P_\tau)| \geq \text{Re}\, \psi(P_\tau X_0 P_\tau) \\ &= \text{Re}\, \psi(X_0) - \text{Re}\, \psi(P_\tau X_0 Q_\tau) - \text{Re}\, \psi(Q_\tau X_0 P_\tau) - \text{Re}\, \psi(Q_\tau X_0 Q_\tau) \\ &\geq \delta_0 - \|\psi\|(\|P_\tau X_0 Q_\tau\| + \|Q_\tau X_0 P_\tau\| + \|Q_\tau X_0 Q_\tau\|) \\ &\geq \delta_0 - 3 \max\{\|X_0 Q_\tau\|, \|Q_\tau X_0\|\}. \end{aligned}$$

But $X_0 Q_\tau = X_0 - X_0 P_\tau$ and $Q_\tau X_0 = X_0 - P_\tau X_0$; consequently both of them converge to 0 in norm because (P_α) is an approximate identity. Thus, for $0 < \eta < \delta_0$, there is a τ_0 such that if $\tau \geq \tau_0$ we have

$$\|P_\tau X_0 P_\tau\| \geq \delta_0 - \eta.$$

In particular, by (3.1), we have, for $\tau \geq \tau_0$,

$$\alpha_\tau(\psi) \leq 3^{1/2}\left(1 - (\delta - \eta)^2\right)^{1/2}. \tag{3.2}$$

Now let $\xi > 0$ be arbitrary. Since the weak*-topology and the norm topology coincide on finite-dimensional subspaces, there is an α_0 such that for all $\alpha, \beta \geq \alpha_0$,

$$\|P_{\tau_0}(\phi_\alpha - \phi_\beta) P_{\tau_0}\| < \xi.$$

Consequently for all $\alpha, \beta \geq \alpha_0$, $\alpha \neq \beta$,

$$\begin{aligned} \|\phi_\alpha - \phi_\beta\| &\leq \|P_{\tau_0}(\phi_\alpha - \phi_\beta) P_{\tau_0}\| + \alpha_{\tau_0}(\phi_\alpha) + \alpha_{\tau_0}(\phi_\beta) \\ &< \xi + 2(3)^{1/2}\left(1 - (\delta_0 - \eta)^2\right)^{1/2}, \end{aligned}$$

by (3.2) and so

$$\underset{inf}{\alpha \neq \beta}, \|\phi_\alpha - \phi_\beta\| < \xi + 2(3)^{1/2}\left(1 - (\delta_0 - \eta)^2\right)^{1/2}.$$

Since ξ and $0 < \eta < \delta_0$ are arbitrary we have

$$\text{sep}\big((\phi_\alpha)\big) \leq 2(3)^{1/2}\left(1 - (\delta_0 - \eta)^2\right)^{1/2} < \varepsilon_1 < \varepsilon_0.$$

This contradicts that $\text{sep}\big((\phi_\alpha)\big) > \varepsilon_0$.

The theorem now follows from [10].

Theorem 3.2 *Let G be a second countable locally compact group. Then the following are equivalent:*

(a) *G is compact*

(b) *$B(G)$ has UKK^**

(c) *$B(G)$ has KK^*.*

Proof That (a)$\Longrightarrow$(b) follows from Theorem 3.1 and the implications (b)$\Longrightarrow$(c) is obvious.

To prove (c)$\Longrightarrow$(a), we first note that if G is second countable, $L^1(G)$ is norm separable. Consequently $C^*(G)$ is also norm separable. Hence the weak*-topology is metrizable. It follows from (c) that the weak*-topology and norm topology agrees on the unit sphere of $B(G)$. Hence by Theorem 3.9 in [3], G must be a compact group.

□

4 Fixed point property for $A(G)$ and $B(G)$

A von Neumann algebra M is called *atomic* if every non-zero projection in M majorizes a non-zero minimal projection. A locally compact group G is called an *[AU]-group* if the enveloping von Neumann algebra of $C^*(G)$ ($= C^*(G)$ with Arens multiplication) is atomic, or equivalently, the von Neumann algebra generated by every continuous unitary representation of G is atomic); it is an *[AR]-group* if $VN(G)$ is atomic. Clearly

$$[\text{compact}] \subseteq [\text{AU}] \subseteq [\text{AR}].$$

However, the inclusions are proper (see [28]).

Bruck [6] shows that if a Banach space E has the FPP then E has the FPP *for commutative semigroup*, i.e. whenever K is a non-empty weakly compact convex subset of E, and $\mathcal{S}$ is a summutative semigroup of nonexpansive mappings from K into K, then K contains a common fixed point for $\mathcal{S}$ (see [12], Appendix]).

Theorem 4.1 *Let G be an* [AP]*-group, then $A(G)$ has the* FPP *for commutative semigroups.*

Proof Let $S = \{\phi \in A(G) \cap P(G);\ \phi(e) = 1\}$ and $\text{ext}\,(S)$ be the set of its extreme points. As in the proof of Theorem 4 in [8], if for each $\phi \in \text{ext}\,(S)$, $\{\pi_\phi, H_\phi\}$ is the representation of the von Neumann algebra $VN(G)$ induced by the GNS-construction, then $VN(G)$ is isomorphic to the direct-sum $\sum \oplus \mathcal{B}(H_\phi)$ taken over all $\phi \in \text{ext}\,(S)$, where $\mathcal{B}(H_\phi)$ denote the set of all bounded linear operators from the Hilbert space H_ϕ into itself. It follows from the uniqueness of the predual of $VN(G)$ that $A(G)$ is linearly isometric to the ℓ_1-sum $\sum \oplus J(H_\phi)$, $\phi \in \text{ext}\,(S)$, where $J(H_\phi)$ denotes the space of trace class operators on H_ϕ. Let H be the Hilbert space direct summand of $\{H_\phi,\ \phi \in \text{ext}\,(S)\}$. Then the space $\sum \oplus J(H_\phi)$ can be isometrically embedded into $J(H)$. By a result of Lennard [21], $J(H)$ has property UKK.

It follows that $J(H)$ and every subspace of it has the FPP. Consequently, by Bruck's theorem [6], $A(G)$ has the FPP for commutative semigroups. □

A similar argument as above will show:

Theorem 4.2 *If G is an* [AU]*-group, then $B(G)$ has the* FPP *for commutative semigroups.*

It follows from above that if G is compact, then $A(G) = B(G)$ has the FPP for commutative semigroup. However, there are non-compact solvable locally compact group (for example, when G is the so-called "$ax+b$" group = affine group) such that $VN(G)$ is an atomic von Neumann algebra (see [2]). For such G, $A(G)$ has the FPP for commutative semigroup by Theorem 4.1. Also, if G is the Fell group (which is the natural semi-direct product of the p-adic numbers with the compact group of p-adic units for a fixed prime p), then G is non-compact, totally disconnected and has countable dual; so it is an [AU]-group [28, Remark 4.6]. So $B(G)$ has the FPP for commutative semigroup by Theorem 4.2.

5 Notes and remarks

Theorem 3.1 and Theorem 3.2 (for amenable groups) are proved in [18, Theorem 5]. Also, it is shown in [20, Corollary 3.3], that if G is an [AR]-group (resp. an [AU]-group), then $A(G)$ (resp. $B(G)$) has the FPP. Note that Theorem 3.1 remains valid for any Banach space E which is the dual of a C^*-algebra $\mathcal{A}$ such that $\mathcal{A}^*$ is an ideal in the enveloping von Neumann algebra $\mathcal{A}^{**}$. This is the case when $E = \mathcal{T}(H) = \mathcal{C}(H)^*$, where $\mathcal{C}(H)$ is the C^*-algebra of compact operators on a Hilbert space or $E = \ell_1 = C_0^*$ (see [18, Lemma 4]).

Problem 1. Is there an example of a non-compact group G such that $B(G)$ has the FPP*, or even the weak*-normal structure?

Problem 2. Does Theorem 3.2 ((c)$\Longrightarrow$(a)) remain valid for non-second countable group?

A locally compact group G is called an *[IN]-group* if there is a compact neighborhood of the identity e in G which is invariant under inner automorphisms; G is called a *[SIN]-group* if there is a base for the neighbourhood system of e consisting of compact sets invariant under inner automorphisms (or equivalently, the left and right uniformities on G are the same). Obviously

$$[\text{compact}] \subseteq [\text{SIN}] \subseteq [\text{IN}]$$

and the inclusions are proper. Furthermore, all [IN]-groups are unimodular.

In [20], it is shown that if G is a connected [IN]-group, then $A(G)$ has the FPP if and only if G is compact. Also, if G is a [SIN]-group, and $A(G)$ has the FPP, then G must contain an open compact normal subgroup. Recently, Michael Leinert and the author are able to show that if G is an [IN]-group and $A(G)$ has the FPP, then G must be compact. Details of this and other results will appear elsewhere.

A Banach space E is said to have the *Radon-Nikodym property* (=RNP) if each closed convex subset D of E is denotable i.e. for any $\varepsilon > 0$ there exists an x in D such that $x \notin \overline{\text{co}}(D \backslash B_\varepsilon(x))$ (closed convex hull of $D \backslash B_\varepsilon(x)$), where $B_\varepsilon(x) = \{y \in X : \|x - y\| < \varepsilon\}$. The proof of Theorem 4.1 shows that if $A(G)$ has the RNP, then $A(G)$ has the FPP (see [8], [19] and [28]).

Problem 3. If G is a locally compact group, and $A(G)$ has the FPP, does $A(G)$ have the RNP? This is the case when G is either abelian or a connected [IN]-group (see [20, Theorem 3.5]).

Note that, as known, if a dual Banach space E has the property:

(**) The weak*-topology and the norm topology agree on the unit sphere of E.

Then E has the RNP.

Note that Property (*) in the proof of Theorem 3.1 implies property (**).

Theorem 5.1 *Let G be a locally compact group, then the followings are equivalent:*

(a) *G has property* (*).

(b) *G has property* (**).

(c) *G is compact.*

The implication (b)$\Longrightarrow$(c) follows from a result in [13] and the implication (b)$\Longrightarrow$(c) of Theorem 5.1 is proved in [3, Theorem 3.9].

The following diagrams summarize the relationship among various geometric properties for

(i) $A(G)$:

G is compact
$\Downarrow \not\Uparrow$
$A(G)$ has the RNP
$\Downarrow$
$A(G)$ has the FPP

(ii) $B(G)$:

$B(G)$ has property (*) $\Longleftrightarrow$ $B(G)$ has property (**) $\Longleftrightarrow$ G is compact

$B(G)$ has property (**) $\Downarrow \not\Uparrow$ $B(G)$ has the RNP $\Downarrow$ $B(G)$ has FPP

G is compact $\Downarrow$ $B(G)$ has UKK^* $\Downarrow$ $B(G)$ has weak*-normal structure $\Downarrow$ $B(G)$ has FPP*

$B(G)$ has FPP $\Nwarrow$ $B(G)$ has FPP*

In [26], Takahashi showed that if S is an amenable semigroup, and $\mathcal{S} = \{T_s : s \in S\}$ is a representation of S as nonexpansive mappings from a compact convex subset K of a Banach space into K, then K contains a common fixed point for $\mathcal{S}$. This result, which generalises a well-known theorem for commutative semigroup establishes the first link between amenability of a semigroup S and a non-linear fixed point property of S on compact convex sets (see [12, Appendix]).

Problem 4. If G is a compact group, does $A(G) = B(G)$ have the FPP for amenable semigroups?

References

[1] D. Alspach, A fixed point free nonexpansive map, *Proc. Amer. Math. Soc.* **82** (1981), 423–424.

[2] L. Baggett and K. Taylor, Groups with completely regular reducible representation, *Proc. Amer. Math. Soc.* **72** (1978), 596–600.

[3] M.B. Bekka, E. Kaniuth, A.T. Lau and G. Schlichting, Weak*-closedness of subspaces of Fourier-Stieltjes algebras and weak*-continuity of the restriction map, *Trans. of Amer. Math. Soc.* **350** (1998), 2277–2296.

[4] A. Belanger and B. Forrest, Geometric properties of coefficient function spaces determined by unitary representations of a locally compact group, *J. of Math. Anal. and Appl.* **193** (1995), 390–405.

[5] F.E. Browder, Nonexpansive nonlinear operators in Banach spaces, *Proc. Nat. Acad. Sci. U.S.A.* **54** (1965), 1041–1044.

[6] R.E. Bruck, A common fixed point theorem for a commuting family of nonexpansive mappings, *Pacific J. Math.* **53** (1974), 59–71.

[7] C. Chou, A.T. Lau and J. Rosenblatt, Approximation of compact operators by sums of translations, *Illinois J. Math.* **29** (1985), 340–350.

[8] C.H. Chu, A note on scattered C^*-algebras and the Radon-Nikodym property, *J. London Math. Society* **24** (1981), 533–536.

[9] M.M. Day, *Normed Linear Spaces.* Springer-Verlag, New York 1973.

[10] D. van Dulst and B. Sims, Fixed points of nonexpansive mappings and Chebyshev centers in Banach spaces with norms of type (KK), Banach space theory and its applications, *Bucharest, Lecture Notes in Math.* **991**, Springer-Verlag, New York (1981).

[11] P. Eymard, L^1 algèbre de Fourier d'un groupe localement compact, *Bull. Soc. Math. France* **92** (1964), 181–236.

[12] K. Goebel and W.A. Kirk, Topics in metric fixed point theory, *Cambridge Studies in Advanced Mathematics* **28**, Cambridge 1990.

[13] E. Graniner and M. Leinert, On some topologies which coincide on the unit sphere of the Fourier-Stieltjes algebra $B(G)$ and of the measure algebra $M(G)$, *Rocky Mountain J. Math.* **11** (1981), 459–472.

[14] E. Hewitt and K. Ross, *Abstract Harmonic Analysis I*, Springer Verlag, New York 1963.

[15] R. Huff, Banach spaces which are nearly uniformly convex, *Rocky Mountain J. Math.* **10** (1980), 743–749.

[16] M.A. Khamsi and W.A. Kirk, *An Introduction to Metric Spaces and Fixed Point Theory*, John Wiley & Sons Inc., New York 2001.

[17] W.A. Kirk, A fixed point theorem for mappings which do not increase distance, *American Math. Monthly* **72** (1965), 1004–1006.

[18] A.T. Lau and P.F. Mah, Normal structure in dual Banach spaces associated with a locally compact group, *Trans. Amer. Math. Soc.* **310** (1988), 341–353.

[19] A.T. Lau and A. Ülger, Some geometric properties on the Fourier and Fourier Stieltjes algebras of locally compact groups, Arens regularity and related problems, *Trans. Amer. Math. Soc.* **337** (1993), 321–359.

[20] A.T. Lau, P.F. Mah and A.Ülger, Fixed point property and normal structure for Banach spaces associated to locally compact groups, *Proc. Amer. Math. Soc.* **125** (1997), 2021–2027.

[21] C. Lennard, C_1 is uniformly Kadec-Klee, *Proc. Amer. Math. Soc.* **109** (1990), 71–77.

[22] T.C. Lim, Asymptotic centres and nonexpansive mappings in some conjugate spaces, *Pacific J. Math.* **90** (1980), 135–143.

[23] B. Murray, Points fixes des contractions sur convexe forme de L^1, *Seminaine d'Analyse Functionnelle 80-81 Ecole Polytechnique.*

[24] A.L.T. Paterson, *Amenability*, Amer. Math. Soc. Survey and Monograph **29** 1988.

[25] J.P. Pier, *Amenable Locally Compact Groups*, John Wiley & Son, New York 1984.

[26] W. Takahashi, A fixed point theorem for amenable semigroups of nonexpansive mappings, *Kodai Math. Sem. Report* **21** (1969), 383–386.

[27] M. Takesaki, *Theory of Operator Algebras 1*, Springer-Verlag, New York 1979.

[28] K. Taylor, Geometry of the Fourier algebras and locally compact groups with atomic representations, *Math. Ann.* **262** (1983), 183–190.

Effective and Averaged Energy Densities in One-Dimensional Wave Propagation through Spatio-Temporal Dielectric Laminates with Negative Effective Values of ϵ and μ

Konstantin A. Lurie[1] and Suzanne L. Weekes[2]

[1] Department of Mathematical Sciences, Worcester Polytechnic Institute, 100 Institute Road, Worcester MA 01609, USA email: klurie@wpi.edu

[2] Department of Mathematical Sciences, Worcester Polytechnic Institute, 100 Institute Road, Worcester MA 01609, USA email: sweekes@wpi.edu

Abstract. This paper introduces spatio-temporal dielectric laminates (dynamic materials) with negative effective permittivity ϵ and permeability μ. We discuss the connection between the averaged energy density of a plane electromagnetic wave propagating through such laminates and the effective energy density associated with the effective material parameters. In general, these energies are not the same, though there are situations in which they may become identical.

AMS Subject Classification. 78A48, 78M40, 78M30, 78M25

Key words. dynamic materials, composites, energy density, homogenization, effective properties

1 Introduction

Spatio-temporal material composites or dynamic materials are composites in which the constituent properties such as permeability, permittivity, density, etc. are distributed on a microscale in time, as well as in space. Optimal material design for static or non-smart applications generally results in the formation of ordinary composites - mixtures where the design variables are position dependent but invariant in time. When it comes to dynamic applications, temporal variability in the material properties is needed in order to adequately match the changing environment. Therefore, we have made it our goal to study such structures in order to understand their behaviour and potential applications.

In [1, 4–7, 11–13], spatio-temporal laminates are studied in both mechanical and electromagnetic contexts. The focus of this paper is on one-dimensional wave propagation through electromagnetic materials, specifically, isotropic dielectrics, and most of the discussion is related to composites assembled in space-time from two original dielectric constituents each characterized by its permittivity ϵ and permeability μ. We use the notion of effective material parameters to describe the performance of the low frequency modulated waves propagating through such composites. These parameters are generated by the variational formulation of the problem of wave motion through material media in general, and through composites in particular.

Physically, we are interested in the energy of modulated waves and we want to know how energy is linked to the effective parameters. The energy is best measured as a result of an averaging procedure applied directly to the energy density within the original materials that constitute a composite. The relation between this averaged energy and the effective parameters is far from obvious and is studied in this paper. After the introduction in Section 1, we give a spatio-temporal dielectric laminar construction that yields negative effective permittivity and permeability. This is done in Section 2. In Section 3, we develop a technique of averaging the energy densities in one-dimensional electromagnetic wave propagation through such structures. We find that this energy is not the same as the energy that is formally associated in a standard way with the effective properties introduced before. However, in a special coordinate frame, both energy densities are indeed the same. Numerical illustration and confirmation of results are given in Section 4.

Dynamic materials may appear in very diverse physical implementations, but we may distinguish two principal ways of making them by the spatio-temporal mixing of ordinary materials via the processes of activation and kinetization [1, 5, 6]. *Activated* dynamic materials are obtained by instantaneous or gradual change of the material parameters in various parts of the system in the absence of relative motion of those parts. *Kinetic* dynamic materials are obtained when various parts of the system are exposed to relative motion that is prearranged and generated in a certain way. The difference between these types is best illustrated by an example.

Consider a transmission line. Its discrete version may be interpreted as an array of LC-cells connected in parallel (Fig. 1). Assume that the properties in each cell can be switched between (L_1, C_1) and (L_2, C_2). If the cells are densely distributed along the line, then, by due switching, the linear inductance L and capacitance C of the line may become almost arbitrary functions of the spatial coordinate z and time t. In particular, we may produce a periodic *LC-laminate* assembled from alternating (L_1, C_1) and (L_2, C_2) segments (Fig. 2). This figure shows the case when the switching occurs so that the pattern of such segments moves along the z-axis at velocity V creating the laminar structure in space-time. We refer to an (L_i, C_i) segment as material i. This is a pure case of activation since the construction does not include any motion of the physical material itself; what moves is the property pattern. *Activated* materials appear as a result of the homogenization procedure applied to this type of construction.

Now consider a dielectric rod assembled from alternating sections of isotropic dielectrics with material constants (ϵ_1, μ_1) and (ϵ_2, μ_2). We refer to these dielectrics as materials 1 and 2 respectively. Within each section, the material may be brought into its individual material motion along the z-axis at velocities v_1 (material 1) and v_2 (material 2). A discontinuous velocity pattern may be implemented either through the use of a special "caterpillar construction" introduced in [7] and shown in Figure 3, or approximately by

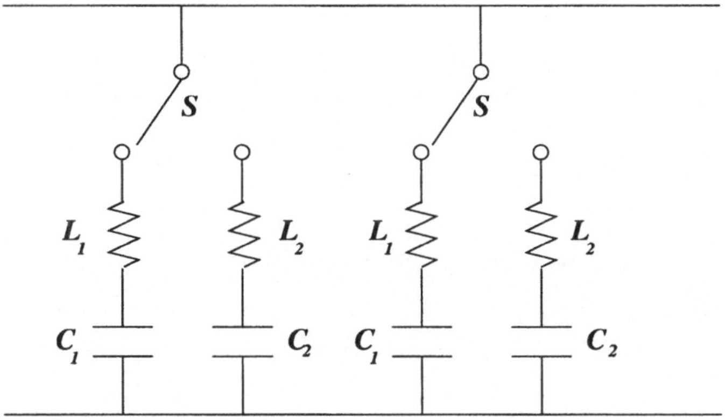

Figure 1: A discrete version of a transmission line.

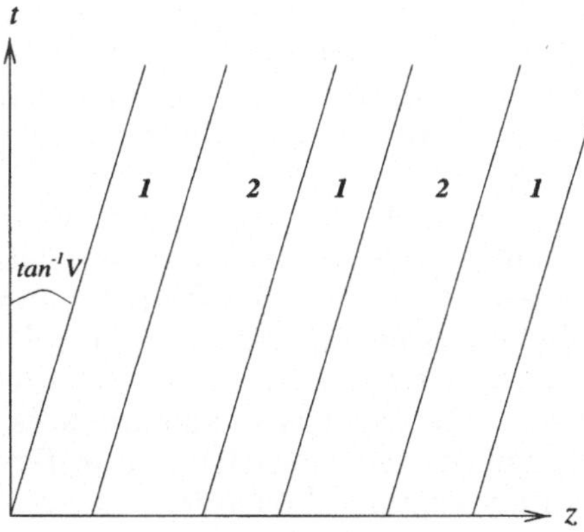

Figure 2: Material laminate in space-time.

a fast periodic longitudinal vibration of a dielectric continuum in the form of a standing wave. The property pattern, i.e. the position of the segments, is stationary in a laboratory frame; what moves is the dielectric material itself within the segments. This is a case of pure kinetization; a *kinetic* material appears after we apply homogenization to this type of construction.

In both activated and kinetic scenarios, homogenization introduces an averaged characterization of the composite material in terms of its effective constants. This characterization is valid for disturbances whose wavelengths are long compared to the period of the material pattern.

The difference between activated and kinetic materials can be formalized in terms of the tensor s of their material properties [5]. If an isotropic dielectric with properties (ϵ, μ) is stationary with respect to a Minkowskian laboratory frame $x_1 = x, \quad x_2 = y, \quad x_3 = z, \quad x_4 = ict$, then its material

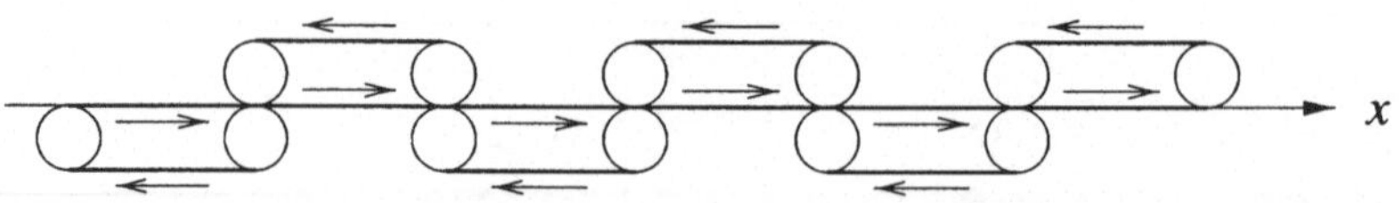

Figure 3: Construction of a kinetic dynamic material

property tensor s is specified by the formula

$$s = -\frac{1}{\mu c}(a_{12}a_{12} + a_{13}a_{13} + a_{23}a_{23}) - \epsilon c(a_{14}a_{14} + a_{24}a_{24} + a_{34}a_{34}), \tag{1.1}$$

where $a_{mn} = (1/2)(e_m e_n - e_n e_m)$ for $m, n = 1, \ldots, 4$ denote the eigentensors defined as skew-symmetric combinations of the unit vectors e_m of the Minkowskian x_m-axes. Here c is the speed of light in a vacuum. If the material is moving with respect to a laboratory frame, then its material tensor is given by equation (1.1) but with a_{mn} replaced by a'_{mn} formed by the vectors e'_m which are linked with e_m by a Lorentz transform. Motion of the material generates rotation of the eigenaxes of a'_{mn}, specifically, of the time axis x'_4, relative to the x_m-axes. As a result, this rotation produces the well-known Minkowski's material relations for a moving dielectric medium.

The tensor language formalizes the said difference between activated and kinetic dynamic composites. For activated composites, the original constituents *differ in the eigenvalues* $1/\mu c, \epsilon c$ *of their material tensors*, but their eigentensors a_{mn} remain the same since there is no relative material motion. If the entire structure experiences a common background motion with respect to a laboratory frame, the identity of eigentensors will continue to hold. For kinetic composites, the original constituents *differ in their eigentensors* while their eigenvalues may remain the same.

The analysis and discussion in this paper is confined to activated dielectric laminates.

2 An activated rank-one laminate in space-time

Consider a laminate (Fig. 2) assembled from two isotropic dielectrics characterized by different pairs of values of $\epsilon = \epsilon(z,t)$ and $\mu = \mu(z,t)$:

$$(\epsilon(z,t), \mu(z,t)) = \begin{cases} (\epsilon_1, \mu_1) & \text{material 1,} \\ (\epsilon_2, \mu_2) & \text{material 2.} \end{cases} \tag{2.1}$$

This formation is periodic in $\xi = z - Vt$ with period δ; we may perceive it as a periodic array of segments of the z-axis carrying materials 1 and 2 and occupying, respectively, portions m_1 and m_2 of the period. The array (property pattern) is assumed moving along the z-axis at the velocity V which specifies the slope of the layers' interfaces in Figure 2, whereas the dielectric materials that fill the segments remain at rest with respect to the laboratory frame.

The slope $V = dz/dt$ satisfies the inequality

$$(V^2 - a_1^2)(V^2 - a_2^2) \geq 0, \tag{2.2}$$

where $a_i = 1/\sqrt{\epsilon_i \mu_i}$ is the phase velocity of light in material i. This inequality is necessary to guarantee smoothness of the relevant solution (the non-appearance of shocks).

The material formation (2.1) is assumed created and maintained by an external activating agent; as an example, we may refer to [3] where a wavelike pattern of linear capacitance is materialized in a periodic array of $p - n$ junction diodes. We are not interested in this paper in the detailed analysis of the activator's work; for our purposes it is sufficient to consider it as a device that maintains a necessary exchange of energy and momentum between the material formation and the environment.

For one-dimensional wave propagation, the Maxwell's system

$$\text{curl}\,\mathbf{E} = -\mathbf{B}_t, \quad \text{div}\,\mathbf{B} = 0, \quad \text{curl}\,\mathbf{H} = \mathbf{D}_t, \quad \text{div}\,\mathbf{D} = 0,$$

is satisfied by the vectors

$$\mathbf{E} = u_t\mathbf{j}, \quad \mathbf{B} = u_z\mathbf{i}, \quad \mathbf{H} = v_t\mathbf{i}, \quad \mathbf{D} = v_z\mathbf{j}, \tag{2.3}$$

representing a plane electromagnetic wave traveling in the z-direction. The material relations $\mathbf{D} = \epsilon\mathbf{E}$, $\mathbf{B} = \mu\mathbf{H}$ generate the system

$$\epsilon u_t = v_z, \quad \frac{1}{\mu}u_z = v_t. \tag{2.4}$$

After homogenization, it is replaced by the system [4]

$$\alpha c u_z + \beta u_t = V v_z + v_t, \tag{2.5}$$

$$V u_z + u_t = \theta(\alpha c v_z + \beta v_t), \tag{2.6}$$

with parameters α, β, θ defined as

$$\alpha = \frac{1}{c}\frac{\left\langle \frac{1}{\varepsilon\mu(V^2-a^2)} \right\rangle}{\left\langle \frac{1}{\varepsilon(V^2-a^2)} \right\rangle}, \quad \beta = V\frac{\left\langle \frac{1}{V^2-a^2} \right\rangle}{\left\langle \frac{1}{\varepsilon(V^2-a^2)} \right\rangle}, \quad \theta = \frac{\left\langle \frac{1}{\varepsilon(V^2-a^2)} \right\rangle}{\left\langle \frac{1}{\mu(V^2-a^2)} \right\rangle}, \tag{2.7}$$

where

$$\langle \cdot \rangle = m_1(\cdot)_1 + m_2(\cdot)_2,$$

and c is the speed of light in a vacuum.

In equations (2.5), (2.6) and below, we preserve the original symbols u, v to denote the weak limits of the relevant quantities, i.e. their values averaged over the period δ of lamination; this period is originally assumed small compared to a typical wavelength of u and v measured in a laboratory frame. The system (2.5), (2.6) represents asymptotic equations valid in the limit $\delta \to 0$. It is not surprising that parameters α, β, θ appear to be dependent on V, the velocity of the property pattern relative to the material. An equivalent form of these equations is given by [5]

$$pu_z - qu_t = v_t, \quad qu_z + ru_t = v_z, \tag{2.8}$$

with parameters p, q, r defined as

$$p = \frac{V^2 - \theta\alpha^2 c^2}{\theta(\beta V - \alpha c)}, \quad q = -\frac{V - \theta\alpha c\beta}{\theta(\beta V - \alpha c)}, \quad r = -\frac{1 - \theta\beta^2}{\theta(\beta V - \alpha c)}. \tag{2.9}$$

Introduce the "primed" coordinate frame z', t' moving with a velocity w with respect to the laboratory frame z, t. Coordinates z', t' are linked with z, t by the Lorentz formulae

$$z' = \gamma^{-1}(z - wt), \quad t' = \gamma^{-1}\left(t - \frac{w}{c^2}z\right), \quad \gamma = \sqrt{1 - w^2/c^2}. \tag{2.10}$$

If w is defined as a root of

$$\frac{q}{c^2}w^2 + \left(\frac{p}{c^2} - r\right)w + q = 0, \tag{2.11}$$

then the system (2.8) becomes diagonalized, i.e. reduced to equations

$$(p + qw)u_{z'} = v_{t'}, \quad \left(\frac{p}{c^2} + \frac{q}{w}\right)u_{t'} = v_{z'}, \tag{2.12}$$

specifying the effective permittivity $\mathcal{E}$ and permeability M (c.f. (2.4)) via the formulae for the eigenvalues of the material tensor of a composite:

$$\mathcal{E}c = \frac{p}{c} + \frac{qc}{w}, \quad \frac{1}{Mc} = \frac{p}{c} + \frac{qw}{c}. \tag{2.13}$$

Applying direct inspection and referring to (2.11), we obtain the following expression for the second invariant of a material tensor:

$$\frac{\mathcal{E}}{M} = \left(\frac{p}{c} + \frac{qc}{w}\right)\left(\frac{p}{c} + \frac{qw}{c}\right) = pr + q^2,$$

so by equations (2.9),

$$\frac{\mathcal{E}}{M} = pr + q^2 = 1/\theta. \tag{2.14}$$

We will also need the formula for the first invariant $\mathcal{E}c + 1/Mc$ of the effective tensors of material parameters. This formula follows from (2.11) and (2.13):

$$\mathcal{E}c + 1/Mc = \frac{2p}{c} + \frac{qc}{w} + \frac{qw}{c} = \frac{p}{c} + rc. \tag{2.15}$$

Given (2.9), we rewrite (2.15) in the form

$$\mathcal{E}c + 1/Mc = \frac{1}{\beta(V/c) - \alpha}\left[\left(\frac{V^2}{c^2} - 1\right)\frac{1}{\theta} - (\alpha^2 - \beta^2)\right]. \tag{2.16}$$

We now summarize restrictions that should be observed when operating with these formulae. We first assume that parameters $\epsilon_1, \ldots, \mu_2$ are all positive. Without a loss of generality, set $a_2^2 > a_1^2$, that is $\epsilon_1\mu_1 > \epsilon_2\mu_2$. This ordering holds true if we are either in the *regular* case where $\epsilon_1 > \epsilon_2, \mu_1 > \mu_2$ or in the *irregular* case where $\epsilon_1 > \epsilon_2, \mu_1 < \mu_2$, (or $\epsilon_1 < \epsilon_2, \mu_1 > \mu_2$). These possibilities will affect the admissible values of some characteristic parameters listed below.

Define the symbol $\bar{\epsilon}$ as

$$\bar{\epsilon} = m_1\epsilon_2 + m_2\epsilon_1,$$

and similarly introduce the symbols $\bar{\mu}$, $\overline{(1/\epsilon)}$, $\overline{(1/\mu)}$. It is easily checked that, for positive ϵ, μ,

$$\begin{aligned} &\bar{\epsilon}\overline{(1/\epsilon)} \geq 1, \\ &\bar{\mu}\overline{(1/\mu)} \geq 1, \\ &a_1^2 \leq (1/\bar{\epsilon})\overline{(1/\mu)} \leq a_2^2, \\ &a_1^2 \leq (1/\bar{\mu})\overline{(1/\epsilon)} \leq a_2^2. \end{aligned} \tag{2.17}$$

Also,

$$\begin{aligned} &(1/\bar{\epsilon})(1/\bar{\mu}) \leq a_2^2, \\ &\overline{(1/\epsilon)}\,\overline{(1/\mu)} \geq a_1^2, \end{aligned} \tag{2.18}$$

for both the regular and irregular cases; and, for the regular case,

$$\begin{aligned} &(1/\bar{\epsilon})(1/\bar{\mu}) \geq a_1^2, \\ &\overline{(1/\epsilon)}\,\overline{(1/\mu)} \leq a_2^2. \end{aligned} \tag{2.19}$$

In the irregular case, there exists [4] a range of m_1 for which

$$(1/\bar{\epsilon})(1/\bar{\mu}) \leq a_1^2, \tag{2.20}$$

and a range of m_1 for which

$$\overline{(1/\epsilon)}\,\overline{(1/\mu)} \geq a_2^2. \tag{2.21}$$

For example, if $\epsilon_1 = \mu_1 = 1,\ \ \epsilon_2 = 9,\ \ \mu_2 = 0.1$, then we have the irregular case, and (2.20) holds if $m_1 \leq 71/72$, and (2.21) holds if $m_1 \geq 1/72$. Both inequalities are satisfied when $1/72 \leq m_1 \leq 71/72$.

Aside from the restrictions in (2.2),

$$V^2 \le a_1^2 \text{ or } V^2 \ge a_2^2, \tag{2.22}$$

specifying the slow and fast ranges for V^2, we note a universal inequality $V^2 \le c^2$. The roots, w, of equation (2.11) should be real. When w is real, (2.12) defines an isotropic effective material. Since the product of the roots equals c^2, one of those roots has absolute value less than or equal to c; this particular root participates in the Lorentz transform (2.10). We thus demand that the discriminant of (2.11) be non-negative:

$$\left(\frac{p}{c^2} - r\right)^2 - 4\frac{q^2}{c^2} \ge 0,$$

or, equivalently,

$$\left(\frac{p}{c} + rc\right)^2 - 4(pr + q^2) \ge 0.$$

Given (2.14) and (2.15), this is

$$\left(\mathcal{E}c + \frac{1}{Mc}\right)^2 - 4\frac{\mathcal{E}}{M} \ge 0. \tag{2.23}$$

When the roots of (2.11) are real, so are $\mathcal{E}$ and M.

We now consider the explicit expressions for the first and the second invariants $I_1 = \mathcal{E}c + 1/Mc$, $I_2 = \mathcal{E}/M$ of the effective material tensor s [5]. Using (2.7), (2.13), and (2.14), we find after some calculation,

$$I_1 = \mathcal{E}c + \frac{1}{Mc} = \frac{\bar{\epsilon}\bar{\mu} + \frac{c^2}{a_1^2 a_2^2}}{\bar{\epsilon}\mu_1\mu_2 c\left[V^2 - \frac{1}{\bar{\epsilon}}\left(\frac{\bar{1}}{\mu}\right)\right]}(V^2 - k), \tag{2.24}$$

$$I_2 = \frac{\mathcal{E}}{M} = \frac{\langle\frac{1}{\mu}\rangle}{\langle\frac{1}{\epsilon}\rangle}\frac{V^2 - \frac{1}{\bar{\mu}}\left(\frac{\bar{1}}{\epsilon}\right)}{V^2 - \frac{1}{\bar{\epsilon}}\left(\frac{\bar{1}}{\mu}\right)}. \tag{2.25}$$

Then inequality (2.23) takes the form

$$\left(\bar{\epsilon}\bar{\mu} + \frac{c^2}{a_1^2 a_2^2}\right)^2 (V^2 - k)^2 \ge 4\frac{c^2}{a_1^2 a_2^2}\bar{\epsilon}\bar{\mu}\left[V^2 - \frac{1}{\bar{\mu}}\left(\frac{\bar{1}}{\epsilon}\right)\right]\left[V^2 - \frac{1}{\bar{\epsilon}}\left(\frac{\bar{1}}{\mu}\right)\right], \tag{2.26}$$

where

$$k = \frac{c^2\left(\frac{\bar{1}}{\epsilon}\right)\left(\frac{\bar{1}}{\mu}\right) + a_1^2 a_2^2}{c^2 + a_1^2 a_2^2 \bar{\epsilon}\bar{\mu}}. \tag{2.27}$$

Define the parameter σ as

$$\sigma = \frac{c^2}{a_1^2 a_2^2 \bar{\epsilon}\bar{\mu}}, \tag{2.28}$$

so (2.27) is rewritten as

$$k = \frac{\sigma\bar{\epsilon}\bar{\mu}\left(\overline{\frac{1}{\epsilon}}\right)\left(\overline{\frac{1}{\mu}}\right)+1}{\sigma+1}\,\frac{1}{\bar{\epsilon}\bar{\mu}}. \tag{2.29}$$

Inequality (2.26) is now rewritten as

$$(1+\sigma)^2(V^2-k)^2 \geq 4\sigma\left[V^2 - \frac{1}{\bar{\mu}}\left(\overline{\frac{1}{\epsilon}}\right)\right]\left[V^2 - \frac{1}{\bar{\epsilon}}\left(\overline{\frac{1}{\mu}}\right)\right]. \tag{2.30}$$

It is easy to see that $k < c^2$. Assuming the contrary, gives the inequality $f(c^2) < 0$ where $f(\lambda)$ is defined as

$$f(\lambda) \equiv \lambda^2 + \lambda\left[a_1^2 a_2^2 \bar{\epsilon}\bar{\mu} - \left(\overline{\frac{1}{\epsilon}}\right)\left(\overline{\frac{1}{\mu}}\right)\right] - a_1^2 a_2^2.$$

Clearly, the equation $f(\lambda) = 0$ has real roots of opposite signs, the product of these roots being $-a_1^2 a_2^2$. We check that $f(a_1^2) = 2m_1 a_1^2(a_1^2 - a_2^2) < 0$, and $f(a_2^2) = 2m_2 a_2^2(a_2^2 - a_1^2) > 0$; this means that a_2^2 exceeds the positive root of $f(\lambda) = 0$, and $f(\lambda) > 0$ for $\lambda \geq a_2^2$. Since $c^2 > a_2^2$, we conclude that $f(c^2) > 0$, and hence our assumption is false.

In this paper, we consider the existence and construction of spatio-temporal dielectric laminates with negative effective permittivity $\mathcal{E}$ and permeability M. First assume that both $\epsilon, \mu > 0$; the product (2.25) is then non-negative given (2.22) and (2.17). The sum (2.24) may be negative if either (i) $V^2 < (1/\bar{\epsilon})(\overline{1/\mu})$ and $V^2 > k$, or (ii) $V^2 > (1/\bar{\epsilon})(\overline{1/\mu})$ and $V^2 < k$. The second possibility can be made consistent with (2.30), as shown by the following argument. Referring to (2.17) and to (2.22), we conclude that V^2 should be taken greater than a_2^2. This may come to agreement with $V^2 < k$ since k may exceed a_2^2 if the value of σ is sufficiently large. In fact, if $\sigma \to \infty$, then k monotonically increases approaching the value $(\overline{1/\epsilon})(\overline{1/\mu})$ *which may exceed a_2^2 for the irregular case* (see (2.21)). Considering this case and choosing V such that $a_2^2 < V^2 < k$, we observe that, for sufficiently large values of σ, inequality (2.30) holds.

This conclusion was based on the assumption that ϵ, μ are positive for the original materials. When these constants are negative, then the same argument shows that the effective parameters $\mathcal{E}, M$ may become positive. This follows directly from the definition of k.

3 The energy considerations

While the effective parameters of an activated spatio-temporal composite may become negative, an averaged value of the electromagnetic energy density measured in a laboratory frame remains positive. For the case of one-dimensional wave propagation through a rank one laminate as considered in Section 2, the electric and magnetic energy densities averaged over the laminate period δ are respectively defined by the formulae

$$\langle w_e\rangle = \frac{1}{2}\langle \mathbf{ED}\rangle = \frac{1}{2}\langle \epsilon u_t^2\rangle, \tag{3.1}$$

$$\langle w_m\rangle = \frac{1}{2}\langle \mathbf{BH}\rangle = \frac{1}{2}\left\langle \frac{1}{\mu}u_z^2\right\rangle. \tag{3.2}$$

Because of the continuity of u and v across the layers' interface, the derivatives

$$\begin{aligned} u_\tau &= u_t + Vu_z,\\ v_\tau &= v_t + Vv_z = \epsilon V u_t + \tfrac{1}{\mu}u_z \end{aligned} \tag{3.3}$$

are also continuous. We use (3.3) to express u_t, u_z as functions of ϵ, μ, V, and the continuous derivatives, u_τ, v_τ:

$$\begin{aligned} u_t &= -\tfrac{a^2}{V^2-a^2}\,u_\tau + \tfrac{V}{\epsilon(V^2-a^2)}\,v_\tau,\\ u_z &= \tfrac{V}{V^2-a^2}\,u_\tau - \tfrac{1}{\epsilon(V^2-a^2)}\,v_\tau. \end{aligned} \tag{3.4}$$

The value of $\langle w_e\rangle$ is thus calculated as

$$\begin{aligned} \langle w_e\rangle = \tfrac{1}{2}\left\langle \epsilon\left(\tfrac{a^2}{V^2-a^2}\right)^2\right\rangle u_\tau^2 - \left\langle \tfrac{a^2}{(V^2-a^2)^2}\right\rangle V u_\tau v_\tau\\ + \tfrac{1}{2}\left\langle \tfrac{1}{\epsilon(V^2-a^2)^2}\right\rangle V^2 v_\tau^2. \end{aligned} \tag{3.5}$$

In this formula, the derivatives u_τ, v_τ remain unaffected by averaging, and are identical with their averaged values. The latter are linked with the averaged values $\langle u_t\rangle, \langle u_z\rangle, \langle v_t\rangle, \langle v_z\rangle$ through the formulae (see (2.8) and (2.9))

(3.6)

$$\begin{aligned} u_\tau &= \langle u_t\rangle + V\langle u_z\rangle,\\ v_\tau &= \langle v_t\rangle + V\langle v_z\rangle = (p+qV)\langle u_z\rangle - (q-rV)\langle u_t\rangle = \alpha c\langle u_z\rangle + \beta\langle u_t\rangle. \end{aligned}$$

Note that these formulae relate the *averaged* values of $u_z, \ldots, v_t$, and are therefore different from those in (3.3) which relate pointwise values and hold along the layers' interfaces. As in Section 2, we preserve below the symbols $u_z, \ldots, v_t$ introduced for the pointwise values to also represent the averaged values; the appropriate meaning will follow from the context.

By eliminating u_τ, v_τ from (3.5) with the aid of (3.6), we arrive at the following expression for $\langle w_e \rangle$:

(3.7)
$$\langle w_e \rangle = \frac{1}{2}\left\langle \epsilon \frac{\left(\frac{V}{\epsilon}\beta - a^2\right)^2}{(V^2-a^2)^2} \right\rangle u_t^2 + \left\langle \frac{\epsilon}{(V^2-a^2)^2}\left(\frac{V}{\epsilon}\beta - a^2\right)\left(\frac{c\alpha}{\epsilon} - a^2\right) \right\rangle V u_t u_z$$
$$+\frac{1}{2}\left\langle \epsilon \frac{\left(\frac{c\alpha}{\epsilon} - a^2\right)^2}{(V^2-a^2)^2} \right\rangle V^2 u_z^2.$$

By a similar argument, we calculate $\langle w_m \rangle$ as

$$\langle w_m \rangle = \frac{1}{2}\left\langle \frac{1}{\mu} \frac{\left(V - \frac{\beta}{\epsilon}\right)^2}{(V^2-a^2)^2} \right\rangle u_t^2 + \left\langle \frac{1}{\mu} \frac{\left(V - \frac{\beta}{\epsilon}\right)\left(V^2 - \frac{c\alpha}{\epsilon}\right)}{(V^2-a^2)^2} \right\rangle u_t u_z$$

(3.8)

$$+\frac{1}{2}\left\langle \frac{1}{\mu} \frac{\left(V^2 - \frac{c\alpha}{\epsilon}\right)^2}{(V^2-a^2)^2} \right\rangle u_z^2.$$

The sum $\langle w_e \rangle + \langle w_m \rangle = \langle w_e + w_m \rangle$ represents an averaged value $\langle W_{tt} \rangle$ of the electromagnetic energy density $W_{tt} = w_e + w_m$ measured in a laboratory frame. The averaged value $\langle W_{zt} \rangle$ of the momentum density $W_{zt} = \epsilon u_t u_z$, originally dependent on pointwise values of u_t, u_z, is given by the formula

$$\langle W_{zt} \rangle = \langle \epsilon u_t u_z \rangle = \left\langle \epsilon \frac{\left(V - \frac{\beta}{\epsilon}\right)\left(V\frac{\beta}{\epsilon} - a^2\right)}{(V^2-a^2)^2} \right\rangle u_t^2$$

(3.9) $$+\left\langle \frac{\epsilon}{(V^2-a^2)^2}\left[\left(V\frac{\beta}{\epsilon} - a^2\right)\left(V^2 - \frac{c\alpha}{\epsilon}\right) + V\left(V - \frac{\beta}{\epsilon}\right)\left(\frac{c\alpha}{\epsilon} - a^2\right)\right] \right\rangle u_t u_z$$

$$+\left\langle \epsilon V \frac{\left(\frac{c\alpha}{\epsilon} - a^2\right)\left(V^2 - \frac{c\alpha}{\epsilon}\right)}{(V^2-a^2)^2} \right\rangle u_z^2.$$

Like equations (3.7) and (3.8), this formula contains the averaged values of u_t, u_z. The energy flux density $W_{tz} = -(1/\mu) u_t u_z$ has an averaged value

(3.10)
$$\langle W_{tz} \rangle = -\left\langle \frac{1}{\mu} u_t u_z \right\rangle = -\left\langle \frac{1}{\mu} \frac{\left(V - \frac{\beta}{\epsilon}\right)\left(V\frac{\beta}{\epsilon} - a^2\right)}{(V^2-a^2)^2} \right\rangle u_t^2$$

$$-\left\langle \frac{1}{\mu(V^2-a^2)^2}\left[\left(V\frac{\beta}{\epsilon} - a^2\right)\left(V^2 - \frac{cd}{\epsilon}\right) + V\left(V - \frac{\beta}{\epsilon}\right)\left(\frac{cd}{\epsilon} - a^2\right)\right] \right\rangle u_t u_z$$

$$-\left\langle \frac{V}{\mu} \frac{\left(\frac{cd}{\epsilon} - a^2\right)\left(V^2 - \frac{cd}{\epsilon}\right)}{(V^2-a^2)^2} \right\rangle u_z^2,$$

and the momentum flux density W_{zz} equals $-W_{tt}$.

The action density is denoted by L:

$$L = w_e - w_m = \frac{1}{2}(u_t v_z - u_z v_t). \tag{3.11}$$

The action density is quasi-affine, i.e. its averaged value $\langle L\rangle$ is equal to the action density of the averaged field $\bar{L}$ (the *effective action density*). This result follows from the chain of equalities

$$\begin{aligned}\langle L\rangle &= \tfrac{1}{2}\langle u_t\, v_z - u_z\, v_t\rangle = \tfrac{1}{2}\langle u_{t'}\, v_{z'} - u_{z'}\, v_{t'}\rangle \\ &= \tfrac{1}{2}[u_{t'}\,\langle v_{z'}\rangle - \langle u_{z'}\rangle\, v_{t'}] = \tfrac{1}{2}[\langle u_{t'}\rangle\,\langle v_{z'}\rangle - \langle u_{z'}\rangle\,\langle v_{t'}\rangle] = \bar{L}.\end{aligned} \tag{3.12}$$

We used here the Lorentz-invariance of action, with the primed frame z', t' moving along with the property interfaces and preserving continuity of tangential fields $\mathbf{E}' = \mathbf{E} + (\mathbf{V}\times\mathbf{B}) = (u_t + V\, u_z)\,\mathbf{j} = u_{t'}\,\mathbf{j}$, and $\mathbf{H}' = \mathbf{H} - (\mathbf{V}\times\mathbf{D}) = (v_t + V\, v_z)\,\mathbf{i} = v_{t'}\,\mathbf{i}$. We confirm this result also by a direct inspection based on (3.7) and (3.8) and using (2.9); the calculation shows that

$$\langle w_e - w_m\rangle = \langle L\rangle = \bar{L} = \frac{1}{2}r u_t^2 + q u_t u_z - \frac{1}{2}p u_z^2. \tag{3.13}$$

The effective action density serves as the integrand (Lagrangian) for the functional

$$\int\!\!\int \bar{L}\; dzdt$$

generating (2.8) as Euler equations.

The effective Lagrangian $\bar{L}$ generates an effective energy-momentum tensor $\bar{W}$; its components are specified by the standard formulae

$$\begin{aligned}\bar{W}_{tt} &= u_t\frac{\partial\bar{L}}{\partial u_t} - \bar{L} = \tfrac{1}{2}(r u_t^2 + p u_z^2),\\ \bar{W}_{tz} &= u_t\frac{\partial\bar{L}}{\partial u_z} = q u_t^2 - p u_t u_z,\\ \bar{W}_{zt} &= u_z\frac{\partial\bar{L}}{\partial u_t} = r u_z u_t + q u_z^2\\ \bar{W}_{zz} &= u_z\frac{\partial\bar{L}}{\partial u_t} - \bar{L} - \tfrac{1}{2}(r u_t^2 + p u_z^2).\end{aligned} \tag{3.14}$$

The component $\bar{W}_{tt}$ may be termed, in a standard way, the effective energy density, the component $\bar{W}_{zt}$, the effective momentum density, and so on. They satisfy the system

$$\begin{aligned}\frac{\partial\bar{W}_{tt}}{\partial t} + \frac{\partial\bar{W}_{tz}}{\partial z} &= 0,\\ \frac{\partial\bar{W}_{zt}}{\partial t} + \frac{\partial\bar{W}_{zz}}{\partial z} &= 0,\end{aligned} \tag{3.15}$$

following from (2.8).

For $V \neq 0$, the quasi-affinity property (3.13) does not extend in an analogous fashion to the components of the energy-momentum tensor, i.e. the energy density, the energy density flux, etc. For example, the averaged energy density $\langle W_{tt} \rangle$ in a laboratory frame is *not equal* to $\bar{W}_{tt}$.

However, if we consider a special frame (2.10) moving at velocity $w = V$ relative to a laboratory, then, in such a frame,

$$\langle W_{t't'} \rangle = \bar{W}_{t't'}, \qquad \langle W_{t'z'} \rangle = \bar{W}_{t'z'}. \tag{3.16}$$

To prove this, we use the formulae for components of the energy-momentum tensor of original constituents:

$$\begin{aligned} W_{tt} &= \tfrac{1}{2}(\epsilon u_t^2 + \tfrac{1}{\mu} u_z^2), \\ W_{tz} &= -\tfrac{1}{\mu} u_t u_z, \\ W_{zt} &= \epsilon u_t u_z, \\ W_{zz} &= -\tfrac{1}{2}(\epsilon u_t^2 + \tfrac{1}{\mu} u_z^2). \end{aligned} \tag{3.17}$$

These components satisfy the system

$$\begin{aligned} \tfrac{\partial}{\partial t} W_{tt} + \tfrac{\partial}{\partial z} W_{tz} &= -\tfrac{1}{2}[\epsilon_t u_t^2 - \left(\tfrac{1}{\mu}\right)_t u_z^2], \\ \tfrac{\partial}{\partial t} W_{zt} + \tfrac{\partial}{\partial z} W_{zz} &= -\tfrac{1}{2}[\epsilon_z u_t^2 - \left(\tfrac{1}{\mu}\right)_z u_z^2]. \end{aligned} \tag{3.18}$$

To simplify calculations, consider a Galilean frame (a special case of (2.10) when $c \to \infty$)

$$z' = z - Vt, \quad t' = t. \tag{3.19}$$

In this frame, the interfaces separating different materials in activated laminates remain immovable. The energy density and the energy density flux are given in the same frame by the formulae

$$\begin{aligned} W_{t't'} &= W_{tt} + V W_{zt}, \\ W_{t'z'} &= -V(W_{tt} + V W_{zt}) + W_{tz} + V W_{zz}. \end{aligned} \tag{3.20}$$

By direct inspection, and with the reference to equations (3.7)–(3.10), (3.13) and (3.17), we observe that

$$\begin{aligned} \langle W_{tt} + V W_{zt} \rangle &= \bar{W}_{tt} + V \bar{W}_{zt}, \\ \langle W_{tz} + V W_{zz} \rangle &= \bar{W}_{tz} + V \bar{W}_{zz}, \end{aligned} \tag{3.21}$$

and, consequently, equations (3.16) hold due to (3.20). This result remains valid also with regard to a Lorentz frame (2.10).

The first equation of (3.18) indicates that the net rate of increase of the energy of the electromagnetic field in a unit segment is equal to the work produced per unit time by an external agent maintaining the variable property pattern. This work is given by the right hand side of the first equation in (3.18). In the case of an activated laminate, both ϵ and μ satisfy the equation $(\cdot)_t + V(\cdot)_z = 0$. From (3.18), we conclude that the term

$$-\frac{1}{2}[\epsilon_t u_t^2 - \left(\frac{1}{\mu}\right)_t u_z^2] \tag{3.22}$$

is equal to the expression

$$-V\left(\frac{\partial}{\partial t}W_{zt} + \frac{\partial}{\partial z}W_{zz}\right).$$

At first glance, the presence of the work term (3.22) might be interpreted as equivalent to the change in the energy density W_{tt} of the field in the laboratory frame by the value VW_{zt}, and to a simultaneous change of the energy flux W_{tz} by the value VW_{zz}:

$$\frac{\partial}{\partial t}(W_{tt} + V\,W_{zt}) + \frac{\partial}{\partial z}(W_{tz} + V\,W_{zz}) = 0.$$

By applying averaging to this equation and by referring to (3.21), we arrive at the equation

$$\frac{\partial}{\partial t}(\bar{W}_{tt} + V\,\bar{W}_{zt}) + \frac{\partial}{\partial z}(\bar{W}_{tz} + V\,\bar{W}_{zz}) = 0,$$

which follows from (3.14). The combination $\bar{W}_{tt}+V\bar{W}_{zt}$, the averaged energy density $\langle W_{t't'}\rangle$ *in the moving frame* (3.19), might then be interpreted *in a laboratory frame* as the averaged net energy density composed of the energy $\langle W_{tt}\rangle$ of the electromagnetic wave minus the energy, $-V\langle W_{zt}\rangle$, needed to support the variable property pattern.

This interpretation is valid only in part, however. The effective energy density $\bar{W}_{tt} + V\,\bar{W}_{zt}$ in a moving frame (3.21) may be calculated by (3.14) for the D'Alembert waves $u_i = u(z - \lambda_i t)$, $i = 1, 2$, where λ_i are the roots of the quadratic equation

$$\lambda^2 r - 2q\lambda - p = 0, \tag{3.23}$$

following from (2.8). After some calculation, we get

$$\bar{W}_{tt} + V\bar{W}_{zt} = (r\lambda - q)(\lambda - V). \tag{3.24}$$

If the original materials are identical ($\epsilon_1 = \epsilon_2 = \epsilon, \mu_1 = \mu_2 = \mu$), then $p = \frac{1}{\mu}, q = 0, r = \epsilon$ by (2.7) and (2.9) and $\lambda_{1,2} = \pm a,\ \ a = 1/\sqrt{\epsilon\mu}$; the energy $\bar{W}_{tt} + V\,\bar{W}_{zt}$ thus becomes equal to

$$\bar{W}_{tt} + V\,\bar{W}_{zt} = \epsilon\lambda(\lambda - V). \tag{3.25}$$

This expression depends on V, a purely kinematic effect related exclusively to the frame and not associated with the variability of the property pattern. When $\lambda = a$ (a "slow" wave in the moving frame), then $\bar{W}_{tt} + V\bar{W}_{zt}$ becomes negative if $V > a$ [10]; when $\lambda = -a$ (a "fast" wave in the moving frame), then $\bar{W}_{tt} + V\bar{W}_{zt}$ remains positive for $V > 0$. We see that, for a *uniform* property pattern, when the work (3.22) is zero, the quantity $\bar{W}_{tt} + V\bar{W}_{zt}$ is not equal to $\bar{W}_{tt}$ which is, in this case, the same as $\langle W_{tt}\rangle$.

If the original substances are different in their material parameters, then the energy (3.24) should be calculated for $\lambda = \lambda_{1,2}$ given by [6]:

$$(3.26)\qquad \lambda_{1,2} = -\frac{V\left[\bar{\epsilon}\bar{\mu} - \left(\frac{\bar{1}}{a^2}\right)\right] \mp \frac{1}{a_1 a_2}\sqrt{\bar{\epsilon}\bar{\mu}\left[V^2 - \frac{1}{\bar{\epsilon}}\left(\frac{\bar{1}}{\mu}\right)\right]\left[V^2 - \frac{1}{\bar{\mu}}\left(\frac{\bar{1}}{\epsilon}\right)\right]}}{\frac{1}{a_1^2 a_2^2}\left[V^2 - \left(\frac{\bar{1}}{\epsilon}\right)\left(\frac{\bar{1}}{\mu}\right)\right]}.$$

In particular, if we consider the irregular case $\epsilon_2 > \epsilon_1, \mu_2 < \mu_1, a_2 > a_1$, then calculations show that $\bar{W}_{tt} + V\bar{W}_{zt}$ *may become negative for both waves* $\lambda = \lambda_1$ and $\lambda = \lambda_2$ if $a_2^2 \le V^2 \le (\overline{1/\epsilon})(\overline{1/\mu})$. The velocity V then also falls into the interval (λ_1, λ_2). This may occur when the D'Alembert waves become coordinated [4], i.e. when λ_1 and λ_2 are both positive, and the effective (averaged) energy densities $\bar{W}_{tt} + V\bar{W}_{zt}$ both become negative.

Waves of negative energy density may propagate through a medium that is not in thermodynamic equilibrium [2, 9, 10], and an activated laminate represents a typical example of such. There is a physical analogy between the build up of waves having negative energy and the anomalous Doppler effect as discussed in [8].

To conclude this section, we note that the coordinate frame (3.19) moving along with the interface can never become proper (i.e. such that (2.8) becomes diagonalized) for a tensor $\bar{W}$, except for the trivial case $V = 0$. To prove this, apply (3.14) to express $\bar{W}_{t't'}$ as

$$\bar{W}_{t't'} = \bar{W}_{tt} + V\bar{W}_{zt} = \frac{1}{2}(ru_t^2 + pu_z^2) + V(ru_z u_t + qu_z^2).$$

In the frame (3.19),

$$u_{t'} = u_t + Vu_z, \qquad u_{z'} = u_z,$$

and $\bar{W}_{t't'}$ takes the form

$$\bar{W}_{t't'} = \frac{1}{2}ru_{t'}^2 + \frac{1}{2}(p - rV^2 + 2qV)u_{z'}^2.$$

The frame (3.19) would become proper if $w = q/r = V$ (see (2.11) with $c \to \infty$); in this case, by (2.7) and (2.9), $p - rV^2 + 2qV = (pr + q^2)/r = \theta^{-1}r^{-1}$, and the effective constants become equal to

$$\mathcal{E} = r, \qquad M = \theta r.$$

However, by (2.9), the equality $V = q/r$ is transformed to

$$\alpha\,\beta\,c = \beta^2\,V,$$

which is impossible by (2.7), unless $V = 0$.

4 Numerical examples and discussion

In this section, we focus on activated laminates made up of two materials, (ϵ_1, μ_1) and (ϵ_2, μ_2), to demonstrate the theoretical results given in the previous sections. We use a numerical method to simulate the heterogeneous problem described by equations (2.4). The effective behaviour of a disturbance through such a laminate is shown in contour plots later in this section.

Directly computing the numerical solution to wave propagation through fast range ($V^2 > a_2^2$) dynamic laminates has proven to be a challenging problem. A more standard conservative finite difference approach analogous to the one taken in [11] for the slow range ($V^2 < a_1^2$) and for static laminates yields an unstable scheme. Numerical results are degraded since accuracy is quickly lost due to the growth of short waves which enter into the computation as truncation and round-off error. In [12], an approach is taken that successfully circumvents the appearance of these instabilities in the case of temporal laminates when $V = \infty$.

For the fast range laminates, we make the following change of coordinates: $\tau = t - \frac{z}{V}, \zeta = z$ yielding the PDE system (c.f. (2.4))

$$\epsilon u_\tau + \frac{1}{V} v_\tau = v_\zeta,$$

$$\mu v_\tau + \frac{1}{V} u_\tau = u_\zeta.$$

In terms of characteristic variables, this is the convection system

$$(u + v/\nu)_\tau - \frac{aV}{V+a}(u + v/\nu)_\zeta = 0 \tag{4.1}$$

$$(u - v/\nu)_\tau + \frac{aV}{V-a}(u - v/\nu)_\zeta = 0, \tag{4.2}$$

where $\nu = \sqrt{\epsilon/\mu}$ is the material impedance, and a is the phase velocity of the material. In ζ, τ coordinates, the fast range dynamic material is as a temporal material where the property pattern depends on τ alone and has period ε. When a wave is incident on the pattern interfaces, $\tau = n\varepsilon$ or $\tau = (n + m_1)\varepsilon$ for n an integer, two new waves arise which both move into the new material. These waves are of the same wave number as the incident wave when looked upon in the new coordinate system. However, short wave modes unavoidably introduced into the computation will grow and destroy the fidelity of the results. We perform a spectral decomposition of the initial data, and at very regular intervals in the course of the numerical computation, we filter out those wave modes that lie without the range initially present. This spectral approach has proved successful and we illustrate some of the results below.

We consider a rank-one activated laminate made up of two isotropic dielectrics. We take the material parameters to be

$$(\epsilon_1, \mu_1) = (1, 1) \qquad (\epsilon_2, \mu_2) = (9, 0.1); \tag{4.3}$$

thus $a_1 = 1, a_2 = 1.0541$ on a scale with $c = 10a_2$. Take the mixing factor to be $m_1 = 0.5$, so

$$(1/\bar{\epsilon})(\overline{1/\mu}) = 1.1, \qquad (\overline{1/\epsilon})(\overline{1/\mu}) = 3.0556,$$

$$k = 2.9757, \qquad \sigma = 36.3636,$$

from (2.27) and (2.28). For fast range laminates, the effective values of permittivity and permeability, $\mathcal{E}$ and M, are real and the effective material isotropic when either

$$a_2 = 1.0541 < V < 1.5853, \qquad \text{or} \qquad V > 1.9755,$$

from (2.30). In the first range, the effective values are both negative as predicted, whereas the values are positive in the second range. Figure 4 represents the plot of $1/M$ versus $\mathcal{E}$ as V varies within the acceptable range. Figure 5 plots the corresponding energy densities computed from (3.24) for each of the D'Alembert waves (3.26) represented by solid and dashed curves.

Figures 6 and 7 show contour plots of u when an initial Gaussian pulse,

$$u(z, 0) = e^{-z^2}, \quad v(z, 0) = 0,$$

propagates through a fast range laminate with material parameters given in (4.3), and $m_1 = 0.5$. We note that these are the results that come from the direct, detailed computation of the unhomogenized equations, not from computing solutions to the effective equations. The figures show the results for u in z, t coordinates when $V = 1.3$ and $V = 4$. The horizontal axis gives the z-values, while time is on the vertical axis.

Calculations show that for $V = 1.3$, the theoretical values for λ_1, λ_2 are 1.09324 and 2.7147 - coordinated wave motion as seen clearly in Figure 6. Looking at the numerical results in the contour plot of Figure 6, we estimate that the slower moving disturbance travels at a velocity 1.0895, and the faster wave has velocity 2.7133. From Figures 4 and 5, we expect that the effective material coefficients when $V = 1.3$ are negative as are the energy densities. We find that $\mathcal{E} = -4.0293$ and $M = -0.3532$. The energy densities $\bar{W}_{tt} + V\bar{W}_{zt}$ given by (3.24) are -0.69835 and -4.77846 using the theoretical values for the effective velocities; the energy densities are $-0.71917, -4.77023$ using the numerical values of λ_i in (3.24).

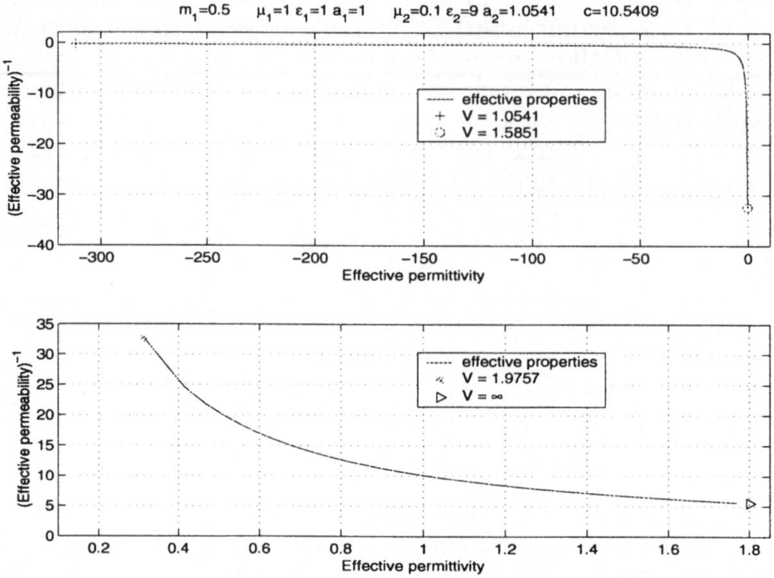

Figure 4: Effective permittivities and permeabilities of dielectric laminates with $(\epsilon_1, \mu_1) = (1, 1), (\epsilon_2, \mu_2) = (9, 0.1), m_1 = 0.5$, for varying V.

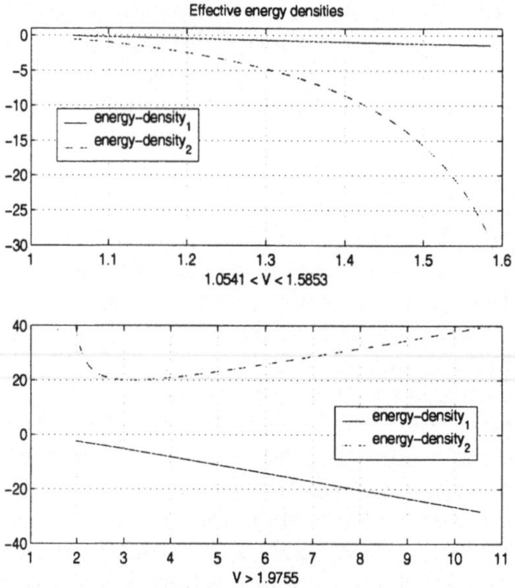

Figure 5: Effective energy densities of dielectric laminates with $(\epsilon_1, \mu_1) = (1, 1), (\epsilon_2, \mu_2) = (9, 0.1), m_1 = 0.5$, for varying V.

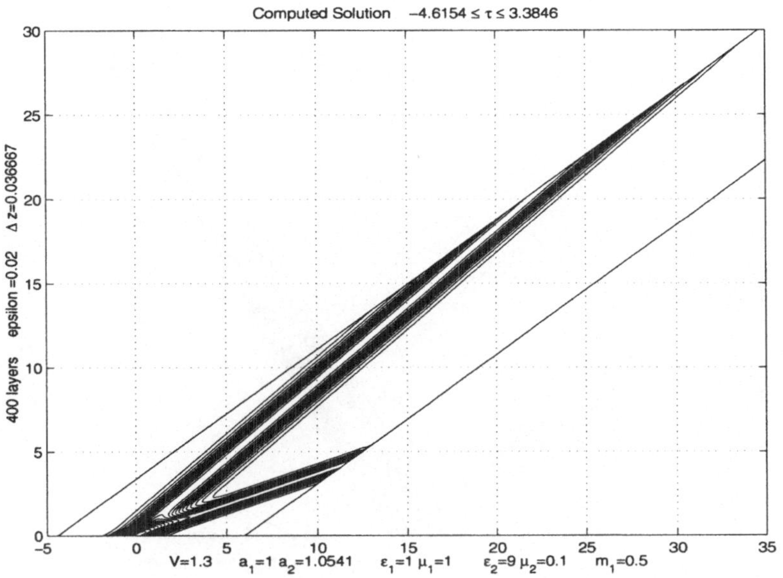

Figure 6: Wave propagation through a fast range laminate where $V = 1.3$ yields a homogenized material with negative effective coefficients.

For $V = 4$, the theoretical values for λ_1, λ_2 are 1.4001 and -2.6362 - no coordination. Looking at the numerical results in the contour plot of Figure 7, we find that the disturbances travel with velocities 1.4 and -2.6296. The effective material properties, $\mathcal{E}$ and M, are both positive as indicated in Figure 4, and take the values 1.5582 and 0.1564 respectively. As given in Figure 5, the energy densities are of opposite signs when $V = 4$ and from (3.24) they take the values -8.20495 and 20.94315 using the theoretical values of λ_i, and -8.20633 and 20.9587, using the numerically computed values of the effective velocities.

The values of the energy densities that have been calculated incorporate contributions due both to the frame motion and to the variable property pattern. To single out the contribution caused specifically by the variable property pattern, one has to subtract the value of the energy density calculated by the formula (3.25) for a pure material from the corresponding value of $\bar{W}_{tt} + V\bar{W}_{zt}$ calculated by (3.24) for the variable material pattern. In Figures 8 and 9, we show how the energy densities of both waves that develop in an activated laminate vary as the fraction of material 1 (i.e. m_1) in the mixture increases from 0 to 1.

For pure materials, as discussed near the end of Section 3, one wave is "fast" and the other "slow". The upper/lower plot is associated with the wave that is "slow"/"fast" for the pure materials of (4.3). For a "slow" wave, the energy density remains negative for all m_1; a "fast" wave has energy density negative in the presence of coordination, and positive in the

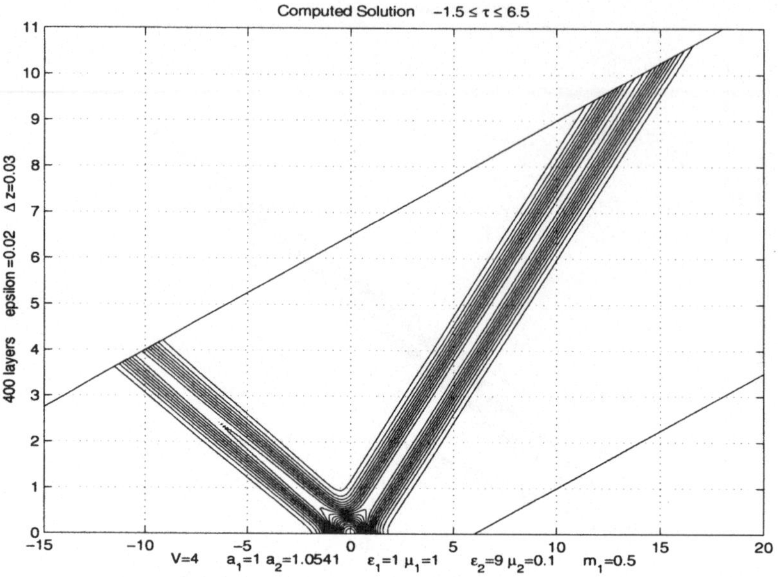

Figure 7: Wave propagation through a fast range laminate where $V = 4$ yields a homogenized material with positive effective coefficients.

absence of it. The energy densities of pure materials are indicated on these plots. It is clearly seen that the energy contribution due to the variable property pattern goes to zero as the property pattern becomes uniform, i.e. $\epsilon_2 \rightarrow \epsilon_1$ and $\mu_2 \rightarrow \mu_1$, or vice versa.

Acknowledgements

The authors acknowledge the support of this work through NSF Grant DMS-0204673.

References

[1] Blekhman, I. I., Lurie, K. A., On dynamic materials. *Proc. Rus. Acad. Sci. (Doklady)* **37** No. 2 (2000), 182-185.

[2] Chu, L. J., A kinetic power theorem, *IRE Conference on Electron Tube Research*, University of New Hampshire (1951).

[3] Louisell, W. H., Coupled mode and parametric electronics, *Wiley*, New York, 1960.

[4] Lurie, K. A., Effective properties of smart elastic laminates and the screening phenomenon, *Int. J. Solids Struct.* **34** (1997), 1633-1643.

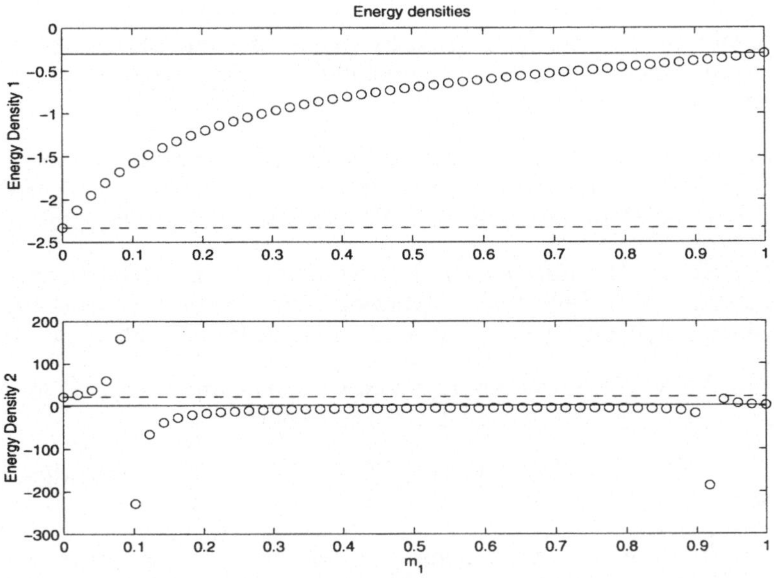

Figure 8: Energy densities of composite vs. m_1, for $V = 1.3$. Solid line is energy density of pure material 1; dashed line is energy density of pure material 2.

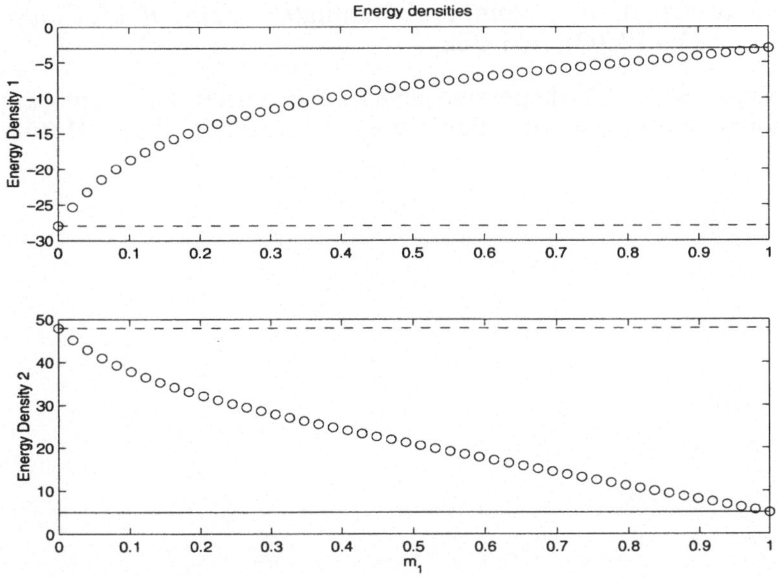

Figure 9: Energy densities of composite vs. m_1, for $V = 4$. Solid line is energy density of pure material 1; dashed line is energy density of pure material 2.

[5] Lurie, K. A., The problem of effective parameters of a mixture of two isotropic dielectrics distributed in space-time and the conservation law for wave impedance in one-dimensional wave propagation, *Proc. R. Soc. Lond.* A**454** (1998), 1767-1779.

[6] Lurie, K. A., Control of the coefficients of linear hyperbolic equations via spatio-temporal composites, in *Homogenization*, V. Berdichevsky, V. Jikov, G. Papanicolaou eds., World Scientific (1999), 285-315.

[7] Lurie, K. A., Bounds for the electromagnetic material properties of a spatio-temporal dielectric polycrystal with respect to one-dimensional wave propagation, *Proc. R. Soc. London* A**456** (2000), 1547-1557.

[8] Nezlin, M. V., Negative-energy waves and the anomalous Doppler effect, *Sov. Phys. Usp.*, **19** No. 11 (1976), 946–954.

[9] Pierce, J. R., All about waves, The MIT Press Cambridge, MA and London, 1974.

[10] Sturrock, P. A., In what sense do slow waves carry negative energy? *Journal of Applied Physics* **31** No. 11 (1960), 2052-2056.

[11] Weekes, S. L., Numerical modelling of wave phenomena in dynamic materials, *Applied Numerical Mathematics* **37** (2001), 417-440.

[12] Weekes, S. L., A stable scheme for the numerical computation of long wave propagation in temporal laminates, *Journal of Computational Physics* **176** (2002), 345–362.

[13] Weekes, S. L., A dispersive effective equation for wave propagation through spatio-temporal laminates, to appear in *Wave Motion*.

A Note on the Uniqueness of Solutions to Nonlinear, Discrete, Vector Boundary Value Problems

J. Mawhin[1] and C. C. Tisdell[2]

[1] Département de Mathématique, Université Catholique de Louvain, B-1348 Louvain-la-Neuve, Belgium e–mail : mawhin@math.ucl.ac.be

[2]Department of Mathematics, The University of Queensland, Brisbane, 4072 Queensland, Australia e–mail : cct@maths.uq.edu.au

Abstract. Discrete, two-point boundary value problems for nonlinear systems of difference equations are researched and conditions are formulated under which solutions to the discrete problem are unique. These discrete boundary value problems may arise as finite difference approximations to two-point boundary value problems for systems of second-order, ordinary differential equations, or, they may occur in their own right. Some existence and convergence theorems for solutions to the discrete problem are also presented.

AMS Subject Classification. 34B15, 39A12

Keywords. Uniqueness of solutions, vector equations, discrete boundary value problems, second order systems of ordinary differential equations.

1 Introduction

Consider the continuous two-point boundary value problem (BVP)

$$y'' = f(t, y) + F(t)y', \quad 0 \le t \le 1, \tag{1.1}$$

$$y(0) = A, \quad y(1) = B, \tag{1.2}$$

and its discrete approximation

$$\Delta^2 y_{k+1}/h^2 = f(t_k, y_k) + F(t_k)\Delta y_k/h, k = 1, \dots, n-1, \tag{1.3}$$

$$y_0 = A, \quad y_n = B, \tag{1.4}$$

where $f(t, y)$ maps $[0, 1] \times \mathbb{R}^d$ into $\mathbb{R}^d$, $F(t)$ is a $d \times d$ matrix, the step size $h = 1/n$ and grid points $t_k = kh$ for $k = 0, \dots, n$. Following the notation of

Lasota [11], the first and second (backward) differences are given respectively by:

$$\Delta y_k = \begin{cases} y_k - y_{k-1}, & \text{for } k = 1, \ldots, n, \\ 0, & \text{for } k = 0, \end{cases}$$

$$\Delta^2 y_{k+1} = \begin{cases} y_{k+1} - 2y_k + y_{k-1}, & \text{for } k = 1, \ldots, n-1, \\ 0, & \text{for } k = 0 \text{ or } k = n. \end{cases}$$

In the literature, uniqueness of solutions to second-order, discrete BVPs has been investigated by Agarwal [2] and [3] (see also references therein), Cheng and Lin [6], Kelley and Peterson [10] (see also references therein), Lasota [11], Tisdell [19] and Usmani and Agarwal [2].

In this note, some simple inequalities are formulated which ensure the solutions to the nonlinear discrete BVP (1.3), (1.4) are unique. The work was motivated by the theorems in [19] which guaranteed the uniqueness of solutions to the (linear) discrete vector BVP (1.3), (1.4) for the special case $f(t, y) = B(t)y$.

The main interest for formulating these results lies in the fact that uniqueness of solutions may not commute from the continuous problem to the discrete problem (see Agarwal [1]). Thus it is of value to study the finite difference approximation (1.3), (1.4) in its own right and then to formulate results which apply to both the continuous and discrete BVP.

The primary benefit of this note is that the results contained herein apply to discrete nonlinear *vector* equations which involve $\Delta y_k/h$. The new theorems extend known uniqueness results from the one-dimensional case (for example, [11]) to the vector case and extend the uniqueness workings on discrete linear vector equations (for example [19]) to the nonlinear case.

Another point of interest is that the new uniqueness and existence results do not rely on "a sufficiently small step-size", therefore the ideas are applicable to purely discrete BVPs such as

$$\Delta^2 y_{k+1} = f(k, y_k) + F(k)\Delta y_k, \qquad k = 1, \ldots, n-1,$$
$$y_0 = A, \quad y_n = B,$$

as well as to finite difference approximations.

2 Notations and preliminary results

Let $\|y\|$ denote the usual norm on $\mathbb{R}^d$ and set

$$\|\bar{y}\| = \max\{\|y_k\| : k = 0, \ldots, n\}.$$

For a bounded open set J let ∂J denote the boundary of J and let $\bar{J}$ denote its closure. Denote the dot product of two vectors y and z by $y \bullet z$. If M is a matrix then denote its transpose by M^T. Denote the space of m times continuously differentiable functions mapping from R to S by $C^m(R;S)$ endowed with the usual maximum norm. If $S = \mathbb{R}$ then simply write $C^m(R)$.

A solution to problem (1.1) is a function $y \in C^2([0,1];\mathbb{R}^d)$ which satisfies (1.1) for all $x \in [0,1]$.

A solution to problem (1.3) is a vector $\bar{y} = (y_0, \ldots, y_n) \in \mathbb{R}^{(n+1)d}$ which satisfies (1.3) for all $k = 1, \ldots, n-1$. The value of the k-th component, y_k, of a solution $\bar{y}$ of (1.3)) is expected to approximate $y(x_k)$, for some solution y of (1.1).

3 Uniqueness of solutions

In this section, some simple inequalities, originally derived from Hartman [8] and Blot et al [5], are used to formulate uniqueness results for solutions of the discrete problem (1.3), (1.4).

The following is an extension of [19,Theorem 1].

Theorem 3.1. *Let $F(t)$ and $f(t,y)$ satisfy*

$$(2(f(t,u) - f(t,v)) - F(t)F^T(t)(u-v)) \bullet (u-v) > 0, \tag{3.1}$$

for $t \in [0,1]$ and all vectors u and v, with $u \neq v$.

Then the discrete problem (1.3) has, at most, one solution satisfying the discrete boundary conditions (1.4).

Proof. The proof follows similar lines to that in [8] and [5]. Let $\bar{u}$ and $\bar{v}$ be two solutions to (1.3), (1.4) and put

$$r_k = \|u_k - v_k\|^2 \quad \text{for} \quad k = 0, \ldots, n.$$

Using the product rule for the difference operator,

$$\begin{aligned}\Delta^2 r_{k+1}/h^2 &= 2(u_k - v_k) \bullet (\Delta^2 u_{k+1} - \Delta^2 v_{k+1})/h^2) \\ &+ \|(\Delta u_k - \Delta v_k)/h\|^2 + \|(\Delta u_{k+1} - \Delta v_{k+1})/h\|^2.\end{aligned}$$

Using the identity $Mb \bullet c = b \bullet M^T c$, it can be verified that

$$\begin{aligned}&2((u_k - v_k) \bullet (\Delta^2 u_{k+1} - \Delta^2 v_{k+1})/h^2) + \|(\Delta u_k - \Delta v_k)/h\|^2 \\ &= \|(\Delta u_k - \Delta v_k)/h + F^T(t_k)(u_k - v_k)\|^2 \\ &+(2(f(t_k,u_k) - f(t_k,v_k)) - F(t_k)F^T(t_k)(u_k - v_k)) \bullet (u_k - v_k),\end{aligned}$$

and therefore from inequality (3.1) it follows that $\Delta^2 r_{k+1}/h^2 > 0$ for each $k = 1, \ldots, n-1$. Since the discrete boundary conditions (1.4) mean $r_0 = r_n = 0$, it follows from a discrete maximum principle that $r_k = 0$ for each $k = 0, \ldots, n$. That is, there is, at most, one solution to (1.3) satisfying the discrete boundary conditions (1.4). This concludes the proof.

Remark 3.1. The inequalities in Theorem 3.1 also guarantee uniqueness of solutions to the continuous problem (1.1), (1.2). Thus, Theorem 3.1 gives a connection between uniqueness for the continuous problem and uniqueness for its associated finite difference approximation.

Remark 3.2. Since the conditions in Theorem 3.1 do not involve any restrictions on the step-size h, the uniqueness property of Theorem 3.1 also applies to those discrete BVPs which do not arise as approximations to continuous BVPs, for example, the case $h = 1$.

4 Existence of a unique solution

Theorem 4.1 *Let $R > 0$ be a constant and let $f(t, y)$ and $F(t)$ be continuous on $[0, 1] \times \mathbb{R}^d$ and $[0, 1]$ respectively. Let*

$$(4.1) \qquad (2f(t, y) - F(t)F^T(t)y) \bullet y > 0, \ \textit{for } t \in [0,1] \textit{ and } \|y\| = R,$$

$$(2(f(t, u) - f(t, v)) - F(t)F^T(t)(u - v)) \bullet (u - v) > 0,$$

$$(4.2) \qquad \textit{for } t \in [0,1] \textit{ and } \|u\|, \|v\| < R \textit{ with } u \neq v.$$

and let $\|A\|$, $\|B\| < R$. Then the discrete problem (1.3), (1.4) has a unique solution $\bar{y}$ satisfying $\|\bar{y}\| < R$.

Proof. *Part (i): Existence.*

It may be checked by direct computation that problem (1.3), (1.4) has a solution $\bar{y}$ if and only if

$$y_k = h \sum_{i=1}^{n-1} G(x_k, s_i) g(s_i, y_i, \Delta y_i/h) + A(1 - x_k) + Bx_k, \ k = 0, \ldots, n,$$

where

$$G(x, t) = \begin{cases} (x-1)t & \text{for } 0 \le t \le x \le 1, \\ (t-1)x & \text{for } 0 \le x \le t \le 1, \end{cases}$$

and, for brevity, $g(x_k, y_k, \Delta y_k/h) = f(t_k, y_k) + F(t_k)\Delta y_k/h$.
Define

$$(4.3) \ T(y)_k = h \sum_{i=1}^{n-1} G(x_k, s_i) g(s_i, y_i, \Delta y_i/h) + A(1-x_k) + Bx_k, \ k = 0, \ldots, n.$$

The problem is thus reduced to showing that $T(\bar{y}) = \bar{y}$ for some $\bar{y} \in \mathbb{R}^{(n+1)d}$. This is done by using degree theory. Let

$$\Omega = \{\bar{y} \in \mathbb{R}^{(n+1)d} : \ \|\bar{y}\| < R\}.$$

From the simple properties of the summation operator, and since f is continuous, see that T is a continuous operator. Now consider

$$(I - \lambda T)(\bar{y}) = 0, \ \lambda \in [0,1]. \tag{4.4}$$

This is equivalent to $\bar{y}$ satisfying

$$\Delta^2 y_{k+1}/h^2 = \lambda f\left(x_k, y_k, \Delta y_k/h\right), \quad k = 1, \ldots, n-1, \tag{4.5}$$

$$y_0 = \lambda A, \ y_n = \lambda B. \tag{4.6}$$

Now show that if $(I - \lambda T)(\bar{y}) = 0$ and $\bar{y} \in \bar{\Omega}$ then $\bar{y} \in \Omega$ (and consequently $\bar{y} \notin \partial\Omega$). Firstly see that this is trivially satisfied for $\lambda = 0$ so assume $\lambda \in (0,1]$.

Suppose that a solution $\bar{y}$ to (4.5), (4.6) exists and define $r_k = \|y_k\|^2$ for $k = 0, \ldots, n$. First show that $r_k < R^2$ for $k = 0, \ldots, n$. Argue by contradiction and assume that r_k has a maximum when $\|y_k\| = R$ for some $k = 0, \ldots, n$. Since $\|A\|$, $\|B\| < R$ then r_k cannot have a maximum for $k = 0$ or $k = n$. Therefore $\Delta r_k/h \geq 0$ and $\Delta r_{k+1}/h \leq 0$ for some $k = 1, \ldots, n-1$, that is, $\Delta^2 r_{k+1}/h^2 \leq 0$. Now, using the product rule for the difference operator,

$$\begin{aligned}\Delta^2 r_{k+1}/h^2 &= 2(y_k \bullet \Delta^2 y_{k+1}/h^2) + \|\Delta y_k/h\|^2 + \|\Delta y_{k+1}/h\|^2, \\ &\geq 2\lambda(y_k \bullet (f(t_k, y_k) + F(t_k)\Delta y_k/h)) + \|\Delta y_k/h\|^2.\end{aligned}$$

As in the proof on Theorem 3.1, using the identity $Mb \bullet c = b \bullet M^T c$, it can be verified that

$$2\lambda(f(t_k, y_k) + F(t_k)\Delta y_k/h) \bullet y_k + \|\Delta y_k/h\|^2 =$$

$$\|\Delta y_k/h + \lambda F^T(t_k) y_k\|^2 + \lambda(2f(t_k, y_k) - F(t_k)F^T(t_k)y_k) \bullet y_k).$$

Therefore

$$\Delta^2 r_{k+1}/h^2 \geq \lambda(2f(t_k, y_k) - F(t_k)F^T(t_k)y_k) \bullet y_k > 0,$$

by inequality (4.1) and thus r_k cannot have a maximum when $\|y_k\| = R$ for all $k = 0, \ldots, n$. Thus $r_k < R^2$, that is, $\|y_k\| < R$ for each $k = 0, \ldots, n$.

Thus every solution $\bar{y}$ to (4.4) which satisfies $\bar{y} \in \bar{\Omega}$ must satisfy $\bar{y} \in \Omega$. Therefore $(I - \lambda T)(\bar{y}) \neq 0$, for all $\lambda \in [0,1]$ and $\bar{y} \in \partial\Omega$. The degree is defined on the bounded, open set Ω and by the invariance of the degree und er homotopy (see [12])

$$d((I - \lambda T)(\bar{y}), \Omega, 0) = d((I - T)(\bar{y}), \Omega, 0) = d(I, \Omega, 0) = 1 (\neq 0),$$

since $0 \in \Omega$. Therefore T has a fixed point and thus there is a solution $\bar{y}$ to (1.3), (1.4).

Part (ii) : Uniqueness.

Inequality (4.2) implies that Theorem 3.1 is applicable to solutions $\bar{y}$ which satisfy $\|\bar{y}\| < R$ and thus the uniqueness property follows. This concludes the proof.

Theorem 4.2. *Let $R > 0$ be a constant and let inequality (4.2) hold. Let $f(t, y)$ and $F(t)$ be continuous on $[0,1] \times \mathbb{R}^d$ and $[0.1]$ respectively. Let*

$$y \bullet (f(t,y) + F(t)y') + \|y'\|^2 > 0, \tag{4.7}$$

$$\text{if } y \bullet y' = 0, \ \|y\| = R, \ \|y'\| \leq N + 1,$$

where N is the constant supplied by Lemma 6.1, and $\|A\|$, $\|B\| < R$. Then (1.3), (1.4) has a unique solution $\bar{y}$ which satisfies $\|\bar{y}\| < R$, for sufficiently small h.

Proof. The main details of the proof are sketched, with the full details on the existence of at least one solution to be found in [16]. The continuity requirements, inequalities (4.7) and $\|A\|$, $\|B\| < R$, in conjunction with Lemma 6.1, are needed to apply Theorem 2 from [16] and it follows that there exists at least one solution $\bar{y}$ to (1.3), (1.4) satisfying $\|\bar{y}\| < R$, for sufficiently small h. Since inequality (4.2) holds, Theorem 3.1 is applicable to solutions $\bar{y}$ and thus solutions are unique. This concludes the proof.

Remark 4.1. If the uniqueness condition of inequality (4.2) is removed from Theorem 4.2, then Theorem 4.2 reduces to Theorem 2 of [16].

Remark 4.2. The primary advantage of Theorem 4.1 over Theorem 4.2 is that, in many cases, the inequalities in Theorem 4.1 are much easier to check than those of Theorem 4.2. Also, there are no restrictions on the step-size h in Theorem 4.1, which means the theory applies to purely discrete equations and not just to those which approximate differential equations.

Remark 4.3. The conditions of both Theorems 4.1 and 4.2 guarantee the existence of a unique solution to the continuous problem, although when examining only the continuous problem the inequalities of these theorems may be relaxed as in Part II, Chapter XII of [8]. For example, in [5], inequality (3.1) is replaced with

$$(4(f(t,u) - f(t,v)) - F(t)F^T(t)(u-v)) \bullet (u-v) > 0, \tag{4.8}$$

$$\text{for } t \in [0,1] \text{ and all vectors } u \text{ and } v, \text{ with } u \neq v.$$

and uniqueness of solutions to (1.1), (1.2) follows. It appears that by arguing along similar lines as in the proof of Theorem 2 in [16], inequality (3.1) may be replaced with inequality (4.8) and uniqueness of solutions to the discrete problem are still guaranteed by considering a sufficiently small step-size.

Remark 4.4. When comparing inequalities (4.8) and (3.1) (and not placing restrictions on the step-size) the distinction occurs because of the difference between the product rule for differential and the product rule for differences.

5 The scalar case

The scalar case is now briefly discussed. For the case $d = 1$, Theorems 3.1 and 4.1 reduce respectively to the following.

Lemma 5.1. *Let $f(t, y)$ and $F(t)$ be real-valued functions satisfying*

$$(2(f(t,u) - f(t,v)) - F(t)^2(u - v))(u - v) > 0, \tag{5.1}$$

$$\textit{for } t \in [0,1] \textit{ and all } u \textit{ and } v, \quad \textit{with } u \neq v.$$

Then the discrete problem (1.3) has, at most, one solution satisfying the discrete boundary conditions (1.4).

Lemma 5.2. *Let $R > 0$ be a constant and let $f(t, y)$ and $F(t)$ be real-valued functions which are continuous on $[0, 1] \times \mathbb{R}$ and $[0, 1]$ respectively. Let*

$$(2f(t,y) - F(t)^2 y)y > 0, \;\; t \in [0,1], \;\; \|y\| = R, \tag{5.2}$$

$$(2(f(t,u) - f(t,v)) - F(t)^2(u - v))(u - v) > 0, \tag{5.3}$$

$$\textit{for } t \in [0,1] \textit{ and } \|u\|, \|v\| < R \textit{ with } u \neq v.$$

and let $\|A\|$, $\|B\| < R$. Then the discrete problem (1.3), (1.4) has a unique solution $\bar{y}$ satisfying $\|\bar{y}\| < R$.

Remark 5.1. It appears that by incorporating inequality (4.2) into the existing theory on discrete lower and discrete upper solution methods, variants of uniqueness and existence results from [7] and [9] follow.

6 Convergence of solutions

In this section the previous results are applied to relate solutions of the discrete problem to solutions of the continuous problem. The following lemma will be needed.

Lemma 6.1. *Let $R > 0$ be a constant. There exists an $N > 0$ such that for every solution $\bar{y}$ to (1.3) which satisfies $\|\bar{y}\| \leq R$ then $\|\Delta y_k\|/h \leq N$ for $k = 1, \ldots, n$ and N is independent of the step-size h.*

Proof. To see this, choose

$$\Phi(\|s\|) = \sup_{(t,y) \in [0,1] \times \bar{B}_R(0)} \|f(t,y)\| + \sup_{t \in [0,1]} \|F(t)\| \|s\|$$

and see that

$$\|f(t,y) + F(t)s\| \leq \Phi(\|s\|), \text{ for } t \in [0,1], \;\; \|y\| \leq R, \;\; s \in R^d,$$

with $\lim_{u \to \infty} u^2/\Phi(u) = \infty$. The result follows by applying Theorem 1 of [16].

Remark 6.1. The fact that N, in Lemma 6.1, is independent of the step-size h is an important property as it allows the formulation of convergence results between solutions to the discrete problem and solutions to the continuous problem. See Gaines [7] for more details. (See Schmitt and Thompson [14] for a discussion on the continuous case).

The following is a generalization of [7, Theorem 2.5].

Theorem 6.1. *Let the assumptions of Theorem 4.1 or 4.2 hold. The unique solution to (1.3), (1.4) converges to the unique solution to (1.1), (1.2) in the following sense : Given $\varepsilon > 0$ there exists a $\delta = \delta(\varepsilon) > 0$ such that if $0 < h < \delta$ and $\bar{y}$ is the solution of (1.3), (1.4) then the solution $y(x)$ of (1.1), (1.2) satisfies :*

$$\max\{\|y(t,\bar{y}) - y(t)\| : 0 \le t \le 1\} \le \varepsilon,$$

and

$$\max\{\|v(t,\bar{y}) - y'(t)\| : 0 \le t \le 1\} \le \varepsilon,$$

where $y(t,\bar{y}) = y_k + (t - t_k)\Delta y_{k+1}/h$, for $t_k \le t \le t_{k+1}$, and

$$v(t,\bar{y}) = \begin{cases} \Delta y_k/h + (t-t_k)\Delta^2 y_{k+1}/h^2, & \text{for} \quad t_k \le t \le t_{k+1}, \\ \Delta y_1/h, \quad \text{for} \quad 0 \le t \le t_1. \end{cases}$$

Proof. The proof is similar to that of [7] and so is omitted.

Acknowledgement.

This paper was written during a stay of the first author at the University of Queensland, as an Ethel Raybould Visiting Fellow.

References

[1] R.P. Agarwal, On Multipoint Boundary Value Problems for Discrete Equations, *J. Math. Anal. Appl.*, 96 (1983), 520-534.

[2] R.P. Agarwal, On Boundary Value Problems of Second Order Discrete Systems, *Appl. Anal.*, 20 (1985), 1-17.

[3] R.P. Agarwal, Difference Equations and Inequalities, *Marcel Dekker*, New York, 1992.

[4] J.V. Baxley, Existence Theorems for Nonlinear Second Order Boundary Value Problems, *J. Differential Equations*, 85 (1990), 125 - 150.

[5] J. Blot, P. Cieutat and J. Mawhin, Almost-Periodic Oscillations of Monotone Second-Order Systems, *Adv. Differential Equations*, 2 (1997), 693-714.

[6] S.S. Cheng and S.S. Lin, Existence and Uniqueness Theorems for Nonlinear Difference Boundary Value Problems, *Utilitas Math.*, 39 (1991), 167-186.

[7] R.E. Gaines, Difference Equations Associated With Boundary Value Problems For Second Order Nonlinear Ordinary Differential Equations, *SIAM J. Numer. Anal.*, 11 (1974), 411-434.

[8] P. Hartman, Ordinary Differential Equations, *John Wiley and Sons*, New York, 1964.

[9] J. Henderson and H.B. Thompson, Difference Equations Associated with Boundary Value Problems for Second Order Nonlinear Ordinary Differential Equations, *J. Differ. Equations Appl.* (in press).

[10] W.G. Kelley and A.C. Peterson, Difference Equations. An Introduction with Applications, *Academic Press*, Boston, 1991.

[11] A. Lasota, A Discrete Boundary Value Problem, *Ann. Polon. Math.*, 20 (1968), 183-190.

[12] N.G. Lloyd, Degree Theory, *Cambridge University Press*, London, 1978.

[13] J.A. Pennline, Constructive Existence and Uniqueness Theorems for Two-Point Boundary Value Problems with a Linear Gradient Term, *Appl. Math. Comput.*, 16 (1984), 233-260.

[14] K. Schmitt and R. Thompson, Boundary Value Problems for Infinite Systems of Second-Order Differential Equations, *J. Differential Equations*, 18 (1975), 277-295.

[15] H.B. Thompson, Topological Methods for Some Boundary Value Problems, *Comput. Math. Appl.*, 42 (2001), 487-495.

[16] H.B. Thompson and C.C. Tisdell, Systems of Difference Equations Associated With Boundary Value Problems for Second Order Systems of Ordinary Differential Equations, *J. Math. Anal. Appl.*, 248 (2000), 333-347.

[17] H.B. Thompson and C.C. Tisdell, Boundary Value Problems for Systems of Difference Equations Associated with Systems of Second-Order Ordinary Differential Equations, *Appl. Math. Lett.* (in press).

[18] H.B. Thompson and C.C. Tisdell, On the Nonexistence of Spurious Solutions to Discrete, Two-Point Boundary Value Problems, *Appl. Math. Lett.* (in press).

[19] C.C. Tisdell, On the Uniqueness of Solutions to Discrete, Vector, Two-Point Boundary Value Problems, (to appear).

[20] R.A. Usmani and R.P. Agarwal, On the Numerical Solution of Two Point Discrete Boundary Value Problems, *Appl. Math. Comput.*, 25 (1988), 247-264.

Absolutely Continuous Invariant Measures on the Sphere

Sylvia Novo and Carmen Núñez[1]

Departamento de Matemática Aplicada a la Ingeniería, ETSII, Universidad de Valladolid, Paseo del Cauce S/N, E-47011 Valladolid, Spain.
E-mails: sylnov@wmatem.eis.uva.es, carnun@wmatem.eis.uva.es.

Abstract. This paper is concerned with the dynamical behavior of the solutions of random linear systems. Conditions which are sufficient for the presence of absolutely continuous dynamics on the spherical flow are analyzed and methods of construction of invariant measures are provided.

AMS Subject Classification. 37C40.

Key words. Spherical flow, invariant measures, absolutely continuous dynamics.

1 Introduction

Approaches based upon ergodic and topological theories are frequently used in the analysis of the behavior of the solutions of non-autonomous linear differential systems: instead of considering a single system, the idea is to study a random family of systems related by a continuous flow on a base set, usually a compact metric space. This setting allows us to consider a linear skew-product flow over the linear bundle, and hence a non-linear skew-product flow over the spherical bundle. Therefore one can use the ideas and methods from topological dynamics, what is not possible when just a single non-autonomous equation is considered. The underline reason is that a collective property should be easier to describe than a single one – although the behavior may differ in a set of null measure: the strong relation between the trajectories on the base provides common qualitative properties of the solutions of "almost all" the systems of the family. Examples of this classical and useful framework can be found in [10, 9, 4, 8, 5, 3] and references therein, among many others.

In particular, Novo and Obaya [8] obtain a complete ergodic classification for two-dimensional systems. The first step of this description is the distinction between the cases in which the spherical flow has or does not have an absolutely continuous invariant measure. Later, Arnold, Cong and Oseledets [2] provide an ergodic classification for lineal discrete cocycles,

[1]Work partially supported by C.I.C.Y.T. under project PB98-0359 and by Junta de Castilla y León and European Community under project VA19/00B.

also based on the analysis of the measures on the sphere which are invariant under the action of these cocycles.

Although the classification provided by these last authors is rather complete, there are still several ergodic and topological aspects which are far away to be well understood. This paper is part of a continuing effort to complete this description (see for instance [7, 5, 6]). The final objectives are to characterize the minimal subsets, to locate the invariant subsets where the ergodic measures are concentrated and to describe the global behavior of the flow. Here we obtain conditions which are sufficient for the presence of absolutely continuous dynamics and provide methods of construction of invariant measures. Our proofs require a careful analysis of the non-lineal functional equation satisfied by the density functions of such measures.

The results here presented constitute an extension (far away form trivial) to the general dimension case of the results of Obaya and Paramio [9] characterizing the density functions of the spherical flow associated to a one-dimensional random Schrödinger equation, which are the starting point for the ergodic classification obtained in [8].

2 Absolutely continuous dynamics of the spherical flow

Let $\sigma : \mathbb{R} \times \Omega \to \Omega$, $(t,\omega) \mapsto \omega{\cdot}t$ be a continuous flow on a compact metric space Ω. Let $S : \Omega \to \mathfrak{sl}(n,\mathbb{R})$ be a continuous function, where $\mathfrak{sl}(n,\mathbb{R})$ represents the algebra of matrices with null trace. The random family of n-dimensional linear systems

$$\mathbf{x}' = S(\omega{\cdot}t)\,\mathbf{x}\,, \quad \omega \in \Omega \tag{2.1}$$

induces a skew-product flow in the linear bundle $V_{\mathbb{R}} = \Omega \times \mathbb{R}^n$. It takes $(\omega, \mathbf{x}_0)$ after time t to $(\omega{\cdot}t, \mathbf{x}(t,\omega,\mathbf{x}_0))$, where $\mathbf{x}(t,\omega,\mathbf{x}_0)$ is the solution of the equation (2.1) corresponding to ω with initial data $\mathbf{x}_0$; i.e. $\mathbf{x}(t,\omega,\mathbf{x}_0) = U(t,\omega)\,\mathbf{x}_0$, where $U(t,\omega)$ is the fundamental matrix solution of (2.1) with $U(0,\omega) = I_n$.

Let us represent by $\mathbb{S}^{n-1}$ the unit sphere of $\mathbb{R}^n$, and by $|\mathbf{x}|$ the euclidean norm of the vector $\mathbf{x} \in \mathbb{R}^n$. By projection on the unit sphere, family (2.1) also induces a skew-product flow τ on the compact bundle $\Omega \times \mathbb{S}^{n-1}$, given by

$$\tau(t,\omega,\mathbf{x}_0) = \left(\omega{\cdot}t, \frac{\mathbf{x}(t,\omega,\mathbf{x}_0)}{|\mathbf{x}(t,\omega,\mathbf{x}_0)|}\right), \quad (\omega,\mathbf{x}_0) \in \Omega \times \mathbb{S}^{n-1}\,.$$

This kind of random family comes frequently from a single non-autonomous real lineal system

$$\mathbf{x}' = S_0(t)\,\mathbf{x}\,, \quad t \in \mathbb{R} \tag{2.2}$$

whose matrix of coefficients is uniformly continuous and bounded, as explained in what follows. Consider the set $C(\mathbb{R}, L(\mathbb{R}^2))$ endowed with the topology of the uniform convergence on the compact sets and define Ω to be the *hull* in $C(\mathbb{R}, L(\mathbb{R}^2))$ of S_0, namely the closure of the set of translated maps $\{S_{0,t} \mid t \in \mathbb{R}\}$ (with $S_{0,t}(s) = S_0(t+s)$. The translation $\mathbb{R} \times \Omega \to \Omega$, $(t,\omega) \to \omega{\cdot}t$, with $\omega{\cdot}t(s) = \omega(t+s)$ defines a continuous flow σ on Ω. Taking $S : \Omega \to L(\mathbb{R}^2)$ as the operator mapping ω to $\omega(0)$, we obtain a family (2.1) which includes the original system (2.2): just take $\omega = S_0$ and note that in this case $S(\omega{\cdot}t) = \omega(t) = S_0(t)$. In this way, almost periodic and recurrent systems are included in the collective formulation given by (2.1) and we obtain a continuous flow from a non-autonomous equation.

Let us consider again the general framework stated at the beginning of this section. As said in the introduction, the strong connection between the orbits on the base Ω provides common properties of the solutions of (2.1) which are satisfied almost everywhere with respect to a fixed σ-ergodic measure m_0 on the base space Ω. If (Ω, σ) is uniquely ergodic, or minimal, it happens frequently that qualitative properties are satisfied for every $\omega \in \Omega$, not just m_0-a.e. But in general, the choice of the ergodic measure m_0 is fundamental in the ergodic and topological description we intend to obtain.

As explained in the introduction, the objective is the description of the measurable structure of the flow σ. We denote by ν the canonical volume form on $\mathbb{S}^{n-1}$ (induced by the one of $\mathbb{R}^n$) and set $m_1 = m_0 \otimes \nu$. The dynamical system $(\Omega \times \mathbb{S}^{n-1}, \tau, m_1)$ is said to be in the *absolutely continuous* case if there exists a τ-invariant measure on the compact bundle which is absolutely continuous with respect to m_1. Otherwise, the dynamics is *singular.*

Our first result characterizes the density functions of those τ-invariant measures which are absolutely continuous with respect to m_1.

Proposition 2.1 *Let $p \in L^1(\Omega \times \mathbb{S}^{n-1}, m_1)$ be a non-zero positive real function. The following statements are equivalent:*

(i) *the measure $d\mu = p\, dm_1$ is τ-invariant;*

(ii) *p coincides m_1-almost everywhere with a measurable solution of the equation*

$$p(\tau(t,\omega,\mathbf{x}_0)) = p(\omega,\mathbf{x}_0)\, |\mathbf{x}(t,\omega,\mathbf{x}_0)|^n . \tag{2.3}$$

Proof. The τ-invariance of the measure μ is equivalent to the equality

$$\int_{\Omega\times\mathbb{S}^{n-1}} h(\omega,\mathbf{x}_0)\, d\mu = \int_{\Omega\times\mathbb{S}^{n-1}} h(\tau(t,\omega,\mathbf{x}_0))\, d\mu \tag{2.4}$$

for every function $h \in C(\Omega \times \mathbb{S}^{n-1})$ and every $t \in \mathbb{R}$, whereas the definition of μ and the σ-invariance of the measure m_0 assure that

$$\int_{\Omega\times\mathbb{S}^{n-1}} h(\omega,\mathbf{x}_0)\, d\mu = \int_\Omega \left(\int_{\mathbb{S}^{n-1}} h(\omega{\cdot}t,\mathbf{x}_0)\, p(\omega{\cdot}t,\mathbf{x}_0)\, d\nu \right) dm_0 . \tag{2.5}$$

For each pair $t \in \mathbb{R}$ and $\omega \in \Omega$ we consider the diffeomorphism

$$\mathbb{S}^{n-1} \to \mathbb{S}^{n-1}, \quad \mathbf{x}_0 \mapsto \tau_2(t,\omega,\mathbf{x}_0),$$

where τ_2 represents the second component of the flow τ; i.e. $\tau_2(t,\omega,\mathbf{x}_0) = \mathbf{x}(t,\omega,\mathbf{x}_0)/|\mathbf{x}(t,\omega,\mathbf{x}_0)|$. The change of variables formula provides

$$\int_{\mathbb{S}^{n-1}} h(\omega{\cdot}t,\mathbf{x}_0)p(\omega{\cdot}t,\mathbf{x}_0)d\nu$$
$$= \int_{\mathbb{S}^{n-1}} h(\tau(t,\omega,\mathbf{x}_0))p(\tau(t,\omega,\mathbf{x}_0))|J_\nu(\tau_2(t,\omega,\mathbf{x}_0))|d\nu$$

where $J_\nu(\tau_2)$ denotes the Jacobian determinant of τ_2 with respect to ν. Our next step is to check that

$$J_\nu(\tau_2(t,\omega,\mathbf{x}_0)) = |\mathbf{x}(t,\omega,\mathbf{x}_0)|^{-n} \tag{2.6}$$

for every $\mathbf{x}_0 \in \mathbb{S}^{n-1}$. To this end, we represent by $\mathcal{X}_\omega(t,\cdot)$ the time-dependent vector field induced on the sphere by the system (2.1) and recall that the Lie derivative formula for this kind of fields (see [1]) assures that

$$J_\nu(\tau_2(t,\omega,\mathbf{x}_0)) = \exp\left(\int_0^t \operatorname{div}_\nu \mathcal{X}_\omega(s,\tau_2(s,\omega,\mathbf{x}_0))\,ds\right). \tag{2.7}$$

In order to compute the divergence of $\mathcal{X}_\omega$ we need to work on a local chart of the sphere. We fix $\mathbf{x}_0 \in \mathbb{S}^{n-1} \subset \mathbb{R}^n$ and assume that it does not belong to the set $\{\mathbf{x} \in \mathbb{R}^n \mid x_1 = 0,\ x_2 \geq 0\}$. Outside this set we can take coordinates $(\boldsymbol{\varphi},\rho)$ with $\boldsymbol{\varphi} = (\varphi_1,\ldots,\varphi_{n-1})$, where

$$\varphi_1 = \theta_1\,, \qquad \varphi_j = \int_{-\pi/2}^{\theta_j} \sin^{j-1} s\,ds \quad \text{for} \quad j = 2,\ldots,n-1\,, \qquad \rho = \frac{1}{n}r^n\,,$$

being $\theta_1,\ldots,\theta_{n-1},r$ defined from the usual coordinates of $\mathbb{R}^n$ by means of the relations

$$\begin{aligned} x_1 &= r\sin\theta_{n-1}\cdots\sin\theta_3\sin\theta_2\sin\theta_1\,,\\ x_2 &= r\sin\theta_{n-1}\cdots\sin\theta_3\sin\theta_2\cos\theta_1\,,\\ x_3 &= r\sin\theta_{n-1}\cdots\sin\theta_3\cos\theta_2\,,\\ &\ \ \vdots\\ x_n &= r\cos\theta_{n-1} \end{aligned}$$

and the conditions $r > 0$, $\theta_1 \in (0,2\pi)$ and $\theta_j \in (0,\pi)$ for $j = 2,\ldots,n-1$. It is easy to check that in these coordinates the systems (2.1) are given by

$$\boldsymbol{\varphi}' = \mathbf{f}(\omega{\cdot}t,\boldsymbol{\varphi})\,, \tag{2.8}$$
$$\rho' = g(\omega{\cdot}t,\boldsymbol{\varphi})\,\rho\,, \tag{2.9}$$

for certain functions $\mathbf{f}$ and g. Besides, $(x_1,\ldots,x_n)\mapsto(\varphi_1,\ldots,\varphi_{n-1},\rho)$ is a symplectic change of coordinates. From this and the fact that $\operatorname{tr} S(\omega)=0$ we conclude that

$$\operatorname{tr}\frac{\partial \mathbf{f}}{\partial \boldsymbol{\varphi}}(\omega,\boldsymbol{\varphi})+g(\omega,\boldsymbol{\varphi})=0\,.$$

In addition, since the flow on the local chart $\boldsymbol{\varphi}$ chosen on the sphere is given by equation (2.8), we obtain

$$\operatorname{div}_\nu \mathcal{X}_\omega(t,\mathbf{x}_0)=\operatorname{tr}\frac{\partial \mathbf{f}}{\partial \boldsymbol{\varphi}}(\omega{\cdot}t,\boldsymbol{\varphi}_0)=-g(\omega{\cdot}t,\boldsymbol{\varphi}_0)\,,$$

where $\boldsymbol{\varphi}_0$ is the point of the chart corresponding to $\mathbf{x}_0$. Hence, for small t,

$$\operatorname{div}_\nu \mathcal{X}_\omega(t,\tau_2(t,\omega,\mathbf{x}_0))=-g(\omega{\cdot}t,\boldsymbol{\varphi}(t,\omega,\boldsymbol{\varphi}_0))\,,$$

where $\boldsymbol{\varphi}(t,\omega,\boldsymbol{\varphi}_0)$ is the solution of (2.8) with initial data $\boldsymbol{\varphi}(0,\omega,\boldsymbol{\varphi}_0)=\boldsymbol{\varphi}_0$. Relation (2.7), equation (2.9) and the definition of ρ provide equality (2.6) for t small enough. Since the same argument works on any chart of the sphere, the equality is valid for any value of t.

Substituting (2.6) in (2.5) and comparing with (2.4) we conclude that the measure μ is τ-invariant if and only if for each $t\in\mathbb{R}$

$$p(\tau(t,\omega,\mathbf{x}_0))\,|\mathbf{x}(t,\omega,\mathbf{x}_0)|^{-n}=p(\omega,\mathbf{x}_0)$$

m_1-a.e. on $\Omega\times\mathbb{S}^{n-1}$. A known argument of ergodic theory (see [9]) shows that a function p with this property coincides m_1-a.e. with a solution everywhere of the functional equation, which completes the proof. □

The above property is fundamental in the proof of our main result, which characterizes the presence of absolutely continuous dynamics and provides a way of construction of density functions. As usual, χ_A represents the characteristic function of the set A.

Theorem 2.2 *The following statement are equivalent:*

(i) *the dynamics of* $(\Omega\times\mathbb{S}^{n-1},\tau,m_1)$ *is absolutely continuous;*

(ii) *there exists a* τ*-invariant set* $C\subset\Omega\times\mathbb{S}^{n-1}$ *with* $m_1(C)>0$ *such that for every* $(\omega,\mathbf{x}_0)\in C$ *the limit*

$$p(\omega,\mathbf{x}_0)=\lim_{T\to\infty}\frac{1}{2T}\int_{-T}^{T}|\mathbf{x}(t,\omega,\mathbf{x}_0)|^{-n}dt$$

exists and is a positive real number;

(iii) *there exists a* τ*-invariant set* $D\subset\Omega\times\mathbb{S}^{n-1}$ *with* $m_1(D)>0$ *and a real constant* $k>0$ *such that for every* $(\omega,\mathbf{x}_0)\in D$,

$$\limsup_{T\to\infty}\frac{1}{2T}\int_{-T}^{T}\chi_{[-k,k]}(|\mathbf{x}(t,\omega,\mathbf{x}_0)|)\,dt>0\,.$$

Moreover, under these conditions, $d\mu = (1/m_1(C))p\,\chi_C\,dm_1$ *defines a normalized* τ*-invariant measure* μ *on* $\Omega \times \mathbb{S}^{n-1}$.

Proof. Taking Proposition 2.1 as starting point, the proofs of (i)$\Leftrightarrow$(ii) and of the last assertion of the theorem follow step by step the proofs of Propositions 2.3(i) and 2.4(i) of [9] (see also [8]), where the analogous results for the one-dimensional Schrödinger equation are established.

In order to show (i)$\Leftrightarrow$(iii) we make use of the following property: the absolutely continuous character of the dynamics on the sphere bundles is simultaneous for the n-dimensional systems (2.1) and the $(n+1)$-dimensional systems

$$\widetilde{\mathbf{x}}' = \widetilde{S}(\omega{\cdot}t)\,\widetilde{\mathbf{x}} = \begin{bmatrix} S(\omega{\cdot}t) & 0 \\ 0 & 0 \end{bmatrix} \widetilde{\mathbf{x}}, \tag{2.10}$$

as we explain now. According to the results of [2], the absolutely continuous character of the dynamics induced on the sphere by a random family of systems with trace 0 is equivalent to the existence of a linear change of variables taking these systems to skew-symmetric form; in other words, to the fact that the corresponding cocycle (determined by the fundamental matrix solution which agrees with the identity at $t=0$) is cohomologous to an orthogonal one. On the other hand, it is proved in [11] that a cocycle $U:\mathbb{R}\times\Omega\to \mathrm{GL}(d,\mathbb{R})$ is cohomologous to an orthogonal one if and only if it is *bounded*, i.e. if for any $\varepsilon>0$ there exists a compact subset $K_\varepsilon \subset \mathrm{GL}(d,\mathbb{R})$ such that $m_0(\{\omega\in\Omega \mid U(t,\omega)\notin K_\varepsilon\})<\varepsilon$ for every $t\in\mathbb{R}$. The obvious relation between the fundamental matrices of (2.1) and the extended systems (2.10) proves our assertion.

Assume that (i) is true, and hence the dynamics induced on $\Omega\times\mathbb{S}^n$ by (2.10) is absolutely continuous. We represent by $\widetilde{\tau}$ and $\widetilde{m}_1$ the flow and the product measure on $\Omega\times\mathbb{S}^n$, and by $\widetilde{\mathbf{x}}(t,\omega,\widetilde{\mathbf{x}}_0)$ the solution of (2.10) with initial data $\widetilde{\mathbf{x}}_0$. Let $\widetilde{C}$ be the $\widetilde{\tau}$-invariant subset of $\Omega\times\mathbb{S}^n$ provided by the corresponding property (ii). Then, for any $(\omega,\widetilde{\mathbf{x}}_0)\in\widetilde{C}$,

$$\lim_{T\to\infty}\frac{1}{2T}\int_{-T}^{T}|\widetilde{\mathbf{x}}(t,\omega,\widetilde{\mathbf{x}}_0)|^{-(n+1)}dt>0\,.$$

From here, the independence of this property with respect to the modulus of the initial data, and the equality $\widetilde{\mathbf{x}}(t,\omega,\widetilde{\mathbf{x}}_0)=(\mathbf{x}(t,\omega,\mathbf{x}_0),a)$ for $\widetilde{\mathbf{x}}_0=(\mathbf{x}_0,a)$, we conclude that there exists a τ-invariant subset $C\subset\Omega\times\mathbb{S}^{n-1}$ with $m_1(C)>0$ such that for any $(\omega,\mathbf{x}_0)\in C$ there is a constant $a\neq0$ with

$$\lim_{T\to\infty}\frac{1}{2T}\int_{-T}^{T}(|\mathbf{x}(t,\omega,\mathbf{x}_0)|^2+a^2)^{-(n+1)/2}dt>0\,.$$

Now assume, by contradiction, that (iii) is false. In particular, for any $k\in\mathbb{N}$

$$\limsup_{T\to\infty}\frac{1}{2T}\int_{-T}^{T}\chi_{[-k,k]}(|\mathbf{x}(t,\omega,\mathbf{x}_0)|)\,dt=0 \tag{2.11}$$

m_1-a.a. Then we can choose $(\omega, \mathbf{x}_0) \in C$ such that (2.11) is satisfied for every $k \in \mathbb{N}$, and a non-null real a with

$$\alpha = \lim_{T\to\infty} \frac{1}{2T} \int_{-T}^{T} (|\mathbf{x}(t,\omega,\mathbf{x}_0)|^2 + a^2)^{-(n+1)/2} dt > 0 .$$

We also choose an integer k with $(k^2 + a^2)^{-(n+1)/2} < \alpha$. Under these conditions,

$$\begin{aligned}
\alpha \leq & \limsup_{T\to\infty} \frac{1}{2T} \int_{\{t\in[-T,T] \,|\, |\mathbf{x}(t,\omega,\mathbf{x}_0)|<k\}} (|\mathbf{x}(t,\omega,\mathbf{x}_0)|^2 + a^2)^{-(n+1)/2} dt \\
& + \limsup_{T\to\infty} \frac{1}{2T} \int_{\{t\in[-T,T] \,|\, |\mathbf{x}(t,\omega,\mathbf{x}_0)|>k\}} (|\mathbf{x}(t,\omega,\mathbf{x}_0)|^2 + a^2)^{-(n+1)/2} dt \\
\leq & \limsup_{T\to\infty} \frac{1}{2T} \int_{-T}^{T} (k^2 + a^2)^{-(n+1)/2} dt = (k^2 + a^2)^{-(n+1)/2} < \alpha ,
\end{aligned}$$

which is impossible. Note that the first limit is 0 as a consequence of (2.11). Consequently, (i)⇒(iii) is proved.

In order to prove the converse assertion, assume that (iii) is true, fix $\varepsilon \in (0,1)$, and define the set $\widetilde{D}$ as follows: the pair $(\omega, \widetilde{\mathbf{x}}) \in \Omega \times \mathbb{S}^n$ belongs to $\widetilde{D}$ if and only if $\widetilde{\mathbf{x}}_0$ agrees with a pair $(\mathbf{x}, a)$, where $|a| \geq \varepsilon$ and $\mathbf{x} = \lambda \mathbf{x}_0$ for a non-null real λ and an element $\mathbf{x}_0 \in \mathbb{S}^{n-1}$ such that $(\omega, \mathbf{x}_0) \in D$.

So defined, the set $\widetilde{D}$ is $\widetilde{\tau}$-invariant and $\widetilde{m}_1(\widetilde{D}) > 0$. Moreover, for any $(\omega, \widetilde{\mathbf{x}}_0) \in \widetilde{D}$ there exist an element $(\omega, \mathbf{x}_0) \in D$ and a constant $a > \varepsilon$ such that

$$\begin{aligned}
\widetilde{p}(\omega, \widetilde{\mathbf{x}}_0) &= \limsup_{T\to\infty} \frac{1}{2T} \int_{-T}^{T} |\widetilde{\mathbf{x}}(t,\omega,\widetilde{\mathbf{x}}_0)|^{-(n+1)}\, dt \\
&= \limsup_{T\to\infty} \frac{1}{2T} \int_{-T}^{T} \frac{(a^2+1)^{(n+1)/2}}{(a^2 + |\mathbf{x}(t,\omega,\mathbf{x}_0)|^2)^{(n+1)/2}}\, dt \\
&\geq \frac{(a^2+1)^{(n+1)/2}}{(a^2+k^2)^{(n+1)/2}} \limsup_{T\to\infty} \frac{1}{2T} \int_{-T}^{T} \chi_{[-k,k]}(|\mathbf{x}(t,\omega,\mathbf{x}_0)|)\, dt > 0 .
\end{aligned}$$

In addition, it is immediate to check that $\widetilde{p}(\omega, \widetilde{\mathbf{x}}_0) \leq (1 + 1/\varepsilon^2)^{(n+1)/2}$ for every $(\omega, \widetilde{\mathbf{x}}_0) \in \widetilde{D}$. Under these conditions, the same arguments used in the results of [9] and [8] cited above show that $d\widetilde{\mu} = (1/\widetilde{m}_1(\widetilde{D}))\widetilde{p}\chi_{\widetilde{D}} dm_1$ defines a $\widetilde{\tau}$-invariant measure on $\Omega \times \mathbb{S}^n$. This is equivalent to the assertion (i) and completes the proof of the theorem. □

References

[1] R. Abraham, J.E. Marsden and T. Ratiu, *Manifolds, Tensor Analysis, and Applications*, Applied Mathematical Sciences **75**, Second Edition, Springer-Verlag, New York, 1988.

[2] L. Arnold, N.D. Cong and V.I. Oseledets, Jordan normal form for linear cocycles, *Random Oper. Stochastic Equation* **7** (4) (1999), 303–358.

[3] R. Fabbri, R. Johnson and C. Núñez, On the Yakubovich Frequency Theorem for linear non-autonomous control processes, *Discrete Contin. Dynam. Systems*, to appear.

[4] R. Johnson and M. Nerurkar, Controllability, stabilization, and the regulator problem for random differential systems, *Mem. Amer. Math. Soc.* **646**, Amer. Math. Soc., Providence, 1998.

[5] R. Johnson, S. Novo and R. Obaya, Ergodic properties and Weyl *M*-functions for random linear Hamiltonian systems, *Proc. Roy. Soc. Edinburgh* **130A** (2000), 1045–1079.

[6] S. Novo and C. Núñez, Linear Hamiltonian systems with absolutely continuous dynamics, *Nonlinear Anal. T.M.A.* **47** (**2**) (2001), 1401-1406.

[7] S. Novo, C. Núñez and R. Obaya, Ergodic properties and rotation number for linear Hamiltonian systems, *J. Differential Equations* **148** (1998), 148–185.

[8] S. Novo and R. Obaya, An ergodic and topological approach to almost periodic bidimensional linear systems, *Contemporary Mathematics* Vol. 215, Amer. Math. Soc., Providence, Rhode Island, 1998, 299–322.

[9] R. Obaya and M. Paramio, Directional differentiability of the rotation number for the almost periodic Schrödinger equation, *Duke Math. J.* **66** (1992), 521–552.

[10] R.J. Sacker and G.R. Sell, A spectral theory for linear differential systems, *J. Differential Equations* **27** (1978), 320–358.

[11] K. Schmidt, Amenability, Kazhdan's property T, strong egodicity and invariant means for ergodic group actions, *Ergod. Th. Dynam. Sys.* **1** (1981), 223–236.

Is Convexity Useful for the Study of Monotonicity?[1]

Jean-Paul Penot

Mathématiques appliquées C.N.R.S. 2070, Faculté des sciences, BP 1155, 64013 PAU cedex, France

Abstract. Recently, some representations of maximal monotone operators by convex functions have been devised. Here we raise the question: can one get results about maximal monotone operators from classical tools of convex analysis? We propose partial answers to this question.

1 Introduction

The links and analogies between convex functions and maximal monotone operators are numerous and striking. Among them are local boundedness on the interiors of their domains [27], the Bronsted-Rockafellar theorem [31], [33], qualification conditions for calculus rules of sums and compositions [1], [29] etc...

It would be of interest to find means to pass from results about convex functions to results about maximal monotone operators. The representation theorems of Fitzpatrick [12], Krauss [13]-[16], Martinez-Legaz and Théra [17] give some hopes in this direction. In [8] a criteria for the maximal monotonicity of an operator is provided in terms of the Fitzpatrick function associated with an operator and of its conjugate. One may wonder whether the main results about maximal monotone operators can be derived from classical facts in convex analysis.

In the first section of the present paper we exhibit a convex function on $X \times X^*$ associated with any operator from a normed vector space (n.v.s.) X to its dual. When the operator is monotone, this function majorizes the coupling function c. In the second section we exhibit functions which can serve to characterize the domains and the ranges of maximal monotone operators. However, these functions are non convex, so that the virtual convexity of these sets cannot be established with the help of such functions. In the third section we study convex functions which are close to functions introduced by Simons and used by him and his collaborators in a number of papers (see [30], [31] for recent references). However, the proofs of the results rely not only on properties of convex functions, but also make use of techniques from the study of monotone operators, so that our query is only partially answered. In the last section, we revisit some results about sums given in [5].

[1]Dedicated to Stephen Simons on the occasion of his retirement

After getting the results of this paper, we discovered they are close to those in the book [30] by Simons, so that we made a comparison with his approach. The main difference lies in the fact that our functions are defined on $X \times X^*$ instead of the free vector space generated by the graph of the operator (or the convex hull of the graph in that space). In this sense, the "big convexification" of Simons can be seen as a linearization rather than a convexification. Although both approaches are instances of relaxation processes, the framework we use is simpler as it does not go beyond the involved spaces. We hope this facility will incite the reader to return to the monograph [30] and to the early references on monotone operators.

2 The convex function associated with an operator

The observation of the following lemma is obvious, but useful.

Lemma 1 *Given a subset S of a vector space Z and a function $f : S \to \mathbb{R}$, the family of convex functions g on Z such that $g \mid S \leq f$ has a largest element g_S. It is given by*

$$g_S(x) = \inf\{\sum_{i=1}^{m} t_i f(x_i) : m \in \mathbb{N},\ x = \sum_{i=1}^{m} t_i x_i,\ x_i \in S,\ t_i \geq 0, \sum_{i=1}^{m} t_i = 1\}. \tag{2.1}$$

Here we use the convention that $\inf \emptyset = +\infty$, so that the domain of g_S is the convex hull $\text{co}(S)$ of S.

Proof. The first assertion stems from the fact that if $(g_i)_{i \in I}$ is a family of convex functions satisfying $g_i \mid S \leq f$ for each $i \in I$, then $g := \sup_{i \in I} g_i$ is convex and such that $g \mid S \leq f$. The second one is a simple verification. ∎

The preceding construction enjoys pleasant rules.

Lemma 2 *For any function f on S, any affine function h on Z and any $\lambda \in \mathbb{R}_+$ the largest convex function g such that $g \mid S \leq \lambda f$ (resp. $g \mid S \leq f + h \mid S$) is λg_S (resp. $g_S + h$).*

Proof. If g is a convex function such that $g \mid S \leq f + h \mid S$ then $(g - h) \mid S \leq f$, so that $g - h \leq g_S$ and $g \leq g_S + h$. The reverse inequality is obtained by changing h in $-h$. The assertion with λf is obtained in a similar way or from the construction. ∎

In the sequel we extend any function h on a subset C of Z by setting $h(z) = +\infty$ for $z \in Z \backslash C$. Viewed in this way, the extension g_S of f does not depend on the ambient space in the following sense: if $S \subset Z \subset W$, where W and Z are two vector spaces, then g_S is the restriction to Z of the convex extension of f to W defined as in the preceding lemma. In order to compare the function we introduce with the "big convexification" used in [30], let us denote by $W := \mathbb{R}^{(Z)}$ the free vector space generated by Z, i.e. the algebraic dual of the product space $\mathbb{R}^Z$, or the sum of Z copies of $\mathbb{R}$: W is the subset of $\mathbb{R}^Z$ formed of those elements $(r_z)_{z\in Z}$ whose components are zero but a finite number of them. There is a natural embedding $\delta : Z \to W$ given by $\delta(\overline{z}) = p_{\overline{z}}$ where $p_{\overline{z}}$ is the projection $(r_z)_{z\in Z} \mapsto r_{\overline{z}}$ (i.e. the evaluation at $\overline{z}$ if $\mathbb{R}^Z$ is identified with the space of functions from Z to $\mathbb{R}$ or the Dirac measure with support $\overline{z}$). There is also a canonical mapping $\pi : W \to Z$ given by

$$\pi(\sum_{y\in Y} r_y\delta(y)) = \sum_{y\in Y} r_y y \tag{2.2}$$

for any finite subset Y of Z and any family $(r_y)_{y\in Y}$ of real numbers. Note that π is linear and surjective and that δ is injective, but nonlinear. Moreover one has $\pi \circ \delta = I_Z$, the identity mapping. Now, any function f on a subset S of Z with values in a vector space has a unique linear extension f^W to W, hence to $\mathrm{co}(\delta(S))$, given for $w := \sum_{y\in Y} r_y\delta(y)$, where Y is a finite subset of Z and $(r_y)_{y\in Y}$ is a family of real numbers, by

$$f^W(w) = \sum_{y\in Y\cap S} r_y f(y) \text{ if } Y\cap S \neq \emptyset, \quad f^W(w) = 0 \text{ if } Y\cap S = \emptyset. \tag{2.3}$$

The following statement, whose proof follows from the definitions (2.1) and (2.3), clarifies the links between the convexification g_S of f and its big convexification (or rather linearization) f^W.

Lemma 3 *For any $w \in \mathrm{co}(\delta(S))$ one has $g_S(\pi(w)) \leq f^W(w)$. Moreover, for any $z \in \mathrm{co}(S)$ one has $g_S(z) = \inf\{f^W(w) : w \in \mathrm{co}(\delta(S)),\ \pi(w) = z\}$.*

When Z is a normed vector space (n.v.s.) one obtains in a similar way the following topological counterpart.

Lemma 4 *There is a greatest element $\overline{g_S}$ of the set H of lower semicontinuous (l.s.c.) convex functions h on Z such that $h \mid S \leq f$; it is the lower semicontinuous hull $\overline{g_S}$ of g_S. When $\overline{g_S}$ does not take the value $-\infty$, it is given by $\overline{g_S} = (f + \iota_S)^{**}$, where ι_S is the indicator function of S given by $\iota_S(z) = 0$ for $z \in S$, $\iota_S(z) = +\infty$ when $z \in Z\backslash S$.*

Note that the domain of $\overline{g_S}$ is contained in the closure of the convex hull co$S$. Also, observe that when X is reflexive, for any $A \subset X \times X^*$, $A^{-1} \subset X^* \times X$ can be considered as an operator from X^* to its dual and then

$$g_{A^{-1}}(x^*, x) = g_A(x, x^*) \qquad \forall (x, x^*) \in X \times X^*.$$

The following result has been inspired by [30] Lemma 9.1; however, note that it takes place in the space $Z := X \times X^*$ which is much smaller than the space in which is taken the "big convexification" of [30] which is the space $\mathbb{R}^{(X\times X^*)}$ i.e. the (algebraic) dual of the space $\mathbb{R}^{X\times X^*}$ or the sum of $X \times X^*$ copies of $\mathbb{R}$. In the sequel, we denote by c the coupling function on $Z := X \times X^*$:

$$c(x, x^*) := \langle x^*, x\rangle.$$

It is quite natural to expect that such a function on Z plays a key role in the study of monotone operators. That is shown in the next proposition. We are indebted to J.E. Martínez-Legaz for the implication (f)⇒(a); this implication is also mentioned in [30] and [35].

Proposition 5 *For a nonempty subset S of $Z := X \times X^*$ the following assertions are equivalent:*

(a) S is monotone;

(b) the greatest convex function g_S on $\mathrm{co}(S)$ such that $g_S \mid S \leq c \mid S$ satisfies $g_S(x,x^*) + g_S(y,y^*) \geq \langle x^*, y\rangle + \langle y^*, x\rangle$ for any $(x,x^*), (y,y^*) \in \mathrm{co}(S)$.

(c) the greatest convex function g_S on $\mathrm{co}(S)$ such that $g_S \mid S \leq c \mid S$ satisfies $g_S \geq c$ on $\mathrm{co}(S)$. In particular $g_S \mid S = c \mid S$.

(d) the greatest l.s.c. convex function $\overline{g_S}$ on $X \times X^*$ such that $\overline{g_S} \mid S \leq c \mid S$ satisfies $\overline{g_S} \geq c$ on $X \times X^*$. In particular $\overline{g_S} \mid S = c \mid S$.

(e) there exists a closed proper convex function $\overline{g}$ on $X \times X^*$ such that $\overline{g} \geq c$ and $\overline{g} \mid S = c \mid S$.

(f) there exists a convex function g on $\mathrm{co}(S)$ such that $g \geq c \mid \mathrm{co}(S)$ and $g \mid S = c \mid S$.

Proof. a)⇒(b) If S is monotone, then for any (x,x^*), $(y,y^*) \in S$ one has

$$\langle x^*, x\rangle \geq -\langle y^*, y\rangle + \langle x^*, y\rangle + \langle y^*, x\rangle,$$

hence, since the right hand side is affine in (x,x^*),

$$g_S(x^*,x) \geq -\langle y^*, y\rangle + \langle x^*, y\rangle + \langle y^*, x\rangle \qquad \forall (x,x^*) \in \mathrm{co}(S),\ \forall (y,y^*) \in S$$

or

$$\langle y^*, y\rangle \geq -g_S(x^*,x) + \langle x^*, y\rangle + \langle y^*, x\rangle.$$

Again, we can deduce from this inequality in which we fix $(x,x^*) \in \mathrm{co}(S)$ that

$$g_S(y^*,y) \geq -g_S(x^*,x) + \langle x^*, y\rangle + \langle y^*, x\rangle \qquad \forall (y,y^*) \in \mathrm{co}(S).$$

(b)⇒(c) Taking $(y,y^*) = (x,x^*) \in \mathrm{co}(S)$ in the relation of (b) we get $g_S(x^*,x) \geq \langle x^*, x\rangle$.

(c)⇒(d) This follows from the continuity of c.

(d)⇒(e) The fact that $\overline{g} := \overline{g_S}$ is bounded below by c ensures that it is proper.

(e)⇒(f) being obvious, let us show that (f)⇒(a). Let g be a convex function g on $\mathrm{co}(S)$ such that $g \geq c$ and $g \mid S = c \mid S$. Then, for any (x,x^*), (y,y^*) in S one has

$$\begin{aligned}\frac{1}{2}\langle x^*, x\rangle + \frac{1}{2}\langle y^*, y\rangle &= \frac{1}{2}g(x,x^*) + \frac{1}{2}g(y,y^*)\\ &\geq g(\frac{1}{2}(x+y), \frac{1}{2}(x^*+y^*)) \geq c(\frac{1}{2}(x+y), \frac{1}{2}(x^*+y^*))\\ &= \frac{1}{4}\langle x^*, x\rangle + \frac{1}{4}\langle y^*, y\rangle + \frac{1}{4}\langle x^*, y\rangle + \frac{1}{4}\langle y^*, x\rangle,\end{aligned}$$

so that $\langle x^* - y^*, x - y\rangle \geq 0$: S is monotone. ■

Corollary 6 *If S and T are monotone subsets of $Z := X \times X^*$ such that $S \subset T$, then one has $g_T \leq g_S$.*

Proof. This follows from the fact that $g_T \mid S = (g_T \mid T) \mid S = (c \mid T) \mid S = c \mid S$, hence $g_T \leq g_S$ as g_T is convex. ■

Corollary 7 *If S is a nonempty monotone subset of $Z := X \times X^*$, then $(g_S + \iota_{coS})^* \leq (c + \iota_{coS})^*$.*

Corollary 8 *If S is a nonempty monotone subset of $Z := X \times X^*$, then, for any $z := (x, x^*) \in X \times X^*$, one has $g_S(z) + \frac{1}{2}\|x\|^2 + \frac{1}{2}\|x^*\|^2 \geq 0$.*

Proof. This follows from assertion (c) of the preceding proposition and the inequality $\langle x^*, x\rangle + \frac{1}{2}\|x\|^2 + \frac{1}{2}\|x^*\|^2 \geq 0$. ■

The function $(x, x^*) \mapsto \frac{1}{2}\|x\|^2 + \frac{1}{2}\|x^*\|^2$ is equal to its conjugate on $X \times X^*$ when $X \times X^*$ is paired with itself by $\langle (x, x^*), (y, y^*)\rangle = \langle x^*, y\rangle + \langle y^*, x\rangle$. One may wonder whether some consequences could be drawn from this observation and from the preceding relation.

Proposition 9 *If S is maximal monotone, then $S = \{z \in Z : g_S(z) = c(z)\}$.*

Proof. We already know that $S \subset S' := \{z \in Z : g_S(z) = c(z)\}$ and the proof of Proposition 5 shows that S' is monotone. Since S is maximal monotone, we have $S' = S$. ■

Proposition 10 *If S is a nonempty monotone subset of $Z := X \times X^*$, then for any $(x, x^*) \in S$ one has*

$$(x^*, x) \in \partial g_S(x, x^*), \quad (x, x^*) \in \partial g_S^*(x^*, x).$$

Proof. The first assertion follows from the inequality

$$\langle w^*, w\rangle \geq \langle x^*, w - x\rangle + \langle w^* - x^*, x\rangle + \langle x^*, x\rangle$$

for any $(x, x^*) \in S^0$, $(w, w^*) \in S$ which shows that for any operator S and any $(x, x^*) \in S \cap S^0$ one has $(x^*, x) \in \partial(c + \iota_S)(x, x^*)$. Thus, when S is monotone, for any $(x, x^*) \in S$ one has $(x^*, x) \in \partial(c + \iota_S)^{**}(x, x^*)$ and $(c+\iota_S)^{**}(x, x^*) = \langle x^*, x\rangle = g_S(x, x^*)$. Since $g_S \geq (c+\iota_S)^{**}$, the first assertion follows. By a general fact, the second one is a consequence of that assertion. ■

Remark. The function g_S^* is precisely the Fitzpatrick's function associated with S and the second assertion is given in [12] Theorem 3.4. This function also satisfies $g_S^* \geq c$ and when S is maximal monotone, one has $S = \{(x, x^*) \in Z : g_S^*(x, x^*) = c(x, x^*)\}$.

3 Characterization of domains and ranges

Given an operator $A : X \rightrightarrows X^*$ (identified with its graph $G(A)$ when there is no risk of confusion) with nonempty domain $D(A)$ and $x \in X$ we set

$$j_A(x) := \sup_{\substack{(w,w^*)\in A \\ w\neq x}} \langle w^*, \frac{x-w}{\|x-w\|}\rangle, \quad j_A(x) := \sup_{\substack{(w,w^*)\in \mathrm{co}\, A \\ w\neq x}} \frac{1}{\|x-w\|}\left(\langle w^*, x\rangle - g_A(w,w^*)\right),$$

$$\ell_A(x) := \lim_{r\to\infty} \sup_{\substack{(w,w^*)\in A \\ \|w^*\|>r}} \frac{\langle w^*, x\rangle - g_A(w,w^*)}{\|x-w\|} = \inf_{r>0} \sup_{\substack{(w,w^*)\in A \\ \|w^*\|>r}} \frac{\langle w^*, x\rangle - g_A(w,w^*)}{\|x-w\|},$$

$$\ell_A^c(x) := \lim_{r\to\infty} \sup_{\substack{(w,w^*)\in \mathrm{co} A \\ \|w^*\|>r}} \frac{\langle w^*, x\rangle - g_A(w,w^*)}{\|x-w\|} = \inf_{r>0} \sup_{\substack{(w,w^*)\in \mathrm{co} A \\ \|w^*\|>r}} \frac{\langle w^*, x\rangle - g_A(w,w^*)}{\|x-w\|}.$$

Although j_A and ℓ_A are not convex functions, they have interesting properties: for $A, B : X \rightrightarrows X^*$

$$j_A \leq j_A + h_B,$$
$$A \subset B \Longrightarrow j_A \leq h_B, \quad \ell_A \leq \ell_B.$$

We also observe that if A is monotone, then, by Proposition 5 (c) we have $\ell_A \leq j_A$,

$$D(A) \subset \mathrm{dom}\, j_A \subset \mathrm{dom}\, \ell_A.$$

In fact, for any $x \in D(A)$ and any $x^* \in A(x)$, we have, for any $(w, w^*) \in A$, $\langle w^*, x - w\rangle \leq \langle x^*, x - w\rangle \leq \|x^*\| \, \|x - w\|$, hence $\ell_A(x) \leq j_A(x) \leq \|x^*\|$.

Moreover, whenever $k \geq j_A(x)$, one has $\langle w^*, x - w\rangle \leq k\, \|x - w\|$ for any $(w, w^*) \in A$, hence, by assertion (c) of Proposition 5

$$g_A(w, w^*) \geq \langle w^*, w\rangle \geq \langle w^*, x\rangle - k\, \|x - w\|.$$

Proposition 11 *For a monotone operator $A \subset X \times X^*$ and $x \in X$, among the following assertions one has the implications (a)⇒(b)⇔(c)⇔(d)⇔(e)⇔(f). If A is maximal monotone, all these assertions are equivalent:*

(a) $x \in D(A)$;

(b) there exists $x^* \in X^*$ *such that* $g_A(w, w^*) \geq \langle w^*, x\rangle + \langle x^*, w - x\rangle$ *for any* $(w, w^*) \in \mathrm{co}\, A$;

(c) there exists $k \in \mathbb{R}_+$ *such that* $g_A(w, w^*) \geq \langle w^*, x\rangle - k\, \|w - x\|$ *for any* $(w, w^*) \in \mathrm{co}\, A$;

(d) $x \in \mathrm{dom}\, j_A$;

(e) there exists $k \in \mathbb{R}_+$ *such that* $g_A(w, w^*) \geq \langle w^*, x\rangle - k\, \|w - x\|$ *for any* $(w, w^*) \in X \times X^*$;

(f) there exists $x^* \in X^*$ *such that* $g_A(w, w^*) \geq \langle w^*, x\rangle + \langle x^*, w - x\rangle$ *for any* $(w, w^*) \in X \times X^*$;

(g) $x \in \mathrm{dom}\, \ell_A^c$.

We will deduce this result from the fact that $g_A = c$ on A when A is monotone and from the following lemma of independent interest in which we set

$$A^0 := \{(x, x^*) \in X \times X^* : \langle w^* - x^*, w - x\rangle \geq 0 \,\forall (w, w^*) \in A\}.$$

We note, following [18] that A is monotone iff $A \subset A^0$ and that A is maximal monotone iff $A = A^0$. These observations show that the proposition is indeed a consequence of the lemma.

Lemma 12 *For any operator $A \subset X \times X^*$ and $x \in X$, the following assertions are equivalent:*

(a^0) $x \in D(A^0)$;

(b) there exists $x^* \in X^*$ such that $g_A(w, w^*) \geq \langle w^*, x\rangle + \langle x^*, w - x\rangle$ for any $(w, w^*) \in \text{co } A$;

(c) there exists $k \in \mathbb{R}_+$ such that $g_A(w, w^*) \geq \langle w^*, x\rangle - k\|w - x\|$ for any $(w, w^*) \in \text{co } A$;

(d) $x \in \text{dom } j_A$;

(e) there exists $k \in \mathbb{R}_+$ such that $g_A(w, w^*) \geq \langle w^*, x\rangle - k\|w - x\|$ for any $(w, w^*) \in X \times X^*$;

(f) there exists $x^* \in X^*$ such that $g_A(w, w^*) \geq \langle w^*, x\rangle + \langle x^*, w - x\rangle$ for any $(w, w^*) \in X \times X^*$;

Moreover, when A is monotone, the preceding assertions are equivalent to the following one:

(g) $x \in \text{dom } \ell_A^c$.

Proof. (a^0)$\Rightarrow$(b) Given $x \in D(A^0)$, $x^* \in A^0(x)$, we see that c majorizes the affine function $(w, w^*) \mapsto \langle w^*, x\rangle + \langle x^*, w - x\rangle$ on A, hence g_A majorizes this function on co A.

(b)$\Rightarrow$(c) The implication follows from $\langle x^*, w - x\rangle \geq -k\|w - x\|$ with $k := \|x^*\|$.

(c)$\Leftrightarrow$(d) is an obvious consequence of the definitions.

(c)$\Rightarrow$(e) stems from the fact that g_A takes the value $+\infty$ on $X \times X^* \backslash \text{co } A$.

(e)$\Rightarrow$(f) Assertion (e) amounts to $g_A(v + x, w^*) - \langle w^*, x\rangle \geq -k\|v\|$ for any $(v, w^*) \in X \times X^*$, so that, by the sandwich theorem, one can find some $(u, x^*) \in X \times X^*$, $r \in \mathbb{R}$ such that for any $(v, w^*) \in X \times X^*$ one has

$$g_A(v + x, w^*) - \langle w^*, x\rangle \geq \langle x^*, v\rangle + \langle w^*, u\rangle + r \geq -k\|v\|.$$

The second inequality shows that $r \geq 0$ and that the linear form $w^* \mapsto \langle w^*, u\rangle$ is bounded below on X^*, hence is null: $u = 0$. Thus, setting $w := x + v$, one sees that (f) holds.

(f)$\Rightarrow$(a^0) When (x, x^*) is such that $g_A(w, w^*) \geq \langle w^*, x\rangle + \langle x^*, w - x\rangle$ for any $(w, w^*) \in A$, since $c \mid A \geq g_A \mid A$, we get $(x, x^*) \in A^0$.

(e)$\Rightarrow$(g) is obvious. Let us prove that (g)$\Rightarrow$(e) when A is monotone. Let $k > \ell_A^c(x)$. There exists $r > 0$ such that

$$\sup_{\substack{(w,w^*)\in \text{co } A \\ \|w^*\| > r}} \frac{\langle w^*, x\rangle - g_A(w, w^*)}{\|x - w\|} < k.$$

Thus, for $(w,w^*) \in \text{co}\, A$ with $\|w^*\| > r$, one has $g_A(w,w^*) \geq \langle w^*, x\rangle - k\,\|w-x\|$. Since for $(w,w^*) \in \text{co}\, A$ with $\|w^*\| \leq r$ one has

$$g_A(w,w^*) \geq \langle w^*, w\rangle \geq \langle w^*, x\rangle - r\,\|w-x\|,$$

assertion (e) holds with k replaced with $k' := \max(k,r)$. ∎

Using the fact that for any operator A one has $R(A) = D(A^{-1})$, and $g_{A^{-1}}(w^*,w) = g_A(w,w^*)$, we get the following characterization of the range of a maximal monotone operator.

Proposition 13 *Suppose X is reflexive. For a maximal monotone operator $A \subset X \times X^*$ and $x^* \in X$, the following assertions are equivalent:*

(a) $x^* \in R(A)$;

(b) there exists $x \in X$ such that $g_A(w,w^*) \geq \langle x^*, w\rangle + \langle w^* - x^*, x\rangle$ for any $(w,w^*) \in \text{co}\, A$;

(c) there exists $k \in \mathbb{R}_+$ such that $g_A(w,w^*) \geq \langle x^*, w\rangle - k\,\|w^* - x^*\|$ for any $(w,w^*) \in \text{co}\, A$;

(d) $x^* \in \text{dom}\, j_A$;

(e) there exists $k \in \mathbb{R}_+$ such that $g_A(w,w^*) \geq \langle x^*, w\rangle - k\,\|w^* - x^*\|$ for any $(w,w^*) \in X \times X^*$;

(f) there exists $x \in X$ such that $g_A(w,w^*) \geq \langle x^*, w\rangle + \langle w^* - x^*, x\rangle$ for any $(w,w^*) \in X \times X^*$.

4 Approximate characterizations of domains and ranges

Although the functions introduced in the preceding sections provide characterizations of the domain and the range of a maximal monotone operator, they cannot be used to prove almost or virtual convexity of these sets. The present section focuses on such a question.

Given $A \subset X \times X^*$, let us introduce the following variants of Simons' functions ψ_A and χ_A on X by setting

$$s_A(x) := \sup_{(w,w^*)\in A} \frac{\langle w^*, x\rangle - g_A(w,w^*)}{\|w\|+1}, \qquad s_A^c(x) := \sup_{(w,w^*)\in \text{co}\, A} \frac{\langle w^*, x\rangle - g_A(w,w^*)}{\|w\|+1}.$$

For an arbitrary operator A these functions do not coincide with the functions ψ_A and χ_A given by

$$\psi_A(x) := \sup_{(w,w^*)\in A} \frac{\langle w^*, x-w\rangle}{\|w\|+1} \qquad \chi_A(x) := \sup_{m\geq 1}\ \sup_{(t_1,\ldots,t_m)\in\Delta_m}\ \sup_{(w_i,w_i^*)\in A} \frac{\sum_{i=1}^m t_i\langle w_i^*, x-w_i\rangle}{\|\sum_{i=1}^m t_i w_i\|+1},$$

where $\Delta_m := \{(t_1,\ldots,t_m) \in \mathbb{R}_+^m : t_1 + \ldots + t_m = 1\}$. However, since $g_A \mid A = c \mid A$ when A is monotone, one has $s_A = \psi_A$ in that case. A comparison between s_A^c and χ_A is included in the following lemma.

Lemma 14 *For any operator A one has $s_A^c \geq \chi_A \geq \psi_A$. When A is monotone one has $s_A = \psi_A \leq \chi_A \leq s_A^c$.*

Proof. Let $m \geq 1$, $(t_1, ..., t_m) \in \Delta_m$, $(w_i, w_i^*) \in A$ for $i = 1, ..., m$. Let $(w, w^*) := \sum_{i=1}^m t_i(w_i, w_i^*) \in \text{co } A$. Then $\|\sum_{i=1}^m t_i w_i\| = \|w\|$ and, by the second part of Lemma 1,

$$\sum_{i=1}^m t_i \langle w_i^*, x - w_i \rangle = \langle w^*, x \rangle - \sum_{i=1}^m t_i \langle w_i^*, w_i \rangle \leq \langle w^*, x \rangle - g_A(w, w^*).$$

The result follows by taking the supremum over m, $(t_1, ..., t_m) \in \Delta_m$, $(w_i, w_i^*) \in A$. ■

The functions s_A and s_A^c are lower semicontinuous convex functions on X. Let us show that they share with χ_A the following property.

Proposition 15 *For any monotone operator one has $D(A) \subset \text{dom } s_A^c \subset \text{dom } \chi_A \subset \text{dom } s_A$, so that s_A, χ_A and s_A^c are closed proper convex functions when A is nonempty. If A is maximal monotone then*

$$\text{core dom } s_A^c = \text{int } D(A) = \text{int dom } s_A^c.$$

Proof. Given $x \in D(A)$, let $x^* \in A(x)$. Then, for any $m \geq 1$, $(t_1, ..., t_m) \in \Delta_m$, $(w_i, w_i^*) \in A$ for $i = 1, ..., m$ and $(w, w^*) := \sum_{i=1}^m t_i(w_i, w_i^*) \in \text{co } A$ we have $\langle w_i^*, x - w_i \rangle \leq \langle x^*, x \rangle - \langle x^*, w_i \rangle$, hence

$$\sum_{i=1}^m t_i \langle w_i^*, x - w_i \rangle \leq \|x^*\| \, \|x\| - \langle x^*, \sum_{i=1}^m t_i w_i \rangle \leq \|x^*\| \left(\|x\| + \|w\| \right)$$

Taking the supremum over the decompositions of (w, w^*) as convex combinations of elements of A, it follows that

$$\langle w^*, x \rangle - g_A(w, w^*) \leq \|x^*\| \left(\|x\| + \|w\| \right) \leq k(1 + \|w\|) \tag{4.1}$$

for $k := \|x^*\| \max(\|x\|, 1)$. Thus $s_A^c(x) \leq k$ and $D(A) \subset \text{dom } s_A^c$.

Now, let $x \in \text{core dom } s_A^c$. Since s_A is a closed proper convex function, there exist some $r > 0$ and some $m > 0$ such that $s_A^c(x + u) \leq m$ for any $u \in rB_X$, the closed ball with center 0 and radius r in X. Thus, by definition of s_A^c one has $\langle w^*, x+u \rangle - g_A(w, w^*) \leq m \left(\|w\| + 1 \right) \leq m \|w - x\| + m \|x\| + m$ for any $u \in rB_X$, $(w, w^*) \in \text{co } A$, hence, by taking the supremum over $u \in rB_X$,

$$\langle w^*, x \rangle + r \|w^*\| \leq g_A(w, w^*) + m \|w - x\| + m \|x\| + m \qquad \forall (w, w^*) \in \text{co } A. \tag{4.2}$$

When $r \|w^*\| \geq m \|x\| + m$, this inequality implies

$$g_A(w, w^*) \geq \langle w^*, x \rangle - m \|w - x\| \qquad \forall (w, w^*) \in \text{co } A,$$

When $\|w^*\| < m' := r^{-1}(m\|x\| + m)$, the inequality $g_A(w,w^*) \geq \langle w^*, w\rangle$ ensures that $g_A(w,w^*) \geq \langle w^*, x\rangle - m'\|w - x\|$. Setting $k := \max(m, m')$ we obtain assertion (c) of Proposition 11, so that $x \in D(A)$. Since core dom s_A^c is open, we get core dom $s_A^c \subset$ int $D(A) \subset$ core $D(A)$. The reverse inclusions are consequences of the inclusion $D(A) \subset$ dom s_A^c. ∎

Let us note in passing that the preceding proof easily yields a famous result of Rockafellar [27]; see also [3] and [23].

Theorem 16 *Any maximal monotone operator A is locally bounded on the interior of its domain D(A).*

Proof. Taking $(w, w^*) \in A$ in relation (4.2), so that $g_A(w,w^*) = \langle w^*, w\rangle$, we have

$$r\|w^*\| \leq \langle w^*, w - x\rangle + m\|w - x\| + m\|x\| + m.$$

Thus, for any $r' \in (0, r)$, for any $(w, w^*) \in A$ with $w \in x + r'B_X$ we get

$$(r - r')\|w^*\| \leq mr' + m\|x\| + m,$$

hence $\|w^*\| \leq (r - r')^{-1}(mr' + m\|x\| + m)$. ∎

Now let us present a characterization of the closure of the range of a maximal monotone operator.

Given a monotone operator $A : X \rightrightarrows X^*$ with nonempty domain, we set, for $x^* \in X^*$,

$$t_A(x^*) := s_{A^{-1}}(x^*) = \sup_{(w,w^*)\in A} \frac{\langle x^* - w^*, w\rangle}{\|w^*\| + 1},$$

$$n_A(x^*) := \limsup_{\|w\|\to\infty,\ (w,w^*)\in A} \frac{\langle x^* - w^*, w\rangle}{\|w^*\| + 1} = \inf_{r\geq 0} \sup_{(w,w^*)\in A,\ \|w\|\geq r} \frac{\langle x^* - w^*, w\rangle}{\|w^*\| + 1}.$$

Clearly, t_A is a closed proper convex function and $n_A \leq t_A$. Since n_A is the limit of a decreasing family of closed convex functions, it is a convex function.

We first observe that if A is monotone one has

$$R(A) \subset \text{dom}\, t_A \subset \text{dom}\, n_A.$$

In fact, for any $x^* \in R(A)$, $x \in A^{-1}(x^*)$, one has for $(w, w^*) \in A$ $\langle x^* - w^*, w\rangle \leq \langle x^* - w^*, x\rangle \leq \|x\|\|w^*\| + \|x\|\|x^*\|$. The following result slightly extends [5] Lemma 1 as the interiors are replaced with the cores (or algebraic interiors).

Lemma 17 *Let A be a maximal monotone operator from X into X*, where X is reflexive. Then*

$$R(A) \subset \text{dom}\, t_A \subset \text{dom}\, n_A \subset \text{cl}(R(A)),$$

$$\text{core dom}\, t_A \subset \text{core dom}\, n_A \subset R(A).$$

Proof. The first two inclusions have been observed above. Let $x^* \in \operatorname{dom} n_A$ and let $c > n_A(x^*)$. There exists $r > 0$ such that

$$(w, w^*) \in A,\ \|w\| > r \Rightarrow \langle x^* - w^*, w\rangle \leq c\,\|w^*\| + c. \tag{4.3}$$

Since A is maximal monotone, for any $\varepsilon \in (0,1)$ there exists $w_\varepsilon \in X$ such that

$$x^* \in \varepsilon J(w_\varepsilon) + A(w_\varepsilon). \tag{4.4}$$

If there exists a sequence (ε_n) in $(0,1)$ such that $\|w_{\varepsilon_n}\| \leq r$, we get that $x^* = \lim_n(x^* - \varepsilon_n J(w_{\varepsilon_n})) \in \operatorname{cl}(R(A))$. Assume that for some $\alpha > 0$, $\|w_\varepsilon\| > r$ for each $\varepsilon \in (0,\alpha)$. Taking $(w,w^*) = (w_\varepsilon, x^* - \varepsilon J(w_\varepsilon))$ in (4.3), we get

$$\varepsilon\,\|w_\varepsilon\|^2 = \langle \varepsilon J(w_\varepsilon), w_\varepsilon\rangle \leq c\,\|x^* - \varepsilon J(w_\varepsilon)\| + c \leq c\varepsilon\,\|w_\varepsilon\| + c\,\|x^*\| + c,$$

hence

$$\varepsilon\,\|w_\varepsilon\|^2 \leq \frac{1}{2}\varepsilon\,\|w_\varepsilon\|^2 + \frac{1}{2}\varepsilon c^2 + c\,\|x^*\| + c,$$

and $\varepsilon\,\|w_\varepsilon\|^2 \leq \varepsilon c^2 + 2c\,\|x^*\| + 2c$. Then $\|\varepsilon J(w_\varepsilon)\| \to 0$ and we get $x^* = \lim_{\varepsilon\to 0}(x^* - \varepsilon J(w_\varepsilon)) \in \operatorname{cl}(R(A))$.

Now let $x^* \in \operatorname{core} \operatorname{dom} n_A$. For any $z^* \in X^*$ there exists some $t > 0$ such that $x^* + tz^* \in \operatorname{dom} n_A$. Let $c > n_A(x^* + tz^*)$ and $r > 0$ be such that

$$(w, w^*) \in A,\ \|w\| > r \Rightarrow \langle x^* + tz^* - w^*, w\rangle \leq c\,\|w^*\| + c.$$

Taking again $(w,w^*) = (w_\varepsilon, x^* - \varepsilon J(w_\varepsilon))$, with w_ε defined by (4.4), and using the fact that $\langle J(w_\varepsilon), w_\varepsilon\rangle \geq 0$, we get either $\|w_\varepsilon\| \leq r$ or $t\langle z^*, w_\varepsilon\rangle \leq c\,\|x^* - \varepsilon J(w_\varepsilon)\| + c \leq c'$ for some constant c' since $\varepsilon J(w_\varepsilon)$ is bounded. The uniform boundedness theorem yields that $(w_\varepsilon)_{\varepsilon\in(0,1)}$ is bounded. If w_0 is a weak cluster point of w_ε as $\varepsilon \to 0$, the closedness of A ensures that $x^* \in A(w_0) : x^* \in R(A)$. ∎

Theorem 18 *Let A be a maximal monotone operator from X into X^*, where X is reflexive. Then*

$$\begin{aligned}
\operatorname{cl}(R(A)) = \operatorname{cl}(\operatorname{co}R(A)) &= \operatorname{cl}(\operatorname{dom} t_A) = \operatorname{cl}(\operatorname{dom} n_A),\\
\operatorname{int}\operatorname{dom} t_A = \operatorname{core}\operatorname{dom} t_A &= \operatorname{core}\operatorname{dom} n_A = \operatorname{int}(R(A)),\\
\operatorname{cl}(D(A)) = \operatorname{cl}(\operatorname{co}D(A)) &= \operatorname{cl}(\operatorname{dom} s_A) = \operatorname{cl}(\operatorname{dom} s_A^c),\\
\operatorname{int}\operatorname{dom} s_A = \operatorname{core}\operatorname{dom} s_A &= \operatorname{core}\operatorname{dom} n_{A^{-1}} = \operatorname{int}(D(A))
\end{aligned}$$

Proof. The assertions concerning the range are direct consequences of the preceding lemma. The ones dealing with the domain can be deduced by replacing A with A^{-1}. ∎

5 The range of a sum

Given an operator $A \subset X \times X^*$ and a subset S^* of X^*, we set

$$M_A(S^*) := \{x \in X : \forall x^* \in S^* \quad \sup_{(w,w^*)\in A} \langle w^* - x^*, x - w\rangle < +\infty\},$$

or, equivalently, using the function m_A given by $m_A(x, x^*) := \sup_{(w,w^*)\in A}\langle w^* - x^*, x - w\rangle$,

$$M_A(S^*) = \{x \in X : \{x\} \times S^* \subset \operatorname{dom} m_A\}.$$

For $S^* := \{x^*\}$, with $x^* \in R(A)$ one has $A^{-1}x^* \subset M_A(S^*)$, as easily seen. Moreover,

$$M_A(S^*) \neq \emptyset \Rightarrow S^* \subset \operatorname{dom} t_A. \tag{5.1}$$

In fact, when $x \in M_A(S^*)$, for any $x^* \in S^*$ there exists some $c \in \mathbb{R}$ such that for any $(w, w^*) \in A$ one has

$$\begin{aligned}\langle x^* - w^*, w\rangle &\leq \langle x^* - w^*, x\rangle + c \\ &\leq \|x\|\,\|x^*\| + \|x\|\,\|w^*\| + c,\end{aligned}$$

hence

$$t_A(x^*) := \sup\{\frac{\langle x^* - w^*, w\rangle}{\|w^*\| + 1} : (w, w^*) \in A\} < +\infty.$$

The following result is close to [30] Theorem 19.4; it extends both [5] Theorem 3 and Theorem 4. In the first of these results one assumes that $D(A) \subset M_A(R(A))$, $D(B) \subset M_B(R(B))$ and $D(A) \cap D(B) \neq \emptyset$; in the second one, one assumes that $D(A) \subset D(B)$ and $D(B) \subset M_B(R(B))$, so that, for $x^* \in R(A)$, one has $A^{-1}x^* \subset M_A(\{x^*\}) \cap D(A) \subset M_A(\{x^*\}) \cap M_B(R(B))$.

Theorem 19 *Let A and B be monotone operators such that $A+B$ is maximal monotone. Let $S^*, T^* \subset X^*$ be such that $M_A(S^*) \cap M_B(T^*)$ is nonempty. Then $S^* + T^* \subset \operatorname{cl}\,(R(A + B))$.*

In particular, when $M_A(R(A)) \cap M_B(R(B))$ is nonempty one has $R(A) + R(B) \subset \operatorname{cl}\,(R(A + B))$.

Proof. Let $x \in M_A(S^*) \cap M_B(T^*)$. For any $x^* \in S^*$ and any $y^* \in T^*$ there exist $a(x^*), b(y^*)$ in $\mathbb{R}$ such that

$$\begin{aligned}&\forall (w, u^*) \in A \;\; \langle x^* - u^*, w\rangle \leq \langle x^* - u^*, x\rangle + a(x^*), \\ &\forall (w, v^*) \in B \;\; \langle y^* - v^*, w\rangle \leq \langle y^* - v^*, x\rangle + b(y^*),\end{aligned}$$

hence, adding sides by sides,

$$\forall w \in D(A+B), \forall w^* \in (A+B)(w) \;\; \langle x^* + y^* - w^*, w\rangle \leq \langle x^* + y^* - w^*, x\rangle + a(x^*) + b(y^*).$$

This means that $x \in M_{A+B}(S^* + T^*)$. Relation (5.1) and the preceding lemma yield the conclusion. ■

We have shown that the functions we studied above form a link between the study of monotone operators and convex analysis. Some questions would require further study.

Questions. 1) What are the relationships between the various functions involved in our study?

2) What about their subdifferentials and conjugates? In particular, can one get the density part of Theorem 18 by using the density of the domain of the subdifferential in the domain of the function?

3) What are the relationships between the various functions involved in our study and the ones which represent maximal monotone operators as in [12]-[16], [17]?

References

[1] H. Attouch, On the maximality of the sum of two maximal monotone operators, Nonlinear Anal. Th. Methods Appl. 5 (2) , 143-147 (1981).

[2] H. Attouch, *Variational convergence for functions and operators*, Pitman, London (1984).

[3] Borwein, J. and Fitzpatrick, S.P., Local boundedness of operators under minimal hypotheses, Bull. Austr. Math. Soc. 39, 439-441 (1989).

[4] Brézis, H., *Opérateurs maximaux monotones et semi-groupes de contractions dans les espaces de Hilbert*, North-Holland, Amsterdam (1971).

[5] Brézis, H. and Haraux, A., Image d'une somme d'opérateurs maximaux monotones et applications, Israël J. Math. 23 (2) (1976), 165-186.

[6] Burachik, R. S., Sagastizábal, C. A., Svaiter, B.F., ε-enlargements of maximal monotone operators: Theory and applications, Fukushima, M. et al. (eds.), *Reformulation: nonsmooth, piecewise smooth, semismooth and smoothing methods*, Kluwer, Boston. Appl. Optim. 22, 25-43 (1999).

[7] Burachik, R.S., Svaiter, B.F., ε-enlargements of maximal monotone operators in Banach spaces, Set-Valued Anal. 7, No.2, 117-132 (1999).

[8] Burachik, R.S., Svaiter, B.F., Maximal monotonicity, conjugation and the duality product, preprint, IMPA Rio, October 2001.

[9] Chu, Liang-Ju, On Brézis-Haraux approximation with applications, Far East J. Math. Sci. 4, (3), 425-442 (1996).

[10] Chu, Liang-Ju, On the sum of monotone operators, Mich. Math. J. 43, No.2, 273-289 (1996).

[11] Coodey, M. and Simons, S., The convex function determined by a multifunction, Bull. Australian Math. Soc. 54 (1996), 87-97.

[12] Fitzpatrick, S., Representing monotone operators by convex functions, *Functional Analysis and Optimization, workshop and miniconference*, Canberra, Australia 1988, 59-65, Proc. Center Math. Anal. Australian Nat. Univ. 20 (1988).

[13] Krauss, E., On the maximality of the sum of monotone operators, Math. Nachr. 101, 199-206 (1981).

[14] Krauss, E., A representation of maximal monotone operators by saddle functions, Rev. Roum. Math. Pures Appl. 30, 823-836 (1985).

[15] Krauss, E., A representation of arbitrary maximal monotone operators via subgradients of skew-symmetric saddle functions, Nonlinear Anal., Theory Methods Appl. 9, 1381-1399 (1985).

[16] Krauss, E., Maximal monotone operators and saddle functions. I., Z. Anal. Anwend. 5, 333-346 (1986).

[17] Martinez-Legaz, J.E. and Théra, M. A convex representation of maximal monotone operators, J. Nonlinear and Convex Anal. 2 (2) 243-247 (2001).

[18] Martinez-Legaz, J.E. and Svaiter, B.F., Work in progress and lecture in the 10th meeting of the Mode group of the SMAI, Montpellier, March 2002.

[19] Minty, G.J., Monotone (nonlinear) operators in Hilbert spaces, Duke Math. J. 29, 341-346 (1962).

[20] Moudafi, A, On the regularization of the sum of two maximal monotone operators, Nonlinear Anal., Theory Methods Appl. 42A, No.7, 1203-1208 (2000).

[21] Okazawa, N. and Yokota, T., Perturbations of maximal monotone operators applied to the nonlinear Schrödinger and complex Ginzburg-Landau equations, RIMS Kokyuroku 1105, 102-120 (1999).

[22] Pennanen, T., On the range of monotone composite mappings, J. Nonlinear and Convex Anal. (2) (2001), 193-202.

[23] Phelps, R.R., Convex Functions, *Monotone Operators and Differentiability*, Lecture Notes in Maths # 1364, Springer-Verlag, Berlin, 1989, (1993).

[24] Revalski, J. P., Théra, M., Generalized sums of monotone operators, C. R. Acad. Sci., Paris, Sér. I, Math. 329, No.11, 979-984 (1999).

[25] Revalski, J. P., Théra, M., Variational and extended sums of monotone operators, Théra, M. (ed.) et al., *Ill-posed variational problems and regularization techniques,* Lect. Notes Econ. Math. Syst. 477, Springer, Berlin, 229-246 (1999).

[26] Robinson, S. M., Composition duality and maximal monotonicity, Math. Program. 85A, No.1, 1-13 (1999).

[27] Rockafellar, R.T., Local boundedness of nonlinear, monotone operators, Michigan J. Math. 16, 397-407 (1969).

[28] Rockafellar, R.T., On the maximal monotonicity of subdifferential mappings, Pacific J. Math. 33, 209-216 (1970).

[29] Rockafellar, R.T., On the maximality of sums of nonlinear monotone operators, Trans. Amer. Math. Soc. 149, 75-88 (1970).

[30] Simons, S., *Minimax and Monotonicity*, Lecture Notes in Maths 1693, Springer, Berlin (1998).

[31] Simons, S., Maximal monotone multifunctions of Brøndsted-Rockafellar type, Set-Valued Anal. 7, No.3, 255-294 (1999).

[32] Simons, S., Sum theorems for monotone operators and convex functions, Trans. Amer. Math. Soc.

[33] Svaiter, B.F, A family of enlargements of maximal monotone operators, Set-Valued Anal. 8, No.4, 311-328 (2000).

[34] Verona, A., Verona, M. E., Regular maximal monotone operators, Set-Valued Anal. 6, No.3, 303-312 (1998).

[35] C. Zălinescu, *Convex Analysis in General Vector Spaces,* World Scientific, Singapore, (2002).

[36] Zeidler, E., *Nonlinear functional analysis and its applications. II/B: Nonlinear monotone operators,* Springer-Verlag, New York (1990).

p-Superlinear Problems with Jumping Nonlinearities

Kanishka Perera[1]

[1]Department of Mathematical Sciences, Florida Institute of Technology, Melbourne, Florida 32901, USA

Abstract. We obtain nontrivial solutions for a class of p-Laplacian problems that are p-superlinear at infinity and have jumping nonlinearities at zero. The proof is based on showing that the associated variational functional has a homological local linking near the origin.

AMS Subject Classification. 35J20

Keywords. p-Laplacian, p-superlinear problems, minimax eigenvalues, Yang index, variational methods, Morse theory, critical groups, homological local linking

1 Introduction

Consider the quasilinear elliptic boundary value problem

$$(1.1) \qquad \begin{cases} -\Delta_p u = f(x,u) & \text{in } \Omega \\ u = 0 & \text{on } \partial\Omega \end{cases}$$

where Ω is a bounded domain in $\mathbb{R}^N$, $N \geq 1$, $\Delta_p u = \operatorname{div}\left(|\nabla u|^{p-2} \nabla u\right)$ is the p-Laplacian, $1 < p < \infty$, and f is a Carathéodory function on $\Omega \times \mathbb{R}$ satisfying the subcritical growth condition

$$(1.2) \qquad |f(x,t)| \leq C\left(|t|^{q-1} + 1\right) \text{ for some } q < \begin{cases} Np/(N-p) & \text{if } p < N \\ \infty & \text{if } p \geq N. \end{cases}$$

As usual, C denotes a generic positive constant. This problem is called p-superlinear if there is a $\mu > p$ such that

$$(1.3) \qquad |f(x,t)| \geq C\left(|t|^{\mu-1} - 1\right).$$

To ensure that the associated variational functional

$$(1.4) \qquad \Phi(u) = \int_\Omega |\nabla u|^p - p\,F(x,u), \quad u \in W = W_0^{1,p}(\Omega),$$

where $F(x,t) = \displaystyle\int_0^t f(x,s)\,ds$, satisfies the Palais-Smale compactness condition (PS), it is customary to strengthen (1.3) to

$$(1.5) \qquad 0 < \mu\,F(x,t) \leq t f(x,t) \text{ for } |t| \text{ large}.$$

We assume that $f(x,0) \equiv 0$, so that $u = 0$ is a solution, and seek others.

Beginning with Ambrosetti and Rabinowitz [1], many authors have obtained nontrivial solutions of superlinear problems, under various assumptions on the behavior of f near zero, in the semilinear case $p = 2$. In contrast, there are only a handful of papers in the literature devoted to the quasilinear case $p \neq 2$. Dinca, Jebelean, and Mawhin [4] and Jiu and Su [6] considered the case

$$pF(x,t) \leq \overline{\lambda}\, |t|^p, \quad |t| \leq \delta \tag{1.6}$$

for some $\overline{\lambda} < \lambda_1$ and $\delta > 0$, while Liu [7] studied

$$\lambda_1\, |t|^p \leq pF(x,t) \leq \overline{\lambda}\, |t|^p, \quad |t| \leq \delta \tag{1.7}$$

with $\overline{\lambda} < \lambda_2$, where λ_1 and λ_2 are the first and the second Dirichlet eigenvalues of $-\Delta_p$, respectively. In this paper we consider the case where (1.1) has a jumping nonlinearity at zero:

$$f(x,t) = a\,(t^+)^{p-1} + b\,(t^-)^{p-1} + o(|t|^{p-1}) \quad \text{as } t \to 0, \text{ uniformly in } x \tag{1.8}$$

where $t^{\pm} = \max\{\pm t, 0\}$. Our main result involves a new sequence of variational eigenvalues $\lambda_l \to \infty$ recently constructed using a minimax scheme involving the Yang index by the author [8].

Theorem 1.1. *If f satisfies* (1.2), (1.5), *and* (1.8) *with $a, b \in (\lambda_l, \lambda_{l+1})$ for some l, then problem* (1.1) *has a nontrivial solution.*

Under (1.6), Dinca, Jebelean, and Mawhin [4] showed that Φ has a mountain-pass geometry, while the proof of Jiu and Su [6] was based on Morse theory. They showed that the critical groups of Φ at zero are given by $C_q(\Phi, 0) = \delta_{q0}\, \mathbb{Z}$, where δ is the Kronecker delta, while (1.5) implies those at infinity are all trivial. Since $C_0(\Phi, 0) \not\approx C_0(\Phi, \infty)$, there must then be a nontrivial critical point (see, e.g., Chang [3]).

Liu [7] showed that when (1.7) holds Φ has a local linking near the origin with respect to $W = W_1 \oplus W_2$ where W_1 is the 1-dimensional eigenspace associated with λ_1 and W_2 is a subspace complementing W_1, i.e.,

$$\begin{cases} \Phi(u) \leq 0 & \text{for } u \in W_1,\ \|u\| \leq r \\ \Phi(u) > 0 & \text{for } u \in W_2,\ 0 < \|u\| \leq r \end{cases} \tag{1.9}$$

for $r > 0$ sufficiently small. This implies that $C_1(\Phi, 0) \neq 0$ and again gives a nontrivial solution as $C_1(\Phi, \infty) = 0$.

The first difficulty that one encounters when attempting to examine the behavior of Φ near the origin under the hypotheses of Theorem 1.1 is that it is not known whether $\{\lambda_l\}$ is a complete list of eigenvalues. Moreover, there are no eigenspaces to work with, and hence the usual definition of local linking is inadequate. We will use the following generalization introduced by the author [9].

Definition 1.2 (1.1 of [9]). Φ has a homological local linking near the origin if there is a neighborhood B of 0 containing no other critical point and subsets S_1, B_1, and B_2 of B with $0 \notin S_1 \subset B_1 \cap B_2^c$ such that the embedding $\widetilde{H}_{l-1}(S_1) \to \widetilde{H}_{l-1}(B \setminus B_2)$ of reduced homology groups is nontrivial for some l, B_1 is contractible, and

$$\begin{cases} \Phi \leq 0 & \text{on } B_1 \\ \Phi > 0 & \text{on } B_2 \setminus \{0\}. \end{cases} \tag{1.10}$$

We will show that Φ has a homological local linking near the origin and hence $C_l(\Phi, 0) \neq 0$ by Theorem 3.1 of [9]. This will prove our theorem since $C_q(\Phi, \infty) = 0$ for all q.

2 Yang Index and Variational Eigenvalues

We briefly recall the definition and some basic properties of the Yang index and the construction of the sequence $\{\lambda_l\}$.

Yang [10] considered compact Hausdorff spaces with fixed-point-free continuous involutions and used the Čech homology theory, but for our purposes here it suffices to work with closed symmetric subsets of Banach spaces that do not contain the origin and singular homology groups. Following Yang, we first construct a special homology theory defined on the category of all pairs of closed symmetric subsets of Banach spaces that do not contain the origin and all continuous odd maps of such pairs. Let (X, A), $A \subset X$ be such a pair and $C(X, A)$ its singular chain complex with $\mathbb{Z}_2$ coefficients, and denote by $T_\#$ the chain map of $C(X, A)$ induced by the antipodal map $T(x) = -x$. We say that a q-chain c is symmetric if $T_\#(c) = c$, which holds if and only if $c = c' + T_\#(c')$ for some q-chain c'. The symmetric q-chains form a subgroup $C_q(X, A; T)$ of $C_q(X, A)$, and the boundary operator ∂_q maps $C_q(X, A; T)$ into $C_{q-1}(X, A; T)$, so these subgroups form a subcomplex $C(X, A; T)$. We denote by

$$Z_q(X, A; T) = \{c \in C_q(X, A; T) : \partial_q c = 0\}, \tag{2.1}$$

$$B_q(X, A; T) = \{\partial_{q+1} c : c \in C_{q+1}(X, A; T)\}, \tag{2.2}$$

and

$$H_q(X, A; T) = Z_q(X, A; T)/B_q(X, A; T) \tag{2.3}$$

the corresponding cycles, boundaries, and homology groups. A continuous odd map $f : (X, A) \to (Y, B)$ of pairs as above induces a chain map $f_\# : C(X, A; T) \to C(Y, B; T)$ and hence homomorphisms

$$f_* : H_q(X, A; T) \to H_q(Y, B; T). \tag{2.4}$$

Example 2.1 (1.8 of [10]). For the l-sphere,

$$H_q(S^l;T) = \begin{cases} \mathbb{Z}_2 & \text{for } 0 \le q \le l \\ 0 & \text{for } q > l. \end{cases} \tag{2.5}$$

Let X be as above, and define homomorphisms $\nu : Z_q(X;T) \to \mathbb{Z}_2$ inductively by

$$\nu(z) = \begin{cases} \mathrm{In}(c) & \text{for } q = 0 \\ \nu(\partial c) & \text{for } q > 0 \end{cases} \tag{2.6}$$

if $z = c + T_{\#}(c)$, where the index of a 0-chain $c = \sum_i n_i \sigma_i$ is defined by $\mathrm{In}(c) = \sum_i n_i$. As in [10], ν is well-defined and $\nu B_q(X;T) = 0$, so we can define the *index homomorphism* $\nu_* : H_q(X;T) \to \mathbb{Z}_2$ by $\nu_*([z]) = \nu(z)$.

Proposition 2.2 (2.8 of [10]). *If F is a closed subset of X such that $F \cup T(F) = X$ and $A = F \cap T(F)$, then there is a homomorphism $\Delta : H_q(X;T) \to H_{q-1}(A;T)$ such that $\nu_*(\Delta[z]) = \nu_*([z])$.*

Taking $F = X$ we see that if $\nu_* H_l(X;T) = \mathbb{Z}_2$, then $\nu_* H_q(X;T) = \mathbb{Z}_2$ for $0 \le q \le l$. We define the *Yang index* of X by

$$i_Y(X) = \inf \{ l \ge -1 : \nu_* H_{l+1}(X;T) = 0 \}, \tag{2.7}$$

taking $\inf \emptyset = \infty$. Clearly, $\nu_* H_0(X;T) = \mathbb{Z}_2$ if $X \ne \emptyset$, so $i_Y(X) = -1$ if and only if $X = \emptyset$.

Example 2.3 (3.4 of [10]). $i_Y(S^l) = l$

Proposition 2.4 (2.4 of [10]). *If $f : X \to Y$ is as above, then $\nu_*(f_*([z])) = \nu_*([z])$ for $[z] \in H_q(X;T)$, and hence $i_Y(X) \le i_Y(Y)$. In particular, this inequality holds if $X \subset Y$.*

Recall that the *Krasnoselskii Genus* of X is defined by

$$\gamma(X) = \inf \{ l \ge 0 : \exists \text{ a continuous odd map } f : X \to S^{l-1} \}. \tag{2.8}$$

By Example 2.3 and Proposition 2.4,

Proposition 2.5. $\gamma(X) \ge i_Y(X) + 1$

Now, the Dirichlet eigenvalues of the p-Laplacian are the critical values of

$$I(u) = \int_\Omega |\nabla u|^p, \quad u \in M = \{u \in W : \|u\|_p = 1\}, \tag{2.9}$$

which satisfies (PS) (see, e.g., Drábek and Robinson [5]). Denote by $\mathcal{A}$ the class of closed symmetric subsets of M, let

$$\mathcal{F}_l = \{A \in \mathcal{A} : i_Y(A) \geq l - 1\}, \tag{2.10}$$

and set

$$\lambda_l := \inf_{A \in \mathcal{F}_l} \sup_{u \in A} I(u). \tag{2.11}$$

Proposition 2.6 (3.1 of [8]). *λ_l is an eigenvalue of $-\Delta_p$ and $\lambda_l \uparrow \infty$.*

Proof. If λ_l is not a critical value of I, then there is an $\varepsilon > 0$ and an odd homeomorphism $\eta : M \to M$ such that $\eta(I_{\lambda_l+\varepsilon}) \subset I_{\lambda_l-\varepsilon}$ by a lemma of Bonnet [2] (the standard first deformation lemma is not sufficient here as the manifold M is not of class $C^{1,1}$ when $p < 2$). Take $A \in \mathcal{F}_l$ with $\max I(A) \leq \lambda_l + \varepsilon$ and set $\widetilde{A} = \eta(A)$. Then $\widetilde{A} \in \mathcal{A}$ since η is an odd homeomorphism and $i_Y(\widetilde{A}) \geq i_Y(A) \geq l - 1$ by Proposition 2.4, so $\widetilde{A} \in \mathcal{F}_l$, but $\max I(\widetilde{A}) \leq \lambda_l - \varepsilon$, a contradiction.

Since $\mathcal{F}_l \supset \mathcal{F}_{l+1}$, $\lambda_l \leq \lambda_{l+1}$. To see that $\lambda_l \to \infty$, recall that this holds for the Ljusternik-Schnirelmann eigenvalues $\mu_l := \inf_{A \in \mathcal{G}_l} \sup_{u \in A} I(u)$ where $\mathcal{G}_l = \{A \in \mathcal{A} : \gamma(A) \geq l\}$. $\mathcal{F}_l \subset \mathcal{G}_l$ by Proposition 2.5, so $\lambda_l \geq \mu_l$. □

3 Proof of Theorem 1.1

As we noted at the end of the introduction, it suffices to show that Φ has a homological local linking near the origin. Denote by $\pi : W \setminus \{0\} \to M$ the radial projection onto M. Taking $\underline{\lambda} \in (\lambda_l, \min\{a, b\})$, $\overline{\lambda} \in (\max\{a, b\}, \lambda_{l+1})$,

$$\underline{\lambda}\, |t|^p - C\, |t|^q \leq p\, F(x, t) \leq \overline{\lambda}\, |t|^p + C\, |t|^q \quad \forall t \tag{3.1}$$

by (1.2) and (1.8). Now taking $\underline{\mu} \in (\lambda_l, \underline{\lambda})$, $\overline{\mu} \in (\overline{\lambda}, \lambda_{l+1})$, it follows that

$$\Phi(u) \leq \|u\|^p - \underline{\lambda}\, \|u\|_p^p + C\, \|u\|_q^q \leq -\left(\underline{\lambda}/\underline{\mu} - 1\right) \|u\|^p + C\, \|u\|^q \tag{3.2}$$

for $u \in \pi^{-1}(I_{\underline{\mu}})$, while

$$\Phi(u) \geq \|u\|^p - \overline{\lambda}\, \|u\|_p^p - C\, \|u\|_q^q > \left(1 - \overline{\lambda}/\overline{\mu}\right) \|u\|^p - C\, \|u\|^q \tag{3.3}$$

for $u \in \pi^{-1}(M \setminus I_{\overline{\mu}})$. Since $q > p$ by (1.2) and (1.3), taking

$$B = \{u \in W : \|u\| \le r\} \tag{3.4}$$

with $r > 0$ sufficiently small, (1.10) holds for

$$S_1 = \partial B \cap \pi^{-1}(I_{\underline{\mu}}), \tag{3.5}$$

$$B_1 = \{tu : u \in S_1,\, t \in [0,1]\}, \tag{3.6}$$

and

$$B_2 = \{tu : u \in \partial B \cap \pi^{-1}(M \setminus I_{\overline{\mu}}),\, t \in [0,1]\}. \tag{3.7}$$

It only remains to show that the embedding $\widetilde{H}_{l-1}(S_1) \to \widetilde{H}_{l-1}(B \setminus B_2)$ is nontrivial. Since the pair $(B \setminus B_2, S_1)$ is homotopic to $(\partial B \setminus B_2, S_1)$ via the radial projection onto ∂B, which in turn is homotopic to $(I_{\overline{\mu}}, I_{\underline{\mu}})$ via π, it suffices to show that the embedding $\widetilde{H}_{l-1}(I_{\underline{\mu}}) \to \widetilde{H}_{l-1}(I_{\overline{\mu}})$ is nontrivial. Since I is even, $I_{\underline{\mu}} \in \mathcal{A}$, and since $\underline{\mu} > \lambda_l$, there is an $A \in \mathcal{F}_l$ such that $A \subset I_{\underline{\mu}}$, so $i_Y(I_{\underline{\mu}}) \ge i_Y(A) \ge l - 1$ by Proposition 2.4 and hence $\nu_* H_{l-1}(I_{\underline{\mu}}; T) \ne 0$ by (2.7). We show that if $[z] \in H_{l-1}(I_{\underline{\mu}}; T)$ is such that $\nu_*([z]) \ne 0$, then $[z] \ne 0$ in $\widetilde{H}_{l-1}(I_{\overline{\mu}})$. Arguing indirectly, assume that $z \in B_{l-1}(I_{\overline{\mu}})$, say, $z = \partial c$. Since $z \in B_{l-1}(I_{\overline{\mu}}; T)$, $T_\#(z) = z$. Let $c' = c + T_\#(c)$. Then $c' \in Z_l(I_{\overline{\mu}}; T)$ since $\partial c' = z + T_\#(z) = 2z = 0 \mod 2$, and $\nu_*([c']) = \nu(c') = \nu(\partial c) = \nu(z) \ne 0$, so $\nu_* H_l(I_{\overline{\mu}}; T) \ne 0$. But then $i_Y(I_{\overline{\mu}}) \ge l$ by (2.7), so $I_{\overline{\mu}} \in \mathcal{F}_{l+1}$ and hence $\lambda_{l+1} \le \overline{\mu}$, a contradiction.

References

[1] A. Ambrosetti and P. H. Rabinowitz. Dual variational methods in critical point theory and applications. *J. Functional Analysis*, 14:349–381, 1973.

[2] A. Bonnet. A deformation lemma on a C^1 manifold. *Manuscripta Math.*, 81(3-4):339–359, 1993.

[3] K.-C. Chang. *Infinite-dimensional Morse theory and multiple solution problems*, volume 6 of *Progress in Nonlinear Differential Equations and their Applications*. Birkhäuser Boston Inc., Boston, MA, 1993.

[4] G. Dinca, P. Jebelean, and J. Mawhin. Variational and topological methods for Dirichlet problems with *p*-Laplacian. *Port. Math. (N.S.)*, 58(3):339–378, 2001.

[5] P. Drábek and S. B. Robinson. Resonance problems for the p-Laplacian. *J. Funct. Anal.*, 169(1):189–200, 1999.

[6] Q. Jiu and J. Su. Existence and multiplicity results for perturbations of the p-Laplacian. preprint.

[7] S. Liu. Existence of solutions to a superlinear p-Laplacian equation. *Electron. J. Differential Equations*, pages No. 66, 6 pp. (electronic), 2001.

[8] K. Perera. Nontrivial critical groups in p-Laplacian problems via the Yang index. to appear in Topol. Methods Nonlinear Anal.

[9] K. Perera. Homological local linking. *Abstr. Appl. Anal.*, 3(1-2):181–189, 1998.

[10] C.-T. Yang. On theorems of Borsuk-Ulam, Kakutani-Yamabe-Yujobô and Dyson. II. *Ann. of Math. (2)*, 62:271–283, 1955.

Preliminary Normal Forms for a Class of Singular Equilibria in Implicit ODEs

Ricardo Riaza

Departamento de Matemática Aplicada a las Tecnologías de la Información, ETSI Telecomunicación, Universidad Politécnica de Madrid, 28040 Madrid, Spain e–mail: rrr@mat.upm.es

Abstract. This paper addresses some qualitative issues concerning singularities of quasilinear ODEs. A local normal form analysis and several dynamic phenomena are extensively reviewed. Singular equilibria, excluded from previous approaches based on normal forms, are framed in this context: a preliminary normal form is presented for a certain class of these singular equilibria. These results provide a framework for the analysis of directional convergence phenomena arising in singular root-finding problems.

AMS Subject Classification. 34A09, 34C20, 34D20

Key words. Quasilinear ODE, normal form, singularity, singular equilibrium, continuous Newton method.

1 Introduction

Quasilinear or *linearly implicit* ODEs are defined by a system of the form $A(x)x' = f(x)$, for sufficiently smooth matrix- and vector-valued functions A, f. The local study of such systems on a neighborhood of a given x^* is strongly influenced by the behavior of $A(x)$ around this point. A non-singular matrix $A(x^*)$ trivially leads to a local vector field analysis, whereas a constant rank deficiency in $A(x)$ around x^* often defines a differential-algebraic equation (DAE) [3, 7, 9]. In contrast, the case in which $A(x^*)$ is singular but arbitrarily close points display a non-singular $A(x)$ defines x^* as a *singularity* of the quasilinear system [13, 14, 15, 20, 22, 25, 28, 29, 33]. Other types of singularities have been considered in [4, 5, 6, 12, 16, 17, 19, 23, 27, 30, 31, 32]. These singular systems come from problems concerning time-varying and/or non-linear electrical circuits, magnetohydrodynamics, optimal control, or power systems. Singularities of quasilinear problems also arise in the description of the reduced flow in semiexplicit singular index 1 DAEs [22, 23].

A detailed study of local normal forms around singularities of quasilinear systems has been recently performed [29]. Previous results in this direction can be found in [13, 14, 20, 33]. On the other hand, several phenomena concerning singular bifurcations, impasse points, singular equilibria or singularity crossing phenomena have been addressed within different frameworks and using different terminological conventions, yielding a certain gap among the various approaches developed in this context: compare e.g. [11, 28, 29, 33] with [15, 16, 17, 25] or [4, 12, 31, 32].

Singular equilibrium points are excluded from the normal form analysis presented in [29]: they would fall in the framework of resonant I-singularities, in the terminology introduced there. Singular equilibria are important concerning certain bifurcations in quasilinear and semiexplicit systems [1, 2, 11, 22, 31, 32]. They also arise when continuous-time systems are used to address singular root-finding problems [21, 23, 24, 25]. Singular roots are displayed, for example, at bifurcation points in PC continuation schemes which use Newton's method as a corrector. The present work is mainly focused on the type of singular equilibria arising in this singular root-finding context.

The link between continuous-time methods for singular root-finding problems and quasilinear ODEs is defined by the quasilinear form of the continuous Newton method $-J(x)x' = f(x)$, where $J(x)$ stands for the derivative $f'(x)$: see references in [10, 24, 25, 26]. In this setting, a phenomenon of directional stability from cone-shaped regions has been proved for a particular class of (so-called *stationary*) singular equilibria [21, 25], extending previous results formulated in the discrete-time context (see [8, 18] and references therein). However, several related issues remain unsolved; the framework presented in this paper should be of help in the analysis of these issues, which are significant not only in the dynamic characterization of singular equilibria in continuous-time systems, but also regarding the behavior of different iterative schemes derived through numerical integration [26].

In this context, the present document is aimed at two main goals. First, we attempt to somehow shorten the gap between the different approaches and terminological conventions developed on this topic: after introducing some terminology in Section 2, Section 3 presents a comprehensive survey of results concerning normal forms and dynamic phenomena around singularities of quasilinear problems. The second purpose of the work is to place stationary singular equilibria in this normal form context, through the discussion of a preliminary normal form for such equilibrium points: this is carried out in Section 4. The main results are presented in Propositions 3, 4 and, particularly, 5. Ultimately, this study should be of help in the analysis of the above-mentioned open questions, and also in the classification of phase portraits around stationary equilibria In turn, this should be relevant in the study of root-finding methods around singular solutions. Concluding remarks are compiled in Section 5.

2 Background: Singularities of quasilinear ODEs

Let us consider a differential system of the form

$$A(x)x' = f(x), \tag{2.1}$$

where $A \in C^k(\Omega, \mathbb{R}^{n\times n})$ and $f \in C^k(\Omega, \mathbb{R}^n)$, $\Omega \subseteq \mathbb{R}^n$, $k \in \{1, 2, ..., \infty, \omega\}$, C^ω meaning real analytic, and $\mathbb{R}^{n\times n}$ standing for the set of $n \times n$ real matrices.

Systems like (2.1) are often called *quasilinear* (see e.g. [9, 16, 17]) or *linearly implicit* [3]. They have also been referred to as *constrained* systems [29, 30]. Several other names such as *singular* systems, *semistate* equations or *descriptor* systems are used for them in different fields. Under additional assumptions, these equations may be considered as *differential-algebraic* problems [3, 9]. The pair (A, f) is sometimes called a *generalized vector field* [7, 13, 14].

As it was indicated in Section 1, if $A(x^*)$ is a regular matrix, equation (2.1) may be locally reduced to an explicit ODE. On the contrary, if $A(x)$ is singular with constant rank around a given x^*, the study may be often addressed within the context of differential-algebraic equations or DAEs. However, we will be interested in cases in which $A(x)$ is singular only on a local hypersurface Ψ which includes x^*. More generally, a singularity may be defined as follows:

Definition 1. *The point x^* is called a* singularity *of (2.1) if $A(x^*)$ is non-invertible but there exists a sequence $\{x_i\}_{i\in\mathbb{N}} \to x^*$ such that $A(x_i)$ is invertible $\forall i \in \mathbb{N}$.*

Singularities are sometimes called in general *impasse points* [29]; we will however reserve this name for a particular type of singular points. Definition 1 makes it possible to exclude cases with a constantly rank-deficient $A(x)$, which often yields well-defined dynamics on lower dimensional manifolds, as it happens in differential-algebraic problems. Other types of singularities are discussed in [1, 2, 4, 5, 6, 12, 16, 17, 19, 23, 27, 31, 32] and references therein.

Definition 1 is a generalization of situations in which $A(x)$ is non-invertible only on a *singular hypersurface* Ψ which includes x^*. This is the case in non-critical problems [15]:

Definition 2. *A singularity x^* of (2.1) is called* non-critical *if* $(\det A)'(x^*) \neq 0$.

Non-critical singularities are termed *regular impasse points* in [29]. It is important to note that definition 2 implies that $\dim \operatorname{Ker} A(x^*) = 1$ (see e.g. [15, 22]), that is, $\operatorname{rk} A(x^*) = n - 1$. If we simply impose this condition at a singularity x^*, without the non-critical assumption, we will say that the system undergoes a rank-1 deficiency at this point.

Definition 3. *A non-critical singularity x^* of (2.1) is called* standard *if $(\det A)'(x^*)v \neq 0$ for any $v \in \operatorname{Ker} A(x^*) - \{0\}$.*

This is taken from [25], and extends a previous concept from [15], which applies only to the below-defined algebraic singular points. Standard singularities exactly correspond to regular impasse points which are not *Kernel-singular* or *K-singular* in the terminology of [29].

The following definition introduces a classification of singularities which describes substantially different dynamic phenomena:

Definition 4. *A singular point x^* of (2.1) will be called an* algebraic singularity *if $f(x^*) \notin \operatorname{Im} A(x^*)$. If, on the contrary, $f(x^*) \in \operatorname{Im} A(x^*)$, x^* will be referred to as a* geometric singularity.

These notions come from a taxonomy introduced in [16, 17]. Standard geometric singularities are *Image-singular* or *I-singular* regular impasse points in [29], and they are essentially the quasilinear analog of the so-called *pseudoequilibria* in [31]. A variety of different phenomena may be displayed at these points, namely, the existence of smooth trajectories crossing the singular set, the presence of equilibria at the singular manifold, or singularity-induced bifurcations in parameterized problems. On the contrary, standard algebraic singularities are shown in [15] to display a true *impasse* behavior, since a pair of trajectories collapse there in finite time. This phenomenon has also been observed in semiexplicit DAEs modeling nonlinear electrical circuits [5, 6, 19].

In the local study of quasilinear systems, it is often useful to premultiply (2.1) by the adjoint (transpose of the matrix of cofactors) $\mathrm{Adj}A(x)$, which yields the so-called *canonical form* [15]

$$\omega(x)x' = g(x), \tag{2.2}$$

with $\omega(x) = \det A(x)$, and $g(x) = \mathrm{Adj}A(x)f(x)$.

Definition 5. *The vector field* $g(x) = \mathrm{Adj}A(x)f(x)$ *will be called the* desingularized field *associated with (2.1).*

For the use of this term see e.g. [10, 23] and references therein. Again, this field has received other names, such as *regularized* [28], *extended* [29] or, in a slightly different context, *transformed* [31] vector field. The importance of this concept stems from the fact that trajectories of the quasilinear system away from the singular set are exactly those of the desingularized field, up to a time reparameterization. This is in turn closely related to the notion of orbital equivalence in quasilinear systems, to be discussed below.

Note that, under the non-critical assumption, geometric singularities are exactly those singular points where the desingularized field g vanishes (see e.g. [25]). Remark that, on the contrary, not every equilibrium point of the desingularized field corresponds to an equilibrium of the vector field f. The main attention in this work will be focused on standard singular equilibria of f, which can be framed within the context of standard geometric singularities (or I-singular regular impasse points in the terminology of [29]).

3 Local normal forms

3.1 Equivalence.

Several interrelated notions of topological or smooth equivalence have been recently proposed for quasilinear systems around singular points [7, 13, 14, 20, 28, 29]. All of them are based on the existence of a local homeomorphism or C^k-diffeomorphism preserving the singular set and the arcs of orbits away from singular points. The basic difference between them concerns the requirement that time parameterization or at least orientation be preserved. In contrast to the classical notion of orbital equivalence for explicit ODEs, the requirement that only orientation be preserved in quasilinear systems

[13, 14, 28] may lead to *singular* reparameterization of trajectories: for instance, desingularized trajectories reaching the singularity in infinite time may yield finite-time singularity crossing phenomena in the original dynamics (see e.g. [12]). Also, allowing for time reversal of trajectories converts the two normal forms for standard algebraic singularities (non-singular impasse points in [29]) discussed in [20] into only one [29].

The present study will be based on the following notion of equivalence:

Definition 6. *System (2.1) will be said C^k-equivalent (locally around a singular point x^*) to*

$$\tilde{A}(v)v' = \tilde{f}(v), \tag{3.1}$$

if there exists a local C^k diffeomorphism $x = \phi(v)$ and a C^k non-singular matrix-valued function $E(x)$, both defined on a neighborhood of x^*, such that

$$\tilde{A}(v) = E(\phi(v))A(\phi(v))\phi'(v), \quad \tilde{f}(v) = E(\phi(v))f(\phi(v)).$$

This definition may be easily proved equivalent to the one introduced in [7] for arbitrary differentiable manifolds. It essentially expresses that the local phase portrait of (3.1) is mapped into that of (2.1) through the diffeomorphism $x = \phi(v)$. Parameterization of trajectories is preserved and so is the singular set, since

$$\det\tilde{A}(v) = \det E(\phi(v)) \ \det A(\phi(v)) \ \det\phi'(v)$$

and then $\det\tilde{A}(v) = 0 \Leftrightarrow \det A(\phi(v)) = 0$. In turn, premultiplication by E does not transform the phase portrait of the system, but it does change the analytical expression describing the dynamics: this allows one to get different formulations for the same dynamical behavior. In a certain sense, the simplest of these formulations defines a *normal form* for the quasilinear system. Sometimes, this term is reserved for expressions which completely characterize the equivalence class of a given system, or for parameterized expressions in which different values of the parameter distinguish different equivalence classes for the dynamics. In this context, an analytical expression which only provides a simplified form for a given system is called a *preliminary* normal form. A survey of results concerning local normal forms is presented in the remainder of this section.

3.2 Rank-1 deficiencies.

Without further assumptions, a preliminary normal form can be given for singular points of quasilinear systems which satisfy $\mathrm{rk}A(x^*) = n-1$: this will be the case of non-critical and, in particular, standard singularities. The reasoning below follows [29, 33].

Proposition 1. *Let x^* be a singularity of (2.1) where $\mathrm{rk}A(x^*) = n - 1$. There exists a local C^k diffeomorphism $x = \phi_1(u)$, with $x^* = \phi_1(0)$, and a*

C^k non-singular matrix-valued function $E_1(x)$, which converts (2.1) into the equivalent form

$$\begin{aligned} u_1' &= \hat{f}_1(u) \\ &\vdots \\ u_{n-1}' &= \hat{f}_{n-1}(u) \\ \beta(u)u_n' &= \hat{f}_n(u), \end{aligned} \tag{3.2}$$

that is, $\hat{A}(u)u' = \hat{f}(u)$, with

$$\begin{aligned} \hat{A}(u) &= E_1(\phi_1(u))A(\phi_1(u))\phi_1'(u) = \begin{pmatrix} I_{n-1} & 0 \\ 0 & \beta(u) \end{pmatrix}, \\ \hat{f}(u) &= E_1(\phi_1(u))f(\phi_1(u)). \end{aligned}$$

Proof. Let us write

$$A(x) = \begin{pmatrix} a_{11}(x) & \dots & a_{1n}(x) \\ \vdots & & \vdots \\ a_{n1}(x) & \dots & a_{nn}(x) \end{pmatrix}.$$

The condition $\mathrm{rk}A(x^*) = n-1$ means that, around x^*, $n-1$ rows of $A(x)$ are linearly independent. If we assume without loss of generality that the first $n-1$ rows are linearly independent, this is equivalent to the linear independence of the $n-1$ linear forms

$$\sum_{j=1}^{n} a_{ij}(x)dx_j, \; i = 1, \dots, n-1,$$

which, in turn, implies the local existence of a vector field which is annihilated by these $n-1$ forms. Let $x = \phi_1(u)$ be a C^k change of coordinates for which this vector field reads $\partial/\partial u_n$ and assume, without loss of generality, that $x^* = \phi_1(0)$. This change of coordinates transforms (2.1) into a system of the form $B(u)u' = f(\phi_1(u))$, with

$$\begin{aligned} B(u) = A(\phi_1(u))\phi_1'(u) &= \begin{pmatrix} b_{11}(u) & \dots & b_{1(n-1)}(u) & 0 \\ \vdots & & \vdots & \vdots \\ b_{(n-1)1}(u) & \dots & b_{(n-1)(n-1)}(u) & 0 \\ b_{n1}(u) & \dots & b_{n(n-1)}(u) & \beta(u) \end{pmatrix} \equiv \\ &\equiv \begin{pmatrix} \tilde{B}(u) & 0 \\ b(u) & \beta(u) \end{pmatrix}. \end{aligned}$$

Remark that, by construction, $\det\tilde{B}(u) \neq 0$, and $\beta(u) = 0$ if and only if u is a singular point. The condition $\det\tilde{B}(u) \neq 0$ implies the existence of $n-1$ linear forms

$$c_i(u) = c_{i1}(u)du_1 + \ldots + c_{i(n-1)}(u)du_{n-1}, \; i = 1, \ldots, n-1,$$

such that $\sum_{j=1}^{n-1} c_{ij}b_{ji} = 1$, $\sum_{j=1}^{n-1} c_{ij}b_{jk} = 0$ if $i \neq k$. Thus, defining

$$C(u) = \begin{pmatrix} c_{11}(u) & \ldots & c_{1(n-1)}(u) & 0 \\ \vdots & & \vdots & \vdots \\ c_{(n-1)1}(u) & \ldots & c_{(n-1)(n-1)}(u) & 0 \\ 0 & 0 & 0 & 1 \end{pmatrix},$$

we have

$$C(u)B(u) = \begin{pmatrix} I_{n-1} & 0 \\ b(u) & \beta(u) \end{pmatrix}.$$

Finally, premultiplying $C(u)B(u)$ by

$$D(u) = \begin{pmatrix} I_{n-1} & 0 \\ -b(u) & 1 \end{pmatrix},$$

we get

$$D(u)C(u)B(u) = D(u)C(u)A(\phi_1(u))\phi_1'(u) = \begin{pmatrix} I_{n-1} & 0 \\ 0 & \beta(u) \end{pmatrix} \equiv \hat{A}(u).$$

Hence, premultiplication by $D(u)C(u)$ transforms $B(u)u' = f(\phi_1(u))$ into (3.2), with $\hat{f}(u) = D(u)C(u)f(\phi_1(u))$. The matrix function $E_1(x)$ is simply $D(\phi_1^{-1}(x))C(\phi_1^{-1}(x))$.

□

The vanishing of the scalar-valued function $\beta(u)$ characterizes the singularities of the quasilinear system. In particular, if β does vanish on a whole neighborhood of $0 = \phi_1^{-1}(x^*)$, we would be led to the constant rank-deficient setting of [7]. Finally, for later use remark that the the condition defining the diffeomorphism $x = \phi(u)$ implies

$$A(\phi_1(0))\frac{\partial x}{\partial u_n}(0) = 0,$$

and $\partial x/\partial u_n(0) \neq 0$ since ϕ_1 is a local diffeomorphism: it then follows that $\partial x/\partial u_n(0) \in \mathrm{Ker}A(\phi_1(0)) - \{0\}$. This means that the u_n-coordinate curve is tangent to Ker$A$ at the origin. This property will be useful in the analysis of standard singularities carried out below.

3.3 Standard singularities.

Standard singularities verify that $\text{rk}A(x^*) = n-1$ (see e.g. [15, 22]) and, therefore, the reasoning above applies to this kind of singular points. Further simplification is however possible (see [33]):

Proposition 2. *Let x^* be a standard singularity of (2.1). There exists a local C^k diffeomorphism $x = \phi(w)$, with $x^* = \phi(0)$, and a C^k non-singular matrix-valued function $E(x)$, which converts (2.1) into the equivalent form*

$$
\begin{aligned}
y_1' &= \bar{f}_1(y,z)\\
&\vdots\\
y_{n-1}' &= \bar{f}_{n-1}(y,z)\\
zz' &= \bar{f}_n(y,z),
\end{aligned}
\tag{3.3}
$$

where $y = (y_1,\ldots,y_{n-1})$ stands for $(w_1,\ldots,w_{n-1})$ and $z = w_n$. System (3.3) may be written as $\bar{A}(w)w' = \bar{f}(w)$ with

$$\bar{A}(w) = E(\phi(w))A(\phi(w))\phi'(w) = \begin{pmatrix} I_{n-1} & 0 \\ 0 & z \end{pmatrix}, \quad \bar{f}(w) = E(\phi(w))f(\phi(w)).$$

Proof. Note that the standard assumption implies that x^* is non-critical and, therefore, $\text{rk}A(x^*) = n-1$. Hence, Proposition 1 is of application. The scalar function $\beta(u)$ introduced there satisfies, by construction,

$$\beta(u) = \det E_1(\phi_1(u))\,\det A(\phi_1(u))\,\det\phi_1'(u).$$

Let $0 = \phi_1^{-1}(x^*)$. Since $\det A(\phi_1(0)) = 0$, we have

$$
\begin{aligned}
\frac{\partial\beta}{\partial u_n}(0) &= \det E_1(x^*)\,\frac{\partial\det A}{\partial u_n}(0)\,\det\phi_1'(0) = \\
&= \det E_1(x^*)\,(\det A)'(x^*)\frac{\partial x}{\partial u_n}(0)\,\det\phi_1'(0).
\end{aligned}
$$

As it was indicated at the end of 3.2, the condition defining the diffeomorphism $x = \phi(u)$ implies $\partial x/\partial u_n(0) \in \text{Ker}A(\phi_1(0)) - \{0\}$. Due to the standard nature of the singularity, we then have

$$(\det A)'(x^*)\frac{\partial x}{\partial u_n}(0) \neq 0$$

and, therefore, $\partial\beta/\partial u_n(0) \neq 0$.

Let us then introduce new coordinates $w = (y, z) = \xi(u)$, through the relations $y = (u_1, \ldots, u_{n-1})$, $z = \beta(u)$. Remark that singular points are now characterized by the condition $z = 0$ and, therefore, $(y_1, \ldots, y_{n-1})$ defines a local parameterization of the singular set. Additionally, by construction the z-coordinate curve is still tangent to Ker$A$. With the notation

$$d(u) = (\frac{\partial \beta}{\partial u_1}(u), \ldots, \frac{\partial \beta}{\partial u_{n-1}}(u)),\ \gamma(u) = \frac{\partial \beta}{\partial u_n}(u) \neq 0,$$

we have

$$\xi' = \begin{pmatrix} I_{n-1} & 0 \\ d & \gamma \end{pmatrix}$$

and, denoting $\phi_2 = \xi^{-1}$,

$$\phi_2' = \begin{pmatrix} I_{n-1} & 0 \\ -d/\gamma & 1/\gamma \end{pmatrix}.$$

Let us then perform in the system $\hat{A}(u)u' = \hat{f}(u)$ represented by (3.2) the coordinate change $u = \phi_2(w)$, together with a premultiplication by

$$E_2(u) = \begin{pmatrix} I_{n-1} & 0 \\ \beta d & \gamma \end{pmatrix}.$$

Some simple computations yield

$$\bar{A}(w) = E_2(\phi_2(w))\hat{A}(\phi_2(w))\phi_2'(w) = \begin{pmatrix} I_{n-1} & 0 \\ 0 & z \end{pmatrix},$$

as required. The matrix-valued function $E(x)$ is given by

$$E(x) = E_2(\phi_1^{-1}(x))E_1(x),$$

whereas the diffeomorphism $x = \phi(w)$ is defined by the relation $\phi = \phi_1 \circ \phi_2$.

□

The transformed system $\bar{A}(w)w' = \bar{f}$ given by (3.3) provides a simple setting for the analysis of different types of singularities. For instance, algebraic and geometric standard singularities are simply characterized by the conditions $\bar{f}_n(y, 0) \neq 0$ and $\bar{f}_n(y, 0) = 0$, respectively. Also, *weak singularities* considered in [21, 24, 25] can be easily characterized in this context. Within this framework, the desingularized vector field $\bar{g} = \text{Adj}\bar{A}\bar{f}$ reads

$$\bar{g}(y, z) = \begin{pmatrix} z\bar{f}_1(y, z) \\ \vdots \\ z\bar{f}_{n-1}(y, z) \\ \bar{f}_n(y, z) \end{pmatrix},$$

and can be easily derived from the original desingularized field $g(x)$ through the relation $\bar{g}(w) = \det E(\phi(w))\ \mathrm{Adj}\,\phi'(w)\ g(\phi(w))$.

3.4 Algebraic and geometric standard singularities

Standard algebraic singularities were shown by Rabier [15] to display an impasse behavior. This means that, if x^* is a standard algebraic singularity, there exists trajectories which reach x^* but cannot be continued beyond that point. More precisely, if x^* is a standard algebraic singularity (a standard singular point, in the terminology of [15]), then there exist exactly two solutions $x(t) \in C^0([0,\tilde{t}], \mathbb{R}^n) \cap C^1((0,\tilde{t}], \mathbb{R}^n)$ verifying $x(0) = x^*$, both defined either for some $\tilde{t} > 0$ or $\tilde{t} < 0$ ($[0,\tilde{t}]$ standing in the latter case for $[\tilde{t}, 0]$). These solutions verify that $||x'(t)|| \to \infty$ as $t \to 0$. The impasse point can be either a *backward* (also called *repelling* and *inaccessible*) or a *forward* (resp. *attracting, accessible*) one if $\tilde{t} > 0$ or $\tilde{t} < 0$, respectively.

Standard algebraic singularities (nonsingular impasse points in [29]) were then addressed in [13, 20], arriving to the normal forms

$$\begin{aligned} y' &= 0 \\ zz' &= \pm 1, \end{aligned} \tag{3.4}$$

where the values $+1$ and -1 correspond to backward and forward impasse points, respectively. If time reversal of trajectories is allowed, both cases are equivalent [29].

The local behavior around standard geometric singularities (I-singularities in [29]) is much richer than that of algebraic points. Some preliminary results can be found in [14]. A taxonomy of the behavior expected in these cases follows from the remark that geometric singularities are equilibria of the desingularized field $g(x)$. Furthermore, the linearization of g at a geometric singularity x^* yields a Jacobian with rank at most 2. This can be directly derived from the definition $g(x) = \mathrm{Adj}A(x)f(x)$ or using (y, z)-coordinates, since in this case

$$\bar{g}'(0,0) = \begin{pmatrix} 0 & \bar{f}_1(0,0) \\ \vdots & \vdots \\ 0 & \bar{f}_{n-1}(0,0) \\ \partial \bar{f}_n/\partial y(0,0) & \partial \bar{f}_n/\partial z(0,0) \end{pmatrix},$$

and rk $\bar{g}'(0,0) \leq 2$. Actually, since the set of geometric singularities is characterized by the conditions $z = 0$, $\bar{f}_n = 0$, it follows that, under the assumption $\partial \bar{f}_n/\partial y \neq 0$, this set is locally a hypersurface of the singular manifold, that is, an $(n-2)$-dimensional submanifold of the singular set.

The rank of $\bar{g}'$ may actually be 2, 1 or 0, each case displaying different phenomena. Problems with rank 1 or 0 typically arise when the system has a singular equilibrium, that is, a singular point x^* where $f(x^*) = 0$. On the other hand, geometric singularities (I-singularities) with rk $\bar{g}'(0,0) = 2$ have been studied in [29]. Under a non-resonance condition and for any finite k, a standard geometric singularity (I-singularity) is shown in [29] to be C^k-orbitally equivalent (in $\mathbb{R}^n$, $n \geq 3$) to one of the normal forms

$$\begin{aligned} y_1' &= 0 \\ &\vdots \\ y_{n-2}' &= 0 \\ y_{n-1}' &= \lambda(y_1, \ldots, y_{n-2}) \\ zz' &= y_{n-1} + z, \end{aligned} \tag{3.5}$$

where $\lambda(y_1, \ldots, y_{n-2})$ can be written as $\lambda + y_1$ or $\lambda \pm y_1^2 \pm y_2^2 \pm \ldots \pm y_{n-2}^2$, with $\lambda \in \mathbb{R}$. Depending on the value of λ, different phase portraits may be displayed: see details in [29]. In the planar case, the classification is simpler and the normal form above amounts to

$$\begin{aligned} y' &= \lambda \\ zz' &= y + z. \end{aligned} \tag{3.6}$$

3.5 Weak singularities.

Certain singular problems can be shown to display a smooth behavior around non-critical geometric singularities. More precisely, if A and f are C^k, x^* is a non-critical singularity, and we assume the condition $f(x) \in \mathrm{Im}A(x)$ for all x in a neighborhood of x^*, then there exists a C^{k-1} extension of the vector field $g(x)/\omega(x)$ around x^* [21, 24, 25]. This defines x^* as a *weak* or *removable* singularity. Note that the requirement above amounts to the condition that x^* is geometric and every singularity in a neighborhood of x^* is also geometric. Although this is a non-generic phenomenon, its study has turned out to be useful in the analysis of several stability and discretization issues [25, 26].

In standard problems, this weak behavior admits a particularly simple description using the normal form (3.3). As it was indicated above, standard geometric singularities are characterized by the conditions $z = 0$, $\bar{f}_n = 0$. The weak assumption may then be expressed as $\bar{f}_n(y, 0) = 0$ for all y in a neighborhood of the origin. It is easy to show that, in this situation, $\bar{f}_n(y, z)$ may be written as $z\varphi(y, z)$ for some $\varphi \in C^{k-1}$ (see [24]). The normal form (3.3) amounts to

$$\begin{aligned} y_1' &= \bar{f}_1(y, z) \\ &\vdots \\ y_{n-1}' &= \bar{f}_{n-1}(y, z) \\ z' &= \varphi(y, z), \end{aligned} \tag{3.7}$$

away from $z = 0$. In fact, this is an explicit ODE which admits a trivial extension to the singular set $z = 0$.

4 Singular equilibria

A singular equilibrium point of (2.1) is defined by the condition $f(x^*) = 0$ at a singularity x^*. Singular equilibria of quasilinear problems fall in the context of resonant geometric singularities or I-singularities and, therefore, are excluded from the normal form analysis in [29]. These equilibrium points are important in the study of singular bifurcations [1, 2, 22, 31, 32], and also in singular root-finding problems addressed via the continuous Newton method (see [10, 24, 25, 26] and references therein):

$$-J(x)x' = f(x), \tag{4.1}$$

where $J(x)$ stands for the derivative $f'(x)$. Note that this system is a particular instance of (2.1) with $A(x) = -J(x)$.

Singular equilibrium points are also equilibria of the desingularized field g. Moreover, since $f(x^*) = 0$, the derivative of the desingularized field reads

$$g'(x^*) = \mathrm{Adj}A(x^*)J(x^*),$$

and, due to the fact that $\mathrm{rkAdj}A(x^*) = 1$ under the non-critical assumption (which implies $\mathrm{rk}A(x^*) = n - 1$), it is $\mathrm{rk}g'(x^*) \leq 1$. This can also be seen in (y, z)-coordinates, since in this case it is

$$\bar{g}'(0,0) = \begin{pmatrix} 0 & 0 \\ \vdots & \vdots \\ 0 & 0 \\ \partial\bar{f}_n/\partial y(0,0) & \partial\bar{f}_n/\partial z(0,0) \end{pmatrix}.$$

Actually, two different cases may happen, depending on the actual value of $\mathrm{rk}g'(x^*)$.

Definition 7. *A non-critical singular equilibrium of (2.1) is called* stationary *if* $\mathrm{Im}J(x^*) \subseteq \mathrm{Im}A(x^*)$.

Non-stationary equilibria are defined by the (generic) condition $\mathrm{Im}J(x^*) \not\subseteq \mathrm{Im}A(x^*)$, that is, $\mathrm{Adj}A(x^*)J(x^*) \neq 0$, which in turn is equivalent to $\mathrm{rk}g'(x^*) = 1$. On the contrary, stationary singular equilibria, characterized by the condition $\mathrm{Im}J(x^*) \subseteq \mathrm{Im}A(x^*)$, satisfy $\mathrm{rk}g'(x^*) = 0$.

Non-stationary singular equilibria arise for instance in the context of the Singularity Induced Bifurcation Theorem [1, 2, 22, 31, 32]. Under generic assumptions, in this situation there exists a codimension-one manifold $\mathcal{C}$, transversal to both the singular set and $\mathrm{Ker}A(x^*)$, along which there is a well-defined vector field displaying an equilibrium at x^*. The stability properties of this equilibrium are characterized by the matrix pencil $\{A(x^*), -J(x^*)\}$ [21]. Moreover, a smooth solution of (2.1), transversal to $\mathcal{C}$, may cross the singular manifold through x^* in finite time [12].

On the other hand, every singular equilibrium is stationary in the particular case of the continuous Newton method (4.1). The term *stationary* follows from the fact that convergence to the singular equilibrium from n-dimensional sets may happen in this kind of problems [21, 24, 25]. In particular, finite-time singularity crossing through x^* cannot happen in this situation. In fact, several issues regarding the actual shape of the attraction domain of standard singular equilibria in the continuous Newton method remain open: although details are beyond the purposes of the present work, these issues might be tackled through the normal form approach here presented.

4.1 A preliminary normal form for stationary singular equilibria.

The stationarity condition $\mathrm{Im}J(x^*) \subseteq \mathrm{Im}A(x^*)$ is preserved under the equivalence notion introduced in definition 6. This makes it possible to simplify the normal form (3.3) at stationary equilibria:

Proposition 3. *Let x^* be a stationary singular equilibrium. If x^* is standard, the normal form (3.3) can be rewritten as*

$$\begin{aligned} y' &= Hy + pz + \text{h.o.t.} \\ zz' &= \lambda z^2 + (q^T y)z + y^T G y + \text{h.o.t.} \end{aligned} \tag{4.2}$$

where $H,\ G \in \mathbb{R}^{(n-1)\times(n-1)}$, G being symmetric, $p,\ q \in \mathbb{R}^{n-1}$, and $\lambda \in \mathbb{R}$.

Proof. With the notation of Proposition 2, the normal form (3.3) for standard singularities is defined by $\bar{A}(w) = E(\phi(w))A(\phi(w))\phi'(w)$ and $\bar{f}(w) = E(\phi(w))f(\phi(w))$, with $w = (y, z)$. Let $\bar{J}(w)$ stand for $\bar{f}'(w)$ and recall that $0 = \phi^{-1}(x^*)$. Since $f(x^*) = 0$ implies $\bar{f}(0) = 0$, it is

$$\bar{J}(0) = E(\phi(0))J(\phi(0))\phi'(0).$$

Now, $\mathrm{Im}J(\phi(0)) \subseteq \mathrm{Im}A(\phi(0))$ implies

$$\mathrm{Im}E(\phi(0))J(\phi(0))\phi'(0) \subseteq \mathrm{Im}E(\phi(0))A(\phi(0))\phi'(0),$$

that is, $\mathrm{Im}\bar{J}(0) \subseteq \mathrm{Im}\bar{A}(0)$. Therefore, since

$$\bar{A}(0) = \begin{pmatrix} I_{n-1} & 0 \\ 0 & 0 \end{pmatrix},$$

the stationary condition reads, in (y, z)-coordinates, $\bar{f}_n'(0,0) = 0$, which means that there are no first-order terms in the expansion of $\bar{f}_n$.

□

The additional assumption $\mathrm{Ker}A(x^*) = \mathrm{Ker}J(x^*)$, which is satisfied in the particular case of the continuous Newton method, implies $p = 0$:

Proposition 4. *Let the assumptions of Proposition 3 hold. If $\mathrm{Ker}A(x^*) = \mathrm{Ker}J(x^*)$, then the normal form (4.2) can be rewritten as*

$$\begin{aligned} y' &= Hy + \text{h.o.t.} \\ zz' &= \lambda z^2 + (q^T y)z + y^T G y + \text{h.o.t.} \end{aligned} \tag{4.3}$$

Proof. In a way similar to the one followed in Proposition 3, it is easy to check that the Kernel identity is preserved under equivalence, that is, $\mathrm{Ker}A(x^*) = \mathrm{Ker}J(x^*)$ implies $\mathrm{Ker}\bar{A}(0) = \mathrm{Ker}\bar{J}(0)$. Now, since

$$\bar{A}(0) = \begin{pmatrix} I_{n-1} & 0 \\ 0 & 0 \end{pmatrix}, \quad \bar{J}(0) = \begin{pmatrix} H & p \\ 0 & 0 \end{pmatrix},$$

it must be $p = 0$.

□

4.2 The continuous Newton method.

The preliminary normal forms (4.2) and (4.3) described in Propositions 3 and 4, respectively, are based on certain relations between the Image and Kernel spaces of A and J at x^*. These relations are trivially satisfied by the continuous Newton method (4.1) at singular equilibria, since the identity $A = -J$ is verified pointwise. Nevertheless, this identity allows for further simplification:

Proposition 5. *Let x^* be a standard singular equilibrium of the continuous Newton method (4.1). Then, the normal form (4.3) can be written as*

$$\begin{aligned} y' &= -y + \text{h.o.t.} \\ zz' &= (-1/2)z^2 + y^T G y + \text{h.o.t.} \end{aligned} \tag{4.4}$$

that is, $H = -I_{n-1}$, $\lambda = -1/2$, $q = 0$.

Proof. In this case, the relation $A(x) = -J(x)$ characterizing the continuous Newton method is not directly transferred to $\bar{A}$, $\bar{J}$. Recall that $\bar{A}(w) = E(\phi(w))A(\phi(w))\phi'(w)$ and $\bar{f}(w) = E(\phi(w))f(\phi(w))$. From the last relation, we have

$$\begin{aligned} \bar{J}(w) &= E(\phi(w))J(\phi(w))\phi'(w) + E'(\phi(w))\phi'(w)f(\phi(w)) \\ &= -\bar{A}(w) + E'(\phi(w))\phi'(w)f(\phi(w)), \end{aligned}$$

that is,

$$\bar{A}(w) = -\bar{J}(w) + E'(\phi(w))\phi'(w)f(\phi(w)),$$

or

$$\bar{A} = -\bar{J} + (E \circ \phi)'(f \circ \phi). \tag{4.5}$$

This means that there is an additional term $(E \circ \phi)'(f \circ \phi)$ due to the change of coordinates $x = \phi(w)$ and the multiplication by the matrix $E(x)$. Nevertheless, this additional term vanishes at the singular equilibrium, since $f(\phi(0)) = f(x^*) = 0$. Hence,

$$\bar{A}(0) = \begin{pmatrix} I_{n-1} & 0 \\ 0 & 0 \end{pmatrix} = -\bar{J}(0) = \begin{pmatrix} -H & 0 \\ 0 & 0 \end{pmatrix},$$

showing that $H = -I_{n-1}$.

The relations $\lambda = -1/2$, $q = 0$ involve second-order terms arising in the expansion of $\bar{f}_n$. Taking the last row of (4.5), we get

$$\bar{A}_n = -\bar{J}_n + (E_n \circ \phi)'(f \circ \phi),$$

and then

$$\bar{A}'_n = -\bar{f}''_n + (E_n \circ \phi)''(f \circ \phi) + (E_n \circ \phi)'(f \circ \phi)'.$$

Focusing on $0 = \phi^{-1}(x^*)$, this relation reads

$$\bar{A}'_n(0) = \begin{pmatrix} 0 & 0 \\ 0 & 1 \end{pmatrix} = -\bar{f}''_n(0) + (E_n \circ \phi)'(0)(f \circ \phi)'(0),$$

since $f(\phi(0)) = f(x^*) = 0$. Taking the last column, and using the fact that $f'(x^*)\partial x/\partial z(0) = 0$ since $\partial x/\partial z(0) \in \mathrm{Ker} A(x^*) = \mathrm{Ker}(-f'(x^*))$ by construction, we get

$$\begin{pmatrix} 0 \\ 1 \end{pmatrix} = -\begin{pmatrix} \partial^2 \bar{f}/\partial y \partial z(0) \\ \partial^2 \bar{f}/\partial z^2(0) \end{pmatrix}.$$

This means

$$q^T = \frac{\partial^2 \bar{f}}{\partial y \partial z}(0) = 0, \quad \lambda = \frac{1}{2}\frac{\partial^2 \bar{f}}{\partial z^2}(0) = -\frac{1}{2}.$$

□

In weak problems, the normal form (4.4) amounts to

$$\begin{aligned} y' &= -y + \text{h.o.t.} \\ z' &= -z/2 + \text{h.o.t.} \end{aligned}$$

This expresses that the continuous Newton field has a hyperbolic linearization at weak singular equilibria, with a simple eigenvalue -1/2 and a multiple (index-one) eigenvalue -1. Note that, at regular equilibria, the -1 eigenvalue is unique (see e.g. [25]): this fact supports the phenomenon of quadratic convergence displayed by the classical Newton iteration at regular roots. Newton's method is obtained after Euler discretization with stepsize 1 of the continuous Newton system. In singular problems, the -1/2 eigenvalue is responsible for the loss of quadratic convergence. This is a well-known behavior in one-dimensional problems, where quadratic convergence may be recovered simply using Euler discretization with stepsize 2. Higher dimensional problems, however, require multi-stage integration schemes to guarantee a quadratic convergence ratio at singular solutions: see details in [26].

On the other hand, a result of directional convergence has been proved for singular equilibria of the continuous Newton method which do not satisfy the weak condition [21, 25]. The example presented below attempts to illustrate this phenomenon. It is worth remarking that the preliminary normal form (4.4) might be of help in the analysis of several related issues concerning the actual shape of the attraction domain of singular equilibria.

4.3 Directional convergence: example.

Standard singular equilibria of the continuous Newton method display a cone-shaped region with axis in the direction of Ker$J$ (therefore, tangent to $\partial/\partial z$, that is, to the z-coordinate curve), which is positively invariant and convergent to the singular solution [21, 25]. A rough explanation of this behavior can be sketched using the notation of (4.4). Such a cone-shaped region is defined by a restriction of the form $\|y\| \leq \theta \|z\|$: for sufficiently small θ, there is a leading term which behaves as $(-1/2)z$ in the last equation of (4.4). This means that the evolution in the z-direction is slower than the dynamics in transversal directions (corresponding to y-coordinates, for which the linearization reads $y' = -y$) and, therefore, trajectories are driven towards the origin "along" the z-coordinate curve. In turn, this invariant region is preserved under appropriate discretization schemes, yielding an n-dimensional convergence region for different iterative root-finding methods [8, 18, 26]. The following example attempts to illustrate this phenomenon in the continuous-time setting.

Consider

$$f(x_1, x_2) = (x_1^2 - x_1^4, x_1^5 + x_2^2 - 1).$$

The Jacobian matrix is

$$J(x_1, x_2) = \begin{pmatrix} 2x_1 - 4x_1^3 & 0 \\ 5x_1^4 & 2x_2 \end{pmatrix}.$$

The singular set is defined by the condition $\omega(x_1, x_2) = \det J(x_1, x_2) = 4x_1 x_2 (1 - 2x_1^2) = 0$ and, therefore, comprises the straight lines $x_1 = 0,\ x_2 = 0$ and $x_1 = \pm\sqrt{1/2}$. In particular, there is a singular equilibrium point located at $(1, 0)$. The desingularized field $g(x_1, x_2) = -\mathrm{Adj}J(x_1, x_2) f(x_1, x_2)$ has the expression

$$g(x_1, x_2) = (x_1^2 x_2 (x_1^2 - 1), x_1(-x_1^7 + 3x_1^5 + 4x_1^2 x_2^2 - 4x_1^2 - 2x_2^2 + 2)),$$

the Newton field being defined by the quotient g/ω.

Figure 1 displays the domain of attraction of the singular equilibrium placed at $(1, 0)$. There is a locally cone-shaped region of convergence with vertical axis defined by the direction of Ker$J(1, 0)$. The singular set $x_2 = 0$ is formed by forward impasse points for x_1 approximately less than 1.85 (with the exception of the equilibrium $(1, 0)$), whereas values of x_1 greater than 1.85 yield backward impasse points. Trajectories emanating from these singularities eventually converge to the singular equilibrium.

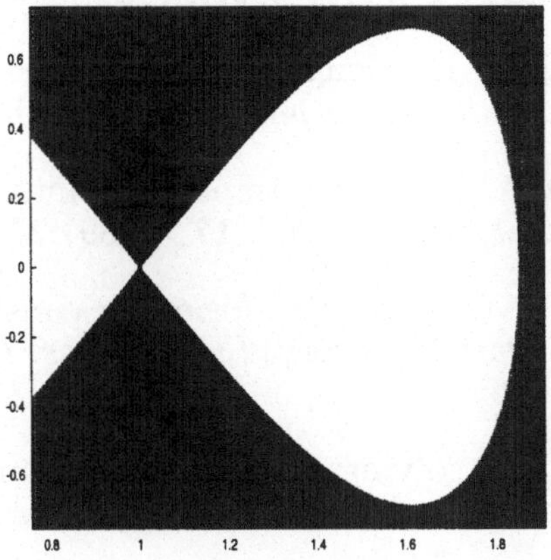

Figure 1: Cone-shaped domain of attraction.

5 Concluding remarks

Several issues concerning local equivalence of quasilinear ODEs around singular points have been considered in this paper. A review of normal forms for these problems has allowed for a simple description of several dynamic phenomena. A preliminary normal form for a family of (so-called stationary) singular equilibria arising in root-finding problems has been presented. This approach should be relevant not only in the topological or smooth classification of phase portraits around such equilibria, but also regarding the characterization of the actual shape of the attraction domain of singular solutions in root-finding methods.

Acknowledgements

The author gratefully acknowledges several stimulating discussions with Professor Rafael Ortega, from Universidad de Granada, Spain.

References

[1] R. E. Beardmore, Stability and bifurcation properties of index-1 DAEs, *Numer. Algorithms* **19** (1998) 43-53.

[2] R. E. Beardmore, The singularity-induced bifurcation and its Kronecker normal form, *SIAM J. Matrix Anal. Appl.* **23** (2001) 126-137.

[3] K. E. Brenan, S. L. Campbell and L. R. Petzold, *Numerical Solution of Initial-Value Problems in Differential-Algebraic Equations*, Classics in Applied Mathematics, SIAM, 1996.

[4] S. L. Campbell and W. Marszalek, DAEs arising from traveling wave solutions of PDEs, *J. Comput. Appl. Math.* **82** (1997) 41-58.

[5] L. O. Chua and A. D. Deng, Impasse points, I: Numerical aspects, *Internat. J. Circuit Theory Appl.* **17** (1989) 213-235.

[6] L. O. Chua and A. D. Deng, Impasse points, II: Analytical aspects, *Internat. J. Circuit Theory Appl.* **17** (1989) 271-282.

[7] L. O. Chua and H. Oka, Normal forms for constrained nonlinear differential equations, Part I: Theory, *IEEE Trans. Circuits and Systems* **35** (1988) 881-901.

[8] A. O. Griewank, On solving nonlinear equations with simple singularities or nearly singular solutions, *SIAM Review* **27** (1985) 537-572.

[9] E. Hairer and G. Wanner, *Solving Ordinary Differential Equations II: Stiff and Differential-Algebraic Problems*, Springer-Verlag, 1996.

[10] J. Th. Jongen, P. Jonker and F. Twilt, The continuous, desingularized Newton method for meromorphic functions, *Acta Appl. Math.* **13** (1988) 81-121.

[11] J. Llibre, J. Sotomayor and M. Zhitomirskii, Impasse bifurcations of constrained systems, in A. Galves, J. K. Hale and C. Rocha, eds., *Differential Equations and Dynamical Systems*, Fields Institute Communications Vol. 31, AMS (2002), 235-255.

[12] W. Marszalek and S. L. Campbell, DAEs arising from traveling wave solutions of PDEs II, *Comp. Math. Appl.* **37** (1999) 15-34.

[13] M. Medved, Normal forms of implicit and observed implicit differential equations, *Riv. Mat. Pura ed Appl.* **10** (1991) 95-107.

[14] M. Medved, Qualitative properties of generalized vector fields, *Riv. Mat. Pura ed Appl.* **15** (1994) 7-31.

[15] P. J. Rabier, Implicit differential equations near a singular point, *J. Math. Anal. Appl.* **144** (1989) 425-449.

[16] P. J. Rabier and W. C. Rheinboldt, On impasse points of quasi-linear differential-algebraic equations, *J. Math. Anal. Appl.* **181** (1994) 429-454.

[17] P. J. Rabier and W. C. Rheinboldt, On the computation of impasse points of quasi-linear differential-algebraic equations, *Math. Comp.* **62** (1994) 133-154.

[18] G. W. Reddien, On Newton's method for singular problems, *SIAM J. Numer. Anal.* **15** (1978) 993-996.

[19] G. Reiszig, Differential-algebraic equations and impasse points, *IEEE Trans. Circuits and Systems, Part I* **43** (1996) 122-133.

[20] G. Reiszig and H. Boche, A normal form for implicit differential equations near singular points, Proc. 1997 Europ. Conf. on Circuit Th. and Design (ECCTD'97), Budapest, Hu., vol. 2, pp. 1048-1053, 1997.

[21] R. Riaza, Stability issues in regular and non-critical singular DAEs, *Acta Appl. Math.* **73** (2002) 301-336.

[22] R. Riaza, On the Singularity-Induced Bifurcation Theorem, *IEEE Trans. Aut. Contr.* **47** (2002) 1520-1523.

[23] R. Riaza, S. L. Campbell and W. Marszalek, On singular equilibria of index-1 DAEs, *Cir. Sys. Sig. Proc.* **19** (2000) 131-157.

[24] R. Riaza and P. J. Zufiria, Weak singularities and the continuous Newton method, *J. Math. Anal. Appl.* **236** (1999) 438-462.

[25] R. Riaza and P. J. Zufiria, Stability of singular equilibria in quasilinear implicit differential equations, *J. Differential Equations* **171** (2001) 24-53.

[26] R. Riaza and P. J. Zufiria, Discretization of implicit ODEs for singular root-finding problems, *J. Comput. Appl. Math.* **140** (2002) 695-712. .

[27] H. von Sosen, *Folds and Bifurcations in the Solutions of Semi-Explicit Differential-Algebraic Equations,* PhD. Thesis, Part I. California Institute of Technology, 1994.

[28] J. Sotomayor, Structurally stable differential systems of the form $A(x)x' = F(x)$, *Diff. Eqs. Dyn. Syst.* **5** (1997) 415-422.

[29] J. Sotomayor and M. Zhitomirskii, Impasse singularities of differential systems of the form $A(x)x' = F(x)$, *J. Differential Equations* **169** (2001) 567-587.

[30] F. Takens, Constrained equations; a study of implicit differential equations and their discontinuous solutions, *Lect. Notes Math.* **525**, 143-234, Springer-Verlag, 1976.

[31] V. Venkatasubramanian, H. Schättler and J. Zaborszky, Local bifurcations and feasibility regions in differential-algebraic systems, *IEEE Trans. Aut. Contr.* **40** (1995) 1992-2013.

[32] L. Yang and Y. Tang, An improved version of the singularity-induced bifurcation theorem, *IEEE Trans. Aut. Contr.* **46** (2001) 1483-1486.

[33] M. Zhitomirskii, Local normal forms for constrained systems on 2-manifolds, *Bol. Soc. Bras. Mat.* **24** (1993) 211-232.

Generic Convergence for a Class of Dynamical Systems

Simeon Reich[1] and Alexander J. Zaslavski[2]

[1]Department of Mathematics, The Technion-Israel Institute of Technology, 32000 Haifa, Israel, e–mail: sreich@tx.technion.ac.il

[2]Department of Mathematics, The Technion-Israel Institute of Technology, 32000 Haifa, Israel, e–mail: ajzasl@tx.technion.ac.il

Abstract. We consider a complete metric space $\mathcal{M}$ of sequences of continuous mappings, acting on a bounded closed convex subset K of a Banach space, which share a common continuous convex Lyapunov function f. We show that for a generic pair $(x, \{A_t\}_{t=1}^{\infty}) \in K \times \mathcal{M}$, the sequence $\{f(A_n \dots A_1 x)\}_{n=1}^{\infty}$ tends to the infimum of f on K.

AMS Subject Classification. 47N10, 49M45, 54E35, 54E50, 54E52, 90C25.

Key words. Banach space, complete metric space, convex function, generic property, infinite product.

1 Introduction

Let $(X, ||\cdot||)$ be a Banach space, $K \subset X$ a nonempty bounded closed convex subset of X, and $f : K \to R^1$ a convex continuous function which is bounded from below. Set

$$\inf(f) = \inf\{f(x) : x \in K\}.$$

We consider the topological subspace $K \subset X$ with the relative topology. Denote by $\mathcal{A}$ the set of all continuous self-mappings $A : K \to K$ such that

$$f(Ax) \le f(x) \text{ for all } x \in K. \tag{1.1}$$

In [10, Section 4] we show how to construct many mappings belonging to $\mathcal{A}$ when f is uniformly continuous.

For the set $\mathcal{A}$ we define a metric $\rho : \mathcal{A} \times \mathcal{A} \to R^1$ by

$$\rho(A, B) = \sup\{||Ax - Bx|| : x \in K\}, \; A, B \in \mathcal{A}. \tag{1.2}$$

Clearly, the metric space $\mathcal{A}$ is complete. Denote by $\mathcal{M}$ the set of all sequences $\{A_t\}_{t=1}^{\infty} \subset \mathcal{A}$. A member $\{A_t\}_{t=1}^{\infty}$ of $\mathcal{M}$ will occasionally be denoted by a boldface $\mathbf{A}$.

For the set $\mathcal{M}$ we will consider two uniformities and the topologies induced by them. The first uniformity is determined by the base

(1.3)
$$E_w(N,\epsilon) = \{(\{A_t\}_{t=1}^{\infty}, \{B_t\}_{t=1}^{\infty}) \in \mathcal{M} \times \mathcal{M} : \rho(A_t, B_t) \le \epsilon, \ t = 1, \dots, N\},$$

where N is a natural number and $\epsilon > 0$. Clearly, the uniform space $\mathcal{M}$ with this uniformity is metrizable (by a metric $\rho_w : \mathcal{M} \times \mathcal{M} \to R^1$) and complete (see [6]). We equip the set $\mathcal{M}$ with the topology induced by this uniformity. This topology will be called weak and denoted by τ_w.

The second uniformity is determined by the base

$$(1.4) \qquad E_s(\epsilon) = \{(\{A_t\}_{t=1}^{\infty}, \{B_t\}_{t=1}^{\infty}) \in \mathcal{M} \times \mathcal{M} : \rho(A_t, B_t) \le \epsilon, \ t \ge 1\},$$

where $\epsilon > 0$. Clearly, this uniformity is metrizable (by a metric $\rho_s : \mathcal{M} \times \mathcal{M} \to R^1$) and complete (see [6]). Denote by τ_s the topology induced by this uniformity in $\mathcal{M}$. Since τ_s is clearly stronger than τ_w, it will be called strong.

We also consider the space $K \times \mathcal{A}$ equipped with the product topology and the space $K \times \mathcal{M}$ which is equipped with a pair of topologies. One of them (which is called the weak topology) is the product of the topology of K and the weak topology of $\mathcal{M}$ and the second one (which is called the strong topology) is the product of the topology of K and the strong topology of $\mathcal{M}$.

From the point of view of the theory of dynamical systems, each element of $\mathcal{M}$ describes a nonstationary dynamical system with a Lyapunov function f. Also, some optimization procedures in Banach spaces can be represented by elements of $\mathcal{M}$ (see [7, 8, 10]). For recent studies of the minimization of convex functions on abstract spaces see, for example, [1], [2] and [5].

In [10] we assume that the function f is uniformly continuous and show that for a generic sequence taken from the space $\mathcal{M}$, the sequence of values of the Lyapunov function f along any trajectory tends to the infimum of f. Thus, instead of considering this convergence property for a single sequence of operators, we investigate it for the whole space $\mathcal{M}$, and show that this property holds for most of them. This approach has already been successfully applied in the theory of dynamical systems ([3], [9]), as well as in optimization theory and the calculus of variations ([4, 5], [10-13]).

In [4] we extend the main result of [10]. We introduce the concept of normality, and show that a generic sequence taken from the space $\mathcal{M}$ is normal and that for each normal element of $\mathcal{M}$, the sequence of values of the Lyapunov function f along any trajectory tends to the infimum of f on K. These results, too, are obtained under the assumption that the function f is uniformly continuous. In the present paper, assuming that f is merely continuous, we are still able to obtain two results in this direction. To achieve this, we change our point of view and consider a new framework. The main feature of this new framework is that the initial point of a trajectory of our dynamical system may also vary.

We now state the two main results of our paper.

Theorem 1.1. *There exists a set $\mathcal{F} \subset K \times \mathcal{M}$ which is a countable intersection of open (in the weak topology) everywhere dense (in the strong topology)* *subsets of $K \times \mathcal{M}$ such that for each $(x, \{A_t\}_{t=1}^{\infty}) \in \mathcal{F}$, the following property holds:*

For each $\epsilon > 0$, there exists a neighborhood $\mathcal{U}$ of $(x, \{A_t\}_{t=1}^{\infty})$ in $K \times \mathcal{M}$ with the weak topology and a natural number N such that for each $(y, \{B_t\}_{t=1}^{\infty}) \in \mathcal{U}$,

$$f(B_N \dots B_1 y) < \inf(f) + \epsilon.$$

Theorem 1.2. *There exists a set $\mathcal{F} \subset K \times \mathcal{A}$ which is a countable intersection of open everywhere dense subsets of $K \times \mathcal{A}$ such that for each $(x, A) \in \mathcal{F}$, the following property holds:*

For each $\epsilon > 0$, there exists a neighborhood $\mathcal{U}$ of (x, A) in $K \times \mathcal{A}$ and a natural number N such that for each $(y, B) \in \mathcal{U}$,

$$f(B^N y) < \inf(f) + \epsilon.$$

2 An auxiliary result

The following auxiliary resut was proved in [4], where it was assumed that f was uniformly continuous. The same proof works in our case too. We present it for the reader's convenience.

Proposition 2.1. *There exists a mapping $A \in \mathcal{A}$ with the following property:*

For each $\epsilon > 0$, there is $\delta(\epsilon) > 0$ such that for each $x \in K$ satisfying $f(x) \geq \inf(f) + \epsilon$, the inequality

$$f(Ax) \leq f(x) - \delta(\epsilon)$$

is true.

Proof. If there is $x_{min} \in K$ for which $f(x_{min}) = \inf(f)$, then we can set $Ax = x_{min}$ for all $x \in K$. Therefore we may assume that

$$\{x \in K : \ f(x) = \inf(f)\} = \emptyset. \tag{2.1}$$

Define a set-valued map $a : \ K \to 2^K$ as follows: for each $x \in K$, denote by $a(x)$ the closure (in the norm topology of X) of the set

$$\{y \in K : \ f(y) < 2^{-1}(f(x) + \inf(f))\}. \tag{2.2}$$

Clearly, for each $x \in K$, the set $a(x)$ is nonempty, closed and convex. We will show that the set-valued map a is lower semicontinuous.

Let $x_0 \in K$, $y_0 \in a(x_0)$, and let $\epsilon > 0$. To prove that a is lower semicontinuous, we need to show that there is $\delta > 0$ such that for each $x \in K$ satisfying $||x - x_0|| < \delta$,

$$a(x) \cap \{y \in K : \ ||y - y_0|| < \epsilon\} \neq \emptyset.$$

By the definition of $a(x_0)$, there exists $y_1 \in K$ such that

$$f(y_1) < 2^{-1}(f(x_0) + \inf(f)) \text{ and } ||y_1 - y_0|| < \epsilon/2.$$

Since the function f is continuous, there is $\delta > 0$ such that for each $x \in K$ satisfying $||x - x_0|| < \delta$,

$$f(y_1) < 2^{-1}(f(x) + \inf(f)).$$

Hence $y_1 \in a(x)$ by definition. Therefore a is indeed lower semicontinuous. By Michael's selection theorem, there exists a continuous mapping $A : K \to K$ such that $Ax \in a(x)$ for all $x \in K$. It follows from the definition of a (see (2.2)) that for each $x \in K$,

$$f(Ax) \leq 2^{-1}(f(x) + \inf(f)).$$

This completes the proof of Proposition 2.1. □

3 Preliminary results

By Proposition 2.1, there exists a mapping $A_* \in \mathcal{A}$ with the following property.

(P1) For each $\epsilon > 0$, there is $\delta > 0$ such that for each $x \in K$ satisfying $f(x) \geq \inf(f) + \epsilon$, the inequality

$$f(A_*x) \leq f(x) - \delta$$

holds.

For each $\{A_t\}_{t=1}^{\infty} \in \mathcal{M}$ and each $\gamma \in (0,1)$, we define a sequence of mappings $\mathbf{A}^\gamma = \{A_t^\gamma\}_{t=1}^{\infty} \in \mathcal{M}$ by

$$A_t^\gamma x = (1-\gamma)A_t x + \gamma A_* x, \quad x \in K, \ t = 1, 2, \ldots. \tag{3.1}$$

It is obvious that for each $\mathbf{A} \in \mathcal{M}$,

$$\mathbf{A}^\gamma \to \mathbf{A} \text{ as } \gamma \to 0^+ \text{ in the strong topology.} \tag{3.2}$$

If $\{A_t\}_{t=1}^{\infty} \in \mathcal{M}$ and $A_t = A$, $t = 1, 2, \ldots$, with $A \in \mathcal{A}$, then $A_t^\gamma = A^\gamma$, $t = 1, 2, \ldots$, with $A^\gamma \in \mathcal{A}$ defined by $A^\gamma x = (1-\gamma)Ax + \gamma A_* x$, $x \in K$.

Lemma 3.1. *Let $\{A_t\}_{t=1}^{\infty} \in \mathcal{M}$, $x \in K$, $\gamma \in (0,1)$, and $\epsilon > 0$. Then there exists a natural number N such that*

$$f(A_N^\gamma \ldots A_1^\gamma x) \leq \inf(f) + \epsilon. \tag{3.3}$$

Proof. Let $\delta > 0$ be as guaranteed by property (P1). Choose a natural number

$$N > 4 + 4(\delta\gamma)^{-1}[f(x) - \inf(f) + 1]. \tag{3.4}$$

We will show that (3.3) holds for this N. Assume the converse. Then

$$f(A_N^\gamma \dots A_1^\gamma x) > \inf(f) + \epsilon$$

and

$$f(x) > \inf(f) + \epsilon,\ f(A_k^\gamma \dots A_1^\gamma x) > \inf(f) + \epsilon,\ k = 1, \dots, N. \tag{3.5}$$

It follows from (3.5), the definition of δ and property (P1) that

$$f(A_* x) \le f(x) - \delta,\ f(A_* A_k^\gamma \dots A_1^\gamma x) \le f(A_k^\gamma \dots A_1^\gamma x) - \delta,\ k = 1, \dots, N. \tag{3.6}$$

By (3.6), (3.1), and the convexity of f,

$$f(A_1^\gamma x) = f((1-\gamma)A_1 x + \gamma A_* x) \le (1-\gamma) f(A_1 x) +$$
$$\gamma f(A_* x) \le (1-\gamma) f(x) + \gamma(f(x) - \delta) =$$
$$f(x) - \gamma\delta,$$

and for $k = 1, \dots, N$,

$$f(A_{k+1}^\gamma A_k^\gamma \dots A_1^\gamma x) =$$
$$f((1-\gamma)A_{k+1} A_k^\gamma \dots A_1^\gamma x + \gamma A_* A_k^\gamma \dots A_1^\gamma x) \le$$
$$(1-\gamma) f(A_{k+1} A_k^\gamma \dots A_1^\gamma x) + \gamma f(A_* A_k^\gamma \dots A_1^\gamma x) \le$$
$$(1-\gamma) f(A_k^\gamma \dots A_1^\gamma x) + \gamma f(A_k^\gamma \dots A_1^\gamma x) - \delta\gamma =$$
$$f(A_k^\gamma \dots A_1^\gamma x) - \delta\gamma.$$

Thus

$$f(A_1^\gamma x) \le f(x) - \gamma\delta$$

and

$$f(A_{k+1}^\gamma A_k^\gamma \dots A_1^\gamma x) \le f(A_k^\gamma \dots A_1^\gamma x) - \delta\gamma,\ k = 1, \dots, N.$$

By these inequalities,

$$f(x) - \inf(f) \ge f(x) - f(A_{N+1}^\gamma A_N^\gamma \dots A_1^\gamma x) =$$

$$f(x) - f(A_1^\gamma x) + \sum_{k=1}^{N} [f(A_k^\gamma \dots A_1^\gamma x) - f(A_{k+1}^\gamma \dots A_1^\gamma x)] \ge (N+1)\delta\gamma$$

and

$$N \le (\delta\gamma)^{-1}[f(x) - \inf(f)],$$

a contradiction (see (3.4)). The contradiction we have reached proves the lemma. □

To continue we also need the following lemma.

Lemma 3.2. *Let* $\{A_t\}_{t=1}^{\infty} \in \mathcal{M}$, $x \in K$, $\epsilon > 0$, *and let* N *be a natural number. Then there exists a neighborhood* $\mathcal{U}$ *of* $(x, \{A_t\}_{t=1}^{\infty})$ *in* $K \times \mathcal{M}$ *with the weak topology such that for each* $(y, \{B_t\}_{t=1}^{\infty}) \in \mathcal{U}$,

$$|f(B_N \dots B_1 y) - f(A_N \dots A_1 x)| < \epsilon.$$

This lemma is, in turn, a consequence of the next one.

Lemma 3.3. *Let* $x \in K$, $\epsilon > 0$, *let* N *be a natural number, and let* $A_1, \dots, A_N \in \mathcal{A}$. *Then there exist* $\delta_1, \delta_2 > 0$ *such that for each* $y \in K$ *satisfying* $||y - x|| \leq \delta_1$ *and each* $\{B_i\}_{i=1}^{N} \subset \mathcal{A}$ *satisfying*

$$\rho(A_k, B_k) \leq \delta_2, \ k = 1, \dots, N,$$

the inequality

$$|f(B_N \dots B_1 y) - f(A_N \dots A_1 x)| \leq \epsilon$$

holds.

Proof.

We prove Lemma 3.3 by induction on N. Consider the case $N = 1$. There is $\delta > 0$ such that for each $y \in K$ satisfying $||y - A_1 x|| \leq \delta$, we have

$$|f(y - A_1 x)| \leq \epsilon. \tag{3.7}$$

Since A_1 is continuous, there is $\delta_1 > 0$ such that for each $y \in K$ satisfying $||y - x|| \leq \delta_1$,

$$||A_1 x - A_1 y|| \leq \delta/4. \tag{3.8}$$

Set

$$\delta_2 = \delta/4. \tag{3.9}$$

Assume that $y \in K$,

$$||y - x|| \leq \delta_1 \text{ and } \rho(A_1, B_1) \leq \delta_2. \tag{3.10}$$

Then

$$||A_1 y - B_1 y|| \leq \delta_2 = \delta/4. \tag{3.11}$$

By (3.10) and the definition of δ_1 (see (3.8)),

$$||A_1 x - A_1 y|| \leq \delta/4.$$

Combined with (3.11), this implies that

$$||B_1 y - A_1 x|| \leq ||B_1 y - A_1 y|| + ||A_1 y - A_1 x|| \leq \delta/2.$$

By this inequality and the definition of δ (see (3.7)),

$$|f(B_1 y) - f(A_1 x)| \le \epsilon.$$

Thus for $N = 1$ the assertion of the lemma holds. Assume that k is a natural number and that the assertion of the lemma holds with $N = k$. We will show that it also holds for $N = k + 1$.

Indeed, assume that $x \in K$, $\epsilon > 0$, and $A_1, \dots, A_k, A_{k+1} \in \mathcal{A}$. Since the assertion of the lemma is true for $N = k$, there is $\delta_0 > 0$ such that

$$(3.12) \qquad |f(A_{k+1}A_k \dots A_1 x) - f(C_{k+1}C_k \dots, C_2 y)| \le \epsilon$$

for each $y \in K$ and $C_2, \dots, C_{k+1} \in \mathcal{A}$ satisfying

$$(3.13) \qquad ||y - A_1 x|| \le \delta_0 \text{ and } \rho(A_i, C_i) \le \delta_0,\ i = 2, \dots, k+1.$$

Since A_1 is continuous, there is a number $\delta_1 > 0$ such that

$$(3.14) \qquad ||A_1 z - A_1 x|| \le \delta_0/4 \text{ for each } z \in K \text{ satisfying } ||z - x|| \le \delta_1.$$

Set

$$\delta_2 = \delta_0/4.$$

Assume that $y \in K$, $B_1, \dots, B_{k+1} \in \mathcal{A}$,

$$(3.15) \qquad ||y - x|| \le \delta_1 \text{ and } \rho(B_i, A_i) \le \delta_2,\ i = 1, \dots, k+1.$$

By (3.15) and (3.14),

$$(3.16) \qquad ||A_1 y - A_1 z|| \le \delta_0/4.$$

By (3.16) and (3.15),

$$(3.17) \quad ||A_1 x - B_1 y|| \le ||A_1 x - A_1 y|| + ||A_1 y - B_1 y|| \le \delta_0/4 + \delta_2 = \delta_0/2.$$

In view of (3.17), (3.15), and the choice of δ_0 (see (3.12) and (3.13)),

$$|f(A_{k+1}A_k \dots A_1 x) - f(B_{k+1}B_k \dots B_1 y)| \le \epsilon.$$

Thus the assertion of the lemma holds with $N = k + 1$. This completes the proof of Lemma 3.3. □

Lemma 3.4. *Let $\{A_t\}_{t=1}^\infty \in \mathcal{M}$, $x \in K$, $\gamma \in (0,1)$, and $\epsilon > 0$. Then there exist a natural number N and a neighborhood $\mathcal{U}$ of $(x, \{A_t^\gamma\}_{t=1}^\infty)$ in $K \times \mathcal{M}$ with the weak topology such that for each $(y, \{B_t\}_{t=1}^\infty) \in \mathcal{U}$,*

$$f(B_N \dots B_1 y) \le \inf(f) + \epsilon.$$

holds.

Proof. By Lemma 3.1, there exists a natural number N such that

$$f(A_N^\gamma \dots A_1^\gamma x) \le \inf(f) + \epsilon/4. \tag{3.18}$$

By Lemma 3.2, there exists a neighborhood $\mathcal{U}$ of $(x, \{A_t^\gamma\}_{t=1}^\infty)$ in $K \times \mathcal{M}$ with the weak topology such that for each $(y, \{B_t\}_{t=1}^\infty) \in \mathcal{U}$,

$$|f(B_N \cdots B_1 y) - f(A_N^\gamma \cdots A_1^\gamma x)| \le \epsilon/4. \tag{3.19}$$

Assume that $(y, \{B_t\}_{t=1}^\infty) \in \mathcal{U}$. Then (3.19) holds. By (3.19) and (3.18),

$$f(B_N \cdots B_1 y) \le f(A_N^\gamma \cdots A_1^\gamma x) + \epsilon/4 \le \inf(f) + \epsilon/2.$$

Lemma 3.4 is proved. □

4 Proof of Theorem 1.1

Let $\{A_t\}_{t=1}^\infty \in \mathcal{M}$, $\gamma \in (0,1)$, let i be a natural number and let $x \in K$. By Lemma 3.4, there exist an open (in the weak topology) neighborhood $\mathcal{U}(x, \{A_t\}_{t=1}^\infty, \gamma, i)$ of $(x, \{A_t^\gamma\}_{t=1}^\infty)$ in $K \times \mathcal{M}$ and a natural number $N(x, \{A_t\}_{t=1}^\infty, \gamma, i)$ such that the following property holds.

(P2) If $(y, \{B_t\}_{t=1}^\infty) \in \mathcal{U}(x, \{A_t\}_{t=1}^\infty, \gamma, i)$, then

$$f(B_{N(x,\mathbf{A},\gamma,i)} \cdots B_1 y) \le \inf(f) + i^{-1}.$$

Define

$$\mathcal{F} = \cap_{q=1}^\infty \cup \{U(x, \{A_t\}_{t=1}^\infty, \gamma, q) : \ x \in K, \ \{A_t\}_{t=1}^\infty \in \mathcal{M}, \ \gamma \in (0,1)\}.$$

It is easy to see that $\mathcal{F}$ is a countable intersection of open (in the weak topology) everywhere dense (in the strong topology) subsets of $K \times \mathcal{M}$.

Let $(y, \{B_t\}_{t=1}^\infty) \in \mathcal{F}$ and $\epsilon > 0$. Choose a natural number q_0 for which

$$q_0^{-1} < \epsilon. \tag{4.1}$$

There exist $x \in K$, $\{A_t\}_{t=1}^\infty \in \mathcal{M}$, and $\gamma \in (0,1)$ such that

$$(y, \{B_t\}_{t=1}^\infty) \in \mathcal{U}(x, \{A_t\}_{t=1}^\infty, \gamma, q_0).$$

By (P2), for each $(z, \{C_t\}_{t=1}^\infty) \in \mathcal{U}(x, \{A_t\}_{t=1}^\infty, \gamma, q_0)$,

$$f(C_{N(x,\mathbf{A},\gamma,q_0)} \cdots C_1 z) \le \inf(f) + q_0^{-1} < \inf(f) + \epsilon.$$

Theorem 1.1 is proved. □

Since Theorem 1.2 can be proved analogously to Theorem 1.1, we omit its proof.

Acknowledgments. The first author was partially supported by the Israel Science Foundation founded by the Israel Academy of Sciences and Humanities (Grant 592/00), by the Fund for the Promotion of Research at the Technion, and by the Technion VPR Fund.

References

[1] Y.I. Alber, A.N. Iusem and M.V. Solodov, Minimization of nonsmooth convex functionals in Banach spaces, *J. Convex Analysis,* **4**(1997), 235-255.

[2] D. Butnariu and A.N. Iusem, Totally convex functions for fixed points computation and infinite dimensional optimization, *Kluwer*, Dordrecht, 2000.

[3] F.S. De Blasi and J. Myjak, Generic flows generated by continuous vector fields in Banach spaces, *Adv. in Math.*, **50**(1983), 266-280.

[4] M. Gabour, S. Reich and A.J. Zaslavski, A class of dynamical systems with a convex Lyapunov function, *Experimental, Constructive and Nonlinear Analysis. Canadian Mathematical Society Conference Proceedings,* **27**(2000), 83-91.

[5] A.D. Ioffe and A.J. Zaslavski, Variational principles and well-posedness in optimization and calculus of variations, *SIAM J. Control Optim.*, **38**(2000), 566-581.

[6] J.L. Kelley, General topology, *D. Van Nostrand*, New York, 1955.

[7] Yu.I. Lyubich and G.D. Maistrovskii, On the stability of relaxation processes, *Soviet Math. Dokl.*, **11**(1970), 311-313.

[8] Yu.I. Lyubich and G.D. Maistrovskii, The general theory of relaxation processes for convex functionals, *Russian Mathematical Surveys,* **25**(1970), 57-117.

[9] S. Reich and A.J. Zaslavski, Convergence of generic infinite products of nonexpansive and uniformly continuous operators, *Nonlinear Analysis: Theory, Methods and Applications,* **36**(1999), 1049-1065.

[10] S. Reich and A.J. Zaslavski, On the minimization of convex functionals, *Calculus of Variations and Differential Equations. Chapman & Hall/CRC Research Notes in Mathematics Series, CRC Press,* **410**(1999), Boca Raton, FL, 200-209.

[11] S. Reich and A.J. Zaslavski, Generic convergence of descent methods in Banach spaces, *Math. Oper. Research,* **25**(2000), 231-242.

[12] A.J. Zaslavski, Optimal programs on infinite horizon, 1 and 2, *SIAM Journal on Control and Optimization,* **33**(1995), 1643-1686.

[13] A.J. Zaslavski, Existence of solutions of optimal control problems without convexity assumptions, *Nonlinear Analysis: Theory, Methods and Applications,* **43**(2001), 339-361.

On the Unique Solvability of Some Linear Boundary Value Problems

A. N. Ronto[1] and A. M. Samoilenko[1]

[1]Institute of Mathematics, National Academy of Sciences of Ukraine, Kiev, Ukraine e–mail: ar@imath.kiev.ua, sam@imath.kiev.ua

Abstract. We establish some unique solvability conditions of linear boundary value problems for a linear functional-differential equation in a Banach space. The main assumption is that the corresponding homogeneous boundary condition should be satisfied by the constant functions.

AMS(MOS) Subject Classification. 4C15, 34B10, 35R30

Key words. Linear functional differential equation, unique solvability, homogeneous problem, periodic type problem.

1 Problem formulation and main assumptions

In this work, we establish certain general conditions under whiboundary value problem

$$u'(t) = (lu)(t) + q(t), \qquad t \in [a, b], \tag{1.1}$$

$$h(u) = d \tag{1.2}$$

is uniquely solvable for arbitrary $q \in L([a, b], X)$ and $d \in X$. Here and below, X is some Banach space, whereas $l : C([a, b], X) \to L([a, b], X)$ and $h : C([a, b], X) \to X$ are a continuous linear operators.

As usual [1, 2], by a solution of Eq. (1.1), we mean an absolutely continuous function $x : [a, b] \to X$ satisfying (1.1) almost everywhere on $[a, b]$.

The following two basic assumptions are made throughout this paper.

Assumption 1. *The mapping $h : C([a, b], X) \to X$ possesses a continuous right inverse operator.*

Assumption 2. $h1_X = 0$.

The property described in Assumption 2, which should be understood in the sense of notation (viii) from Section 2 below, resembles the so-called "strong singularity condition" considered in [3].

Remark 1.1. In other words, Assumption 2 means that we have the following inclusion understood in the sense of Remark 6.6 from Section 6:

$$X \subset \ker h. \tag{1.3}$$

A typical example of h satisfying Assumptions 1 and 2 is provided by the operator

$$C([a,b],X) \ni u \longmapsto h(u) := u(b) - u(a), \tag{1.4}$$

which determines the periodic-type boundary condition

$$u(b) - u(a) = d. \tag{1.5}$$

Our solvability conditions are formulated in terms of the spectrum of a certain linear operator in $C([a,b],X)$ associated with the given boundary-value problem (1.1), (1.2).

The paper is arranged as follows. Sections 3—10 contain a number of auxiliary statements referred to in the sequel.

In Section 11, we prove our main result (Theorem 11.1) on the unique solvability of the linear inhomogeneous boundary-value problem (1.1), (1.2) and give a number of corollaries.

Section 12 contains some comments and remarks.

2 Notation

Let $\langle X, \|\cdot\|\rangle$ be a Banach space. The following basic notation is used throughout the paper.

(i) $\mathcal{B}(X)$ is the algebra of all linear bounded operators on X.

(ii) 1_X is the unity in $\mathcal{B}(X)$.

(iii) $\sigma(A)$ is the spectrum of $A \in \mathcal{B}(X)$.

(iv) $r(A)$ is the spectral radius of $A \in \mathcal{B}(X)$.

(v) $\ker A$ (resp., $\operatorname{im} A$) is the kernel (resp., image) of a linear continuous operator A.

(vi) $C([a,b],X)$ is the Banach space of all the continuous mappings from $[a,b]$ to X with the norm

$$C([a,b],X) \ni x \longmapsto \|x\|_{C([a,b],X)} := \max_{t\in[a,b]} \|x(t)\|;$$

(vii) $L([a,b],X)$ is the Banach space of all the Bochner integrable mappings from $[a,b]$ to X with the norm

$$L([a,b],X) \ni x \longmapsto \|x\|_{L([a,b],X)} := \int_a^b \|x(t)\|\,\mathrm{d}t.$$

(viii) When $h : C([a,b],X) \to X$ and $Q : [a,b] \to \mathcal{B}(X)$, the symbol hQ stands for the mapping $X \ni x \longmapsto h(Qx)$.

3 Linear surjections from $C([a,b],X)$ to X

First of all, we introduce a definition.

Definition 3.1. For a continuous linear mapping $\Lambda : X \to C([a,b],X)$ [resp., summable linear mapping $\Lambda : X \to L([a,b],X)$], the symbol $\Lambda^{\diamond}$ will stand for the function $\Lambda^{\diamond} : [a,b] \to \mathcal{B}(X)$ defined by the relation

$$\Lambda^{\diamond}(t)\,x = (\Lambda x)(t), \qquad t \in [a,b], \tag{3.1}$$

for all $x \in X$. We shall call the continuous [resp., summable] abstract function $\Lambda^{\diamond}$ the *mapping associated with* Λ.

Remark 3.1. It is easy to see that, for every continuous linear mapping $\Lambda : X \to C([a,b],X)$, there exists a unique abstract function $\Lambda^{\diamond} : [a,b] \to \mathcal{B}(X)$ having property (3.1). This function is, obviously, continuous. Conversely, given the function $\Lambda^{\diamond}$, we can use equality (3.1) to reconstruct the original mapping Λ in a unique way.

Let X be a Banach space and $h : C([a,b],X) \to X$ be a continuous linear operator.

Lemma 3.1. *The following conditions are equivalent.*

(a) $\operatorname{im} h = X$.

(b) $h : C([a,b],X) \to X$ *possesses a continuous right inverse.*

(c) *There exists a continuous function* $\Phi : [a,b] \to \mathcal{B}(X)$ *such that*

$$h\Phi = 1_X. \tag{3.2}$$

Proof. The equivalence of items (a) and (b) is obvious.

Assume that $\Lambda : X \to C([a,b],X)$ is a continuous right inverse for h. This means that

$$h(\Lambda x) = x \tag{3.3}$$

for every $x \in X$. According to Definition 3.1 and Remark 3.1, it follows from (3.3) that we can put $\Phi := \Lambda^{\diamond}$ in (3.2). It is also easy to show that the existence of a continuous solution of equation (3.2) implies that h is a surjection, which proves our lemma. □

Remark 3.2. When $X = \mathbb{R}^n$, the operator $h : C([a,b],X) \to X$ is given by the formula $C([a,b],X) \ni u \longmapsto \operatorname{col}(h_1(u), h_2(u), \dots, h_n(u))$ where $(h_k)_{k=1}^n$ are continuous linear functionals on the space $C([a,b],X)$. In this case, each of conditions (a)—(c) in Lemma 3.1 is equivalent to the following one:

(d) *The functionals* $(h_k)_{k=1}^n$ *are linearly independent.*

Corollary 3.1. *Let the linear operator* $h : C([a,b],\mathbb{R}^n) \to \mathbb{R}^n$ *be given by the formula* $C([a,b],X) \ni u \longmapsto \sum_{\nu=1}^{m} M_\nu u(t_\nu)$, *where* $\{t_\nu\}_{\nu=1}^{m} \subset [a,b]$ *and* $\{M_\nu\}_{\nu=1}^{m} \subset \mathcal{B}(\mathbb{R}^n)$. *Then each of conditions (a)—(d) formulated above is equivalent to the equality* $\operatorname{rank}[M_1, M_2, \ldots, M_m] = n$.

Proof. This fact is a consequence of the Kronecker–Capelli theorem and the statement given in Remark 3.2. □

4 An equivalent form of problem (1.1), (1.2)

Let us fix a Banach space X and consider the differential equation

$$u'(t) = (gu)(t), \qquad t \in [a,b], \tag{4.1}$$

where $g : C([a,b],X) \to L([a,b],X)$ is a (possibly, non-linear) mapping.

Remark 4.1. Proposition 4.1 below, which is applied in the sequel to the linear equations only (namely, as Proposition 6.2), is formulated here in the general, non-linear case for the notational convenience.

The aim of this section is to formally reduce the boundary value problem (4.1), (1.2) to a certain system of equations which, due their specific form, can be replaced by their restrictions to $\ker h$. In doing so, we shall use an idea similar to that suggested in [4].

As in Section 1, we make Assumption 1 (Assumption 2, generally speaking, is not needed here) and fix a continuous mapping $\Phi : [a,b] \to \mathcal{B}(X)$ such that equality (3.2) holds. Recall that, in view of Assumption 1, the existence of such a function is guaranteed by Lemma 3.1.

Example 4.1. In case h is given by relation (1.4), one can put, e.g., $\Phi(t) := \left(\frac{t-a}{b-a}\right)^{1+\alpha} \cdot 1_X$ for all $t \in [a,b]$ and some positive α.

Given a function Φ with the above properties, we introduce the linear operator $\Pi_{\Phi,h} : C([a,b],X) \to C([a,b],X)$ by setting

$$[\Pi_{\Phi,h} y](t) := y(t) - \Phi(t)h(y), \qquad t \in [a,b] \tag{4.2}$$

for all $y \in C([a,b],X)$.

By virtue of (3.2) and (4.2), the following lemma is obvious.

Lemma 4.1. $\Pi_{\Phi,h} C([a,b],X) \subset \ker h$.

Proof. For every y from $C([a,b],X)$, equality (4.2) yields

$$h\Pi_{\Phi,h} y = [1_X - h\Phi] hy,$$

whence, by virtue of (3.2), it follows immediately that $h\Pi_{\Phi,h} y = 0$. □

Remark 4.2. Identity (3.2) implies, in particular, that $\Pi^2_{\Phi,h} = \Pi_{\Phi,h}$ and, therefore, $\Pi_{\Phi,h}$ is a continuous projector of $C([a,b],X)$ onto $\ker h$.

Let us fix a certain number $\tau \in [a,b]$ and, for every $u \in L([a,b],X)$, set

$$[I_\tau u](t) := \int_\tau^t u(\sigma)\,\mathrm{d}\sigma, \qquad t \in [a,b]. \tag{4.3}$$

Proposition 4.1. *If $u \in C([a,b],X)$ is a solution of the boundary value problem* (4.1), (1.2), *then the equalities*

$$u = \Pi_{\Phi,h} I_\tau gu + [1_X - \Phi h 1_X]\,\xi + \Phi d, \tag{4.4}$$

$$h\,(\xi + I_\tau gu) = d \tag{4.5}$$

hold with $\xi = u(\tau)$.

Conversely, if equations (4.4) *and* (4.5) *hold with some $\xi \in X$, then u is a solution of problem* (4.1), (1.2). *If, furthermore,*

$$\Phi(\tau) = 0. \tag{4.6}$$

then

$$u(\tau) = \xi. \tag{4.7}$$

Proof. It follows from (4.2) and (4.3) that equation (4.4), in fact, has the form

$$\begin{aligned} u(t) = \xi + \int_\tau^t [gu](\sigma)\,\mathrm{d}\sigma - \Phi(t)h\Big(\int_\tau^\cdot [gu](\sigma)\,\mathrm{d}\sigma\Big) \\ + [1_X - \Phi(t)h1_X]\,\xi + \Phi\,(t)\,d, \qquad t \in [a,b]. \end{aligned} \tag{4.8}$$

If u satisfies (1.1) and (1.2), then

$$u(t) = u(\tau) + \int_\tau^t [gu](\sigma)\,\mathrm{d}\sigma, \qquad t \in [a,b]. \tag{4.9}$$

Therefore, by (1.2), it follows that (4.5) is true. Moreover, equality (4.8) [or, which is the same, (4.4)] holds with $\xi = u(\tau)$. Thus, we have shown that (1.1), (1.2) yield (4.4), (4.5).

Conversely, if equalities (4.4) and (4.5) hold for some $u \in C([a,b],X)$ and $\xi \in X$, then u satisfies equation (4.9) and, hence, equation (1.1). By virtue of Lemma 4.1, every solution of (4.8) satisfies (1.2) and, therefore, (1.2) also holds. Equality (4.7) is an immediate consequence of (4.8) and (4.5). This completes the proof of our proposition. □

Remark 4.3. By virtue of Lemma 4.1, for arbitrary fixed d and ξ from X, the mapping

$$C\big([a,b]\,, X\big) \ni u \longmapsto \Pi_{\Phi,h} I_\tau g u + [1_X - \Phi h 1_X]\,\xi + \Phi d$$

transforms $C\big([a,b]\,,X\big)$ to $h^{-1}(d)$ and, therefore, (4.4) can be regarded as an equation on $h^{-1}(d)$. From this viewpoint, Proposition 4.1 is a generalization of the principle described, e.g., in Chapt. I of [5] and Chapt. II of [6].

In the case where the mapping $g : C\big([a,b], X\big) \to L\big([a,b], X\big)$ in (4.1) is given by the formula

$$C([a,b], X) \ni u \longmapsto gu := lu + q, \tag{4.10}$$

i.e., when equation (4.1) coincides with (1.1), we have the following

Corollary 4.1. *If u is a solution of the boundary value problem* (1.1), (1.2), *then*

$$u = \Pi_{\Phi,h} I_\tau l u + [1_X - \Phi h 1_X]\,\xi + \Phi d + \Pi_{\Phi,h} I_\tau q, \tag{4.11}$$

$$h\,(\xi + I_\tau [lu + q]) = d \tag{4.12}$$

with $\xi = u(\tau)$. Conversely, if equalities (4.11) *and* (4.12) *hold with some $\xi \in X$, then u is a solution of problem* (1.1), (1.2).

Proof. This statement is an easy consequence of Proposition 4.1. □

Remark 4.4. Similarly to Proposition 4.1, condition (4.6) for Φ implies that every solution of (4.11), (4.12) satisfies (4.7).

When the operator $h : C([a,b], X) \to X$ in the boundary condition (1.2) satisfies Assumption 2, Corollary 4.1 implies the following statement.

Corollary 4.2. *Under Assumptions 1 and 2, a function $u \in C([a,b], X)$ is a solution of problem* (1.1), (1.2) *if, and only if $(\xi, u) \in X \times C([a,b], X)$ satisfies the equations*

$$u = \Pi_{\Phi,h} I_\tau l u + \Phi d + \Pi_{\Phi,h} I_\tau q, \tag{4.13}$$

$$h\,(I_\tau [lu + q]) = d \tag{4.14}$$

Proof. In view of Remark 1.1, Assumption 2 leads us to (1.3). Therefore, equality (4.11) can be brought to the form (4.13), whereas (4.12) can be rewritten as (4.14). □

5 The kernel of the operator $\Pi_{\Phi,h} I_\tau$

In our subsequent applications, the mapping $\Phi : [a,b] \to \mathcal{B}(X)$ from Proposition 4.1 is absolutely continuous and such that (4.6) holds. For this case, we have the following

Lemma 5.1. *If the mapping $\Phi : [a,b] \to \mathcal{B}(X)$ satisfying* (3.2) *and* (4.6) *is absolutely continuous, then*

$$\ker \Pi_{\Phi,h} I_\tau = \left\{ \Phi'(\cdot)\xi \mid \xi \in X \right\}.$$

Proof. Obviously, the relation $u \in \ker \Pi_{\Phi,h} I_\tau$ and the absolute continuity of Φ imply

$$u(t) = \Phi'(t) h u \qquad \text{for } t \in [a,b] \tag{5.1}$$

and, therefore,

$$\ker \Pi_{\Phi,h} I_\tau \subset \left\{ \Phi'(\cdot)\xi \mid \xi \in X \right\}.$$

Conversely, if $u(\cdot) = \Phi'(\cdot)\xi$ with some $\xi \in X$, then, according to (4.6),

$$\int_\tau^t u(s) ds = [\Phi(t) - \Phi(\tau)]\xi = \Phi(t)\xi$$

for all $t \in [a,b]$, whence, by (3.2), it follows that

$$h I_\tau u = h \Phi \xi = \xi.$$

Therefore, u satisfies (5.1). Integrating (5.1) and taking into account (4.6), we show that $u \in \ker \Pi_{\Phi,h} I_\tau$, which proves the lemma. □

6 Operator $\mathcal{A}_{l,h,\tau}$ and its properties

Let us consider equations (1.1), (1.2) determined by certain continous linear operators $l : C\big([a,b],X\big) \to L\big([a,b],X\big)$ and $h : C\big([a,b],X\big) \to X$.

Definition 6.1. Given a linear operator $l : C\big([a,b],X\big) \to L\big([a,b],X\big)$ and a certain fixed τ from $[a,b]$, we introduce the sequence of operator-valued functions $l^{[k,\tau]} : [a,b] \to \mathcal{B}(X)$, $k = 0,1,2,\ldots$, by putting

$$l^{[0,\tau]} := \left[l|_X \right]^\diamond \tag{6.1}$$

and

$$l^{[k,\tau]} := l \int_\tau^{\cdot} l^{[k-1,\tau]}(s)\, ds \tag{6.2}$$

for $k = 1,2,\ldots$.

Here, the diamond symbol denotes the mapping associated to $l|_X$ according to Definition 3.1.

Remark 6.1. Obviously, the function $l^{[0,\tau]}$ is independent of τ and, for this reason, we write in the sequel $l^{[0]}$ instead of $l^{[0,\tau]}$.

Remark 6.2. According to notation (viii) from Section 2, relation (6.1) can be rewritten as

$$l^{[0]} = l1_X. \tag{6.3}$$

In other words, the action of the restriction of l to the subspace of constant functions is nothing but the multiplication by $l^{[0]}$.

From now on, in addition to Assumptions 1 and 2, we make the following

Assumption 3. *The linear operator $l_h : X \to X$ defined by the formula*

$$l_h := h\left(\int_\tau^{\cdot} l^{[0]}(\sigma)\,d\sigma\right) \tag{6.4}$$

is invertible.

The notation introduced above is justified by the following

Proposition 6.1. *The mapping $l_h : X \to X$ is independent of the choice of τ in relation* (6.4).

Proof. This statement is an immediate consequence of Lemma 6.1 below. □

Lemma 6.1. *Let Assumption 2 hold for the linear operator $h : C([a,b],X) \to X$. Then, for an arbitrary u from $C([a,b],X)$, the function*

$$[a,b] \ni \tau \longmapsto f_u(\tau) := h\left(\int_\tau^{\cdot} (lu)(\sigma)\,d\sigma\right) \tag{6.5}$$

is constant.

Proof. Let us fix an arbitrary u from $C([a,b],X)$. Then, for all $\{\tau_1,\tau_2\}$ from $[a,b]$, formula (6.5) yields

$$hI_{\tau_2}lu - hI_{\tau_1}lu = h\left(\int_{\tau_1}^{\tau_2} (lu)(s)\,ds\right). \tag{6.6}$$

However, $\int_{\tau_1}^{\tau_2}(lu)(s)\,ds$ is a constant vector from X and therefore, in view of Assumption 2 and Remark 1.1, relation (6.6) implies $h\big[I_{\tau_1} - I_{\tau_2}\big]lu = 0$. In view of (4.3) and (6.3), this means that $f_u(\tau_1) = f_u(\tau_2)$, which, in view of the arbitrariness of τ_1 and τ_2, proves our lemma. □

The invertibility of operator (6.4) allows us to put

$$\Phi_{l,h,\tau}(t) := \int_\tau^t l^{[0]}(\sigma)\,d\sigma\, l_h^{-1}, \qquad t \in [a,b] \tag{6.7}$$

and consider the mapping $\Pi_{\Phi,h} : C\big([a,b],X\big) \to C\big([a,b],X\big)$ defined by (4.5) with $\Phi := \Phi_{l,h,\tau}$.

Remark 6.3. Relation (6.4), as is easy to see, guarantees that function (6.7) satisfies conditions (3.2) and (4.6). Furthermore, the function $\Phi_{l,h,\tau}: [a,b] \to \mathcal{B}(X)$ is, obviously, absolutely continuous.

Remark 6.4. It is obvious from Remark 6.3 and Lemma 3.1 that the multiplication by function (6.7) gives a right inverse operator for h. Thus, Assumption 3 implies, in partucular, that Assumption 1 is true.

Let us introduce the mapping

$$[\mathcal{A}_{l,h,\tau}y](t) := \int_\tau^t y(\sigma)\,\mathrm{d}\sigma - \int_\tau^t l^{[0]}(\sigma)\,\mathrm{d}\sigma\, l_h^{-1} h\Big(\int_\tau^{\cdot} y(\sigma)\,\mathrm{d}\sigma\Big) \tag{6.8}$$

for $y \in L([a,b],X)$ and $t \in [a,b]$. The integral in (6.4) is understood in the Bochner sense, whereas $l^{[0]}$ is related to l according to formula (6.1). Recall that we use function (3.1) from Definition 3.1.

It is easy to see that formula (6.8) determines a linear, continuous operator $\mathcal{A}_{l,h,\tau}$ from $L([a,b],X)$ to $C([a,b],X)$.

The reason why we need mapping (6.8) in relation to problem (1.1), (1.2) is explained by Corollary 4.2 and the following

Lemma 6.2. *The equality*

$$\mathcal{A}_{l,h,\tau} = \Pi_{\Phi_{l,h,\tau}} I_\tau \tag{6.9}$$

holds with $\Phi_{l,h,\tau}$ given by (6.7).

Proof. In view of (4.2) and (6.7), the validity of equality (6.9) is readily verified by direct computation. □

Proposition 6.2. *Let l and h in* (1.1), (1.2) *satisfy Assumptions 1—3.*

If $u \in C([a,b],X)$ is a solution of the boundary value problem (1.1), (1.2), *then the equalities*

$$u = \mathcal{A}_{l,h,\tau}(lu+q) + \xi + \int_\tau^{\cdot} l^{[0]}(\sigma)\,\mathrm{d}\sigma\, l_h^{-1} d, \tag{6.10}$$

$$h\Big[\int_\tau^{\cdot} [(lu)(s) + q(s)]\,\mathrm{d}s\Big] = d \tag{6.11}$$

hold with with $\xi = u(\tau)$.

Conversely, if equations (6.10) *and* (6.11) *hold with some $\xi \in X$, then u is a solution of problem* (1.1), (1.2) *and, furthermore, equality* (4.7) *holds.*

Proof. It suffices to fix a τ from $[a,b]$, put $\Phi := \Phi_{l,h,\tau}$ according to equality (6.7), apply Corollary 4.2 from Section 4, and take into account Lemma 6.2. □

Remark 6.5. It follows immediately from Lemma 6.2 and Remark 4.3 that every solution, u, of equation (6.10), if there are any, satisfies condition (1.2).

Lemma 6.3. *The inclusion*

$$X \subset \ker \mathcal{A}_{l,h,\tau} l \tag{6.12}$$

holds.

Remark 6.6. In (1.3), (6.12), and similar relations, we identify X with the subspace of $C([a,b],X)$ constituted by the constant functions.

Proof of Lemma 6.3. By Lemma 6.2, the linear mapping $\mathcal{A}_{l,h,\tau} : L([a,b],X) \to C([a,b],X)$ can be represented in the form (6.9) with $\Phi_{l,h,\tau} : [a,b] \to \mathcal{B}(X)$ defined according to (6.7). It is obvious from (6.7) that, for the latter function,

$$\Phi'_{l,h,\tau}(t) = l^{[0]}(t) l_h^{-1}, \qquad t \in [a,b],$$

and, therefore, by virtue of Lemma 5.1, $\mathcal{A}_{l,h,\tau} lu = 0$ if, and only if

$$[lu](t) = l^{[0]}(t) l_h^{-1} \xi = l(l_h^{-1}\xi), \qquad t \in [a,b], \tag{6.13}$$

with some $\xi \in X$.

It is easy to see that, when $u(t) = c$ for all $t \in [a,b]$, relation (6.13) holds with $\xi := l_h c$. This means that the required assertion holds true. □

Remark 6.7. In other words, Lemma 6.3 claims that $\operatorname{im} l|_X \subset \ker \mathcal{A}_{l,h,\tau}$.

7 A lemma on the "iterated shifts"

Let us introduce the "shift" mapping

$$u \longmapsto \Sigma_f u := u + f, \tag{7.1}$$

where f is a fixed element of a certain Banach space E.

Lemma 7.1. *Let E, F be two Banach spaces, G be a subspace of F, and $l : E \to F$ and $A : F \to E$ be two linear mappings such that*

$$G \subset \ker Al. \tag{7.2}$$

Then, for all $k \geq 1$ and $g \in G$,

$$(\Sigma_g Al)^k = \Sigma_g (Al)^k, \tag{7.3}$$

where the mapping $\Sigma_g : E \to E$ is defined according to (7.1).

Proof. Let us fix arbitrary $g \in G$ and $u \in E$, and set

$$u_k := (\Sigma_g Al)^k u, \qquad k = 0, 1, 2, \ldots.$$

According to definition (7.1), we have $u_1 = g + Alu$ and, by virtue of assumption (7.2),

$$u_2 = g + Al[g + Alu] = g + A[lg + lAlu] = g + (Al)^2 u.$$

Arguing similarly, we can show that

$$u_k = g + (Al)^k u$$

for all $u \in E$ and $k \in \mathbb{N}$. In view of the arbitrariness of u, this yields immediately equality (7.3). Lemma is proved. □

8 Iteration method for equation (6.10)

It follows from Proposition 6.2 that problem (1.1), (1.2) can be rewritten as the corresponding system (6.10), (6.11). In this section, we are interested in establishing the unique solvability of equation (6.10) for an arbitrary ξ from X and, more specifically, in conditions under which the unique solvability can be proved by using the iteration method

$$u_{k+1} := \xi + \mathcal{A}_{l,h,\tau}(lu_k + q) + \int_\tau^{\cdot} l^{[0]}(\sigma)\,\mathrm{d}\sigma\, l_h^{-1} d, \quad k = 0, 1, \ldots \tag{8.1}$$

Surprisingly enough, application of Lemma 7.1 allows one to claim that, under our conditions, the successive approximation method (8.1) for equation (6.10) is completely determined by the iterations of the operator $\mathcal{A}_{l,h,\tau} l$ for an *arbitrary* value of the parameter ξ. More precisely, we have the following

Lemma 8.1. *Under Assumptions 1—3, sequence* (8.1) *can be represented alternatively as*

$$u_k = \xi + \left(\mathcal{A}_{l,h,\tau} l\right)^k u_0 + \sum_{\nu=0}^{k-1} \left(\mathcal{A}_{l,h,\tau} l\right)^\nu \left[\mathcal{A}_{l,h,\tau} q + \int_\tau^{\cdot} l^{[0]}(\sigma)\,\mathrm{d}\sigma\, l_h^{-1} d\right] \tag{8.2}$$

for all $u_0 \in C\big([a,b], X\big)$, $\xi \in X$, *and* $k = 1, 2, \ldots$.

Proof. To obtain equality (8.2), Lemma 7.1 can be applied with

$$A := \mathcal{A}_{l,h,\tau}. \tag{8.3}$$

Indeed, we can put $E := C([a,b], X)$, $F := L([a,b], X)$, and $G := X$ (see Remark 6.6) in Lemma 7.1. Then, by virtue of Lemma 6.3, operator (8.3) satisfies condition (7.2) with respect to the given operator $l : C([a, b], X) \to L([a, b], X)$.

Obviously, sequence (8.1) can be represented in the form

$$u_k = \Sigma_\xi \mathcal{A}_{l,\,h,\,\tau} l u_{k-1} + r, \qquad k = 1, 2, \dots,$$

where

$$r := \mathcal{A}_{l,\,h,\,\tau} q + \int_\tau^{\cdot} l^{[0]}(\sigma)\, d\sigma\, l_h^{-1} d \tag{8.4}$$

and the mapping $\Sigma_\xi : C([a, b], X) \to C([a, b], X)$ is given by equality (7.1). Hence, it follows that

$$u_k = \Sigma_\xi \mathcal{A}_{l,\,h,\,\tau} l \big[\Sigma_\xi \mathcal{A}_{l,\,h,\,\tau} l u_{k-2} + r\big] + r, \qquad k = 2, 3, \dots. \tag{8.5}$$

Rewriting (8.5) as

$$u_k = \big(\Sigma_\xi \mathcal{A}_{l,\,h,\,\tau} l\big)^2 u_{k-2} + \Sigma_\xi \mathcal{A}_{l,\,h,\,\tau} l r + r$$

and arguing similarly, we arrive to the formula

$$u_k = \big(\Sigma_\xi \mathcal{A}_{l,\,h,\,\tau} l\big)^k u_0 + \sum_{\nu=0}^{k-1} \big(\Sigma_\xi \mathcal{A}_{l,\,h,\,\tau} l\big)^\nu r, \qquad k = 1, 2, \dots.$$

By virtue of Lemma 7.1, equality (8.6) can be brought to the form

$$u_k = \Sigma_\xi \Big[\big(\mathcal{A}_{l,\,h,\,\tau} l\big)^k u_0 + \sum_{\nu=0}^{k-1} \big(\mathcal{A}_{l,\,h,\,\tau} l\big)^\nu r\Big], \qquad k = 1, 2, \dots. \tag{8.6}$$

Taking into account definition (7.1) and notation (8.4), we conclude that (8.6) is nothing but (8.2). Lemma is proved. □

It follows from Lemma 8.1 that, under Assumption 2, the structure of operator (6.8) guarantees the simplest possible dependence on the parameter ξ in the method of successive approximations corresponding to the parametrised equation (4.4).

9 A lemma on the perturbation of spectrum

The purpose of this section is to establish a statement used in the proof of Proposition 10.1.

Lemma 9.1. *If $\{A, B\} \subset \mathcal{B}(X)$ are linear operators such that*

$$\sigma(B) \subset \sigma(A) \tag{9.1}$$

and

$$\operatorname{im} B \subset \ker A, \tag{9.2}$$

then

$$\sigma(A + B) = \sigma(A). \tag{9.3}$$

Proof. Indeed, let $\lambda \notin \sigma(A)$ be a regular point for A. Then the equation

$$Ax - \lambda x = y - \phi$$

has the unique solution

$$x(y - \phi, \lambda) := -\lambda^{-1}[y - \phi + \lambda^{-1}A(y - \phi) + \dots]$$

for all y and ϕ from X. Let us consider the equation

$$\phi = Bx(y - \phi, \lambda), \tag{9.4}$$

or, which is the same,

$$\phi = \lambda^{-1}B\sum_{\nu=0}^{+\infty}\lambda^{-\nu}A^{\nu}(\phi - y).$$

Since, obviously, we are seeking for a ϕ in $\operatorname{im} B$, condition (9.2) implies

$$\sum_{\nu=0}^{+\infty}\lambda^{-\nu}A^{\nu}\phi = \phi$$

and, therefore, equation (9.4) rewrites as

$$B\phi - \lambda\phi = B\sum_{\nu=0}^{+\infty}\lambda^{-\nu}A^{\nu}y. \tag{9.5}$$

By virtue of (9.1), the assumption that $\lambda \notin \sigma(A)$ yields $\lambda \notin \sigma(B)$. Therefore, equation (9.5) or, which is the same, (9.4), has a unique solution, say $\phi(y, \lambda)$. Thus, for every y from X, the equation

$$Ax - \lambda x = y - \phi(y, \lambda) \tag{9.6}$$

has a unique solution. Moreover, taking into account the form of equation (9.4), we see that the unique solution

$$\Xi(y, \lambda) := x\,(y - \phi(y, \lambda), \lambda)$$

of (9.6) also satisfies the equation

$$Ax - \lambda x = y - Bx. \tag{9.7}$$

Let us prove that (9.7) cannot have any other solutions. Indeed, in the contrary case, if (9.7) has another solution, say z, then the difference $\delta := \Xi(y, \lambda) - z$ satisfies the equality

$$A\delta - \lambda\delta = -B\delta. \tag{9.8}$$

Since, by (9.2), $\operatorname{im} B$ is contained in $\ker A$, relation (9.8) implies that $A^2\delta = \lambda A\delta$. Therefore, $A\delta = 0$, because otherwise $A\delta$ would be an eigenvector of A with the eigen-value λ, which has been assumed to be regular for A. The same equality (9.8) then yields $B\delta = \lambda\delta$, which can be the case only when $\delta = 0$, because $\lambda \notin \sigma(B)$. Hence, z and $\Xi(y, \lambda)$ coincide.

The argument above shows that, for $\lambda \notin \sigma(A)$ and arbitrary y from X, equation (9.7) has a unique solution and, obviously, this solution continuously depends upon y. Therefore,

$$\sigma(A) \supset \sigma(A + B). \tag{9.9}$$

Conversely, if $\lambda \notin \sigma(A + B)$ and $\lambda \neq 0$, then there exists a bounded inverse operator $(A + B - \lambda 1_X)^{-1}$. Since, in view of condition (9.2), the equality $AB = 0$ holds, we have

$$(A - \lambda 1_X)(B - \lambda 1_X) = -\lambda[A + B - \lambda 1_X], \tag{9.10}$$

an invertible operator. Let us assume that $\lambda \in \sigma(B)$. Then condition (9.1) implies that λ also belongs to $\sigma(A)$. Therefore, the linear mapping $(A - \lambda 1_X)(B - \lambda 1_X)$ is non-invertible, which, in view of equality (9.10), contradicts to our assumption that λ is a regular value for $A + B$.

The contradiction obtained above shows that, in fact, $\lambda \notin \sigma(B)$. Thus, $B - \lambda 1_X$ is invertible and, by (9.10), so does $A - \lambda 1_X$, i. e., $\lambda \notin \sigma(A)$. Hence,

$$\sigma(A + B) \supset \sigma(A). \tag{9.11}$$

Finally, combining (9.11) with (9.9), we arrive at (9.3). □

10 Convergence of the successive approximations for equation (6.10)

We are now in position to establish a statement on the convergence of sequence (8.1), for which purpose we first introduce a definition. Let us fix some $\tau \in [a, b]$ and put

$$\rho(l, h) := r\left(\mathcal{A}_{l,h,\tau} l\right). \tag{10.1}$$

It is clear from (6.8) that $\mathcal{A}_{l,h,\tau} l$ is a continuous linear self-mapping of $C([a, b], X)$ and, therefore, number (10.1) is well-defined.

Proposition 10.1. *The number $\rho(l, h)$ is independent of τ in* (6.8).

Proof. Let us fix arbitrary $\{\tau_1, \tau_2\} \subset [a, b]$ and put

$$A := \mathcal{A}_{l,h,\tau_1} l \tag{10.2}$$

and

$$B := \mathcal{A}_{l,h,\tau_2} l - \mathcal{A}_{l,h,\tau_1} l. \tag{10.3}$$

It follows from (6.8) that, for all $u \in C([a,b],X)$ and $t \in [a,b]$,

$$(Bu)(t) = \int_{\tau_2}^{\tau_1} (lu)(\sigma)\,d\sigma - \int_{\tau_2}^{\tau_1} l^{[0]}(\sigma)\,d\sigma\; l_h^{-1} p_{l,h}(u), \tag{10.4}$$

where

$$p_{l,h}(u) := h\left(\int_{\tau_1}^{\cdot} (lu)(\sigma)\,d\sigma\right). \tag{10.5}$$

Note that, in view of Lemma 6.1, the linear mapping $p_{l,h} : C([a,b],X) \to X$ given by (10.5) is independent of the value of τ_1.

Identity (10.4) implies, in particular, that mapping (10.3) has range in the subspace of constant functions, i. e., $\operatorname{im} B \subset X$. By virtue of Lemma 6.3, operator (10.2) vanishes on the image of B. Furthermore, the spectral radius of B is equal to zero. Indeed, since the mapping B is compact, $\sigma(B)$ consists of its eigenvalues. If λ is an eigenvalue of B, then

$$\lambda u(t) = \int_{\tau_2}^{\tau_1} (lu)(\sigma)\,d\sigma - \int_{\tau_2}^{\tau_1} l^{[0]}(\sigma)\,d\sigma\; l_h^{-1} p_{l,h}(u) \tag{10.6}$$

for all $t \in [a,b]$. This relation implies, in particular, that u is constant. Hence, in view of (10.5) and (6.4), $p_{l,h}(u) = l_h\, u(a)$. Combining this equality with (10.6), we obtain that $u \equiv 0$. Thus, $\sigma(B) = \{0\}$.

The above-stated properties of the operators A and B given by relations (10.2) and (10.3) respectively, allow us to apply Lemma 9.1 from Section 9. It follows from the lemma mentioned that, for all $\{\tau_1, \tau_2\} \subset [a,b]$, we have $r\left(\mathcal{A}_{l,\,h,\,\tau_1} l\right) = r\left(\mathcal{A}_{l,\,h,\,\tau_2} l\right)$, which, by virtue of (10.1), leads us to the required assertion. □

Recall that conditions of Lemma 7.1 are satisfied, in particular, when $A = \mathcal{A}_{l,\,h,\,\tau}$ and l satisfies Assumptions 1, 2, and 3. This, in view of Lemma 8.1, allows us to claim that the unique solvability of equation (6.10) can be proven by finding conditions sufficient for the convergence of sequence (8.1). To formulate such conditions, we need the following

Assumption 4. *The operators l and h are such that the corresponding number* (10.1) *satisfies the inequality*

$$\rho(l,\, h) < 1. \tag{10.7}$$

Remark 10.1. By virtue of Proposition 10.1, the choice of τ does not play any rôle when speaking of the spectral properties of the operator $\mathcal{A}_{l,\,h,\,\tau} l$. From now on, we thus pay no attention to the concrete value of τ, which can be chosen as it may prove convenient.

In case condition (10.7) holds, we have the following result concerning the iteration sequence (8.1) associated with equation (6.10).

Lemma 10.1. *Under Assumptions 1—4, equation* (6.10) *possesses a unique solution u^ξ for every ξ from X, and this solution u^ξ is given by the formula*

$$u^\xi = \xi + \sum_{\nu=0}^{+\infty} (\mathcal{A}_{l,h,\tau} l)^\nu \left[\mathcal{A}_{l,h,\tau} q + \int_\tau^\cdot l^{[0]}(\sigma)\, d\sigma \, l_h^{-1} d \right]. \tag{10.8}$$

Proof. By virtue of Lemma 8.1, the iteration sequence (8.1) corresponding to equation (6.10) admits representation (8.2). In view of (10.1), condition (10.7) guarantees that the series $\sum_{\nu=0}^{+\infty} (\mathcal{A}_{l,h,\tau} l)^\nu \phi$ converges and, in particular, $\lim_{k\to+\infty} (\mathcal{A}_{l,h,\tau} l)^k \phi = 0$ uniformly on $[a, b]$ for every ϕ from $C([a, b], X)$. This implies that $\lim_{k\to+\infty} u_k = u^\xi$, where u^ξ is the function given by (10.8). Obviously, u^ξ is a solution of (6.10).

Let w be another solution of equation (6.10) with the same value of ξ. Then the difference $z := u^\xi - w$ satisfies the relation

$$z = \mathcal{A}_{l,h,\tau} l z. \tag{10.9}$$

However, condition (10.7) yields that equation (10.9) is uniquely solvable. This means that (10.9) implies $z = 0$, i.e., $w = u^\xi$. Thus, (10.8) is the unique solution of (6.10). □

11 Unique solvability of problem (1.1), (1.2)

Here, we use the results of Sections 3—10 to establish several statements concerning the unique solvability of the boundary-value problem (1.1), (1.2).

11.1 Main Theorem

Lemma 10.1 of the preceding section allows us to obtain the following

Theorem 11.1. *If the operators l and h satisfy Assumptions 1—4, then problem* (1.1), (1.2) *has a unique solution for arbitrary $q \in L([a, b], X)$ and $d \in X$. Furthermore, this solution can be represented by the formula*

$$u = l_h^{-1} \left[d - h\left(\int_\tau^\cdot \left[(l\phi_{q,d})(s) + q(s) \right] ds \right) \right] + \phi_{q,d}, \tag{11.1}$$

where

$$\phi_{q,d} := \sum_{\nu=0}^{+\infty} (\mathcal{A}_{l,h,\tau} l)^\nu \left[\mathcal{A}_{l,h,\tau} q + \int_\tau^\cdot l^{[0]}(s)\, ds \, l_h^{-1} d \right] \tag{11.2}$$

and $\tau \in [a, b]$ is arbitrary.

Proof. According to Proposition 6.2, every solution u of problem (1.1), (1.2), if there are any, satisfies equation (6.10) for some $\xi \in X$. Lemma 10.1 guarantees the unique solvability of equation (6.10) for an arbitrary $\xi \in X$. Furthermore, this solution u^{ξ} is represented by formula (10.8).

Let us denote the unique solution of (6.10) by u^{ξ}. Since, by Remark 6.5, the function u^{ξ} satisfies condition (1.2), it remains to find out when u^{ξ} satisfies equation (1.1). However, by virtue of Proposition 4.1, u^{ξ} is a solution of (1.1) if, and only if the parameter ξ satisfies equation (6.11), i.e.,

$$l_h\xi + hI_\tau\Big[l\sum_{\nu=0}^{+\infty}(\mathcal{A}_{l,h,\tau}l)^{\nu}\big[\mathcal{A}_{l,h,\tau}q + \Phi_{l,h,\tau}d\big] + q\Big] = d. \tag{11.3}$$

It is obvious from Assumption 3 that equation (11.3) is uniquely solvable with respect to ξ. In view of notation (11.2), the solution of (11.3) is given by the formula

$$\xi = l_h^{-1}\Big[d - hI_\tau\big(l\phi_{q,d} + q\big)\Big]. \tag{11.4}$$

Inserting (11.4) into (10.8) and taking into account (11.2), we arrive at the conclusion desired. □

Remark 11.1. It is obvious from (11.2) that

$$\phi_{q,0} = \sum_{\nu=0}^{+\infty}(\mathcal{A}_{l,h,\tau}l)^{\nu}\mathcal{A}_{l,h,\tau}q, \quad \phi_{0,d} = \sum_{\nu=0}^{+\infty}(\mathcal{A}_{l,h,\tau}l)^{\nu}\mathcal{A}_{l,h,\tau}\int_\tau^{\cdot} l^{[0]}(\sigma)\,\mathrm{d}\sigma\, l_h^{-1}d$$

and, in particular,

$$\phi_{0,0} = 0. \tag{11.5}$$

In the case of the homogeneous boundary-value problem, Theorem 11.1 yields immediately the following

Corollary 11.1. *Under Assumptions 1— 4, the homogeneous boundary-value problem*

$$u'(t) = (lu)(t), \qquad t \in [a,b], \tag{11.6}$$

$$h(u) = 0 \tag{11.7}$$

has only the trivial solution.

Proof. In view of equality (11.5) from Remark 11.1, the required assertion is an immediate consequence of formula (11.1) of Theorem 11.1 with $q \equiv 0$ and $d = 0$. □

11.2 A Partucular Case

For special kinds of perturbation terms q in equation (1.1), the solution of problem (1.1), (1.2) is necessarily constant. In particular, the following statement holds.

Corollary 11.2. *If $q \in L([a,b],X)$ in* (1.1) *is a solution of the equation*

$$q(t) = l^{[0]}(t)\, l_h^{-1}\Big[h\Big(\int_\tau^{\cdot} q(s)\,ds\Big) - d\Big], \qquad t \in [a,b], \tag{11.8}$$

then the solution of problem (1.1), (1.2), *whose existence and uniqueness is guaranteed by Assumptions 1—4, is the constant function given by the formula*

$$u(t) = l_h^{-1}\Big[d - h\Big(\int_\tau^{\cdot} q(s)\,ds\Big)\Big], \qquad t \in [a,b]. \tag{11.9}$$

Proof. It follows from Theorem 11.1 that, under our conditions, problem (1.1), (1.2) is uniquely solvable, and the solution is given by formula (11.1).

Equality (11.8) implies that, for $t \in [a,b]$,

$$\int_\tau^t q(s)\,ds = \int_\tau^t l^{[0]}(s)\,ds\, l_h^{-1}\Big[h\Big(\int_\tau^{\cdot} q(s)\,ds\Big) - d\Big] + \gamma, \tag{11.10}$$

with some $\gamma \in X$. Taking into account definition (6.8), one can easily verify that (11.10) yields

$$\mathcal{A}_{l,h,\tau} q + \int_\tau^{\cdot} l^{[0]}(\sigma)\,d\sigma\, l_h^{-1} d = \gamma. \tag{11.11}$$

Recalling formula (11.2) and taking into account Lemma 6.3, we conclude that (11.11) implies that $\phi_{q,d}(t) = 0$ for $t \in [a,b]$. This, in view of (11.1) and (4.3), leads us to equality (11.9). □

11.3 System (1.1), (1.2) and Related Cauchy Problems

It is obvious that the operator $C([a,b],X) \to X$ determining an initial value problem cannot possess the property specified in Assumption 2. However, there is still a certain relation between the unique solvability of (1.1), (1.2) and that of a special Cauchy problem.

Theorem 11.2. *Let the operators l and h in* (11.6), (11.7) *satisfy Assumptions 1—3 and, moreover, the Cauchy problem*

$$z'(t) = \mu\big[(lz)(t) - l^{[1,\tau]}(t)\, l_h^{-1} h(z)\big], \qquad t \in [a,b],$$
$$z(\tau) = 0$$

have only the trivial solution for an arbitrary μ with $|\mu| \le 1$ and some τ from $[a,b]$.

Then the homogeneous boundary-value problem (11.6), (11.7) *has only the trivial solution.*

Remark 11.2. It follows from Proposition 10.1 and the proof given below that the concrete value of the initial time τ in Theorem 11.2 is, in fact, inessential.

Proof of Theorem 11.2. Let us take some μ such that

$$|\mu| \leq 1 \tag{11.12}$$

and $\mu \neq 0$, put

$$\lambda := \frac{1}{\mu}, \tag{11.13}$$

and consider the equation

$$\int_\tau^t (lu)(s)\,\mathrm{d}s - \int_\tau^t l^{[0]}(\sigma)\,\mathrm{d}\sigma\, l_h^{-1} h\Big[\int_\tau^{\cdot} (lu)(s)\,\mathrm{d}s\Big] = \lambda u(t) \tag{11.14}$$

for $t \in [a,b]$. Let $u \in C([a,b],X)$ be a solution of (11.14). Then the function

$$z(t) := \int_\tau^t (lu)(s)\,\mathrm{d}s, \qquad t \in [a,b],$$

as can be readily verified, satisfies the initial-value problem

$$z'(t) = \frac{1}{\lambda}\big[(lz)(t) - l^{[1,\tau]}(t)\, l_h^{-1} h(z)\big], \qquad t \in [a,b], \tag{11.15}$$

$$z(\tau) = 0. \tag{11.16}$$

Here, we use the mapping $l^{[1,\tau]} : [a,b] \to \mathcal{B}(X)$ given by formula (6.2).

In view of (11.12) and (11.13), our assumption implies that $z \equiv 0$ is the unique solution of (11.15), (11.16). Hence, every solution of (11.14) satisfies the relation

$$\int_\tau^t (lu)(s)\,\mathrm{d}s = 0, \qquad t \in [a,b]. \tag{11.17}$$

However, in view of the form of equation (11.14), identity (11.17) yields

$$u \equiv 0, \tag{11.18}$$

i.e., (11.14) has only the trivial solution.

One can show [7] that operator

$$\mathcal{A}_{l,h,\tau} l : C([a,b],X) \to C([a,b],X) \tag{11.19}$$

with $\mathcal{A}_{l,h,\tau}$ given by (6.8), is compact. Its spectrum, therefore, consists of eigenvalues and, as is easy to see, a function u from $C([a,b],X)$ is an eigenfunction of operator (11.19) corresponding to the eigenvalue λ if, and only if (11.14) holds.

However, it have been proved above that, under our conditions, function (11.18) is the unique solution of equation (11.14). Hence, every λ with $|\lambda| \geq 1$ is a regular value for mapping (11.19) or, in other words, Assumption 4 holds.

Applying Theorem 11.1 with $q \equiv 0$ and $d = 0$, we conclude that problem (11.6), (11.7) is uniquely solvable. By Remark 11.1, the unique solution of (11.6), (11.7) is given by formula (11.18). □

The Fredholm property of the homogeneous problem (11.6), (11.7) yields the following statement.

Theorem 11.3. *If, under Assumptions 1—3, the Cauchy problem* (11.15), (11.16) *has only the trivial solution for all λ with $|\lambda| \geq 1$, then, for arbitrary $q \in L([a,b], X)$ and $d \in X$, formula* (11.1) *gives the unique solution of the inhomogeneous boundary-value problem* (1.1), (1.2).

11.4 Semi-Homogeneous Problems

By using Theorem 11.1, one can describe certain particular classes of uniquely solvable semi-homogeneous problems (1.1), (11.7) and (11.6), (1.2), the solutions of which are represented by finite sums of expressions containing known functions.

Corollary 11.3. *If the forcing term q in equation* (1.1) *is such that*

$$\mathcal{A}_{l,h,\tau} q \in \ker\left(\mathcal{A}_{l,h,\tau} l\right)^N \tag{11.20}$$

for some natural N, then the solution of (1.1), (11.7), *whose existence and uniqueness is guaranteed by Proposition 11.1, has the form*

$$u(t) = \psi_{q,N}(t) - l_h^{-1} h I_\tau\left[l\psi_{q,N} + q\right], \qquad t \in [a,b], \tag{11.21}$$

where

$$\psi_{q,N} := \sum_{\nu=0}^{N-1} \left(\mathcal{A}_{l,h,\tau} l\right)^\nu \mathcal{A}_{l,h,\tau} q. \tag{11.22}$$

Proof. Considering the Neumann series in the right-hand side of (11.2), we conclude that (11.20) yields

$$\sum_{\nu=0}^{+\infty} \left(\mathcal{A}_{l,h,\tau} l\right)^\nu \mathcal{A}_{l,h,\tau} q = \sum_{\nu=0}^{N-1} \left(\mathcal{A}_{l,h,\tau} l\right)^\nu \mathcal{A}_{l,h,\tau} q.$$

The latter equality leads us immediately to formula (11.21) for the unique solution of problem (1.1), (11.7). □

Quite similarly to Corollary 11.3, we have the following

Corollary 11.4. *When the operators l and h satisfy Assumptions 1—4 and, additionally, $d \in X$ is such that the vector*

$$l_h^{-1} d \tag{11.23}$$

belongs to the kernel of

$$(\mathcal{A}_{l,\,h,\,\tau} l)^N \int_\tau^{\cdot} l^{[0]}(s)\,\mathrm{d}s, \tag{11.24}$$

then the unique solution of problem (1.1), (11.7) *is given by the formula*

$$u(t) = B_{d,N}(t) + l_h^{-1}\left[d - h\left(\int_\tau^{\cdot} (lB_{d,N})(s)\,\mathrm{d}s\right)\right], \qquad t \in [a,b] \tag{11.25}$$

with

$$B_{d,N} := \sum_{\nu=0}^{N-1} (\mathcal{A}_{l,\,h,\,\tau} l)^\nu \int_\tau^{\cdot} l^{[0]}(s)\,\mathrm{d}s\, l_h^{-1} d.$$

In some cases, formula (11.25) from Corollary 11.4 provides a constant solution of problem (1.1), (11.7).

Corollary 11.5. *Let Assumptions 1—4 hold. If, moreover, d in* (1.2) *is such that vector* (11.23) *belongs to the kernel of*

$$\int_\tau^t \left[l^{[1,\tau]}(s) - l^{[0]}(s)\, l_h^{-1} h\left(\int_\tau^{\cdot} l^{[1]}(\sigma)\,\mathrm{d}\sigma\right)\right] \mathrm{d}s \tag{11.26}$$

for every t from $[a,b]$, then the unique solution of problem (11.6), (1.2) *is given by the equality*

$$u(t) = l_h^{-1} d, \qquad t \in [a,b]. \tag{11.27}$$

Proof. It is not difficult to verify by computation that, for all $t \in [a,b]$, the value of the operator $X \to X$ given by formula (11.26) on vector (11.23) coincides with

$$\mathcal{A}_{l,\,h,\,\tau} \int_\tau^{\cdot} l^{[0]}(\sigma)\,\mathrm{d}\sigma\, l_h^{-1} d.$$

Therefore, in view our assumption, vector (11.23) lies in the kernel of operator (11.24) with $N = 1$. Application of Corollary 11.4 completes the proof of our assertion. □

Corollary 11.6. *If Assumptions 1—4 hold and, moreover, the operator l in equation* (11.6) *satisfies the additional condition*

$$l\left(\int_\tau^{\cdot} l^{[0]}(\sigma)\,\mathrm{d}\sigma\right) = l^{[0]} \tag{11.28}$$

for some $\tau \in [a,b]$, then, for an arbitrary d from X, the unique solution of the semi-homogeneous problem (11.6), (1.2) *is the constant function given by the formula* (11.27).

Proof. It suffices to notice that, in view of (6.2), equality (11.28) is nothing but

$$l^{[1,\tau]}(t) = l^{[0]}(t), \qquad t \in [a, b].$$

Therefore, as can be readily verified, formula (11.26), in fact, defines the zero operator. This means that assumptions of Corollary 11.5 are satisfied for an arbitrary $d \in [a, b]$. Applying this corollary, we obtain the required assertion. □

11.5 Autonomous Equations

In this subsection, we establish several statements concerning the "autonomous," in a sense, functional-differential equation (1.1) of the form

$$u'(t) = (lu)(t) + q, \qquad t \in [a, b], \tag{11.29}$$

where q is a constant vector from X. Prior to this, and following [8], we introduce a definition.

Definition 11.1. A mapping g from $C([a, b], X)$ to $L([a, b], X)$ will be called *autonomous* if $g(X) \subset X$.

Remark 11.3. Here and in similar cases, the set of constant functions $[a, b] \to X$ is identified with X and considered as a subspace of both $C([a, b], X)$ and $L([a, b], X)$.

Remark 11.4. The notion introduced by Definition 11.1 is precisely what one usually implies when saying that one or another functional differential equation is "autonomous" (see, e.g., [9]). However, very much surprisingly, we cannot name any paper (except [8]) which would contain the abstract definition given above.

In the sequel, we also need the following

Definition 11.2. For an arbitrary linear mapping $h : C([a, b], X) \to X$, the operator $h_{\#} : X \to X$ is associated with h according to the formula

$$h_{\#} := h(\alpha 1_X), \tag{11.30}$$

where $\alpha(t) := t$ for $t \in [a, b]$.

Remark 11.5. Equality (11.30) should be inderstood in the sense of notation (viii) from Section 2. Thus, $h_{\#}$ is an element from $\mathcal{B}(X)$.

In case l in (11.29) is an autonomous operator, the corresponding function

$$[a, b] \ni t \longmapsto l^{[0]}(t) \in \mathcal{B}(X) \tag{11.31}$$

from Definition 6.1 is constant,

$$l^{[0]}(t) \equiv l^{[0]}, \qquad t \in [a, b] \tag{11.32}$$

and we are allowed to we introduce the following

Assumption 5. *The mappings $l^{[0]} : X \to X$ and $h_{\#} : X \to X$ are invertible.*

Corollary 11.7. *Let the operator $l : C([a,b],X) \to L([a,b],X)$ in equation (11.29) be autonomous.*

Then, under Assumptions 1—3 and 5, the unique solution of problem (11.29), (11.7) *is the constant function given by the formula*

$$u(t) = -\left[l^{[0]}\right]^{-1} q, \qquad t \in [a,b]$$

for an arbitrary q from X.

Proof. Since function (11.31) is constant, it follows from Lemma 6.1 that operator (6.4) has the form

$$l_h = h_{\#} l^{[0]}. \tag{11.33}$$

By virtue of this relation, Assumption 5 implies that Assumption 4 holds.

Furthermore, in view of (11.30), (11.32), and (11.33), we have

$$l^{[0]} l_h^{-1} h\Big(\int_\tau^{\cdot} q \, \mathrm{d}s\Big) = l^{[0]} \big(l^{[0]}\big)^{-1} h_{\#}^{-1} h_{\#} = 1_X,$$

whence it follows that condition (11.8) holds for every constant q. Applying Corollary 11.2 and putting $d = 0$ in (11.9), we arrive at the required assertion. □

Remark 11.6. The assertion of Corollary 11.7, generally speaking, is not true for the inhomogeneous problem (1.1), (1.2).

11.6 Ordinary Differential Systems with Constant Coefficients

Here, we consider the simplest autonomous equation (11.6) of the form

$$u'(t) = P u(t), \qquad t \in [a,b], \tag{11.34}$$

where $P \in \mathcal{B}(X)$.

We need a notation. For every non-zero λ, we set

$$H_\lambda := h(\alpha_\lambda 1_X), \tag{11.35}$$

where

$$\alpha_\lambda(t) := \exp\left(t\lambda^{-1}\right), \qquad t \in [a,b]. \tag{11.36}$$

Here, as above, the value of h on a function $[a,b] \to \mathcal{B}(X)$ is computed accoriding to notation (viii) from Sect. 2. Obviously, $H_\lambda \in \mathcal{B}(X)$ for all non-zero λ.

Furthermore, we introduce the following mapping $Q_{h,\tau} : C([a,b],X) \to C([a,b],X)$:

$$[Q_{h,\tau} u](t) := \int_\tau^t u(s) \, \mathrm{d}s - (t-\tau) h_{\#}^{-1} h\Big(\int_\tau^{\cdot} u(\sigma) \, \mathrm{d}\sigma\Big), \quad t \in [a,b]. \tag{11.37}$$

Of course, we assume the existence of $h_{\#}^{-1}$.

Lemma 11.1. *Let $h : C([a,b],X) \to X$ satisfy Assumption 2. Then a non-zero λ, for which the corresponding operator* (11.35) *is invertible, belongs to the set of regular points of operator* (11.37).

Remark 11.7. It is easy to see from (6.8) that

$$Q_{h,\tau} = \mathcal{A}_{l,\,h,\,\tau} \tag{11.38}$$

with $l : C([a,b],X) \to L([a,b],X)$ given by formula

$$C([a,b],X) \ni u \longmapsto lu := Pu. \tag{11.39}$$

Therefore, it follows from Proposition 10.1 that the spectral properties of operator (11.37) are independent of the value of τ.

Proof of Lemma 11.1. Let us fix a certain non-zero λ and assume that, on the contrary, the corresponding operator (11.35) is invertible but $\lambda \in \sigma(Q_{h,\tau})$. Then, since $Q_{h,\tau} : C([a,b],X) \to C([a,b],X)$ is compact, there exists a non-trivial u from $C([a,b],X)$ such that

$$\int_\tau^t u(s)\,\mathrm{d}s - (t-\tau)h_{\#}^{-1}h\Big(\int_\tau^{\cdot} u(\sigma)\,\mathrm{d}\sigma\Big) = \lambda u(t), \quad t \in [a,b]. \tag{11.40}$$

It follows immediately from (11.40) that the function

$$z(t) := \int_\tau^t u(s)\,\mathrm{d}s, \qquad t \in [a,b] \tag{11.41}$$

satisfies the homogeneous initial-value problem

$$\lambda z'(t) = z(t) - (t-\tau)h_{\#}^{-1}h(z), \qquad t \in [a,b], \tag{11.42}$$

$$z(\tau) = 0. \tag{11.43}$$

By using the Cauchy formula, one can readily show that problem (11.42), (11.43) is equivalent to the equation

$$z(t) = -\frac{1}{\lambda}\int_\tau^t (s-\tau)\exp\left[\lambda^{-1}(t-s)\right]\mathrm{d}s\, h_{\#}^{-1}h(z), \qquad t \in [a,b],$$

which, after computation, can be rewritten as

$$z(t) = \Big[t - \lambda\exp(\lambda^{-1}(t-\tau)) + \lambda - \tau\Big]h_{\#}^{-1}h(z), \quad t \in [a,b]. \tag{11.44}$$

Let us apply h to both sides of relation (11.44). Then, taking into account Remark 1.1, we obtain

$$-\lambda\exp(-\tau/\lambda)H_\lambda h_{\#}^{-1}h(z) = 0. \tag{11.45}$$

By assumption, operator (11.35) is invertible and $\lambda \neq 0$. Hence, equality (11.45) yields

$$h(z) = 0. \tag{11.46}$$

Finally, recalling notation (11.41) and taking into account the form of equation (11.40), we see that (11.46) implies $u \equiv 0$, contrary to our assumption that u is an eigenfunction of operator (11.37). The contradiction obtained proves the lemma. □

Let us introduce a condition on the operators P and h in (11.34), (11.7).

Assumption 6. *The operators $P \in \mathcal{B}(X)$ and $h : C([a,b],X) \to X$ are such that*

$$h_{\#}^{-1}h(Pz) = Ph_{\#}^{-1}h(z) \tag{11.47}$$

for all z from $C([a,b],X)$ such that $z(\tau) = 0$.

We have the following

Theorem 11.4. *Let h in (11.7) satisfy Assumptions 1 and 2. Furthermore, let the operator P in equation (11.34) and the mapping $h_{\#}$ associated with h according to (11.30) be invertible and such that Assumption 6 holds with some $\tau \in [a,b]$.*

Then the condition

$$r(P)\sup\{|\lambda| : \textit{operator } (11.35) \textit{ is non-invertible}\} < 1 \tag{11.48}$$

guarantees the absence of non-trivial solutions of the homogeneous boundary-value problem (11.34), (11.7).

Remark 11.8. In view of (11.35), the expression in braces in condition (11.48) does not involve P. This strange, on the first glance, fact is explained by the presence of the additional condition (11.47) in Assumption 6.

Proof of Theorem 11.4. It is not difficult to verify that, in view of Assumption 6, the mapping $C([a,b],X) \to C([a,b],X)$ given by formula

$$C([a,b],X) \ni u \longmapsto Pu$$

commutes with operator (11.37). Therefore, as is well-known (see, e.g., §149 in [10]), it follows that

$$r(PQ_{h,\tau}) \le r(P)r(Q_{h,\tau}) \tag{11.49}$$

However, by virtue of Lemma 11.1, condition (11.48) implies the inequality $r(P)r(Q_{h,\tau}) < 1$, which, combined with (11.49), yields

$$r(PQ_{h,\tau}) < 1. \tag{11.50}$$

On the other hand, according to Remark 11.7, equality (11.38) holds with $l : C([a,b],X) \to L([a,b],X)$ given by (11.39). Thus, we see that, under our conditions, Assumption 4 holds. Since, morever, Assumption 3 is true in view of the invertibility of P and $h_{\#}$, application of Corollary 11.1 leads us to the conclusion desired. Theorem is proved. □

Similarly to Theorem 11.3, the last statement implies the following

Corollary 11.8. *If P and $h_{\#}$ are invertible and, furthermore, Assumptions 1, 2, and 6 hold, then condition* (11.48) *guarantees the unique solvability of the inhomogeneous boundary-value problem*

$$u'(t) = Pu(t) + q(t), \qquad t \in [a,b], \tag{11.51}$$
$$h(u) = d$$

for arbitrary $q \in L([a,b],X)$ and $d \in X$.

Remark 11.9. It should be noted that, in order to apply Theorem 11.4, there is no need to compute the value of the operator h on the mapping

$$[a,b] \ni t \longmapsto \exp(tP); \tag{11.52}$$

it is *unnecessary* to know mapping (11.52) explicitly. One has only to find the value of h on $\alpha_\lambda 1_X$, where α_λ is the *scalar* function given by (11.36).

We conclude the section with some examples. Consider, e.g., the multipoint boundary-value problem with the conditions

$$\sum_{k=1}^{m} M_k u(t_k) = 0, \tag{11.53}$$

where $(t_k)_{k=1}^m \subset [a,b]$ and $(M_k)_{k=1}^m \subset \mathcal{B}(X)$ are such that the operator

$$X^m \ni (u_1, u_2, \ldots, u_n) \longmapsto \sum_{k=1}^{m} M_k u_k$$

is a surjection.

Corollary 11.9. *Let us suppose that P in* (11.34) *and, moreover, the operator $\mu := \sum_{k=1}^m t_k M_k$ are invertible. Furthermore, assume that*

$$r(P)\sup\Big\{|\lambda| : \textit{the operator} \sum_{k=1}^{m} M_k \exp\left(t_k\lambda^{-1}\right) \textit{ is non-invertible}\Big\} < 1$$

and there exists a $\nu_* \in \{1,2,\ldots,n\}$ such that

$$M_\nu P = \mu P \mu^{-1} M_\nu \tag{11.54}$$

for all $\nu \neq \nu_*$.

Then problem (11.34), (11.53) *has only the trivial solution.*

Proof. In view of (11.30), μ is nothing but the operator $h_{\#}$ associated with the mapping

$$C([a,b],X) \ni u \longmapsto h(u) := \sum_{k=1}^{m} M_k u(t_k) \tag{11.55}$$

It is not difficult to verify that (11.54) guarantees the fulfillment of Assumption 6 with $\tau = t_{\nu_*}$. Similarly to Corollary 3.1 of Lemma 3.1 and, one can show that, under our conditions, Assumption 1 holds for the operator h given by (11.55). Thus, Theorem 11.4 is applicable, whence our assertion follows. □

In particular, for the homogeneous boundary-value problem (11.34), (11.53) with $X = \mathbb{R}^n$ and the "multiperiodic" condition of the form

$$u_k(t_k) = u_k(a), \qquad k = 1, 2, \ldots, n, \tag{11.56}$$

where $(t_k)_{k=1}^{n} \subset [a,b]$ are such that $\prod_{k=1}^{n}(t_k - a) \neq 0$, we have

Corollary 11.10. *Assume that P is an invertible matrix and, moreover,*

$$r(P) \max_{1 \le k \le n} (t_k - a) < 2\pi.$$

Then problem (11.34), (11.56) *has only the trivial solution.*

Proof. It is easy to see that, in our case,

$$\mu = \operatorname{diag}\{t_1 - a, t_2 - a, \ldots, t_n - a\}$$

and $\det H_\lambda = \prod_{k=1}^{n}\left[\exp\left(t_k\lambda^{-1}\right) - \exp\left(a\lambda^{-1}\right)\right]$. Therefore, Corollary 11.9 can be applied. □

Corollary 11.10 agrees, in particular, with the following standard

Proposition 11.1. *Under the condition*

$$r(P) < \frac{2\pi}{b-a},$$

the problem

$$u'(t) = Pu(t), \qquad t \in [a,b],$$
$$u(a) = u(b)$$

has only the trivial solution.

The last statement, together with the classical example [11] of the system

$$x_1' = \nu^2 x_2, \qquad x_2' = -\nu^2 x_1,$$

in particular, shows that the unique solvability conditions established in this section are optimal even in the class of ordinary differential equations.

12 Comments

1. By the Riesz characterization theorem, every linear continuous mapping $h : C([a,b],X) \to X$ can be given by the formula

$$C([a,b],X) \ni u \longmapsto h(u) = \int_a^b \mathrm{d}R(s)u(s),$$

where $R : [a,b] \to \mathcal{B}(X)$ has bounded variation on $[a,b]$ and is such that $R(a) = 0$. In terms of the Riesz function R corresponding to the mapping h, our basic Assumption 2 means that the equality

$$\int_a^b \mathrm{d}R(s) = 0 \tag{12.1}$$

holds. Condition (12.1) excludes, in particular, the "non-degenerate" case considered in [12].

2. Proposition 4.1 is a generalisation of the reduction principle extensively used in [5, 6].

3. Theorem 11.1 and its corollaries can be considered as generalizations of some results of [3, 2] (e.g., Corollary 6.2 of [2] and Corollary 4.2.1 of [3]).

Acknowledgement

This work was partially supported by the Ukrainian State Foundation for Fundamental Research through grant No. 01.07/00109.

References

[1] I. T. Kiguradze, *Some Singular Boundary Value Problems for Ordinary Differential Equations.* Tbilisi: Izd. Tbil. Univ., 1975. In Russian.

[2] I. T. Kiguradze, *Initial and Boundary Value Problems for Systems of Ordinary Differential Equations. I. Linear Theory.* Tbilisi: Metsniereba, 1997. In Russian.

[3] A. M. Samoilenko, V. N. Laptinskii, and K. K. Kenzhebaev, *Constructive Methods of Investigating Periodic and Multipoint Boundary Value Problems.* Inst. Math., Kiev, 1999. In Russian.

[4] A. Ronto, "On the boundary value problems with linear multipoint restrictions," *Publ. Univ. of Miskolc, Ser. D., Natur. Sc. Math.*, vol. 36, no. 1, pp. 81–89, 1995.

[5] A. M. Samoilenko and N. I. Ronto, *Numerical-analytic methods of investigating periodic solutions.* Moscow: Mir, 1979.

[6] M. Ronto and A. M. Samoilenko, *Numerical-analytic methods in the theory of boundary-value problems.* River Edge, NJ: World Scientific Publishing Co. Inc., 2000.

[7] N. Dunford and J. T. Schwartz, *Linear operators. Part I.* New York: John Wiley & Sons Inc., 1988.

[8] A. Ronto, "A note on the periods of periodic solutions of some autonomous functional differential equations," *El. J. Qualit. Theory Differential Eqns.*, no. 25, pp. 1–15, 2000.

[9] J. K. Hale, *Theory of Functional Differential Equations.* New York–Heidelberg–Berlin: Springer-Verlag, 1977.

[10] F. Riesz and B. Sz.-Nagy, *Leçons d'Analyse Fonctionelle.* Budapest: Akadémiai Kiadó, 1952.

[11] J. A. Yorke, "Periods of periodic solutions and the Lipschitz constant," *Proc. Amer. Math. Soc.*, vol. 22, pp. 509–512, 1969.

[12] E. P. Trofimchuk and A. V. Kovalenko, "The A. M. Samoilenko numerical-analytic method without a determining equation," *Ukrainian Math. J.*, vol. 47, no. 1, pp. 163–166, 1995.

Weak and Strong Convergence Theorems for Nonlinear Operators of Accretive and Monotone Type and Applications

Wataru Takahashi

Department of Mathematical and Computing Sciences, Tokyo Institute of Technology, 2-12-1, Ohokayama, Meguro-ku, Tokyo 152-8552, Japan, e-mail: wataru@is.titech.ac.jp

Abstract. In this article, motivated by iterative methods for approximation of fixed points of nonexpansive mappings, we discuss weak and strong convergence theorems for nonlinear operators in a Hilbert space or a Banach space. In particular, we state weak and strong convergence theorems for maximal monotone operators in a Hilbert space or a Banach space. Using these results, we also consider the convex minimization problem and the variational inequality problem.

1 Introduction

Let H be a real Hilbert space with inner product $\langle \cdot, \cdot \rangle$ and norm $\|\cdot\|$. Let C be a closed convex subset of H and let P_C be the metric projection of H onto C. A mapping T of C into itself is called nonexpansive if $\|Tx-Ty\| \leq \|x-y\|$ for all $x, y \in C$. We denote by $F(T)$ the set of fixed points of T. A mapping A of C into H is called monotone if for all $x, y \in C$, $\langle x-y, Ax-Ay \rangle \geq 0$. The variational inequality problem for such an A is to find a $z \in C$ such that $\langle Az, x-z \rangle \geq 0$ for all $x \in C$. The set of solutions of the problem is denoted by $VI(C, A)$.

We consider the following problem: Let $f_0, f_1, f_2, \ldots, f_m$ be convex continuous functions of H into $\mathbb{R}$. Then, the problem is to find a $z \in C$ such that

$$f_0(z) = \min\{f_0(x) : x \in C\}, \tag{1.1}$$

where $C = \{x \in H : f_1(x) \leq 0, f_2(x) \leq 0, \ldots, f_m(x) \leq 0\}$. Such a problem is called the convex minimization problem. Let us define a function $g : H \to (-\infty, \infty]$ as follows:

$$g(x) = \begin{cases} f_0(x), & x \in C, \\ \infty, & x \notin C. \end{cases}$$

Then, g is a proper lower semicontinuous convex function and a minimizer $z \in H$ of g is a solution of the convex minimization problem (1). So, let

$g : H \to (-\infty, \infty]$ be a proper convex lower semicontinuous function and consider the convex minimization problem:

$$\min\{g(x) : x \in H\}. \tag{1.2}$$

For such a g, we can define a multivalued operator ∂g on H by

$$\partial g(x) = \{x^* \in H : g(y) \geq g(x) + \langle x^*, y - x\rangle, y \in H\}$$

for all $x \in H$. Such a ∂g is said to be the subdifferential of g. A monotone operator $A \subset H \times H$ is called maximal if its graph

$$G(A) = \{(x, y) : y \in Ax\}$$

is not properly contained in the graph of any other monotone operator. We know that if A is a maximal monotone operator, then $R(I + \lambda A) = H$ for all $\lambda > 0$. A monotone operator A is also called m-accretive if $R(I + \lambda A) = H$ for all $\lambda > 0$. So, we can define, for each positive λ, the resolvent $J_\lambda : R(I + \lambda A) \to D(A)$ by $J_\lambda = (I + \lambda A)^{-1}$. We know that J_λ is a nonexpansive mapping. If $g : H \to (-\infty, \infty]$ is a proper lower semicontinuous convex function, then ∂g is an m-accretive operator.

We know that one method for solving (2) is the proximal point algorithm first introduced by Martinet [20]. The proximal point algorithm is based on the notion of resolvent J_λ, i.e.,

$$J_\lambda x = \arg\min\left\{g(z) + \frac{1}{2\lambda}\|z - x\|^2 : z \in H\right\}.$$

The proximal point algorithm is an iterative procedure, which starts at a point $x_1 \in H$, and generates recursively a sequence $\{x_n\}$ of points $x_{n+1} = J_{\lambda_n} x_n$, where $\{\lambda_n\}$ is a sequence of positive numbers, see, for instance, Rockafellar [30].

On the other hand, Halpern [7] and Mann [19] introduced the following iterative schemes to approximate a fixed point of a nonexpansive mapping T of H into itself:

$$x_{n+1} = \alpha_n x + (1 - \alpha_n) T x_n, \quad n = 1, 2, \ldots$$

and

$$x_{n+1} = \alpha_n x_n + (1 - \alpha_n) T x_n, \quad n = 1, 2, \ldots,$$

respectively, where $x_1 = x \in H$ and $\{\alpha_n\}$ is a sequence in $[0, 1]$. Recently, Nakajo and Takahashi [22] also introduced an iterative scheme of finding a fixed point of a nonexpansive mapping in a Hilbert space by using the hybrid method in mathematical programming.

In this article, motivated by iterative methods for approximation of fixed points of nonexpansive mappings, we discuss weak and strong convergence theorems for nonlinear operators in a Hilbert space or a Banach space. In Section 3, we first state three convergence theorems for nonexpansive mappings in a Hilbert space. They are convergence theorems of Halpern's type,

Mann's type and Nakajo-Takahashi's type. The theorems of Halpern's type and Mann's type are extended to Banach spaces. In Section 4, we consider weak and strong convergence theorems for inverse-strongly-monotone mappings in a Hilbert space. In Section 5, we deal with proximal point algorithms in a Hilbert space. In Section 6, we prove weak and strong convergence theorems for resolvents of accretive operators in a Banach space. In Section 7, we consider the weak and strong convergence of sequences defined by resolvents of maximal monotone operators in a Banach space. Finally, in Section 8, we discuss the convex minimization problem of finding a minimizer of a proper lower semicontinuous convex function and the variational inequality problem in a Banach space.

2 Preliminaries

Let E be a real Banach space with norm $\|\cdot\|$ and let E^* denote the dual of E. We denote the value of $y^* \in E^*$ at $x \in E$ by $\langle x, y^* \rangle$. When $\{x_n\}$ is a sequence in E, we denote the strong convergence of $\{x_n\}$ to $x \in E$ by $x_n \to x$ and the weak convergence by $x_n \rightharpoonup x$. The modulus of convexity of E is defined by

$$\delta(\epsilon) = \inf \left\{ 1 - \frac{\|x+y\|}{2} : \|x\| \leq 1, \|y\| \leq 1, \|x-y\| \geq \epsilon \right\}$$

for every ϵ with $0 \leq \epsilon \leq 2$. A Banach space E is said to be uniformly convex if $\delta(\epsilon) > 0$ for every $\epsilon > 0$. If E is uniformly convex, then δ satisfies that $\delta(\epsilon/r) > 0$ and

$$\left\| \frac{x+y}{2} \right\| \leq r \left(1 - \delta \left(\frac{\epsilon}{r} \right) \right)$$

for every $x, y \in E$ with $\|x\| \leq r$, $\|y\| \leq r$ and $\|x-y\| \geq \epsilon$. Let C be a nonempty closed convex subset of a uniformly convex Banach space E. Then we know that for any $x \in E$, there exists a unique element $z \in C$ such that $\|x - z\| \leq \|x - y\|$ for all $y \in C$. Putting $z = P_C(x)$, we call P_C the metric projection of E onto C. The duality mapping J from E into 2^{E^*} is defined by

$$Jx = \{x^* \in E^* : \langle x, x^* \rangle = \|x\|^2 = \|x^*\|^2\}$$

for every $x \in E$. Let $U = \{x \in E : \|x\| = 1\}$. The norm of E is said to be Gâteaux differentiable if for each $x, y \in U$, the limit

$$\lim_{t \to 0} \frac{\|x + ty\| - \|x\|}{t} \tag{2.1}$$

exists. In the case, E is called smooth. The norm of E is said to be uniformly Gâteaux differentiable if for each $y \in U$, the limit (2.1) is attained uniformly for $x \in U$. It is also said to be Fréchet differentiable if for each $x \in U$, the limit (2.1) is attained uniformly for $y \in U$. A Banach space E is called uniformly smooth if the limit (2.1) is attained uniformly for $x, y \in U$. It is known that if the norm of E is uniformly Gâteaux differentiable, then the duality mapping J is single valued and uniformly norm to weak* continuous on each bounded subset of E. We know the following result: Let E be a uniformly convex Banach space with a Gâteaux differentiable norm. Let C be a nonempty closed convex subset of E and $x_1 \in E$. Then, $x_0 = P_C(x_1)$ if and only if

$$\langle x_0 - y, J(x_1 - x_0)\rangle \geq 0$$

for all $y \in C$, where J is the duality mapping of E.

A Banach space E is said to satisfy Opial's condition [24] if for any sequence $\{x_n\} \subset E$, $x_n \rightharpoonup y$ implies

$$\liminf_{n\to\infty} \|x_n - y\| < \liminf_{n\to\infty} \|x_n - z\|$$

for all $z \in E$ with $z \neq y$. A Hilbert space satisfies Opial's condition.

Let C be a closed convex subset of E. A mapping $T\colon C \to C$ is said to be nonexpansive if $\|Tx - Ty\| \leq \|x - y\|$ for all $x, y \in C$. We denote the set of all fixed points of T by $F(T)$. A closed convex subset C of E is said to have the fixed point property for nonexpansive mappings if every nonexpansive mapping of a bounded closed convex subset D of C into itself has a fixed point in D. We know that a closed convex subset of a uniformly convex Banach space has the fixed point property for nonexpansive mappings. Let D be a subset of C and let P be a mapping of C into D. Then P is said to be sunny if

$$P(Px + t(x - Px)) = Px$$

whenever $Px + t(x - Px) \in C$ for $x \in C$ and $t \geq 0$. A mapping P of C into C is said to be a retraction if $P^2 = P$. We denote the closure of the convex hull of D by $\overline{\text{co}}D$.

Let I denote the identity operator on E. An operator $A \subset E \times E$ with domain $D(A) = \{z \in E : Az \neq \emptyset\}$ and range $R(A) = \bigcup\{Az : z \in D(A)\}$ is said to be accretive if for each $x_i \in D(A)$ and $y_i \in Ax_i$, $i = 1, 2$, there exists $j \in J(x_1 - x_2)$ such that $\langle y_1 - y_2, j\rangle \geq 0$. If A is accretive, then we have

$$\|x_1 - x_2\| \leq \|x_1 - x_2 + r(y_1 - y_2)\|$$

for all $r > 0$. An accretive operator A is said to satisfy the range condition if $\overline{D(A)} \subset \bigcap_{r>0} R(I + rA)$. If A is accretive, then we can define, for each $r > 0$, a nonexpansive single valued mapping $J_r\colon R(I + rA) \to D(A)$ by $J_r = (I + rA)^{-1}$. It is called the resolvent of A. We also define the Yosida approximation A_r by $A_r = (I - J_r)/r$. We know that $A_r x \in AJ_r x$ for all $x \in R(I + rA)$ and $\|A_r x\| \leq \inf\{\|y\| : y \in Ax\}$ for all $x \in D(A) \cap R(I + rA)$. We also know that for an accretive operator A satisfying the range condition, $A^{-1}0 = F(J_r)$ for all $r > 0$. An accretive operator A is said to be m-accretive if $R(I + rA) = E$ for all $r > 0$. Reich [27] proved the following result: Let E be a uniformly convex and uniformly smooth Banach space and let $A \subset E \times E$ be an m-accretive operator such that $A^{-1}0$ is nonempty. Then, for any $x \in E$, the strong limit $\lim_{t\to\infty} J_t x$ exists and belongs to $A^{-1}0$. In this case, putting $Px = \lim_{t\to\infty} J_t x$, we have that P is a sunny nonexpansive retraction of E onto $A^{-1}0$.

A multi-valued operator $A\colon E \to 2^{E^*}$ with domain $D(A) = \{z \in E : Az \neq \emptyset\}$ and range $R(A) = \bigcup\{Az : z \in D(A)\}$ is said to be monotone if $\langle x_1 - x_2, y_1 - y_2\rangle \geq 0$ for each $x_i \in D(A)$ and $y_i \in Ax_i$, $i = 1, 2$. A monotone operator A is said to be maximal if its graph $G(A) = \{(x, y) : y \in Ax\}$ is not properly contained in the graph of any other monotone operator. The following theorems are well known; for instance, see [40].

Theorem 1. *Let E be a reflexive, strictly convex and smooth Banach space and let $A\colon E \to 2^{E^*}$ be a monotone operator. Then A is maximal if and only if $R(J + rA) = E^*$ for all $r > 0$.*

Theorem 2. *Let E be a strictly convex and smooth Banach space and let $x, y \in E$. If $\langle x - y, Jx - Jy\rangle = 0$, then $x = y$.*

These theorems are used essentially in this article. A duality mapping J of a smooth Banach space is said to be weakly sequentially continuous if $x_n \rightharpoonup x$ implies that $Jx_n \overset{*}{\rightharpoonup} J_n x$, where $\overset{*}{\rightharpoonup}$ means the weak* convergence.

3 Approximating fixed points of nonexpansive mappings

Let C be a nonempty closed convex subset of a Hilbert space H and let T be a nonexpansive mapping of C into itself. Then the set $F(T)$ of fixed points of T is closed and convex. Further, if C is bounded, then $F(T)$ is nonempty. There are three iterative methods for approximation of fixed points of nonexpansive mappings in a Hilbert space which are related to the problem of finding a minimizer of a convex function.

Halpern [7] introduced the following iterative scheme to approximate a fixed point of a nonexpansive mapping in a Hilbert space. For the proof, see Wittmann [44] and Takahashi [40].

Theorem 3 ([44]). *Let C be a closed convex subset of a Hilbert space H and let T be a nonexpansive mapping of C into itself such that $F(T)$ is nonempty. Let P be the metric prjection of H onto $F(T)$. Let $x \in C$ and let $\{x_n\}$ be a sequence defined by $x_1 = x$ and*

$$x_{n+1} = \alpha_n x + (1 - \alpha_n)Tx_n, \quad n = 1, 2, \ldots,$$

where $\{\alpha_n\} \subset [0, 1]$ satisfies

$$\lim_{n\to\infty} \alpha_n = 0, \ \sum_{n=1}^{\infty} \alpha_n = \infty \ \text{and} \ \sum_{n=1}^{\infty} |\alpha_{n+1} - \alpha_n| < \infty.$$

Then, $\{x_n\}$ converges strongly to $Px \in F(T)$.

Mann [19] also introduced the iterative scheme for finding a fixed point of a nonexpansive mapping. For the proof, see Takahashi [40].

Theorem 4 ([19]). *Let C be a closed convex subset of a Hilbert space and let T be a nonexpansive mapping of C into itself such that $F(T)$ is nonempty. Let P be the metric projection of H onto $F(T)$. Let $x \in C$ and let $\{x_n\}$ be a sequence defined by $x_1 = x$ and*

$$x_{n+1} = \alpha_n x_n + (1 - \alpha_n)Tx_n, \quad n = 1, 2, \dots,$$

where $\{x_n\} \subset [0, 1]$ satisfies

$$0 \leq \alpha_n < 1 \text{ and } \sum_{n=1}^{\infty} \alpha_n(1 - \alpha_n) = \infty.$$

Then, $\{x_n\}$ converges weakly to $z \in F(T)$, where $z = \lim_{n\to\infty} Px_n$.

Recently, Nakajo and Takahashi [22] proved the following theorem for nonexpansive mappings in a Hilbert space by using the hybrid method in mathematical programming.

Theorem 5 ([22]). *Let C be a closed convex subset of a Hilbert space and let T be a nonexpansive mapping of C into itself such that $F(T)$ is nonempty. Let P be the metric projection of H onto $F(T)$. Let $x_1 = x \in C$ and*

$$\begin{cases} y_n = \alpha_n x_n + (1 - \alpha_n)Tx_n, \\ C_n = \{z \in C : \|y_n - z\| \leq \|x_n - z\|\}, \\ Q_n = \{z \in C : \langle x_n - z, x_1 - x_n \rangle \geq 0\}, \\ x_{n+1} = P_{C_n \cap Q_n}(x_1), \quad n = 1, 2, \dots, \end{cases}$$

where $\{\alpha_n\} \subset [0, 1]$ satisfies $\liminf_{n\to\infty} \alpha_n < 1$ and $P_{C_n \cap Q_n}$ is the metric projection of H onto $C_n \cap Q_n$. Then, $\{x_n\}$ converges strongly to $Px_1 \in F(T)$.

Shioji and Takahashi [32] extended Theorem 3 to that of a Banach space whose norm is uniformly Gâteaux differentiable.

Theorem 6 ([32]). *Let E be a uniformly convex Banach space with a uniformly Gâteaux differentiable norm. Let C be a nonempty closed convex subset of E and let T be a nonexpansive mapping of C into itself such that $F(T)$ is nonempty. Let $\{\alpha_n\}$ be a sequence of real numbers such that*

$$0 \leq \alpha_n < 1, \ \lim_{n\to\infty} \alpha_n = 0, \ \sum_{n=1}^{\infty} \alpha_n = \infty, \text{ and } \sum_{n=1}^{\infty} |\alpha_{n+1} - \alpha_n| < \infty.$$

Suppose $x_1 = x \in C$ and $\{x_n\}$ is given by

$$x_{n+1} = \alpha_n x + (1 - \alpha_n)Tx_n, \quad n = 1, 2, \dots.$$

Then, $\{x_n\}$ converges strongly to $Px \in F(T)$, where P is a unique sunny nonexpansive retraction of C onto $F(T)$.

Reich [26] extended also Mann's result to that of Banach space whose norm is Fréchet differentiable.

Theorem 7 ([26]). *Let C be a closed convex subset of a uniformly convex Banach space E with a Fréchet differentiable norm, let $T : C \to C$ be a nonexpansive mapping such that $F(T)$ is nonempty, and let $\{\alpha_n\}$ be a real sequence such that $0 \leq \alpha_n \leq 1$ and $\sum_{n=1}^{\infty} \alpha_n(1 - \alpha_n) = \infty$. If $x_1 = x \in C$ and*

$$x_{n+1} = \alpha_n T x_n + (1 - \alpha_n)x_n, \quad n = 1, 2, \ldots,$$

then $\{x_n\}$ converges weakly to a fixed point of T.

Problem. Is a Hilbert space in Theorem 5 replaced by a uniformly convex and smooth Banach space?

4 Approximating solutions of valiational inequalities

Let C be a closed convex subset of a Hilbert space H. Then, a mapping A of C into H is called inverse-strongly-monotone if there exists a positive real number α such that

$$\langle x - y, Ax - Ay \rangle \geq \alpha \|Ax - Ay\|^2$$

for all $x, y \in C$; see [5] and [18]. For such a case, A is called α-inverse-strongly-monotone. If a mapping T of C into itself is nonexpansive, then $A = I - T$ is $\frac{1}{2}$-inverse-strongly-monotone and $F(T) = VI(C, A)$; for example, see [9]. A mapping A of C into H is called strongly monotone if there exists a positive number η such that

$$\langle x - y, Ax - Ay \rangle \geq \eta \|x - y\|^2$$

for all $x, y \in C$. In such a case, A is called η-strongly monotone. If A is η-stronly monotone and k-Lipschitz continuous, i.e., $\|Ax - Ay\| \leq k\|x - y\|$ for all $x, y \in C$, then A is $\frac{\eta}{k^2}$-inverse-stronglly-monotone; see [18]. Let f be a continuously Fréchet differentiable convex function H and let ∇f be the gradient of f. If ∇f is $\frac{1}{\alpha}$-Lipschitz continuous, then ∇f is an α-inverse-strongly-monotonoe mapping of C into H; see [2]. We also have that for all $x, y \in C$ and $\lambda > 0$,

$$\begin{aligned}
\|(I - \lambda A)x - (I - \lambda A)y\|^2 &= \|(x - y) - \lambda(Ax - Ay)\|^2 \\
&= \|x - y\|^2 - 2\lambda\langle x - y, Ax - Ay \rangle + \lambda^2 \|Ax - Ay\|^2 \\
&\leq \|x - y\|^2 + \lambda(\lambda - 2\alpha)\|Ax - Ay\|^2.
\end{aligned}$$

So, if $\lambda \leq 2\alpha$, then $I - \lambda A$ is a nonexpansive mapping of C into H. Iiduka and Takahashi [8] proved the following strong convergence theorem for inverse-strongly-monotone mappings and nonexpansive mappings.

Theorem 8 ([8]). *Let C be a closed convex subset of a Hilbert space H. Let A be an α-inverse-strongly-monotone mapping of C into H and let S be a nonexpansive mapping of C into itself such that $F(S) \cap VI(C,A) \neq \phi$. Let $x_1 = x \in C$ and let $\{x_n\}$ be a sequence defined by*

$$x_{n+1} = \alpha_n x + (1-\alpha_n) S P_C(x_n - \lambda_n A x_n), \quad n = 1, 2, \ldots,$$

where $\{\alpha_n\} \subset [0,1)$ and $\{\lambda_n\} \subset [a,b] \subset (0, 2\alpha)$ satisfy

$$\lim_{n\to\infty} \alpha_n = 0, \ \sum_{n=1}^{\infty} \alpha_n = \infty, \ \sum_{n=1}^{\infty} |\alpha_{n+1} - \alpha_n| < \infty, \text{ and } \sum_{n=1}^{\infty} |\lambda_{n+1} - \lambda_n| < \infty.$$

Then, $\{x_n\}$ converges strongly to $z = P_{F(S)\cap VI(C,A)} x$.

As a direct consequence of Theorem 8, we obtain the following result.

Theorem 9. *Let C be a closed convex subset of a Hilbert space H. Let A be an α-inverse-strongly-monotone mapping of C into H such that $VI(C,A) \neq \phi$. Let $x_1 = x \in C$ and let $\{x_n\}$ be a sequence defined by*

$$x_{n+1} = \alpha_n x + (1-\alpha_n) P_C(x_n - \lambda_n A x_n), \quad n = 1, 2, \ldots,$$

where $\{\alpha_n\} \subset [0,1)$ and $\{\lambda_n\} \subset [a,b] \subset (0, 2\alpha)$ satisfy

$$\lim_{n\to\infty} \alpha_n = 0, \ \sum_{n=1}^{\infty} \alpha_n = \infty, \ \sum_{n=1}^{\infty} |\alpha_{n+1} - \alpha_n| < \infty, \text{ and } \sum_{n=1}^{\infty} |\lambda_{n+1} - \lambda_n| < \infty.$$

Then, $\{x_n\}$ converges strongly to $z = P_{VI(C,A)} x$.

Takahashi and Toyoda [42] proved the following weak convergence theorem for inverse-strongly-monotone mappings and nonexpansive mappings.

Theorem 10 ([42]). *Let C be a closed convex subset of a Hilbert space H. Let A be an α-inverse-strongly-monotone mapping of C into H and let S be a nonexpansive mapping of C into itself such that $F(S) \cap VI(C,A) \neq \phi$. Let $x_1 = x \in C$ and let $\{x_n\}$ be a sequence defined by*

$$x_{n+1} = \alpha_n x_n + (1-\alpha_n) S P_C(x_n - \lambda_n A x_n), \quad n = 1, 2, \ldots,$$

where $\{\alpha_n\}$ and $\{\lambda_n\}$ satisfy

$$0 < c \leq \alpha_n \leq d < 1 \text{ and } 0 < a \leq \lambda_n \leq b < 2\alpha.$$

Then, $\{x_n\}$ converges weakly to $z \in F(S) \cap VI(C,A)$.

As a direct consequence of Theorem 10, we obtain the following result.

Theorem 11. *Let C be a closed convex subset of a Hilbert space H. Let A be an α-inverse-strongly-monotone mapping of C into H such that $VI(C, A) \neq \phi$. Let $x_1 = x \in C$ and let $\{x_n\}$ be a sequence defined by*

$$x_{n+1} = \alpha_n x_n + (1 - \alpha_n) P_C(x_n - \lambda_n A x_n), \quad n = 1, 2, \dots,$$

where $\{\alpha_n\}$ and $\{\lambda_n\}$ satisfy

$$0 < c \leq \alpha_n \leq d < 1 \text{ and } 0 < a \leq \lambda_n \leq b < 2\alpha.$$

Then, $\{x_n\}$ converges weakly to $z \in VI(C, A)$.

Next, we can prove a strong convergence theorem for inverse-strongly-monotone mapping by using an idea of Nakajo and Takahashi [22].

Theorem 12 ([9]). *Let C be a closed convex subset of a Hilbert space. Let A be an α-inverse- strongly-monotone mapping of C into H such that $VI(C, A)$ is nonempty. Let $x_1 = x \in C$ and*

$$\begin{cases} y_n = P_C(x_n - \lambda_n A x_n), \\ C_n = \{z \in C : \|y_n - z\| \leq \|x_n - z\|\}, \\ Q_n = \{z \in C : \langle x_n - z, x_1 - x_n \rangle \geq 0\}, \\ x_{n+1} = P_{C_n \cap Q_n}(x_1), \quad n = 1, 2, \dots, \end{cases}$$

where $\{\lambda_n\} \subset [a, \alpha]$ for some $a \in (0, 2\alpha)$. Then $\{x_n\}$ converges strongly to Px_1, where P is the metric projection of H onto $VI(C, A)$.

5 Proximal point algorithms in Hilbert spaces

We consider two proximal point algorithms for sloving (2) in Section 1, with parameters $\{r_n\}$, starting at an initial point x_1 in a Hilbert space H. Kamimura and Takahashi [11] obtained the following strong convergence theorem for resolvents of maximal monotone operators.

Theorem 13 ([11]). *Let H be a Hilbert space and let $A \subset H \times H$ be a maximal monotone operator. Let $x_1 = x \in H$ and let $\{x_n\}$ be a sequence defined by*

$$x_{n+1} = \alpha_n x + (1 - \alpha_n) J_{r_n} x_n, \ n = 1, 2, \dots,$$

where $\{\alpha_n\} \subset [0, 1]$ and $\{r_n\} \subset (0, \infty)$ satisfy

$$\lim_{n \to \infty} \alpha_n = 0, \ \sum_{n=1}^{\infty} \alpha_n = \infty \text{ and } \lim_{n \to \infty} r_n = \infty.$$

If $A^{-1}0 \neq \phi$, then $\{x_n\}$ converges strogly to $Px \in A^{-1}0$, where P is the metric projection of H onto $A^{-1}0$.

Kamimura and Takahashi [11] obtained the following weak convergence theorem for resolvents of maximal monotone operators.

Theorem 14 ([11]). *Let H be a Hilbert space and let $A \subset H \times H$ be a maximal monotone operator. Let $x_1 = x \in H$ and let $\{x_n\}$ be a sequence defined by*

$$x_{n+1} = \alpha_n x_n + (1 - \alpha_n) J_{r_n} x_n, \ n = 1, 2, \dots,$$

where $\{\alpha_n\} \subset [0,1]$ and $\{r_n\} \subset (0, \infty)$ satisfy $\alpha_n \in [0, k]$ for some k with $0 < k < 1$ and $\lim_{n\to\infty} r_n = \infty$. If $A^{-1}0 \neq \phi$, then $\{x_n\}$ converges weakly to $v \in A^{-1}0$, where $v = \lim_{n\to\infty} Px_n$ and P is the metric projection of H onto $A^{-1}0$.

Using Theorems 13 and 14, we obtain the following theorems.

Theorem 15 ([11]). *Let H be a Hilbert space and let $f : H \to (-\infty, \infty]$ be a lower semicontinuous proper convex function. Let $x_1 = x \in H$ and let $\{x_n\}$ be a sequence defined by*

$$x_{n+1} = \alpha_n x + (1 - \alpha_n) J_{r_n} x_n, \ n = 1, 2, \dots,$$
$$J_{r_n} x_n = \arg\min \left\{ f(z) + \frac{1}{2r_n} \|z - x_n\|^2 : z \in H \right\},$$

where $\{\alpha_n\} \subset [0,1]$ and $\{r_n\} \subset (0, \infty)$ satisfy

$$\lim_{n\to\infty} \alpha_n = 0, \ \sum_{n=1}^{\infty} \alpha_n = \infty \text{ and } \lim_{n\to\infty} r_n = \infty.$$

If $(\partial f)^{-1}0 \neq \phi$, then $\{x_n\}$ converges strongly to $v \in H$, which is the minimizer of f nearest to x. Further

$$f(x_{n+1}) - f(v) \leq \alpha_n (f(x) - f(v)) + \frac{1 - \alpha_n}{r_n} \|J_{r_n} x_n - v\| \|J_{r_n} x_n - x_n\|.$$

Theorem 16 ([11]). *Let H be a Hilbert space and let $f : H \to (-\infty, \infty]$ be a lower semicontinuous proper convex function. Let $x_1 = x \in H$ and let $\{x_n\}$ be a sequence defined by*

$$x_{n+1} = \alpha_n x_n + (1 - \alpha_n) J_{r_n} x_n, \ n = 1, 2, \dots,$$
$$J_{r_n} x_n = \arg\min \left\{ f(z) + \frac{1}{2r_n} \|z - x_n\|^2 : z \in H \right\},$$

where $\{\alpha_n\} \subset [0,1]$ and $\{r_n\} \subset (0, \infty)$ satisfy $\alpha_n \in [0, k]$ for some k with $0 < k < 1$ and $\lim_{n\to\infty} r_n = \infty$. If $(\partial f)^{-1}0 \neq \phi$, then $\{x_n\}$ converges weakly to $v \in H$, which is a minimizer of f. Further

$$f(x_{n+1}) - f(v) \leq \alpha_n (f(x_n) - f(v)) + \frac{1 - \alpha_n}{r_n} \|J_{r_n} x_n - v\| \|J_{r_n} x_n - x_n\|.$$

Solodov and Svaiter [34] also proved the following strong convergence theorem.

Theorem 17 ([34]). *Let H be a Hilbert space and let $A \subset H \times H$ be a maximal monotone operator. Let $x \in H$ and let $\{x_n\}$ be a sequence defined by*

$$\begin{cases} x_1 = x \in H, \\ 0 = v_n + \dfrac{1}{r_n}(y_n - x_n), \ v_n \in Ay_n, \\ H_n = \{z \in H : \langle z - y_n, v_n \rangle \leq 0\}, \\ W_n = \{z \in H : \langle z - x_n, x_1 - x_n \rangle \leq 0\}, \\ x_{n+1} = P_{H_n \cap W_n} x_1, \ n = 1, 2, \ldots, \end{cases}$$

where $\{r_n\}$ is a sequence of positive numbers. If $A^{-1}0 \neq \phi$ and $\liminf_{n\to\infty} r_n > 0$, then $\{x_n\}$ converges strongly to $P_{A^{-1}0}x_1$.

6 Convergence theorems for accretive operators

In this section, we first study a strong convergence theorem of Halpern's type for accretive operators in a Banach space. We need the following lemma for the proof of our theorem.

Lemma 18 ([43]). *Let E be a reflexive Banach space whose norm is uniformly Gâteaux differentiable and let $A \subset E \times E$ be an accretive operator which satisfies the range condition. Suppose that every weakly compact convex subset of E has the fixed point property for nonexpansive mappings. Let C be a nonempty closed convex subset of E such that $\overline{D(A)} \subset C \subset \bigcap_{r>0} R(I + rA)$. If $A^{-1}0 \neq \emptyset$, then the strong $\lim_{t\to\infty} J_t x$ exists and belongs to $A^{-1}0$ for all $x \in C$.*

See also Reich [27]. Using this result, we can prove the following theorem. The proof is mainly due to Wittmann [44] and Shioji and Takahashi [32].

Theorem 19 ([12]). *Let E be a uniformly convex Banach space with a uniformly Gâteaux differentiable norm, let $A \subset E \times E$ be an accretive operator which satisfies the range condition, and let C be a nonempty closed convex subset of E such that $\overline{D(A)} \subset C \subset \bigcap_{r>0} R(I + rA)$. Let $x_1 = x \in C$ and let $\{x_n\}$ be a sequence generated by*

$$x_{n+1} = \alpha_n x + (1 - \alpha_n) J_{r_n} x_n, \ n = 1, 2, \ldots,$$

where $\{\alpha_n\} \subset [0, 1]$ and $\{r_n\} \subset (0, \infty)$ satisfy

$$\lim_{n\to\infty} \alpha_n = 0, \ \sum_{n=0}^{\infty} \alpha_n = \infty \text{ and } \lim_{n\to\infty} r_n = \infty.$$

If $A^{-1}0 \neq \emptyset$, then $\{x_n\}$ converges strongly to an element of $A^{-1}0$.

As a direct consequence of Theorem 19, we have the following:

Theorem 20. *Let E be a uniformly convex Banach space with a uniformly Gâteaux differentiable norm and let $A \subset E \times E$ be an m-accretive operator. Let $x_1 = x \in C$ and let $\{x_n\}$ be a sequence generated by*

$$x_{n+1} = \alpha_n x + (1 - \alpha_n) J_{r_n} x_n, \ n = 1, 2, \dots,$$

where $\{\alpha_n\} \subset [0,1]$ and $\{r_n\} \subset (0, \infty)$ satisfy

$$\lim_{n\to\infty} \alpha_n = 0, \ \sum_{n=0}^{\infty} \alpha_n = \infty \text{ and } \lim_{n\to\infty} r_n = \infty.$$

If $A^{-1}0 \neq \emptyset$, then $\{x_n\}$ converges strongly to an element of $A^{-1}0$.

Next, we prove a weak convergence theorem for Mann's type for accretive operators in a Banach space. Before proving the theorem, we need the following two lemmas.

Lemma 21 ([4]). *Let C be a closed bounded convex subset of a uniformly convex Banach space E and let T be a nonexpansive mapping of C into itself. If $\{x_n\}$ converges weakly to $z \in C$ and $\{x_n - Tx_n\}$ converges strongly to 0, then $Tz = z$.*

Lemma 22 ([26]). *Let E be a uniformly convex Banach space whose norm is Fréchet differentiable, let C be a closed convex subset of E and let $\{T_0, T_1, T_2, \dots\}$ be a sequence of nonexpansive mappings of C into itself such that $\bigcap_{n=0}^{\infty} F(T_n)$ is nonempty. Let $x \in C$ and $S_n = T_n T_{n-1} \cdots T_0$ for all $n = 1, 2, \dots$. Then the set $\bigcap_{n=0}^{\infty} \overline{\mathrm{co}}\{S_m x : m \geq n\} \cap U$ consists of at most one point, where $U = \bigcap_{n=0}^{\infty} F(T_n)$.*

For the proof of Lemma 22, see Takahashi and Kim [41]. Now we can prove the following weak convergence theorem.

Theorem 23 ([12]). *Let E be a uniformly convex Banach space whose norm is Fréchet differentiable or which satisfies Opial's condition, let $A \subset E \times E$ be an accretive operator which satisfies the range condition, and let C be a nonempty closed convex subsetof E such that $\overline{D(A)} \subset C \subset \bigcap_{r>0} R(I + rA)$. Let $x_1 = x \in C$ and let $\{x_n\}$ be a sequence generated by*

$$x_{n+1} = \alpha_n x_n + (1 - \alpha_n) J_{r_n} x_n, \ n = 1, 2, \dots,$$

where $\{\alpha_n\} \subset [0,1]$ and $\{r_n\} \subset (0, \infty)$ satisfy

$$\limsup_{n\to\infty} \alpha_n < 1 \text{ and } \liminf_{n\to\infty} r_n > 0.$$

If $A^{-1}0 \neq \emptyset$, then $\{x_n\}$ converges weakly to an element of $A^{-1}0$.

As a direct consequence of Theorem 23, we have the following:

Theorem 24. *Let E be a uniformly convex Banach space whose norm is Fréchet differentiable or which satisfies Opial's condition and let $A \subset E \times E$ be an m-accretive operator. Let $x_1 = x \in C$ and let $\{x_n\}$ be a sequence generated by*

$$x_{n+1} = \alpha_n x_n + (1 - \alpha_n) J_{r_n} x_n, \ n = 1, 2, \ldots,$$

where $\{\alpha_n\} \subset [0, 1]$ and $\{r_n\} \subset (0, \infty)$ satisfy

$$\limsup_{n \to \infty} \alpha_n < 1 \text{ and } \liminf_{n \to \infty} r_n > 0.$$

If $A^{-1}0 \neq \emptyset$, then $\{x_n\}$ converges weakly to an element of $A^{-1}0$.

Problem. Does Theorem 17 hold for accretive operators in a Banach space?

7 Convergence theorems for maximal monotone operators

In this section, we study weak and strong convergence theorems for resolvents of maximal monotone operators in a Banach space. Let E be a uniformly convex and smooth Banach space and let A be a maximal monotone operator from E into E^* such that $A^{-1}0 \neq \phi$. For $x \in E$ and $r > 0$, we consider the following equation

$$0 \in J(x_r - x) + rAx_r.$$

By Theorems 1 and 2, this equation has a unique solution x_r. We denote J_r by $x_r = J_r x$ and such J_r, $r > 0$ are called resolvents of A. Now, we extend Solodov and Svaiter's result [34] to that of a Banach space.

Theorem 25 ([23]). *Let E be a uniformly convex and smooth Banach space and let A be a maximal monotone operator from E into E^*. Suppose $\{x_n\}$ is the sequence generated by*

$$\begin{cases} x_1 \in E, \\ y_n = J_{r_n} x_n, \\ H_n = \{z \in E : \langle y_n - z, J(x_n - y_n) \rangle \geq 0\}, \\ W_n = \{z \in E : \langle x_n - z, J(x_1 - x_n) \rangle \geq 0\}, \\ x_{n+1} = P_{H_n \cap W_n} x_1, \ n = 1, 2, \ldots, \end{cases}$$

where $\{r_n\}$ is a sequence of positive numbers. If $A^{-1}0 \neq \phi$ and $\liminf_{n \to \infty} r_n > 0$, then $\{x_n\}$ converges strongly to $P_{A^{-1}0} x_1$, where $P_{A^{-1}0}$ is the metric projection of E onto $A^{-1}0$.

Next, we establish another extension of Solodov and Svaiter's result [34]. Before establishing it, we give a definition. Let E be a reflexive, strictly convex and smooth Banach space. The function ϕ: $E \times E \to (-\infty, \infty)$ is defined by

$$\phi(x, y) = \|x\|^2 - 2\langle x, Jy\rangle + \|y\|^2$$

for $x, y \in E$, where J is the duality mapping of E; see [1] and [13]. Let C be a nonempty closed convex subset of E and let $x \in E$. Then there exists a unique element $x_0 \in C$ such that

$$\phi(x_0, x) = \inf\{\phi(z, x) : z \in C\}. \tag{7.1}$$

Now, we define the mapping Q_C of E onto C by $Q_C x = x_0$, where x_0 is defined by (4). Such a Q_C is called the generalized projection of E onto C. It is easy to see that in a Hilbert space, the mapping Q_C is coincident with the metric projection.

Lemma 26. *Let E be a smooth Banach space, let C be a nonempty closed convex subset of E, let $x \in E$ and let $x_0 \in C$. Then, the following (1) and (2) are equivalent:*

(1) $\phi(x_0, x) = \min_{y \in C} \phi(y, x)$;

(2) $\langle x_0 - y, Jx - Jx_0\rangle \geq 0$ *for all* $y \in C$.

Kamimura and Takahashi [13] obtained the following strong convergence theorem by using Lemma 26.

Theorem 27 ([13]). *Let E be a uniformly convex and uniformly smooth Banach space and let A be a maximal monotone operator from E into E^* such that $A^{-1}0 \neq \phi$. Let $Q_r = (J + rA)^{-1}J$ for all $r > 0$ and let $\{x_n\}$ be a sequence generated by*

$$\begin{cases} x_1 \in E, \\ y_n = Q_{r_n} x_n, \\ H_n = \{z \in E : \langle z - y_n, Jx_n - Jy_n\rangle \leq 0\}, \\ W_n = \{z \in E : \langle z - x_n, Jx_1 - Jx_n\rangle \leq 0\}, \\ x_{n+1} = Q_{H_n \cap W_n} x_1, \ n = 1, 2, \ldots, \end{cases}$$

where $\{r_n\}$ is a sequence of positive numbers such that $\liminf_{n\to\infty} r_n > 0$. *Then, $\{x_n\}$ converges strongly to $Q_{A^{-1}0} x_1$, where $Q_{A^{-1}0}$ is the generalized projection of E onto $A^{-1}0$.*

Recently, Kohsaka and Takahashi [14] proved a strong convergence theorem of Halpen's type for maximal monotone operators in a Banach space.

Theorem 28 ([14]). *Let E be a smooth and uniformly convex Banach space and let $A \subset E \times E^*$ be a maximal monotone operator. Let $Q_r = (J+rA)^{-1}J$ for all $r > 0$ and let $\{x_n\}$ be a sequence difined as follows:*

$$x_1 = x \in E,$$
$$x_{n+1} = J^{-1}(\alpha_n J(x) + (1-\alpha_n)J(Q_{r_n}x_n)), \quad n = 1, 2, \ldots,$$

where $\{\alpha_n\} \subset [0,1]$ and $\{r_n\} \subset (0,\infty)$ satisfy

$$\lim_{n\to\infty} \alpha_n = 0, \ \sum_{n=1}^{\infty} \alpha_n = \infty \ \text{and} \ \lim_{n\to\infty} r_n = \infty.$$

If $A^{-1}0 \neq \phi$, then $\{x_n\}$ converges strongly to $Q_{A^{-1}0}x$, where $Q_{A^{-1}0}$ is the generalized projection of E onto $A^{-1}0$.

For the sake of getting a weak convergence theorem of Mann's type for maximal monotone operators in a Banach space, we need the following strong convergence theorem.

Theorem 29 ([10]). *Let E be a smooth and uniformly convex Banach space. Let $A \subset E \times E^*$ be a maximal monotone operator such that $A^{-1}0$ is nonempty, let $Q_r = (J + rA)^{-1}J$ for all $r > 0$ and let $Q_{A^{-1}0}$ be the generalized projection of E onto $A^{-1}0$. Let $\{x_n\}$ be a sequence defined as follows: $x_1 = x \in E$ and*

$$x_{n+1} = J^{-1}(\alpha_n J(x_n) + (1-\alpha_n)J(Q_{r_n}x_n)), \quad n = 1, 2, \ldots,$$

where $\{\alpha_n\} \subset [0,1]$ and $\{r_n\} \subset (0,\infty)$. Then, the sequence $\{Q_{A^{-1}0}(x_n)\}$ converges strongly to an element of $A^{-1}0$, which is a unique element $v \in A^{-1}0$ such that

$$\lim_{n\to\infty} \phi(v, x_n) = \min_{y\in A^{-1}0} \lim_{n\to\infty} \phi(y, x_n).$$

Using Theorem 29, we can prove the following theorem in a Banach space which generalizes the results of Rockafellar [30] and Kamimura and Takahashi [11] in a Hilbert space.

Theorem 30 ([10]). *Let E be a smooth and uniformly convex Banach space whose duality mapping J is weakly sequentially continuous. Let $A \subset E \times E^*$ be a maximal monotone operator such that $A^{-1}0$ is nonempty, let $Q_r = (J + rA)^{-1}J$ for all $r > 0$ and let $Q_{A^{-1}0}$ be the generalized projection of E onto $A^{-1}0$. Let $\{x_n\}$ be a sequence defined as follows: $x_1 = x \in E$ and*

$$x_{n+1} = J^{-1}(\alpha_n J(x_n) + (1-\alpha_n)J(Q_{r_n}x_n)), \quad n = 1, 2, \ldots,$$

where $\{\alpha_n\} \subset [0,1]$ and $\{r_n\} \subset (0,\infty)$ satisfy

$$\limsup_{n\to\infty} \alpha_n < 1 \quad \text{and} \quad \liminf_{n\to\infty} r_n > 0.$$

Then, the sequence $\{x_n\}$ converges weakly to an element v of $A^{-1}0$, where $v = \lim_{n\to\infty} Q_{A^{-1}0}(x_n)$.

As a direct consequence of Theorem 30, we obtain the following:

Theorem 31. *Let E be a smooth and uniformly convex Banach space whose duality mapping J is weakly sequentially continuous. Let $A \subset E \times E^*$ be a maximal monotone operator such that $A^{-1}0$ is nonempty, let $Q_r = (J + rA)^{-1}J$ for all $r > 0$ and let $Q_{A^{-1}0}$ be the generalized projection of E onto $A^{-1}0$. Let $\{x_n\}$ be a sequence defined as follows: $x_1 = x \in E$ and*

$$x_{n+1} = Q_{r_n}x_n, \quad n = 1, 2, \dots,$$

where $\{r_n\} \subset (0, \infty)$ satisfies $\liminf_{n\to\infty} r_n > 0$. Then, the sequence $\{x_n\}$ converges weakly to an element v of $A^{-1}0$, where $v = \lim_{n\to\infty} Q_{A^{-1}0}(x_n)$.

Probelm. If E and E^* are uniformly convex Banach spaces, does Theorem 30 hold without assumming that J is weakly sequentially continuous ?

8 Applications

We can apply Theorems 25, 27, 28 and 30 to find a minimizer of a convex function f. Let E be a real Banach space and let $f : E \to (-\infty, \infty]$ be a proper convex lower semicontinuous function. Then the subdifferential ∂f of f is as follows:

$$\partial f(z) = \{v^* \in E^* : f(y) \geq f(z) + \langle y - z, v^* \rangle, \forall y \in E\}, \quad \forall z \in E.$$

Theorem 32 ([23]). *Let E be a uniformly convex and smooth Banach space and let $f : E \to (-\infty, \infty]$ be a proper convex lower semicontinuous function. Assume that $\{r_n\} \subset (0, \infty)$ satisfies $\liminf_{n\to\infty} r_n > 0$ and let $\{x_n\}$ be the sequence generated by*

$$\begin{cases} x_1 \in E, \\ y_n = \arg\min_{z\in E}\{f(z) + \frac{1}{2r_n}\|z - x_n\|^2\}, \\ H_n = \{z \in E : \langle y_n - z, J(x_n - y_n\rangle \geq 0\}, \\ W_n = \{z \in E : \langle x_n - z, J(x_1 - x_n)\rangle \geq 0\}, \\ x_{n+1} = P_{H_n \cap W_n}x_1, \quad n = 1, 2, \dots. \end{cases}$$

If $(\partial f)^{-1}0 \neq \phi$, then $\{x_n\}$ converges strongly to the minimizer of f nearest to x_1.

Proof. Since $f : E \to (-\infty, \infty]$ is a proper convex lower semicontinuous function, by Rockafellar [28], the subdifferential ∂f of f is a maximal monotone operator. We also know that

$$y_n = \arg\min_{z\in E}\{f(z) + \frac{1}{2r_n}\|z - x_n\|^2\}$$

is equivalent to

$$0 \in \partial f(y_n) + \frac{1}{r_n}J(y_n - x_n).$$

So, we have

$$0 \in J(y_n - x_n) + r_n \partial f(y_n).$$

Using Theorem 25, we get the conclusion. □

Theorem 33 ([13]). *Let E be a uniformly convex and uniformly smooth Banach space and let $f : E \to (-\infty, \infty]$ be a proper convex lower semicontinuous function. Assume that $\{r_n\} \subset (0, \infty)$ satisfies $\liminf_{n\to\infty} r_n > 0$ and let $\{x_n\}$ be a sequence generated by*

$$\begin{cases} x_1 \in E, \\ y_n = \arg\min_{z\in E}\{f(z) + \frac{1}{2r_n}\|z\|^2 - \frac{1}{r_n}\langle z, Jx_n\rangle\}, \\ 0 = v_n + \frac{1}{r_n}(Jy_n - Jx_n), \ v_n \in \partial f(y_n), \\ H_n = \{z \in E : \langle z - y_n, v_n\rangle \le 0\}, \\ W_n = \{z \in E : \langle z - x_n, Jx_1 - Jx_n\rangle \le 0\}, \\ x_{n+1} = Q_{H_n \cap W_n} x_1, \quad n = 1, 2, \ldots. \end{cases}$$

If $(\partial f)^{-1}0 \neq \phi$, then $\{x_n\}$ converges strongly to the minimizer of f nearest to x_1.

Proof. We also know that

$$y_n = \arg\min_{z\in E}\{f(z) + \frac{1}{2r_n}\|z\|^2 - \frac{1}{r_n}\langle z, Jx_n\rangle\}$$

is equivalent to

$$0 \in \partial f(y_n) + \frac{1}{r_n}Jy_n - \frac{1}{r_n}Jx_n.$$

So, we have $v_n \in \partial f(y_n)$ such that $0 = v_n + \frac{1}{r_n}(Jy_n - Jx_n)$. Using Theorem 27, we get the conclusion. □

Using Theorem 28, we get the following theorem.

Theorem 34 ([14]). *Let E be a smooth and uniformly convex Banach space and let $f : E \to (-\infty, \infty]$ be a proper lower semicontinuous convex function such that $(\partial f)^{-1}0$ is nonempty. Let $\{x_n\}$ be a sequence defined as follows:*

$$x_1 = x \in E,$$
$$y_n = \arg\min_{y\in E}\{f(y) + \frac{1}{2r_n}\|y\|^2 - \frac{1}{r_n}\langle y, Jx_n\rangle\},$$
$$x_{n+1} = J^{-1}(\alpha_n Jx + (1-\alpha_n)Jy_n), \quad n = 1, 2, \ldots,$$

where $\{\alpha_n\} \subset [0,1]$ and $\{r_n\} \subset (0,\infty)$ satisfy

$$\lim_{n\to\infty} \alpha_n = 0, \ \sum_{n=1}^{\infty} \alpha_n = \infty \text{ and } \lim_{n\to\infty} r_n = \infty.$$

Then, $\{x_n\}$ converges strongly to $Q_{(\partial f)^{-1}0}x$.

Using Theorem 30, we can prove the following weak convergence theorem:

Theorem 35 ([10]). *Let E be a smooth and uniformly convex Banach space whose duality mapping J is weakly sequentially continuous. Let $f : E \to (-\infty, \infty]$ be a proper lower semicontinuous convex function such that $(\partial f)^{-1}0$ is nonempty. Let $\{x_n\}$ be a sequence defined as follows:*

$$\begin{aligned} x_1 &= x \in E, \\ y_n &= \arg\min_{y \in E}\{f(y) + \frac{1}{2r_n}\|y\|^2 - \frac{1}{r_n}\langle y, Jx_n\rangle\}, \\ x_{n+1} &= J^{-1}(\alpha_n Jx_n + (1-\alpha_n)Jy_n), \quad n = 1, 2, \ldots, \end{aligned}$$

where $\{\alpha_n\} \subset [0,1]$ and $\{r_n\} \subset (0,\infty)$ satisfy

$$\limsup_{n\to\infty} \alpha_n < 1 \text{ and } \liminf_{n\to\infty} r_n > 0.$$

Then, $\{x_n\}$ converges weakly to $v \in (\partial f)^{-1}0$. Further $v = \lim Q_{(\partial f)^{-1}0}(x_n)$, where $Q_{(\partial f)^{-1}0}$ is the generalized projection of E onto $(\partial f)^{-1}0$.

We next apply Theorem 30 to the variational inequality problem. Let C be a nonempty closed convex subset of a Banach space and let $A : C \to E^*$ be a single-valued monotone operator which is hemicontinous, that is, continuous along each line segument in C with respect to the *weak** topology of E^*. Then a point $u \in C$ is said to be a solution of the variational inequality for A if

$$\langle y - u, Au\rangle \geq 0$$

holds for all $y \in C$. We denote by $VI(C, A)$ the set of all solutions of the variational inequality for A. We also denote by $N_C(x)$ the normal cone for C at a point $x \in C$, that is,

$$N_C(x) = \{x^* \in E^* : \langle y - x, x^*\rangle \leq 0, \quad y \in C\}.$$

Rockafellar [29] proved that the mapping $A \subset E \times E^*$ defined by

$$Tx = \begin{cases} Ax + N_C(x), & x \in C \\ \phi, & x \notin C \end{cases}$$

is a maximal monotone operator. It is easy to verify that $T^{-1}0 = VI(C, A)$. In this case, if E is smooth, strictly convex and reflexive, then for each $x \in E$ and $r > 0$, the equation

$$J(x) \in J(x_r) + rT(x_r)$$

is equivalent to the following:

$$x_r = VI(C, A + \frac{1}{r}(J - Jx)).$$

So, we can obtain the following weak convergence theorem.

Theorem 36 ([10]). *Let C be a nonempty closed convex subset of a uniformly smooth and uniformly convex Banach space E whose duality mapping J is weakly sequentially continuous. Let $A : C \to E^*$ be a single-valed, monotone and hemicontinuous operator and let $\{x_n\}$ be a sequence defined as follows: $x_1 = x \in E$ and*

$$\begin{cases} y_n = VI(C, A + \frac{1}{r_n}(J - Jx_n)), \\ x_{n+1} = J^{-1}(\alpha_n J(x_n) + (1 - \alpha_n)J(y_n)), \quad n = 1, 2, \dots, \end{cases}$$

where $\{\alpha_n\} \subset [0, 1)$ and $\{r_n\} \subset (0, \infty)$ satisfy

$$\limsup_{n\to\infty} \alpha_n < 1 \text{ and } \liminf_{n\to\infty} r_n > 0.$$

If $VI(C, A) \neq \phi$, then the sequence $\{x_n\}$ converges weakly to an elemnt v of $VI(C, A)$. Further $v = \lim_{n\to\infty} Q(x_n)$, where Q is the generalized projection of E onto $VI(C, A)$.

References

[1] Y. I. Alber, *Metric and generalized projections in Banach spaces: Properties and applications, in Theory and Applications of Nonlinear Operators of Accretive and Monotone Type* (A. G. Kartsatos Ed.), Marcel Dekker, New York, 1996, pp. 15–20.

[2] J. B. Baillon and G. Haddad, *Quelques propriétésdes operateurs angle-bornés et n-cycliquement monotones*, Israel J. Math. **26** (197), 137–150.

[3] H. Brézis and P. L. Lions, *Produits infinis de resolvants*, Israel J. Math. **29** (1978), 329–345.

[4] F. E. Browder, *Semicontractive and semiaccretive nonlinear mappings in Banach spaces*, Bull. Amer. Math. Soc. **74** (1968), 660–665.

[5] F. E. Browder and W. V. Petryshym, *Construction of fixed points of nonlinear mappings in Hilbert space*, J. Math. Anal. Appl. **20** (1967), 197–228.

[6] O. Güler, *On the convergence of the proximal point algorithm for convex minimization*, SIAM J. Control and Optim. **29** (1991), 403–419.

[7] B. Halpern, *Fixed points of nonexpanding maps*, Bull. Amer. Math. Soc. **73** (1967), 957–961.

[8] H. Iiduka and W. Takahashi, *Strong convergence theorems for nonexpansive mappings and monotone mappings*, to appear.

[9] H. Iiduka, W. Takahashi and M. Toyoda, *Approximation of solutions of variational inequalities for monotone mappings*, to appear.

[10] S. Kamimura, F. Kohsaka and W. Takahashi, *Weak and strong convergence theorems for maximal monotone operators in a Banach space*, to appear.

[11] S. Kamimura and W. Takahashi, *Approximating solutions of maximal monotone operators in Hilbert spaces*, J. Approx. Theory **106** (2000), 226–240.

[12] S. Kamimura and W. Takahashi, *Weak and strong convergence of solutions to accretive operator inclusions and applications*, Set-Valued Anal. **8** (2000), 361–374.

[13] S. Kamimura and W. Takahashi, , *Strong convergence of a proximal-type algorithm in a Banach apace*, SIAM. J. Optim., to appear.

[14] F. Kohsaka and W. Takahashi, *Strong convergence of an iterative sequence for maximal monotone operators in a Banach space*, to appear.

[15] A. T. Lau, N. Shioji and W. Takahashi, *Existence of nonexpansive retractions for amenable semigroups of nonexpansive mappings and nonlinear ergodic theorems in Banach spaces*, J. Func. Anal. **161** (1999), 62-75.

[16] A. T. Lau and W. Takahashi, *Invariant submeans and semigroups of nonexpansive mappings on Banach spaces with normal structure*, J. Func. Anal. **142** (1996), 79-88.

[17] P. L. Lions, *Une methode iterative de resolution d'une inequation variationnelle*, Israel J. Math. **31** (1978), 204–208.

[18] F. Liu and M. Z. Nashed, *Regularization of nonlinear ill-posed variational inequalities and convergence rates*, Set-Valued Anal. **6** (1998), 313-344.

[19] W. R. Mann, *Mean value methods in iteration*, Proc. Amer. Math. Soc. **4** (1953), 506–510.

[20] B. Martinet, *Regularisation d'inequations variationnelles par approximations successives*, Rev. Franc. Inform. Rech. Oper. **4** (1970), 154–159.

[21] J. J. Moreau, *Proximité et dualité dans un espace Hilbertien*, Bull. Soc. Math., France **93** (1965), 273–299.

[22] K. Nakajo and W. Takahashi, *Strong convergence theorems for nonexpansive mappings and nonexpansive semigroups*, J. Math. Anal. Appl., to appear.

[23] S. Ohsawa and W. Takahashi, *strong convergence theorems for resolvents of maximal monotone operators*, Arch. Math., to appear.

[24] Z. Opial, *Weak convergence of the sequence of successive approximations for nonexpansive mappings*, Bull. Amer. Math. Soc. **73** (1967), 591–597.

[25] G. B. Passty, *Ergodic convergence to a zero of the sum of monotone operators in Hilbert space*, J. Math. Anal. Appl. **72** (1979), 383–390.

[26] S. Reich, *Weak convergence theorems for nonexpansive mappings in Banach spaces*, J. Math. Anal. Appl. **67** (1979), 274–276.

[27] S. Reich, *Strong convergence theorems for resolvents of accretive operators in Banach spaces*, J. Math. Anal. Appl. **75** (1980), 287–292.

[28] R. T. Rockafellar, *Characterization of the subdifferentials of convex functions*, Pacific J. Math. **17** (1966), 497-510.

[29] R. T. Rockafellar, *On the maximality of sums of nonlinear monotone operators*, Trans. Amer. Math. Soc. **149** (1970), 75–88.

[30] R. T. Rockafellar, *Monotone operators and the proximal point algorithm*, SIAM J. Control and Optim. **14** (1976), 877–898.

[31] N. Shimizu and W. Takahashi, *Strong convergence to common fixed points of families of nonexpansive mappings*, J. Math. Anal. Appl. **211** (1997), 71-83.

[32] N. Shioji and W. Takahashi, *Strong convergence theorems of approximated sequences for nonexpansive mappings in Banach spaces*, Proc. Amer. Math. Soc. **125** (1997), 3641–3645.

[33] M. V. Solodov and B. F. Svaiter, *A hybrid projection – proximal point algorithm*, J. Convex Anal. **6** (1999), 59–70.

[34] M. V. Solodov and B. F. Svaiter, *Forcing strong convergence of proximal point iterations in a Hilbert space*, Math. Program. **87** (2000), 189–202.

[35] W. Takahashi, *A nonlinear ergodic theorem for an amenable semigroup of nonexpansive mappings in a Hilbert space*, Proc. Amer. Math. Soc. **81** (1981), 253-256.

[36] W. Takahashi, *Fixed point theorems for families of nonexpansive mappings on unbounded sets*, J. Math. Soc. Japan **36** (1984), 543-553.

[37] W. Takahashi, *A nonlinear ergodic theorem for a reversible semigroup of nonexpansive mappings in a Hilbert space*, Proc. Amer. Math. Soc. **97** (1986), 55-58.

[38] W. Takahashi, *Fixed point theorems and nonlinear ergodic theorems for nonlinear semigroups and their applications*, Nonlinear Anal. **30** (1997), 1283-1293.

[39] W. Takahashi, *Nonlinear Functional Analysis*, Yokohama Publishers, Yokohama, 2000.

[40] W. Takahashi, *Convex Analysis and Approximation of Fixed Points*, Yokohama Publishers, Yokohama, 2000 (Japanese).

[41] W. Takahashi and G. E. Kim, *Approximating fixed points of nonexpansive mappings in Banach spaces*, Math. Japon. **48** (1998), 1–9.

[42] W. Takahashi and M. Toyoda, *Weak convergence theorems for nonexpansive mappings and monotone mappings*, J. Optim. Theory Appl., to appear.

[43] W. Takahashi and Y. Ueda, *On Reich's strong convergence theorems for resolvents of accretive operators*, J. Math. Anal. Appl. **104** (1984), 546–553.

[44] R. Wittmann, *Approximation of fixed points of nonexpansive mappings*, Arch. Math. **58** (1992), 486–491.

[45] H. K. Xu, *Inequalities in Banach spaces with applications*, Nonlinear Anal. **16** (1991), 1127–1138.

Global Existence and Blow Up for a Wave Equation with a Potential and a Cubic Convolution

Kimitoshi Tsutaya

Department of Mathematics, Hokkaido University, Sapporo 060-0810, Japan
e–mail: tsutaya@math.sci.hokudai.ac.jp

Abstract. Consider a wave equation with a cubic convolution together with a potential in three space dimensions. We show the borderline value of the decay rate at infinity of the initial data between global existence and blow up.

AMS Subject Classification. 35L70, 35B05, 35B30

Key words. Global existence, blow up, wave equations, potential, cubic convolution.

1 Introduction

We consider the wave equation

$$\partial_t^2 u - \Delta u = V_1(x)u + (V_2 * u^2)u \tag{1.1}$$

for $x \in \mathbf{R}^3$, where $V_1(x) = O(|x|^{-\gamma_1})$ as $|x| \to \infty$, $V_2(x) = \nu_2|x|^{-\gamma_2}$, $\nu_2 \in \mathbf{R}$, and $*$ denotes spatial convolution. The Schrödinger equation with the interaction term $V_1(x)u + (V_2 * u^2)u$ were studied by Hayashi and Ozawa [5]. We study the global existence of smooth solutions of (1.1) for small initial data. Moreover, in this paper the potential V_1 is assumed to be small since the solution may blow up in a finite time unless V_1 is small, as proved by Strauss and Tsutaya [10]. In case $V_1(x) \equiv 0$, the initial data have small amplitude, and $V_2(x)$ satisfies some conditions, it is known that smooth solutions exist for all time. Hidano [3] proved not only the global existence but also the existence of scattering operator for $2 < \gamma_2 < 5/2$ using the Lorentz invariance method. See Menzala and Strauss [7], Mochizuki and Motai [8], and Mochizuki [9] for other dimensional cases and the Klein-Gordon equation with a cubic convolution.

One aim of this paper is to permit $V_1(x)$ which is small and decays like $|x|^{-\gamma_1}$. In Section 2, we show that if $\gamma_1 > 2$, $2 < \gamma_2 < 3$, and the small initial data decay like $|x|^{-1-k}$ with $k > 1 + (3 - \gamma_2)/2$, then smooth solutions exist for all time, which improves on the requirement $2 < \gamma_2 < 5/2$ in Hidano [3].

In Section 3 we show that if any of these three conditions are relaxed, then there exist arbitrarily small initial data such that the corresponding solutions blow up in a finite time. If $\gamma_2 < 2$, the blow up is due to Hidano [3]. We prove the blow up in Theorem 3.3 if $\gamma_1 < 2$ combining results obtained in [3] and [10] by the method of Glassey [2]. Then we prove it in Theorem 3.4 by the method of John [4] if $k < 1 + (3 - \gamma_2)/2$.

Finally, it seems interesting to compare the global existence and blow up results in Sections 2 and 3 with those for the wave equation

$$\partial_t^2 u - \Delta u = V_1(x)u + |u|^{p-1}u$$

for $x \in \mathbf{R}^3$ in Strauss and Tsutaya [10]. It is proved that if $p > 1 + \sqrt{2}$, the potential $V_1(x)$ and initial data which are all small and decay like $|x|^{-\gamma_1}$ with $\gamma_1 > 2$ and $|x|^{-1-k}$ with $k > 2/(p-1)$, respectively, then global solutions exist. On the other hand, if any of these conditions are relaxed, then solutions blow up in a finite time even for small initial data.

2 Global existence

We consider the Cauchy problem

$$\begin{cases} \partial_t^2 u - \Delta u = V_1(x)u + (V_2 * u^2)u, & t > 0,\ x \in \mathbf{R}^3, \\ u(0,x) = \varphi(x),\ \partial_t u(0,x) = \psi(x), & x \in \mathbf{R}^3, \end{cases} \tag{2.1}$$

where $V_2(x) = \nu_2 |x|^{-\gamma_2}$ with $\nu_2 \in \mathbf{R}$ and $\gamma_2 > 0$.

Before stating our results, we give some hypotheses :

(H1) The potential $V_1(x)$ satisfies

$$\sum_{|\alpha| \le 2} |\partial_x^\alpha V_1(x)| \le \frac{\nu_1}{(1 + |x|)^{\gamma_1}}$$

with $\gamma_1 > 0$, where $\nu_1 > 0$ is a small parameter.

(H2) $\varphi(x) \in C^3(\mathbf{R}^3)$, $\psi(x) \in C^2(\mathbf{R}^3)$ satisfy

$$\sum_{|\alpha| \le 3} |\partial_x^\alpha \varphi(x)| + \sum_{|\beta| \le 2} |\partial_x^\beta \psi(x)| \le \frac{\varepsilon}{(1 + |x|)^{1+k}}$$

with $k > 0$, where $\varepsilon > 0$ is a small parameter.

Theorem 2.1. *Consider the problem (2.1). Assume the hypotheses (H1) and (H2). If $\gamma_1 > 2$, $2 < \gamma_2 < 3$, $k > 1 + (3 - \gamma_2)/2$, and if ε and ν_1 are sufficiently small, depending on k, γ_1, γ_2 and ν_2, then there exists a unique global C^2 solution of (2.1).*

To prove Theorem 2.1 we consider the linear problem

$$\begin{cases} \partial_t^2 u - \Delta u = f(t,x), & t > 0,\ x \in \mathbf{R}^3, \\ u(0,x) = \varphi(x),\ \partial_t u(0,x) = \psi(x), & x \in \mathbf{R}^3. \end{cases} \tag{2.2}$$

The solution $u(t,x)$ of (2.2) is given by

$$u(t,x) = u^0(t,x) + L[f](t,x), \tag{2.3}$$

where $u^0(t,x)$ is the solution of $\partial_t^2 u - \Delta u = 0$ with the initial data $\varphi(x)$, $\psi(x)$, and

$$L[f](t,x) = \frac{1}{4\pi}\int_0^t (t-s)\int_{|\omega|=1} f(s, x+(t-s)\omega)\,d\omega ds \tag{2.4}$$

is the solution of $\partial_t^2 u - \Delta u = f$ with zero data, where $d\omega$ is the surface measure on the unit sphere $|\omega| = 1$. Note that $f \geq 0$ implies $L[f] \geq 0$.

The following lemma is proved in Asakura [1].

Lemma 2.2. *Suppose that $\varphi(x)$, $\psi(x)$ satisfy (H2). Let $k > 1$. Then the solution u^0 of $\partial_t^2 u^0 - \Delta u^0 = 0$ with the data $\varphi(x)$, $\psi(x)$ satisfies*

$$\sum_{|\alpha|\leq 2} |\partial_x^\alpha u^0(t,x)| \leq \frac{C_k \varepsilon}{(1+t+|x|)(1+|t-|x||)^{k-1}}, \tag{2.5}$$

where the constant C_k depends only on k.

We next introduce the weight function

$$w(t,x) = (1+t+|x|)(1+|t-|x||)^m, \tag{2.6}$$

where $m = \min\{1, k-1\}$, and define the norm for functions $u(t,x)$

$$\|u\| = \sup_{\substack{t\geq 0 \\ x\in \mathbf{R}^3}} w(t,x)|u(t,x)|. \tag{2.7}$$

The following lemma is the basic estimate for the existence proof.

Lemma 2.3. *Let $V_1(x)$ satisfy (H1). If $\gamma_1 > 2$, $2 < \gamma_2 < 3$ and $k > 1+(3-\gamma_2)/2$, then there exists a constant $C > 0$ depending only on k, γ_1, γ_2 and ν_2 such that*

$$\|L[|V_1 u| + |(V_2 * u^2)u|]\| \leq C(\nu_1\|u\| + \|u\|^3), \tag{2.8}$$

where L is defined in (2.4) and $\| \quad \|$ is defined in (2.7).

Before proving Lemma 2.3, we make some preparations. If f is spherically symmetric, i.e., $f(t,x) = f(t,r)$, $r = |x|$, then from (2.4) we can write the solution $L[f]$ in the form

$$u(t,r) = \frac{1}{2r}\int_0^t \int_{|r-t+s|}^{r+t-s} \lambda f(s,\lambda)\,d\lambda ds, \tag{2.9}$$

where $\lambda = |x+(t-s)\omega|$. We denote (2.9) by $u = Pf$. Defining the function

$$\tilde{u}(t,r) = \sup_{|x|=r} |u(t,x)|,$$

we have

$$|(L[f])(t,x)| \le (P\tilde{f})(t,r).$$

Thus, it suffices to estimate

$$P(\tilde{V}_1\tilde{u} + (\widetilde{V_2 * u^2})\tilde{u})(t,r) = \frac{1}{2r}\int_0^t \int_{|r-t+s|}^{r+t-s} \lambda(\tilde{V}_1\tilde{u} + (\widetilde{V_2 * u^2})\tilde{u})d\lambda ds. \tag{2.10}$$

In the course of calculations below, the various constants are simply denoted by C. To prove Lemma 2.3, we need the following proposition.

Proposition 2.4. *Suppose that*

$$|u(t,x)| \le \frac{M}{(1+t+|x|)(1+|t-|x||)^m}$$

with constants $M > 0$, $m > 0$ for all $t \ge 0$ and $x \in \mathbf{R}^3$. If $2 < \gamma_2 < 3$, then

(2.11)

$$|\widetilde{V_2 * u^2}(s,\lambda)| \le \begin{cases} \dfrac{CM^2}{(1+\lambda)(1+s)(1+|\lambda-s|)^{\gamma_2-2}} & (m > 1/2), \\[2ex] \dfrac{CM^2(1+\log(1+|\lambda-s|))}{(1+\lambda)(1+s)(1+|\lambda-s|)^{\gamma_2-2}} & (m = 1/2), \\[2ex] \dfrac{CM^2}{(1+\lambda)(1+s)(1+|\lambda-s|)^{2m+\gamma_2-3}} & (m < 1/2 \text{ and } 2m+\gamma_2-3 > 0), \end{cases}$$

where the constant C depends only on m, γ_2 and ν_2, independent of M.

Proof. We first have

$$\begin{aligned} & |V_2 * u^2(s,y)| \\ = \; & |\nu_2| \int |y-z|^{-\gamma_2} u^2(s,z)dz \\ \le \; & |\nu_2| M^2 \int |y-z|^{-\gamma_2}(1+s+|z|)^{-2}(1+|s-|z||)^{-2m}dz \\ \le \; & \frac{2\pi|\nu_2|M^2}{\lambda}\int_0^\infty \int_{|\lambda-\eta|}^{\lambda+\eta} \rho^{1-\gamma_2}d\rho(1+s+\eta)^{-2}(1+|s-\eta|)^{-2m}\eta d\eta, \end{aligned}$$

where $\lambda = |y|$. Using an elementary fact (see [1])

$$\int_{|\lambda-\eta|}^{\lambda+\eta} \rho^{1-\gamma_2}d\rho \le \frac{C\min\{\lambda,\eta\}}{(\lambda+\eta)|\lambda-\eta|^{\gamma_2-2}}, \tag{2.12}$$

we have

$$
\begin{aligned}
& |V_2 * u^2(s,y)| \\
(2.13) \qquad \leq\ & \frac{CM^2}{\lambda}\int_0^\infty \frac{\min\{\lambda,\eta\}\eta}{(\lambda+\eta)|\lambda-\eta|^{\gamma_2-2}(1+s+\eta)^2(1+|s-\eta|)^{2m}}d\eta \\
\equiv\ & I.
\end{aligned}
$$

To prove (2.11), we have to distinguish seven cases as follows :

1. $\lambda \geq 1$ and $\lambda - s \geq 1$,
2. $\lambda \geq 1$ and $0 \leq \lambda - s \leq 1$,
3. $\lambda \geq 1$ and $s - \lambda \geq 1$,
4. $\lambda \geq 1$ and $0 \leq s - \lambda \leq 1$,
5. $\lambda \leq 1$ and $\lambda \geq s$,
6. $\lambda \leq 1$ and $s - \lambda \geq 1$,
7. $\lambda \leq 1$ and $0 \leq s - \lambda \leq 1$.

<u>Case 1. $\lambda \geq 1$ and $\lambda - s \geq 1$.</u> From (2.13), we have
(2.14)

$$
\begin{aligned}
I \ \leq\ & \frac{CM^2}{1+\lambda}\int_0^\infty \frac{1}{(1+s+\eta)(1+|s-\eta|)^{2m}|\lambda-\eta|^{\gamma_2-2}}d\eta \\
=\ & \frac{CM^2}{1+\lambda}\left(\int_0^s + \int_s^\lambda + \int_\lambda^\infty\right)\frac{1}{(1+s+\eta)(1+|s-\eta|)^{2m}|\lambda-\eta|^{\gamma_2-2}}d\eta \\
\equiv\ & I_1 + I_2 + I_3.
\end{aligned}
$$

We first estimate I_1. If $m > 1/2$, then

$$
(2.15a) \qquad I_1 \leq \frac{CM^2}{1+\lambda}\int_0^s \frac{1}{(1+s+\eta)(1+s-\eta)^{2m}(\lambda-\eta)^{\gamma_2-2}}d\eta,
$$

or

$$
\begin{aligned}
I_1 \ \leq\ & \frac{CM^2}{(1+\lambda)(1+s)(\lambda-s)^{\gamma_2-2}}\int_0^s \frac{1}{(1+s-\eta)^{2m}}d\eta \\
(2.15b) \qquad \leq\ & \frac{CM^2}{(1+\lambda)(1+s)(1+\lambda-s)^{\gamma_2-2}}.
\end{aligned}
$$

If $m \leq 1/2$, then we have to consider two cases. First let $\lambda \geq 2s$. From (2.15b), we have

$$
I_1 \ \leq\ \begin{cases} \dfrac{CM^2(1+s)^{1-2m}}{(1+\lambda)(1+s)(1+\lambda-s)^{\gamma_2-2}} & (m < 1/2), \\[2ex] \dfrac{CM^2\log(1+s)}{(1+\lambda)(1+s)(1+\lambda-s)^{\gamma_2-2}} & (m = 1/2), \end{cases}
$$

$$\leq \begin{cases} \dfrac{CM^2}{(1+\lambda)(1+s)(1+\lambda-s)^{2m+\gamma_2-3}} & (m<1/2), \\[2ex] \dfrac{CM^2\log(1+\lambda-s)}{(1+\lambda)(1+s)(1+\lambda-s)^{\gamma_2-2}} & (m=1/2). \end{cases}$$

On the other hand, if $2s \geq \lambda \geq s$, then by (2.15a) and integration by parts, we have

$$\begin{aligned} I_1 &\leq \frac{CM^2}{(1+\lambda)(1+s)}\left(\int_0^{2s-\lambda}+\int_{2s-\lambda}^{s}\right)\frac{1}{(1+s-\eta)^{2m}(\lambda-\eta)^{\gamma_2-2}}d\eta \\ &\leq \frac{CM^2}{(1+\lambda)(1+s)}\left\{\frac{\lambda^{3-\gamma_2}}{(1+s)^{2m}}+C\int_0^{2s-\lambda}\frac{(\lambda-\eta)^{3-\gamma_2}}{(1+s-\eta)^{2m+1}}d\eta \right. \\ &\quad \left. +\frac{1}{(\lambda-s)^{\gamma_2-2}}\int_{2s-\lambda}^{s}\frac{1}{(1+s-\eta)^{2m}}d\eta\right\}. \end{aligned}$$

Since $\lambda-\eta \leq 2(s-\eta)$ for $0 \leq \eta \leq 2s-\lambda$, and $2m+\gamma_2-3>0$,

$$\begin{aligned} I_1 &\leq \frac{CM^2}{(1+\lambda)(1+s)}\left\{\frac{\lambda^{3-\gamma_2}}{(1+s)^{2m}}+C\int_0^{2s-\lambda}\frac{1}{(1+s-\eta)^{2m+\gamma_2-2}}d\eta \right. \\ &\quad \left. +\frac{1}{(\lambda-s)^{\gamma_2-2}}\int_{2s-\lambda}^{s}\frac{1}{(1+s-\eta)^{2m}}d\eta\right\} \\ &\leq \begin{cases} \dfrac{CM^2}{(1+\lambda)(1+s)(1+\lambda-s)^{2m+\gamma_2-3}} \\ \qquad\qquad\qquad (m<1/2 \text{ and } 2m+\gamma_2-3>0), \\[2ex] \dfrac{CM^2(1+\log(1+\lambda-s))}{(1+\lambda)(1+s)(1+\lambda-s)^{\gamma_2-2}} \qquad (m=1/2). \end{cases} \end{aligned}$$

We next estimate I_2. From (2.14), we obtain

$$\begin{aligned} I_2 &\leq \frac{CM^2}{1+\lambda}\left(\int_s^{(\lambda+s)/2}+\int_{(\lambda+s)/2}^{\lambda}\right)\frac{1}{(1+s+\eta)(1+\eta-s)^{2m}(\lambda-\eta)^{\gamma_2-2}}d\eta \\ &\leq \frac{CM^2}{(1+\lambda)(1+s)(\lambda-s)^{\gamma_2-2}}\int_s^{(\lambda+s)/2}\frac{1}{(1+\eta-s)^{2m}}d\eta \\ &\quad +\frac{CM^2}{(1+\lambda)(1+\lambda+s)(1+\lambda-s)^{2m}}\int_{(\lambda+s)/2}^{\lambda}\frac{1}{(\lambda-\eta)^{\gamma_2-2}}d\eta \end{aligned}$$

$$
\leq \begin{cases} \dfrac{CM^2}{(1+\lambda)(1+s)(1+\lambda-s)^{\gamma_2-2}} & (m > 1/2), \\[2ex] \dfrac{CM^2(1+\log(1+\lambda-s))}{(1+\lambda)(1+s)(1+\lambda-s)^{\gamma_2-2}} & (m = 1/2), \\[2ex] \dfrac{CM^2}{(1+\lambda)(1+s)(1+\lambda-s)^{2m+\gamma_2-3}} & (m < 1/2), \end{cases}
$$

since $2 < \gamma_2 < 3$.

We now estimate I_3. Since $\lambda \geq s$ and $2m+\gamma_2-3>0$, from (2.14) we have

$$
I_3 \leq \frac{CM^2}{(1+\lambda)(1+s)} \left(\int_\lambda^{2\lambda-s+1} + \int_{2\lambda-s+1}^\infty \right) \frac{1}{(1+\eta-s)^{2m}(\eta-\lambda)^{\gamma_2-2}} d\eta.
$$

Integrating by parts for the first integral on the right-hand side, we obtain

$$
\begin{aligned}
I_3 \quad &\leq \quad \frac{CM^2}{(1+\lambda)(1+s)} \left\{ \frac{C}{(1+\lambda-s)^{2m+\gamma_2-3}} \right. \\
&\qquad + C\int_\lambda^{2\lambda-s+1} \frac{1}{(1+\eta-s)^{2m+\gamma_2-2}} d\eta \\
&\qquad \left. + \int_{2\lambda-s+1}^\infty \frac{1}{(\eta-\lambda)^{2m+\gamma_2-2}} d\eta \right\} \\
&\leq \quad \frac{CM^2}{(1+\lambda)(1+s)(1+\lambda-s)^{2m+\gamma_2-3}},
\end{aligned}
$$

since $\lambda \geq s$, $2m+\gamma_2-3>0$, and $2<\gamma_2<3$.

Case 2. $\lambda \geq 1$ and $0 \leq \lambda - s \leq 1$. We proceed as in Case 1. We first estimate I_1. By (2.14) and integrating by parts, we have

$$
\begin{aligned}
I_1 \quad &\leq \quad \frac{CM^2}{(1+\lambda)(1+s)} \int_0^s \frac{1}{(1+s-\eta)^{2m}(\lambda-\eta)^{\gamma_2-2}} d\eta \\
&\leq \quad \frac{CM^2}{(1+\lambda)(1+s)} \left\{ \frac{\lambda^{3-\gamma_2}}{(1+s)^{2m}} + C\int_0^s \frac{1}{(1+s-\eta)^{2m+\gamma_2-2}} d\eta \right\} \\
&\leq \quad \frac{CM^2}{(1+\lambda)(1+s)(1+\lambda-s)^{2m+\gamma_2-3}},
\end{aligned}
$$

since $2m+\gamma_2-3>0$ and $2<\gamma_2<3$.

We next have

$$
\begin{aligned}
I_2 \quad &\leq \quad \frac{CM^2}{(1+\lambda)(1+s)} \int_s^\lambda \frac{1}{(\lambda-\eta)^{\gamma_2-2}} d\eta \\
&\leq \quad \frac{CM^2}{(1+\lambda)(1+s)(1+\lambda-s)^{2m+\gamma_2-3}}.
\end{aligned}
$$

We now estimate I_3. We see that

$$\begin{aligned} I_3 &\leq \frac{CM^2}{1+\lambda}\left(\int_\lambda^{2\lambda-s+1} + \int_{2\lambda-s+1}^\infty\right)\frac{1}{(1+s+\eta)(1+\eta-s)^{2m}(\eta-\lambda)^{\gamma_2-2}}d\eta \\ &\leq \frac{CM^2}{(1+\lambda)(1+s)}\int_\lambda^{2\lambda-s+1}\frac{1}{(\eta-\lambda)^{\gamma_2-2}}d\eta \\ &\quad + \frac{CM^2}{(1+\lambda)(1+s)}\int_{2\lambda-s+1}^\infty\frac{1}{(\eta-\lambda)^{2m+\gamma_2-2}}d\eta \\ &\leq \frac{CM^2}{(1+\lambda)(1+s)(1+\lambda-s)^{2m+\gamma_2-3}}. \end{aligned}$$

Case 3. $\lambda \geq 1$ and $s-\lambda \geq 1$. From (2.13), we have

(2.16)
$$\begin{aligned} I &\leq \frac{CM^2}{1+\lambda}\left(\int_0^\lambda + \int_\lambda^s + \int_s^\infty\right)\frac{1}{(1+s+\eta)(1+|s-\eta|)^{2m}|\lambda-\eta|^{\gamma_2-2}}d\eta \\ &\equiv I_1 + I_2 + I_3. \end{aligned}$$

To estimate I_1, we use integration by parts. Then

$$\begin{aligned} I_1 &\leq \frac{CM^2}{(1+\lambda)(1+s)}\int_0^\lambda\frac{1}{(1+s-\eta)^{2m}(\lambda-\eta)^{\gamma_2-2}}d\eta \\ &\leq \frac{CM^2}{(1+\lambda)(1+s)}\left\{\frac{\lambda^{3-\gamma_2}}{(1+s)^{2m}} + C\int_0^\lambda\frac{1}{(1+s-\eta)^{2m+\gamma_2-2}}d\eta\right\} \\ &\leq \frac{CM^2}{(1+\lambda)(1+s)(1+s-\lambda)^{2m+\gamma_2-3}}, \end{aligned}$$

since $s > \lambda$ and $2m+\gamma_2-3 > 0$.

We next estimate I_2.

$$\begin{aligned} I_2 &\leq \frac{CM^2}{1+\lambda}\left(\int_\lambda^{(\lambda+s)/2} + \int_{(\lambda+s)/2}^s\right)\frac{1}{(1+s+\eta)(1+s-\eta)^{2m}(\eta-\lambda)^{\gamma_2-2}}d\eta \\ &\leq \frac{CM^2}{(1+\lambda)(1+s+\lambda)(1+s-\lambda)^{2m}}\int_\lambda^{(\lambda+s)/2}\frac{1}{(\eta-\lambda)^{\gamma_2-2}}d\eta \\ &\quad + \frac{CM^2}{(1+\lambda)(1+s+\lambda)(s-\lambda)^{\gamma_2-2}}\int_{(\lambda+s)/2}^s\frac{1}{(1+s-\eta)^{2m}}d\eta \end{aligned}$$

$$\leq \begin{cases} \dfrac{CM^2}{(1+\lambda)(1+s)(1+s-\lambda)^{\gamma_2-2}} & (m > 1/2), \\[2ex] \dfrac{CM^2(1+\log(1+s-\lambda))}{(1+\lambda)(1+s)(1+s-\lambda)^{\gamma_2-2}} & (m = 1/2), \\[2ex] \dfrac{CM^2}{(1+\lambda)(1+s)(1+s-\lambda)^{2m+\gamma_2-3}} & (m < 1/2), \end{cases}$$

since $s-\lambda \geq 1$ and $2 < \gamma_2 < 3$.

We now estimate I_3.

$$\begin{aligned} I_3 &\leq \frac{CM^2}{(1+\lambda)(1+s)} \left(\int_s^{2s-\lambda+1} + \int_{2s-\lambda+1}^{\infty} \right) \frac{1}{(1+\eta-s)^{2m}(\eta-\lambda)^{\gamma_2-2}} d\eta \\ &\leq \frac{CM^2}{(1+\lambda)(1+s)(s-\lambda)^{\gamma_2-2}} \int_s^{2s-\lambda+1} \frac{1}{(1+\eta-s)^{2m}} d\eta \\ &\quad + \frac{CM^2}{(1+\lambda)(1+s)} \int_{2s-\lambda+1}^{\infty} \frac{1}{(\eta-s)^{2m+\gamma_2-2}} d\eta \\ &\leq \begin{cases} \dfrac{CM^2}{(1+\lambda)(1+s)(1+s-\lambda)^{\gamma_2-2}} & (m > 1/2), \\[2ex] \dfrac{CM^2(1+\log(1+s-\lambda))}{(1+\lambda)(1+s)(1+s-\lambda)^{\gamma_2-2}} & (m = 1/2), \\[2ex] \dfrac{CM^2}{(1+\lambda)(1+s)(1+s-\lambda)^{2m+\gamma_2-3}} & (m < 1/2), \end{cases} \end{aligned}$$

since $s \geq \lambda+1$ and $2m+\gamma_2-3 > 0$.

Case 4. $\lambda \geq 1$ and $0 \leq s-\lambda \leq 1$. We proceed as in Case 3. From (2.16), we have

$$I_1 \leq \frac{CM^2}{(1+\lambda)(1+s)(1+s-\lambda)^{2m+\gamma_2-3}},$$

and

$$\begin{aligned} I_2 &\leq \frac{CM^2}{(1+\lambda)(1+s)} \int_\lambda^s \frac{1}{(\eta-\lambda)^{\gamma_2-2}} d\eta \\ &\leq \frac{CM^2}{(1+\lambda)(1+s)(1+s-\lambda)^{2m+\gamma_2-3}}, \end{aligned}$$

since $2 < \gamma_2 < 3$.

Integrating by parts, we have

$$\begin{aligned}
I_3 &\le \frac{CM^2}{(1+\lambda)(1+s)}\left(\int_s^{2s-\lambda+1} + \int_{2s-\lambda+1}^{\infty}\right)\frac{1}{(1+\eta-s)^{2m}(\eta-\lambda)^{\gamma_2-2}}d\eta \\
&\le \frac{CM^2}{(1+\lambda)(1+s)}\left\{\frac{1}{(1+s-\lambda)^{2m+\gamma_2-3}} + C\int_s^{2s-\lambda+1} d\eta\right. \\
&\quad \left. + \int_{2s-\lambda+1}^{\infty}\frac{1}{(\eta-s)^{2m+\gamma_2-2}}d\eta\right\} \\
&\le \frac{CM^2}{(1+\lambda)(1+s)(1+s-\lambda)^{2m+\gamma_2-3}}.
\end{aligned}$$

Case 5. $\lambda \le 1$ and $\lambda \ge s$. From (2.13), we obtain

$$\begin{aligned}
I &\le CM^2\left(\int_0^{\lambda} + \int_{\lambda}^{2\lambda+1} + \int_{2\lambda+1}^{\infty}\right)\frac{1}{(1+s+\eta)^2|\eta-\lambda|^{\gamma_2-2}}d\eta \\
&\le CM^2\left\{\lambda^{3-\gamma_2} + \frac{1}{(\lambda+1)^{\gamma_2-2}}\int_{2\lambda+1}^{\infty}\frac{1}{(1+s+\eta)^2}d\eta\right\} \\
&\le \frac{CM^2}{(1+\lambda)(1+s)(1+\lambda-s)^{2m+\gamma_2-3}},
\end{aligned}$$

since $s \le \lambda \le 1$.

Case 6. $\lambda \le 1$ and $s-\lambda \ge 1$. By (2.13),

$$\begin{aligned}
I &\le CM^2\left(\int_0^{\lambda} + \int_{\lambda}^{s} + \int_s^{\infty}\right)\frac{1}{(1+s+\eta)^2(1+|s-\eta|)^{2m}|\lambda-\eta|^{\gamma_2-2}}d\eta \\
&\equiv I_1 + I_2 + I_3.
\end{aligned}$$

Since $2 < \gamma_2 < 3$, we have

$$\begin{aligned}
I_1 &\le \frac{CM^2}{(1+s)^2(1+s-\lambda)^{2m}}\int_0^{\lambda}\frac{1}{(\lambda-\eta)^{\gamma_2-2}}d\eta \\
&\le \frac{CM^2}{(1+\lambda)(1+s)(1+s-\lambda)^{2m+\gamma_2-3}}
\end{aligned}$$

and

$$\begin{aligned}
I_2 &\le CM^2\left(\int_{\lambda}^{(\lambda+s)/2} + \int_{(\lambda+s)/2}^{s}\right)\frac{1}{(1+s+\eta)^2(1+s-\eta)^{2m}(\eta-\lambda)^{\gamma_2-2}}d\eta \\
&\le \frac{CM^2}{(1+s+\lambda)^2(1+s-\lambda)^{2m}}\int_{\lambda}^{(\lambda+s)/2}\frac{1}{(\eta-\lambda)^{\gamma_2-2}}d\eta
\end{aligned}$$

$$
\begin{aligned}
&+\frac{CM^2}{(1+s+\lambda)^2(s-\lambda)^{\gamma_2-2}}\int_{(\lambda+s)/2}^{s}\frac{1}{(1+s-\eta)^{2m}}d\eta \\
\leq\ &\begin{cases}
\dfrac{CM^2}{(1+\lambda)(1+s)(1+s-\lambda)^{\gamma_2-2}} & (m>1/2), \\[2ex]
\dfrac{CM^2(1+\log(1+s-\lambda))}{(1+\lambda)(1+s)(1+s-\lambda)^{\gamma_2-2}} & (m=1/2), \\[2ex]
\dfrac{CM^2}{(1+\lambda)(1+s)(1+s-\lambda)^{2m+\gamma_2-3}} & (m<1/2).
\end{cases}
\end{aligned}
$$

We next estimate I_3. If $m > 1/2$, then

$$
\begin{aligned}
I_3 &\leq \frac{CM^2}{(1+s)^2(s-\lambda)^{\gamma_2-2}}\int_s^\infty \frac{1}{(1+\eta-s)^{2m}}d\eta \\
&\leq \frac{CM^2}{(1+\lambda)(1+s)(1+s-\lambda)^{\gamma_2-2}}.
\end{aligned}
$$

If $m \leq 1/2$, then

$$
\begin{aligned}
I_3 &\leq \frac{CM^2}{(1+s)^2}\left(\int_s^{2s-\lambda+1}+\int_{2s-\lambda+1}^\infty\right)\frac{1}{(1+\eta-s)^{2m}(\eta-\lambda)^{\gamma_2-2}}d\eta \\
&\leq \frac{CM^2}{(1+s)^2(s-\lambda)^{\gamma_2-2}}\int_s^{2s-\lambda+1}\frac{1}{(1+\eta-s)^{2m}}d\eta \\
&\quad+\frac{CM^2}{(1+s)^2}\int_{2s-\lambda+1}^\infty\frac{1}{(\eta-s)^{2m+\gamma_2-2}}d\eta \\
&\leq \begin{cases}
\dfrac{CM^2(\log(1+s-\lambda)+1)}{(1+\lambda)(1+s)(1+s-\lambda)^{\gamma_2-2}} & (m=1/2), \\[2ex]
\dfrac{CM^2}{(1+\lambda)(1+s)(1+s-\lambda)^{2m+\gamma_2-3}} & (m<1/2).
\end{cases}
\end{aligned}
$$

<u>Case 7. $\lambda \leq 1$ and $0 \leq s-\lambda \leq 1$.</u> By (2.13), we easily see that

$$
\begin{aligned}
I &\leq CM^2\left\{\int_0^{\lambda+1}\frac{1}{|\lambda-\eta|^{\gamma_2-2}}d\eta+\int_{\lambda+1}^\infty\frac{1}{(1+s+\eta)^2(\eta-\lambda)^{\gamma_2-2}}d\eta\right\} \\
&\leq \frac{CM^2}{(1+\lambda)(1+s)(1+s-\lambda)^{2m+\gamma_2-3}}.
\end{aligned}
$$

This completes the proof of Proposition 2.4. ■

Proof of Lemma 2.3. It follows from [10] that

$$\|L[|V_1 u|]\| \le C\nu_1 \|u\|$$

holds. Thus, we have only to prove

$$\|L[|(V_2 * u^2)u|]\| \le C\|u\|^3.$$

We shall use the following lemma repeatedly.

Lemma 2.5. *Let $k > 1$. Then*

(2.17)
$$\frac{1}{2r}\int_{|r-t|}^{r+t} \frac{1}{(1+\lambda)^k} d\lambda \le \frac{C_k}{(1+t+r)(1+|t-r|)^{k-1}} \qquad \text{for all } t,\ r \ge 0,$$

where the constant C_k depends only on k.

See [1] for a proof.

Since we use Proposition 2.4 to prove Lemma 2.3, we distinguish three cases $m > 1/2$, $m = 1/2$ and ($m < 1/2$ and $2m + \gamma_2 - 3 > 0$).

Case 1. $m > 1/2$. We have to consider two subcases, $r \ge t$ and $t \ge r$.

(i) $r \ge t$. From (2.6), (2.7), (2.10) and (2.11), we have

$$\begin{aligned}
P &\equiv P(\widetilde{(V_2 * u^2)\tilde{u}})(t,r) \\
&\le \frac{C\|u\|^3}{r}\int_0^t \int_{r-t+s}^{r+t-s} \frac{\lambda}{(1+\lambda)(1+s)(1+\lambda+s)(1+\lambda-s)^{m+\gamma_2-2}} d\lambda ds \\
&= \frac{C\|u\|^3}{r}\int_{D_1} \frac{\lambda}{(1+\lambda)(1+s)(1+\lambda+s)(1+\lambda-s)^{m+\gamma_2-2}} d\lambda ds \\
&\quad + \frac{C\|u\|^3}{r}\int_{D_2} \frac{\lambda}{(1+\lambda)(1+s)(1+\lambda+s)(1+\lambda-s)^{m+\gamma_2-2}} d\lambda ds \\
&\equiv \mathrm{I} + \mathrm{II},
\end{aligned}$$

where the domains of integration D_1 and D_2 are

(2.18a) $$D_1 = \{(s,\lambda) : r - t + s \le \lambda \le r + t - s,\ \lambda \le 2s,\ 0 \le s \le t\},$$

(2.18b) $$D_2 = \{(s,\lambda) : r - t + s \le \lambda \le r + t - s,\ \lambda \ge 2s,\ 0 \le s \le t\}.$$

Note that if $r \ge 2t$, then D_1 is an empty set. To estimate I, we note that $s \ge (\lambda + s)/3$ on the domain D_1. Changing variables by

(2.19) $$\alpha = \lambda + s, \quad \beta = \lambda - s,$$

we have

$$\mathrm{I} \le \frac{C\|u\|^3}{r}\int_{r-t}^{r+t} \frac{1}{(1+\alpha)^2}\int_{r-t}^{\alpha} \frac{1}{(1+\beta)^{m+\gamma_2-2}} d\beta d\alpha$$

If $m+\gamma_2-2>1$, then by Lemma 2.5

$$\begin{aligned} \mathrm{I} &\le \frac{C\|u\|^3}{r}\int_{r-t}^{r+t}\frac{1}{(1+\alpha)^2}d\alpha \\ &\le \frac{C\|u\|^3}{(1+t+r)(1+r-t)}. \end{aligned} \tag{2.20}$$

If $m+\gamma_2-2=1$, then

$$\begin{aligned} \mathrm{I} &\le \frac{C\|u\|^3}{r}\int_{r-t}^{r+t}\frac{\log(1+\alpha)}{(1+\alpha)^{m+\gamma_2-1}}d\alpha \\ &\le \frac{C\|u\|^3}{r}\int_{r-t}^{r+t}\frac{1}{(1+\alpha)^{m+1}}d\alpha \\ &\le \frac{C\|u\|^3}{(1+t+r)(1+r-t)^m}. \end{aligned} \tag{2.21}$$

If $m+\gamma_2-2<1$, then

$$\begin{aligned} \mathrm{I} &\le \frac{C\|u\|^3}{r}\int_{r-t}^{r+t}\frac{1}{(1+\alpha)^{m+\gamma_2-1}}d\alpha \\ &\le \frac{C\|u\|^3}{(1+t+r)(1+r-t)^m}. \end{aligned} \tag{2.22}$$

To estimate II, we note that $\lambda - s \ge (\lambda+s)/3$ on the domain D_2. Changing variables by (2.19), we have

$$\begin{aligned} \mathrm{II} &\le \frac{C\|u\|^3}{r}\int_{r-t}^{r+t}\frac{1}{(1+\alpha)^{m+\gamma_2-1}}\int_{r-t}^{\alpha}\frac{1}{1+\alpha-\beta}d\beta d\alpha \\ &\le \frac{C\|u\|^3}{r}\int_{r-t}^{r+t}\frac{\log(1+\alpha)}{(1+\alpha)^{m+\gamma_2-1}}d\alpha \\ &\le \frac{C\|u\|^3}{r}\int_{r-t}^{r+t}\frac{1}{(1+\alpha)^{m+1}}d\alpha \\ &\le \frac{C\|u\|^3}{(1+t+r)(1+r-t)^m}. \end{aligned}$$

Thus, we obtain in the case $r \ge t$

$$P \le \frac{C\|u\|^3}{(1+t+r)(1+r-t)^m}.$$

(ii) $t \ge r$. From (2.6), (2.7), (2.10) and (2.11), we have

$$\begin{aligned} P &\le \frac{C\|u\|^3}{r}\int_{D_1}\frac{\lambda}{(1+\lambda)(1+s)(1+\lambda+s)(1+|\lambda-s|)^{m+\gamma_2-2}}d\lambda ds \\ &\quad + \frac{C\|u\|^3}{r}\int_{D_2}\frac{\lambda}{(1+\lambda)(1+s)(1+\lambda+s)(1+|\lambda-s|)^{m+\gamma_2-2}}d\lambda ds \\ &\equiv \mathrm{I}+\mathrm{II}, \end{aligned}$$

where the domains of integration D_1 and D_2 are

(2.23a) $$D_1 = \{(s,\lambda) : |r-t+s| \le \lambda \le r+t-s,\ \lambda \le 2s,\ 0 \le s \le t\},$$

(2.23b) $$D_2 = \{(s,\lambda) : t-r-s \le \lambda \le r+t-s,\ \lambda \ge 2s,\ s \ge 0\}.$$

To estimate I, we note that $s \ge (\lambda+s)/3$ on the domain D_1. As before, changing variables by (2.19), we have

$$\begin{aligned} \mathrm{I} &\le \frac{C\|u\|^3}{r}\int_{t-r}^{t+r}\frac{1}{(1+\alpha)^2}\int_{r-t}^{\alpha/3}\frac{1}{(1+|\beta|)^{m+\gamma_2-2}}d\beta d\alpha \\ &= \frac{C\|u\|^3}{r}\int_{t-r}^{t+r}\frac{1}{(1+\alpha)^2}\left(\int_0^{\alpha/3}\frac{1}{(1+\beta)^{m+\gamma_2-2}}d\beta\right. \\ &\qquad\left. + \int_0^{t-r}\frac{1}{(1+\beta)^{m+\gamma_2-2}}d\beta\right)d\alpha. \end{aligned}$$

Proceeding as in (2.20)–(2.22), we obtain

$$\mathrm{I} \le \frac{C\|u\|^3}{(1+t+r)(1+t-r)^m}.$$

To estimate II, we note that $\lambda - s \ge (\lambda+s)/3$ on the domain D_2. Since $\gamma_2 > 2$, using Lemma 2.5, we obtain

$$\begin{aligned} \mathrm{II} &\le \frac{C\|u\|^3}{r}\int_{t-r}^{t+r}\frac{1}{(1+\alpha)^{m+\gamma_2-1}}\int_{\alpha/3}^{\alpha}\frac{1}{1+\alpha-\beta}d\beta d\alpha \\ &\le \frac{C\|u\|^3}{r}\int_{t-r}^{t+r}\frac{1}{(1+\alpha)^{m+1}}d\alpha \\ &\le \frac{C\|u\|^3}{(1+t+r)(1+t-r)^m}. \end{aligned}$$

Thus, we obtain in Case 1 the desired estimate

$$P \le \frac{C\|u\|^3}{(1+t+r)(1+|t-r|)^m}.$$

Case 2. $m = 1/2$.

(i) $r \ge t$. From (2.6), (2.7), (2.10) and (2.11), we have

$$\begin{aligned} P &\le \frac{C\|u\|^3}{r}\int_{D_1}\frac{\lambda(1+\log(1+\lambda-s))}{(1+\lambda)(1+s)(1+\lambda+s)(1+\lambda-s)^{1/2+\gamma_2-2}}d\lambda ds \\ &\quad + \frac{C\|u\|^3}{r}\int_{D_2}\frac{\lambda(1+\log(1+\lambda-s))}{(1+\lambda)(1+s)(1+\lambda+s)(1+\lambda-s)^{1/2+\gamma_2-2}}d\lambda ds \\ &\equiv \mathrm{I} + \mathrm{II}, \end{aligned}$$

where the domains of integration D_1 and D_2 are defined in (2.18). Note that $s \geq (\lambda+s)/3$ on the domain D_1. Changing variables by (2.19), we have by Lemma 2.5,

$$\begin{aligned} \mathrm{I} &\leq \frac{C\|u\|^3}{r}\int_{r-t}^{r+t}\frac{1}{(1+\alpha)^2}\int_{r-t}^{\alpha}\frac{1+\log(1+\beta)}{(1+\beta)^{1/2+\gamma_2-2}}d\beta d\alpha \\ &\leq \frac{C\|u\|^3}{r}\int_{r-t}^{r+t}\frac{1}{(1+\alpha)^2}\int_{r-t}^{\alpha}\frac{1}{(1+\beta)^{1/2}}d\beta d\alpha \\ &\leq \frac{C\|u\|^3}{r}\int_{r-t}^{r+t}\frac{1}{(1+\alpha)^{3/2}}d\alpha \\ &\leq \frac{C\|u\|^3}{(1+t+r)(1+r-t)^{1/2}}, \end{aligned}$$

since $\gamma_2 > 2$.

Noting that $\lambda - s \geq (\lambda+s)/3$ on the domain D_2, we have

$$\begin{aligned} \mathrm{II} &\leq \frac{C\|u\|^3}{r}\int_{r-t}^{r+t}\frac{1}{(1+\alpha)^{1/2+\gamma_2-1}}\int_{r-t}^{\alpha}\frac{1+\log(1+\beta)}{1+\alpha-\beta}d\beta d\alpha \\ &\leq \frac{C\|u\|^3}{r}\int_{r-t}^{r+t}\frac{\log(1+\alpha)(1+\log(1+\alpha))}{(1+\alpha)^{\gamma_2-1/2}}d\alpha \\ &\leq \frac{C\|u\|^3}{r}\int_{r-t}^{r+t}\frac{1}{(1+\alpha)^{3/2}}d\alpha \\ &\leq \frac{C\|u\|^3}{(1+t+r)(1+r-t)^{1/2}}. \end{aligned}$$

Thus, we obtain in the case $r \geq t$ the desired estimate

$$P \leq \frac{C\|u\|^3}{(1+t+r)(1+r-t)^{1/2}}.$$

(ii) $t \geq r$. We have

$$\begin{aligned} P &\leq \frac{C\|u\|^3}{r}\int_{D_1}\frac{\lambda(1+\log(1+|\lambda-s|))}{(1+\lambda)(1+s)(1+\lambda+s)(1+|\lambda-s|)^{1/2+\gamma_2-2}}d\lambda ds \\ &\quad+\frac{C\|u\|^3}{r}\int_{D_2}\frac{\lambda(1+\log(1+|\lambda-s|))}{(1+\lambda)(1+s)(1+\lambda+s)(1+|\lambda-s|)^{1/2+\gamma_2-2}}d\lambda ds \\ &\equiv \mathrm{I}+\mathrm{II}, \end{aligned}$$

where D_1 and D_2 are defined in (2.23). Noting that $s \geq (\lambda+s)/3$ on the domain D_1, we have

$$\mathrm{I} \leq \frac{C\|u\|^3}{r}\int_{t-r}^{t+r}\frac{1}{(1+\alpha)^2}\int_{r-t}^{\alpha/3}\frac{1+\log(1+|\beta|)}{(1+|\beta|)^{1/2+\gamma_2-2}}d\beta d\alpha$$

$$\begin{aligned}
&\leq \frac{C\|u\|^3}{r}\int_{t-r}^{t+r}\frac{1}{(1+\alpha)^2}\left(\int_0^{\alpha/3}\frac{1}{(1+\beta)^{1/2}}d\beta+\int_0^{t-r}\frac{1}{(1+\beta)^{1/2}}d\beta\right)d\alpha\\
&\leq \frac{C\|u\|^3}{r}\int_{t-r}^{t+r}\frac{1}{(1+\alpha)^{3/2}}d\alpha\\
&\leq \frac{C\|u\|^3}{(1+t+r)(1+t-r)^{1/2}}.
\end{aligned}$$

Proceeding as in the previous case (i) $r \geq t$, we obtain

$$\begin{aligned}
\text{II} &\leq \frac{C\|u\|^3}{r}\int_{t-r}^{t+r}\frac{1}{(1+\alpha)^{1/2+\gamma_2-1}}\int_{\alpha/3}^{\alpha}\frac{1+\log(1+\beta)}{1+\alpha-\beta}d\beta d\alpha\\
&\leq \frac{C\|u\|^3}{(1+t+r)(1+t-r)^{1/2}},
\end{aligned}$$

so that

$$P \leq \frac{C\|u\|^3}{(1+t+r)(1+|t-r|)^{1/2}}.$$

Case 3. $m < 1/2$ and $2m+\gamma_2-3>0$.
(i) $r \geq t$.

$$\begin{aligned}
P &\leq \frac{C\|u\|^3}{r}\int_0^t\int_{r-t+s}^{r+t-s}\frac{\lambda}{(1+\lambda)(1+s)(1+\lambda+s)(1+\lambda-s)^{3m+\gamma_2-3}}d\lambda ds\\
&\leq \frac{C\|u\|^3}{r}\int_{r-t}^{r+t}\frac{1}{(1+\alpha)^2}\int_{r-t}^{\alpha}\frac{1}{(1+\beta)^{3m+\gamma_2-3}}d\beta d\alpha\\
&\quad+\frac{C\|u\|^3}{r}\int_{r-t}^{r+t}\frac{1}{(1+\alpha)^{3m+\gamma_2-2}}\int_{r-t}^{\alpha}\frac{1}{1+\alpha-\beta}d\beta d\alpha\\
&\leq \frac{C\|u\|^3}{r}\int_{r-t}^{r+t}\frac{1}{(1+\alpha)^2}\int_{r-t}^{\alpha}\frac{1}{(1+\beta)^{m}}d\beta d\alpha\\
&\quad+\frac{C\|u\|^3}{r}\int_{r-t}^{r+t}\frac{\log(1+\alpha)}{(1+\alpha)^{3m+\gamma_2-2}}d\alpha\\
&\leq \frac{C\|u\|^3}{r}\int_{r-t}^{r+t}\frac{1}{(1+\alpha)^{m+1}}d\alpha\\
&\leq \frac{C\|u\|^3}{(1+t+r)(1+r-t)^{m}}.
\end{aligned}$$

We have used $2m+\gamma_2-3>0$.
(ii) $t \geq r$.

P

$$
\begin{aligned}
&\le \frac{C\|u\|^3}{r}\int_0^t\int_{|r-t+s|}^{r+t-s}\frac{\lambda}{(1+\lambda)(1+s)(1+\lambda+s)(1+|\lambda-s|)^{3m+\gamma_2-3}}d\lambda ds \\
&\le \frac{C\|u\|^3}{r}\int_{t-r}^{t+r}\frac{1}{(1+\alpha)^2}\int_{r-t}^{\alpha/3}\frac{1}{(1+|\beta|)^{3m+\gamma_2-3}}d\beta d\alpha \\
&\quad +\frac{C\|u\|^3}{r}\int_{t-r}^{t+r}\frac{1}{(1+\alpha)^{3m+\gamma_2-2}}\int_{\alpha/3}^{\alpha}\frac{1}{1+\alpha-\beta}d\beta d\alpha \\
&\le \frac{C\|u\|^3}{r}\int_{t-r}^{t+r}\frac{1}{(1+\alpha)^{m+1}}d\alpha \\
&\le \frac{C\|u\|^3}{(1+t+r)(1+t-r)^m}.
\end{aligned}
$$

This completes the proof of Lemma 2.3. ■

Proof of Theorem 2.1. Let X be the linear space defined by

$$X=\{u(t,x):\partial_x^\alpha u(t,x)\in C([0,\infty)\times\mathbf{R}^3),\ \|\partial_x^\alpha u\|<\infty \text{ for } |\alpha|\le 2\},$$

where $\|\quad\|$ is defined in (2.7). We can verify that X is complete with respect to the norm

$$\|u\|_X=\sum_{|\alpha|\le 2}\|\partial_x^\alpha u\|.$$

We denote by X_0 the closed subset of X given by

$$X_0=\{u\in X:\|u\|_X\le 2\varepsilon C_k\},$$

where C_k is the constant in Lemma 2.2. Then it follows from Lemma 2.2 that $\|u^0\|_X\le C_k\varepsilon$. Hence $u^0\in X_0$.

We define the map T by

$$Tu=u^0+L[F(u)] \qquad \text{for } u\in X,$$

where

$$F(u)=V_1(x)u+(V_2*u^2)u.$$

Let $u,\ v\in X_0$. Lemma 2.3 gives us

$$
\begin{aligned}
&\|L[F(u)-F(v)]\| \\
&= \|L[V_1(u-v)+(V_2*u^2)(u-v)+\{V_2*(u+v)(u-v)\}v]\| \\
&\le C\nu_1\|u-v\|+C\|u\|^2\|u-v\|+C(\|u\|+\|v\|)\|u-v\|\|v\| \\
&\le C\nu_1\|u-v\|+C(2C_k\varepsilon)^2\|u-v\|+2C(2C_k\varepsilon)^2\|u-v\| \\
&\le C(\nu_1+(2C_k\varepsilon)^2)\|u-v\|.
\end{aligned}
\tag{2.24}
$$

Let $\partial_j = \partial/\partial x_j$, $j = 1, 2, 3$. Similarly, for $\partial_j L[F(u)-F(v)]$ and $\partial_j\partial_k L[F(u)-F(v)]$, we obtain

(2.25)

$$\begin{aligned}
&\|\partial_j L[F(u) - F(v)]\| \\
\leq\ & C\nu_1(\|u - v\| + \|\partial_j u - \partial_j v\|) \\
&+C\{\|u\|\|\partial_j u\|\|u - v\| + \|u\|^2\|\partial_j u - \partial_j v\| \\
&+(\|\partial_j u\| + \|\partial_j v\|)\|u - v\|\|v\| + (\|u\| + \|v\|)\|\partial_j u - \partial_j v\|\|v\| \\
&+(\|u\| + \|v\|)\|u - v\|\|\partial_j v\|\} \\
\leq\ & C\nu_1\|u - v\|_X + C\|u - v\|_X(\|u\|_X^2 + \|v\|_X^2) \\
\leq\ & C(\nu_1 + (2C_k\varepsilon)^2)\|u - v\|_X,
\end{aligned}$$

and

(2.26)

$$\begin{aligned}
\|\partial_j\partial_k L[F(u) - F(v)]\| &\leq C\nu_1\|u - v\|_X + C\|u - v\|_X(\|u\|_X^2 + \|v\|_X^2) \\
&\leq C(\nu_1 + (2C_k\varepsilon)^2)\|u - v\|_X.
\end{aligned}$$

Combining (2.24)–(2.26), we have

$$\|L(F(u) - F(v))\|_X \leq C(\nu_1 + (2C_k\varepsilon)^2)\|u - v\|_X.$$

Hence, if we choose ε and ν_1 so small that

$$C(\nu_1 + (2C_k\varepsilon)^2) \leq \frac{1}{2}, \tag{2.27}$$

then we obtain

$$\begin{aligned}
\|Tu\|_X &\leq 2C_k\varepsilon, \\
\|Tu - Tv\|_X &\leq C(\nu_1 + (2C_k\varepsilon)^2)\|u - v\|_X \\
&\leq \frac{1}{2}\|u - v\|_X
\end{aligned}$$

for u, $v \in X_0$. Thus, T is a contraction mapping of X_0 into itself. Therefore, by the contraction mapping principle it follows that there exists a unique global solution of (2.1) if we choose ε and ν_1 sufficiently small by (2.27). ∎

3 Blow up

We shall prove an approximate converse of Theorem 2.1. We shall show that if conditions $2 < \gamma_2 < 3$, (H1) with $\gamma_1 > 2$ and (H2) with $k > 1 + (3 - \gamma_2)/2$ are not satisfied, then there exist arbitrarily small solutions that blow up in a finite time. The solutions that we construct are everywhere non-negative.

First we state local existence and uniqueness of solutions of (2.1), and existence of non-negative solutions. The following lemma is proved in Menzala-Strauss [7] and Keller [6].

Lemma 3.1. *(i) Let $0 < \gamma_2 < 3$. Let $V_1 \in L^\infty$, $\varphi \in H^1(\mathbf{R}^3)$ and $\psi \in L^2(\mathbf{R}^3)$. Then there exists a maximal interval $I = [0, T)$ and a unique solution $u \in C(I, H^1)$, $\partial_t u \in C(I, L^2)$ satisfying (2.1). If $T < \infty$, then $\|u(t)\|_{H^1} + \|\partial_t u(t)\|_{L^2} \to \infty$ $(t \to T)$.*
(ii) Let $V_1(x)$ be a non-negative function and $\nu_2 > 0$. Let $\varphi \equiv 0$ and $\psi \geq 0$. Then $u(t,x) \geq 0$ for all $(t,x) \in I \times \mathbf{R}^3$.

We now state our blow up theorems.

Theorem 3.2. *Suppose that $V_1 \geq 0$, $\nu_2 > 0$ and $0 < \gamma_2 < 2$. Let $\varphi \equiv 0$, $\psi \in C(\mathbf{R}^3) \cap L^2(\mathbf{R}^3)$, $\psi \geq 0$ and $\psi \not\equiv 0$. If u is the unique solution of Lemma 3.1, then the existence time T is finite.*

Proof. First, let $\tilde{\psi}(r) = \inf_{r=|x|} \psi(x)$. Then the solution with the initial data $(0, \tilde{\psi}(r))$ is spherically symmetric. Let $u(t,r)$ be the solution with the data $(0, \tilde{\psi}(r))$. If $\tilde{\psi}(r)$ has compact support, then we see from Hidano [3] that the solution $u(t,r)$ has a finite existence time.

Next, if $\tilde{\psi}(r)$ does not have compact support, then the existence time of the solution $u(t,r)$ is finite by comparison argument (see page 185 of [10]). Moreover, using the same argument again, we conclude that the solution $u(t,x)$ with the initial data $(0, \psi(x))$ has a finite existence time. ∎

Theorem 3.3. *Suppose that $V_1(x) \geq C_1 < x >^{-\gamma_1} = C_1(1+|x|^2)^{-\gamma_1/2}$ for all $x \in \mathbf{R}^3$, where $C_1 > 0$ and $0 < \gamma_1 < 2$. Let $\nu_2 > 0$, $0 < \gamma_2 < 3$ and let $\varphi \equiv 0$, $\psi \in C(\mathbf{R}^3) \cap L^2(\mathbf{R}^3)$ with $\psi \geq 0$, and $\psi \not\equiv 0$. If u is the unique solution of Lemma 3.1, then $T < \infty$.*

Proof. The theorem is proved by combining some results obtained in [3] and [10]. To prove Theorem 3.3, we use the same comparison argument in the proof of Theorem 3.2. Thus, it is sufficient to show the theorem for the case where the support of $\tilde{\psi}(r)$ is compact, say in $\{|x| < R\}$. Then the solution $u(t,x)$ has support in $\{|x| \leq t + R\}$. Let

$$J(t) = \int_{\mathbf{R}^3} u(t,x)dx.$$

Integrating Eq. (2.1) on the support $\{|x| \leq t + R\}$, we have

$$J''(t) = \int_{\mathbf{R}^3} V_1(x)u(t,x)dx + \int_{\mathbf{R}^3} (V_2 * u^2)u(t,x)dx. \tag{3.1}$$

See page 184 of [10]. Using only the first term on the right-hand side of (3.1), we obtain

$$J(t) \geq C\exp(Ct^{1-\gamma_1/2}) \quad \text{for } 1 \leq t \leq T. \tag{3.2}$$

See (3.11) of [10]. Using the second term on the right-hand side of (3.1), we have

$$J''(t) \geq C(t+1)^{-(3+\gamma_2)} J(t)^3 \quad \text{for } 1 \leq t \leq T. \tag{3.3}$$

In fact, for $0 < \gamma_2 < 1$, see (6.11) of [3]. When $1 \le \gamma_2 < 3$, we can derive (3.3) from (6.13) of [3] by Hölder's inequality since the solution has support in $\{|x| \le t + R\}$. Combining (3.2) and (3.3) we obtain, for $1 \le t \le T$,

$$J'' \ge CJ^2.$$

This inequality implies $T < \infty$. ■

The following theorem shows that the solution blows up in a finite time if the decay rate of the initial data at infinity is slower than the number depending on γ_2.

Theorem 3.4. *Suppose that $V_1 \ge 0$, $\nu_2 > 0$, $0 < \gamma_2 < 3$ and $\varphi \equiv 0$. Let $\psi \in C(\mathbf{R}^3) \cap L^2(\mathbf{R}^3)$ satisfy*

$$\psi(x) \ge \varepsilon < x >^{-1-k} \tag{3.4}$$

for $\varepsilon > 0$ and k with $1/2 < k < 1 + (3 - \gamma_2)/2$. If u is the unique solution given in Lemma 3.1, then $T < \infty$.

Remark 3.1. We note that $\psi \in L^2(\mathbf{R}^3)$ and (3.4) imply $k > 1/2$.

Proof. We prove Theorem 3.4 following [1] and [4]. Since the solution u given in Lemma 3.1 and V_1 are non-negative, we have

$$u \ge u^0 + L[(V_2 * u^2)u], \tag{3.5}$$

where $u^0(t, x)$ is the solution of $(\partial_t^2 - \Delta)u = 0$ with the initial data $(0, \psi(x))$ satisfying the assumption of Theorem 3.4. The solution $u^0(t, x)$ satisfies

$$u^0(t, x) \ge \frac{\varepsilon t}{(1 + t + |x|)^{1+k}} \qquad \text{for all } t \ge 0 \text{ and } x \in \mathbf{R}^3.$$

In fact, we have by elementary calculation,

$$\begin{aligned} u^0(t, x) &= \frac{t}{4\pi} \int_{|\omega|=1} \psi(x + t\omega) d\omega \\ &\ge \frac{\varepsilon t}{4\pi} \int_{|\omega|=1} \frac{1}{(1 + |x + t\omega|)^{1+k}} d\omega \\ &\ge \frac{\varepsilon t}{4\pi(1 + |x| + t)^{1+k}} \int_{|\omega|=1} d\omega \\ &= \frac{\varepsilon t}{(1 + |x| + t)^{1+k}}. \end{aligned}$$

Hence, we have

$$u(t, x) \ge \frac{\varepsilon t}{(1 + |x| + t)^{1+k}} \qquad \text{for all } t \ge 0 \text{ and } x \in \mathbf{R}^3 \tag{3.6}$$

since $u \geq u_0$.

We now assume that the following estimate holds:

$$u(t,x) \geq \frac{Dt^a}{(1+|x|+t)^b} \tag{3.7}$$

with constants $D > 0$, $a \geq 1$ and $b \geq 0$. Note that (3.7) with $a = 1,\ b = 1+k$ and $D = \varepsilon$ corresponds to (3.6). From (3.5),

(3.8)

$$\begin{aligned} u(t,x) &= \frac{1}{4\pi}\int_0^t \frac{1}{t-s}\int_{|y-x|=t-s}\left(\int_{\mathbf{R}^3}|y-z|^{-\gamma_2}u^2(s,z)dz\right) \\ &\qquad\qquad\qquad\qquad \times u(s,y)dS_y ds \\ &\geq \frac{D^3}{4\pi}\int_0^t \frac{s^{3a}}{t-s}\int_{|y-x|=t-s}\frac{1}{(1+s+|y|)^b} \\ &\qquad \times\left(\int_{\mathbf{R}^3}\frac{1}{|y-z|^{\gamma_2}(1+s+|z|)^{2b}}dz\right)dS_y ds. \end{aligned}$$

Here we need the following lemma:

Lemma 3.5. *Let* $0 < \gamma_2 < 3$. *Then*

$$\int_{\mathbf{R}^3}\frac{1}{|y-z|^{\gamma_2}(1+s+|z|)^{2b}}dz \geq \frac{C|y|^{3-\gamma_2}}{(1+s+|y|)^{2b}} \tag{3.9}$$

holds for $0 \leq s \leq t$ *and all* $y \in \mathbf{R}^3$, *where the constant* C *depends only on* γ_2.

Proof. We have

$$\begin{aligned} &\int_{\mathbf{R}^3}\frac{1}{|y-z|^{\gamma_2}(1+s+|z|)^{2b}}dz \\ &= \frac{2\pi}{|y|}\int_0^\infty\left(\int_{||y|-\eta|}^{|y|+\eta}\rho^{1-\gamma_2}d\rho\right)\frac{\eta}{(1+s+\eta)^{2b}}d\eta \\ &\geq \frac{2\pi}{|y|}\int_0^{|y|}\left(\int_{|y|-\eta}^{|y|+\eta}\rho^{1-\gamma_2}d\rho\right)\frac{\eta}{(1+s+\eta)^{2b}}d\eta. \end{aligned} \tag{3.10}$$

If $2 < \gamma_2 < 3$, then

$$\int_{|y|-\eta}^{|y|+\eta}\rho^{1-\gamma_2}d\rho = \frac{C}{(|y|-\eta)^{\gamma_2-2}}\left\{1-\left(\frac{|y|-\eta}{|y|+\eta}\right)^{\gamma_2-2}\right\}.$$

Note that

$$1-\rho^k \geq \min(1,k)(1-\rho) \quad \text{for } 0 \leq \rho \leq 1. \tag{3.11}$$

Then we have

$$\int_{|y|-\eta}^{|y|+\eta} \rho^{1-\gamma_2} d\rho \geq \frac{C\eta}{(|y|-\eta)^{\gamma_2-2}(|y|+\eta)} \geq C\eta(|y|+\eta)^{1-\gamma_2}. \tag{3.12}$$

If $\gamma_2 = 2$, then

$$\int_{|y|-\eta}^{|y|+\eta} \rho^{1-\gamma_2} d\rho \geq \frac{2\eta}{|y|+\eta}. \tag{3.13}$$

If $0 < \gamma_2 < 2$, then

$$\int_{|y|-\eta}^{|y|+\eta} \rho^{1-\gamma_2} d\rho \geq C\eta(|y|+\eta)^{1-\gamma_2}. \tag{3.14}$$

Combining (3.12)–(3.14), we obtain

$$\int_{|y|-\eta}^{|y|+\eta} \rho^{1-\gamma_2} d\rho \geq C\eta(|y|+\eta)^{1-\gamma_2}, \tag{3.15}$$

where the constant C depends only on γ_2.

From (3.10) and (3.15), we obtain

$$\begin{aligned}\int_{\mathbf{R}^3} \frac{1}{|y-z|^{\gamma_2}(1+s+|z|)^{2b}} dz &\geq \frac{C}{|y|}\int_0^{|y|} \frac{\eta^2}{(|y|+\eta)^{\gamma_2-1}(1+s+\eta)^{2b}} d\eta \\ &\geq \frac{C|y|^{3-\gamma_2}}{(1+s+|y|)^{2b}},\end{aligned}$$

which completes the proof of Lemma 3.5. ∎

We now go back to the proof of Theorem 3.4. From (3.8) and Lemma 3.5 we have

$$\begin{aligned} u(t,x) &\geq \frac{CD^3}{4\pi}\int_0^t \frac{s^{3a}}{t-s}\int_{|y-x|=t-s} \frac{|y|^{3-\gamma_2}}{(1+s+|y|)^{3b}} dS_y ds \\ &= \frac{CD^3}{2r}\int_0^t \int_{|r-t+s|}^{r+t-s} \frac{s^{3a}\lambda^{4-\gamma_2}}{(1+s+\lambda)^{3b}} d\lambda ds, \end{aligned} \tag{3.16}$$

where $r = |x|$. We divide into two cases $t \geq r$ and $r \geq t$. If $t \geq r$, then

$$\begin{aligned} & u(t,x) \\ \geq\ & \frac{CD^3}{r(1+t+r)^{3b}}\left\{\int_{t-r}^t s^{3a}[(r+t-s)^{5-\gamma_2} - (r-t+s)^{5-\gamma_2}]ds \right. \\ & \left. + \int_0^{t-r} s^{3a}[(r+t-s)^{5-\gamma_2} - (t-r-s)^{5-\gamma_2}]ds \right\} \end{aligned}$$

$$= \frac{CD^3}{r(1+t+r)^{3b}}\left\{\int_{t-r}^{t} s^{3a}(r+t-s)^{5-\gamma_2}\left[1-\left(\frac{r-t+s}{r+t-s}\right)^{5-\gamma_2}\right]ds\right.$$
$$\left.+\int_0^{t-r} s^{3a}(r+t-s)^{5-\gamma_2}\left[1-\left(\frac{t-s-r}{t-s+r}\right)^{5-\gamma_2}\right]ds\right\}.$$

Using (3.11), we obtain

$$\begin{aligned} u(t,x) &\geq \frac{CD^3}{r(1+t+r)^{3b}}\left\{\int_{t-r}^{t} s^{3a}(t-s)(r+t-s)^{4-\gamma_2}ds\right. \\ &\qquad \left.+\int_0^{t-r} s^{3a}r(r+t-s)^{4-\gamma_2}ds\right\} \\ &\geq \frac{CD^3}{(1+t+r)^{3b}}\int_0^t s^{3a}(r+t-s)^{3-\gamma_2}(t-s)ds \\ &\geq \frac{CD^3}{(r+t)^{\gamma_2}(1+t+r)^{3b}}\int_0^t s^{3a}(t-s)^4ds \\ &\geq \frac{CD^3t^{3a+5}}{(3a+5)^5(1+t+r)^{3b+\gamma_2}}. \end{aligned}$$

If $r \geq t$, then similarly we have from (3.16)

$$\begin{aligned} u(t,x) &\geq \frac{CD^3}{2r}\int_0^t\int_{r-t+s}^{r+t-s}\frac{s^{3a}\lambda^{4-\gamma_2}}{(1+s+\lambda)^{3b}}d\lambda ds \\ &\geq \frac{CD^3t^{3a+5}}{(3a+5)^5(1+t+r)^{3b+\gamma_2}}. \end{aligned}$$

Thus, we obtain

$$u(t,x) \geq \frac{CD^3t^{3a+5}}{(3a+5)^5(1+t+r)^{3b+\gamma_2}}. \tag{3.17}$$

We define the sequences a_n, b_n and D_n for $n = 0, 1, 2, \cdots$ by

$$a_{n+1} = 3a_n + 5, \quad b_{n+1} = 3b_n + \gamma_2, \quad D_{n+1} = \frac{CD_n^3}{(3a_n+5)^5}, \tag{3.18}$$

$$a_0 = 1, \quad b_0 = 1+k, \quad D_0 = \varepsilon. \tag{3.19}$$

Solving (3.18) and (3.19), we have

$$a_n = \frac{3^n 7}{2} - \frac{5}{2}, \qquad b_n = 3^n\left(1+k+\frac{\gamma_2}{2}\right) - \frac{\gamma_2}{2},$$

and

$$D_{n+1} \geq \frac{CD_n^3}{4^5 3^{5(n+1)}}.$$

Then

$$\begin{aligned}\log D_n &\geq \sum_{j=0}^{n-1} 3^j \log C - 5\sum_{j=0}^{n-1} 3^j \log 4 - 5\sum_{j=1}^{n} j3^{n-j}\log 3 + 3^n \log D_0 \\ &= (\log C - 5\log 4)\frac{3^n-1}{2} - 3^n 5\sum_{j=1}^{n}\frac{j}{3^j}\log 3 + 3^n \log D_0.\end{aligned}$$

For n sufficiently large, we have

$$D_n \geq \exp(E3^n),$$

where

$$E = \min\left(0, \frac{\log C}{2} - 5\log 4\right) - 5\sum_{j=1}^{\infty}\frac{j}{3^j}\log 3 + \log D_0.$$

Replacing a, b, D by a_n, b_n, D_n, respectively in (3.7), we obtain
(3.20)

$$\begin{aligned}u(t,x) &\geq t^{-5/2}(1+t+|x|)^{\gamma_2/2} \\ &\times \exp\left[\left\{E + \frac{7}{2}\log t - (1+k+\frac{\gamma_2}{2})\log(1+t+|x|)\right\}3^n\right].\end{aligned}$$

For t large enough, we can choose a positive δ such that

$$E + \frac{7}{2}\log t - (1+k+\frac{\gamma_2}{2})\log(1+t+|x|) \geq \delta > 0,$$

since $k < 1 + (3-\gamma_2)/2$. Then it follows from (3.20) that $u(t,x) \longrightarrow \infty$ as $n \to \infty$ for t sufficiently large. This is a contradiction, which completes the proof. ■

References

[1] F. Asakura, Existence of a global solution to a semi-linear wave equation with slowly decreasing initial data in three space dimensions, Comm. Partial Differential Equations, **11**, No.13 (1986), 1459–1487.

[2] R.T. Glassey, Finite-time blow-up for solutions of nonlinear wave equations, Math. Z., **177** (1981), 323–340.

[3] K. Hidano, Small data scattering and blow-up for a wave equation with a cubic convolution, Funkcialaj Ekvacioj, **43** (2000), 559–588.

[4] F. John, Blow-up of solutions of nonlinear wave equations in three space dimensions, Manuscripta Math., **28** (1979), 235–268.

[5] N. Hayashi and T. Ozawa, Smoothing effect for some Schrödinger equations, J. Functional Analysis, **85** (1989), 307–348.

[6] J. B. Keller, On solutions of nonlinear wave equations, Comm. Pure Appl. Math., **10** (1957), 523–530.

[7] G. Perla Menzala and W. A. Strauss, On a wave equation with a cubic convolution, J. Diff. Eq., **43** (1982), 93–105.

[8] K. Mochizuki and T. Motai, On small data scattering for some nonlinear wave equations, "Patterns and waves–qualitative analysis of nonlinear differential equations-", Stud. Math. Appl. No.18, 543–560, North-Holland, Amsterdam, 1986.

[9] K. Mochizuki, On small data scattering with cubic convolution nonlinearity, J. Math. Soc. Japan, **41** (1989), 143–160.

[10] W. A. Strauss and K. Tsutaya, Existence and blow up of small amplitude nonlinear waves with a negative potential, Discrete and Conti. Dynamical Systems, **3** (1997), 175–188.

Monotonicity and Compactness Methods Applied to the Nonlinear Schrödinger and Related Equations

Tomomi Yokota

Department of Mathematics, Tokyo University of Science, 26 Wakamiya-cho, Shinjuku-ku, Tokyo 162-0827, Japan e–mail: yokota@rs.kagu.tus.ac.jp

Abstract. Global existence of weak, strong and C^1-solutions is proved for the nonlinear Schrödinger and related equations. The proof is based on monotonicity and compactness methods.

AMS Subject Classification. 35Q55

Key words. Monotonicity methods, accretive operators, Yosida approximations, compactness methods, nonlinear Schrödinger equations.

1 Introduction

Let Ω be a bounded domain in $\mathbf{R}^N$ ($N \in \mathbf{N}$) with C^2-boundary $\partial\Omega$. This paper is concerned with the global existence of weak, strong and C^1-solutions to the following initial-boundary value problem:

$$\text{(P)} \quad \begin{cases} \dfrac{\partial u}{\partial t} - (\lambda + i\alpha)\Delta u + (\kappa + i\beta)|u|^{q-2}u - \gamma u = 0 & \text{on } \Omega \times (0,\infty), \\ u = 0 & \text{on } \partial\Omega \times (0,\infty), \\ u(x,0) = u_0(x), & x \in \Omega, \end{cases}$$

where $\lambda \geq 0$, $\kappa \geq 0$ with "$\lambda\kappa = 0$", $\alpha, \beta, \gamma \in \mathbf{R}$ and $q > 1$ are constants, $i = \sqrt{-1}$, and $u = u(x,t)$ is a complex-valued unknown function on $\Omega \times [0,\infty)$.

In particular, if $\lambda > 0$ and $\kappa > 0$, then (P) is a problem for the complex Ginzburg-Landau equation:

$$\text{(CGL)} \quad \frac{\partial u}{\partial t} - (\lambda + i\alpha)\Delta u + (\kappa + i\beta)|u|^{q-2}u - \gamma u = 0 \quad (\lambda > 0,\ \kappa > 0)$$

and there is much literature on (CGL) (for the global existence of solutions see, e.g., Temam [24], Levermore-Oliver [11], Ginibre-Velo [4, 5] and Okazawa-Yokota [14]–[18]; for the large time behavior of solutions with $\gamma = 0$ see Hayashi-Kaikina-Naumkin [6]–[9]).

This paper deals with the limiting case where $\lambda = 0$ or $\kappa = 0$ so that the equation in (P) can be rewritten as the nonlinear Schrödinger *type* equation

$$\text{(1.1)} \quad \frac{\partial u}{\partial t} - i\alpha\Delta u + (\kappa + i\beta)|u|^{q-2}u - \gamma u = 0$$

or the *dissipative* nonlinear Schrödinger equation (cf. [6])

$$\frac{\partial u}{\partial t} - (\lambda + i\alpha)\Delta u + i\beta|u|^{q-2}u - \gamma u = 0. \tag{1.2}$$

Concerning (1.1), the special case where $\beta = \gamma = 0$ and $\alpha = \kappa = 1$, i.e.,

$$\frac{\partial u}{\partial t} - i\Delta u + |u|^{q-2}u = 0 \tag{1.3}$$

is studied by Lions [12, Chapitre 1, Section 10], Pecher-von Wahl [21], Shigeta [22], Bahuguna-Raghavendra [1] and Okazawa-Yokota [14]. In [12] the global existence of unique weak solutions for (1.3) has been established. On the other hand, the global existence of unique strong solutions for (1.3) has been proved by [21, 22, 1] under the additional condition that $q - 1 \leq (N+2)/(N-2)$ for $N \geq 3$. This restriction on q has been removed by [14]. Unfortunately, we cannot extend these results on (1.3) directly to (1.1) and (1.2), much less to the *usual* nonlinear Schrödinger equation

$$\frac{\partial u}{\partial t} - i\alpha\Delta u + i\beta|u|^{q-2}u = 0. \tag{NLS}$$

It is well-known that (NLS) is studied very extensively; however, most of them are done under the strict restriction on the dimension N and the exponent q. Also, we would like to emphasize that (1.1) and (1.2) are essentially different from (NLS) in the sense that the L^2-norm (or energy) conservation law does not hold for (1.1) and (1.2).

The purpose of this paper is to establish the global existence of weak, strong and C^1-solutions to (P) in the case where $\lambda = 0$ or $\kappa = 0$ *without any upper restriction on the exponent* q. Setting

$$c(q) := \frac{|q-2|}{2\sqrt{q-1}}, \tag{1.4}$$

we discuss the following three kinds of solvability of (P):

1. Global existence of *weak* solutions to (P) in the following three cases.

Case i	$u_0 \in H_0^1(\Omega) \cap L^q(\Omega)$	$\lambda \geq 0,\ \kappa \geq 0$	$\alpha\beta \geq 0$
Case ii	$u_0 \in H_0^1(\Omega)$	$\lambda = 0,\ \kappa > 0$	$\lvert\beta\rvert/\kappa \leq c(q)^{-1}$
Case iii	$u_0 \in L^q(\Omega) \cap L^2(\Omega)$	$\lambda > 0,\ \kappa = 0$	$\lvert\alpha\rvert/\lambda \leq c(q)^{-1}$

2. Global existence of *unique strong* solutions to (P) with $u_0 \in H^2(\Omega) \cap H_0^1(\Omega) \cap L^{2(q-1)}(\Omega)$ and $q \geq 2$ in the following two cases.

Case i	$\lambda = 0,\ \kappa > 0$	$\alpha\beta \geq 0,\ \lvert\beta\rvert/\kappa \leq c(q)^{-1}$
Case ii	$\lambda = 0,\ \kappa > 0$	$\alpha\beta < 0,\ \lvert\beta\rvert/\kappa < c(q)^{-1}$

3. Global existence of *unique* C^1-solutions to (P) with $u_0 \in H^2(\Omega) \cap H_0^1(\Omega)$ and $q \geq 2$ in the following three cases.

Case i	$N = 1$	$\lambda \geq 0,\ \kappa \geq 0$	$\alpha\beta \geq 0$
Case ii	$N = 2, 3$	$\lambda = 0,\ \kappa > 0$	$\alpha\beta \geq 0,\ \lvert\beta\rvert/\kappa \leq c(q)^{-1}$
Case iii	$N = 1, 2, 3$	$\lambda = 0,\ \kappa > 0$	$\alpha\beta < 0,\ \lvert\beta\rvert/\kappa < c(q)^{-1}$

Here we note that the global existence of weak and strong solutions to (P) is discussed without any dimension constraint. The existence result on strong solutions to (P) is partially announced in [18]. Our method here is based on the idea developed in our previous works [14]–[18]. In [14]–[18] we have proved the global existence of weak and strong solutions for (CGL) (and its generalized equations) without any upper restriction on N and q by using the complex space version of monotonicity and compactness methods. In this paper we give a new application of monotonicity and compactness methods to the limiting case where $\lambda = 0$ or $\kappa = 0$.

Before stating our results we define weak, strong and C^1-solutions to (P).

Definition. Let $u(\cdot) \in C([0,\infty); L^2(\Omega))$. Then

(a) $u(\cdot)$ is called a *weak solution* to (P) if
  (a1) $u(0) = u_0$ in $L^2(\Omega)$;
  (a2) $u(\cdot) \in L^2(0,T; H_0^1(\Omega)) \cap L^q(0,T; L^q(\Omega))$ for all $T > 0$;
  (a3) for $T > 0$ and $\varphi(\cdot) \in C_0^\infty((0,T); H_0^1(\Omega) \cap L^q(\Omega))$,

$$
\begin{aligned}
&-\int_0^T \left(u(t), \frac{\partial \varphi}{\partial t}(t)\right)_{L^2} dt + (\lambda + i\alpha)\int_0^T (\nabla u(t), \nabla\varphi(t))_{L^2}\, dt \\
&+(\kappa + i\beta)\int_0^T \langle |u(t)|^{q-2}u(t), \varphi(t)\rangle_{L^{q'},L^q}\, dt - \gamma\int_0^T (u(t), \varphi(t))_{L^2}\, dt = 0,
\end{aligned}
$$

where q' is the conjugate exponent to q: $1/q + 1/q' = 1$.

(b) $u(\cdot)$ is called a *strong solution* to (P) if
  (b1) $u(t) \in H^2(\Omega) \cap H_0^1(\Omega) \cap L^{2(q-1)}(\Omega)$ for all $t \geq 0$;
  (b2) $u(\cdot) \in W^{1,\infty}(0,T; L^2(\Omega))$ for all $T > 0$;
  (b3) $u(\cdot)$ satisfies the equation in (P) a.e. on $(0,\infty)$ as well as the initial condition.

(c) $u(\cdot)$ is called a C^1-*solution* to (P) if
  (c1) $u(\cdot) \in C^1([0,\infty); L^2(\Omega)) \cap C([0,\infty); H^2(\Omega) \cap H_0^1(\Omega))$;
  (c2) $u(\cdot)$ satisfies the equation in (P) on $[0,\infty)$ as well as the initial condition.

Now we state our main results in this paper.

First we present the existence theorem for weak solutions to (P). Part (i) includes the case where $\lambda = \kappa = \gamma = 0$ which corresponds to (NLS). Recently Bechouche-Jüngel [2] constructed a weak solution for (NLS) by taking the inviscid limit of strong solutions for (CGL) as $\lambda \downarrow 0$ and $\kappa \downarrow 0$, while our proof here does not depend on the solvability of (CGL). Part (ii) not only generalizes but also improves the result on (1.3) established by Lions [12]. In fact, we assume in part (ii) that $u_0 \in H_0^1(\Omega)$, while it is assumed in [12, p. 131, Théorème 10.1] that $u_0 \in H_0^1(\Omega) \cap L^{2(q-1)}(\Omega)$.

Theorem 1.1. *Let $N \in \mathbf{N}$, $q > 1$ and $\gamma \in \mathbf{R}$.*

(i) *If $\lambda \geq 0$, $\kappa \geq 0$ with $\lambda\kappa = 0$ and $\alpha\beta \geq 0$, then for any $u_0 \in H_0^1(\Omega) \cap L^q(\Omega)$ there exists a weak solution $u(\cdot) \in C([0,\infty); L^2(\Omega))$ to* (P) *such that*

$$u(\cdot) \in L^\infty(0,T; H_0^1(\Omega) \cap L^q(\Omega)) \ \ \forall T > 0,$$
$$(\partial u/\partial t)(\cdot) \in L^\infty(0,T; H^{-1}(\Omega) + L^{q'}(\Omega)) \ \ \forall T > 0$$

with the estimates

(1.5) $\|u(t)\|_{L^2} \leq e^{\gamma t}\|u_0\|_{L^2} \ \ \forall t \geq 0,$

(1.6) $\frac{1}{2}\|\nabla u(t)\|_{L^2}^2 + \frac{\delta}{q}\|u(t)\|_{L^q}^q \leq e^{M\gamma_+ t}(\frac{1}{2}\|\nabla u_0\|_{L^2}^2 + \frac{\delta}{q}\|u_0\|_{L^q}^q) \ \ a.a.\ t \geq 0,$

where $M := \max\{q, 2\}$, $\gamma_+ := \max\{\gamma, 0\}$ and $\delta > 0$ is a constant depending only on $\lambda + i\alpha$, $\kappa + i\beta$ and q.

(ii) *If $\lambda = 0$, $\kappa > 0$ and $|\beta|/\kappa \leq c(q)^{-1}$ (see (1.4)), then for any $u_0 \in H_0^1(\Omega)$ there exists a "unique" weak solution $u(\cdot) \in C([0,\infty); L^2(\Omega))$ to* (P) *such that*

$$u(\cdot) \in L^\infty(0,T; H_0^1(\Omega)) \cap L^q(0,T; L^q(\Omega)) \ \ \forall T > 0,$$
$$(\partial u/\partial t)(\cdot) \in L^\infty(0,T; H^{-1}(\Omega)) + L^{q'}(0,T; L^{q'}(\Omega)) \ \ \forall T > 0$$

with the estimates

(1.7) $\|\nabla u(t)\|_{L^2} \leq e^{\gamma t}\|\nabla u_0\|_{L^2} \ \ \forall t \geq 0,$

(1.8) $\int_0^T \|u(t)\|_{L^q}^q \, dt \leq \frac{1}{2\kappa} e^{2\gamma_+ T}\|u_0\|_{L^2}^2 \ \ \forall T > 0,$

(1.9) $\|u(t) - v(t)\|_{L^2} \leq e^{\gamma t}\|u_0 - v_0\|_{L^2} \ \ \forall t \geq 0,$

where $v(\cdot)$ is a unique weak solution to (P) *with $v(0) = v_0 \in H_0^1(\Omega)$.*

(iii) *If $\lambda > 0$, $\kappa = 0$ and $|\alpha|/\lambda \leq c(q)^{-1}$, then for any $u_0 \in L^q(\Omega) \cap L^2(\Omega)$ there exists a weak solution $u(\cdot) \in C([0,\infty); L^2(\Omega))$ to* (P) *such that*

$$u(\cdot) \in L^2(0,T; H_0^1(\Omega)) \cap L^\infty(0,T; L^q(\Omega)) \ \ \forall T > 0,$$
$$(\partial u/\partial t)(\cdot) \in L^2(0,T; H^{-1}(\Omega)) + L^\infty(0,T; L^{q'}(\Omega)) \ \ \forall T > 0.$$

Also, $u(\cdot)$ satisfies (1.5) *and*

(1.10) $\int_0^T \|\nabla u(t)\|_{L^2}^2 \, dt \leq \frac{1}{2\lambda} e^{2\gamma_+ t}\|u_0\|_{L^2}^2 \ \ \forall T > 0,$

(1.11) $\|u(t)\|_{L^q} \leq e^{\gamma t}\|u_0\|_{L^q} \ \ \forall t \geq 0.$

Next we present the existence theorem for unique strong solutions to (P), which is a slight generalization of [14, Theorem 1.2].

Theorem 1.2. *Let $N \in \mathbf{N}$, $q \geq 2$, $\lambda = 0$, $\kappa > 0$ and $\gamma \in \mathbf{R}$. Assume that either*

(i) $\alpha\beta \geq 0$ *and* $|\beta|/\kappa \leq c(q)^{-1}$ *(see* (1.4))

or

(ii) $\alpha\beta < 0$ *and* $|\beta|/\kappa < c(q)^{-1}$.

Then for any $u_0 \in H^2(\Omega) \cap H_0^1(\Omega) \cap L^{2(q-1)}(\Omega)$ there exists a unique strong solution $u(\cdot) \in C([0,\infty); L^2(\Omega))$ to (P) *such that*

$$u(\cdot) \in C^{0,1}([0,T]; L^2(\Omega)) \ \forall T > 0, \tag{1.12}$$

$$u(\cdot) \in C^{0,1/2}([0,T]; H_0^1(\Omega)) \cap C^{0,1/q}([0,T]; L^q(\Omega)) \ \forall T > 0, \tag{1.13}$$

$$u(\cdot) \in L^\infty(0,T; H^2(\Omega) \cap L^{2(q-1)}(\Omega)) \ \forall T > 0. \tag{1.14}$$

Also, $u(\cdot)$ satisfies (1.7)–(1.9) *and*

$$\|\Delta u(t)\|_{L^2} \leq K_1 e^{\gamma t} k(u_0) \ \forall t \geq 0, \tag{1.15}$$

$$\|u(t)\|_{L^{2(q-1)}}^{q-1} \leq K_2 e^{\gamma t} k(u_0) \ \forall t \geq 0, \tag{1.16}$$

where K_1, K_2 and $k(u_0)$ are given by

$$K_1 := |\alpha|^{-1}(1 + K_2|\kappa + i\beta|),$$

$$K_2 := \frac{1 + c(q)}{\kappa - \frac{(\alpha\beta)_-}{|\alpha|} c(q)}, \quad (\alpha\beta)_- := -\min\{\alpha\beta, 0\},$$

$$k(u_0) := \| -i\alpha\Delta u_0 + (\kappa + i\beta)|u_0|^{q-2}u_0 - \gamma u_0\|_{L^2} + |\gamma| \cdot \|u_0\|_{L^2}.$$

Furthermore, let $v(\cdot)$ be a unique strong solution to (P) *with $v(0) = v_0 \in H^2(\Omega) \cap H_0^1(\Omega) \cap L^{2(q-1)}(\Omega)$. Then for every $t \geq 0$,*

$$\|\nabla u(t) - \nabla v(t)\|_{L^2}^2 \leq K_1 e^{2\gamma t}(k(u_0) + k(v_0))\|u_0 - v_0\|_{L^2}, \tag{1.17}$$

$$\|u(t) - v(t)\|_{L^q}^q \leq 2^{q-2} K_2 e^{2\gamma t}(k(u_0) + k(v_0))\|u_0 - v_0\|_{L^2}. \tag{1.18}$$

Remark 1. Theorem 1.2 remains true when $1 < q < 2$. However, we should replace $L^{2(q-1)}(\Omega)$ and $\|u\|_{L^{2(q-1)}}$ with $\{u \in L^2(\Omega);\ |u|^{q-2}u \in L^2(\Omega)\}$ and $\||u|^{q-2}u\|_{L^2}$, respectively; note that $0 < q-1 < 1$. So we have assumed for simplicity that $q \geq 2$.

Finally, we present the existence theorem for unique C^1-solutions to (P), which improves partially the result on (1.3) established by Pecher-von Wahl [21]. In fact, we remove the upper restriction of $q \leq 6$ assumed in [21] when $N = 3$.

Theorem 1.3. *Let $q \geq 2$ and $\gamma \in \mathbf{R}$. Assume that any one of the following three conditions is satisfied:*

(i) $N = 1$; $\lambda \geq 0$, $\kappa \geq 0$ *with* $\lambda\kappa = 0$ *and* $\alpha\beta \geq 0$.
(ii) $N = 2, 3$; $\lambda = 0$, $\kappa > 0$, $\alpha\beta \geq 0$ *and* $|\beta|/\kappa \leq c(q)^{-1}$ (*see* (1.4)).
(iii) $N = 1, 2, 3$; $\lambda = 0$, $\kappa > 0$, $\alpha\beta < 0$ *and* $|\beta|/\kappa < c(q)^{-1}$.

Then for any $u_0 \in H^2(\Omega) \cap H_0^1(\Omega)$ *there exists a unique* C^1*-solution* $u(\cdot)$ *to* (P) *such that* $u(\cdot) \in C^1([0,\infty); L^2(\Omega)) \cap C([0,\infty); H^2(\Omega) \cap H_0^1(\Omega))$.

Remark 2. The boundedness of Ω is not essential in Theorems 1.2 and 1.3.

This paper is organized as follows. In Section 2 we prepare some basic inequalities and existence results on abstract evolution equations (monotonicity methods). In Section 3 we prove Theorem 1.1 by using compactness methods. Section 4 is devoted to the proof of Theorem 1.2. The proof is based on monotonicity methods. In Section 5 we prepare the local existence theorem for C^1-solutions to (P) and then prove the global existence theorem (Theorem 1.3) by using the estimates for weak and strong solutions to (P).

2 Preliminaries

In this section we prepare some basic inequalities and global existence results on two kinds of abstract evolution equations.

First we shall give two basic inequalities in $\mathbf{C}$. One is the Liskevich-Perelmuter inequality [13, Lemma 2.2] (for the Hilbert space case see [16, Lemma 2.1]).

Lemma 2.1. *Let $q > 1$ and let $c(q)$ be as in* (1.4). *Then for $z, w \in \mathbf{C}$,*

$$|\mathrm{Im}\,(|z|^{q-2}z - |w|^{q-2}w)(\overline{z} - \overline{w})| \leq c(q)\,\mathrm{Re}\,(|z|^{q-2}z - |w|^{q-2}w)(\overline{z} - \overline{w}).$$

The other one is the following elementary inequality.

Lemma 2.2. *Let $q \geq 2$. Then for $z, w \in \mathbf{C}$,*

$$||z|^{q-2}z - |w|^{q-2}w| \leq (q-1)\max\{|z|^{q-2}, |w|^{q-2}\}|z - w|.$$

Next we define two operators in the complex Hilbert space $H := L^2(\Omega)$ with inner product $(\cdot,\cdot) := (\cdot,\cdot)_{L^2}$ and norm $\|\cdot\| := \|\cdot\|_{L^2}$ as follows:

$$Su := -\Delta u \quad \text{for } u \in D(S) := H^2(\Omega) \cap H_0^1(\Omega), \tag{2.1}$$

$$Bu := |u|^{q-2}u \quad \text{for } u \in D(B) := \{u \in H;\ |u|^{q-2}u \in H\}, \tag{2.2}$$

where $q > 1$. It is well-known that S is a nonnegative selfadjoint operator in H. On the other hand, as shown in [14, Lemma 3.1], B is a nonlinear m-accretive operator in H:

$$\begin{cases} \mathrm{Re}(Bu - Bv, u - v) \geq 0 & \text{for } u, v \in D(B), \\ R(1 + n^{-1}B) = H & \text{for some (and hence every) } n \in \mathbf{N}. \end{cases}$$

Moreover, S and B are represented by subdifferential operators. In fact, define two proper lower semi-continuous convex functions on H as follows:

$$\phi(u) := \begin{cases} (1/2)\|\nabla u\|_{L^2}^2 & \text{if } u \in D(\phi) := H_0^1(\Omega), \\ \infty & \text{otherwise,} \end{cases} \tag{2.3}$$

$$\psi(u) := \begin{cases} (1/q)\|u\|_{L^q}^q & \text{if } u \in D(\psi) := L^q(\Omega) \cap H, \\ \infty & \text{otherwise.} \end{cases} \tag{2.4}$$

Then it is known that the subdifferentials of ϕ and ψ are given by

$$\partial\phi = S, \quad \partial\psi = B.$$

These facts are effectively used in [16]–[18] (for subdifferential operators see, e.g., Showalter [23, Sections IV.1 and IV.2]).

Now let $n \in \mathbf{N}$ and denote by J_n and B_n the resolvent and Yosida approximation of B, respectively:

$$J_n := (1 + n^{-1}B)^{-1}, \tag{2.5}$$
$$B_n := n(1 - J_n) = BJ_n. \tag{2.6}$$

Then B_n is also represented by subdifferential operators. In fact, setting

$$\begin{aligned} \psi_n(u) &:= \min_{v \in H} \left\{ \frac{n}{2}\|v - u\|^2 + \psi(v) \right\} \\ &= \frac{1}{2n}\|B_n u\|^2 + \psi(J_n u) \quad \text{for } u \in H, \end{aligned} \tag{2.7}$$

we see that $\partial\psi_n = B_n$ (see [23, Proposition IV.1.8]).

The next lemma is proved in [16, Lemma 3.2] in more general situations. Here we give a direct proof.

Lemma 2.3. *Let $q > 1$. Then for $u \in H$ and $n \in \mathbf{N}$,*

$$q\psi(J_n u) \le \mathrm{Re}(B_n u, u) \le M\psi_n(u), \quad \mathrm{Im}(B_n u, u) = 0, \tag{2.8}$$

where $M := \max\{q, 2\}$.

Proof. Let $q > 1$, $u \in H$ and $n \in \mathbf{N}$. Writing as

$$(B_n u, u) = (B(J_n u), J_n u) + n^{-1}\|B_n u\|^2 = q\psi(J_n u) + n^{-1}\|B_n u\|^2,$$

we can obtain (2.8). □

Lemma 2.4. *Let $q > 1$. Then for $u \in D(S)$ and $n \in \mathbf{N}$,*

$$|\mathrm{Im}(Su, B_n u)| \le c(q)\mathrm{Re}(Su, B_n u), \tag{2.9}$$

where $c(q)$ is the same as in (1.4).

Proof. When $q \geq 2$, (2.9) has already been proved in [16, Lemma 6.2] (see also [15, Lemma 5.5]). When $1 < q < 2$, we employ the usual procedure which is outlined as follows. Let $\varepsilon > 0$ and define

$$B^{\varepsilon}u := (|u|^2 + \varepsilon)^{(q-2)/2}u \quad \text{for } u \in H.$$

Then it follows that B^{ε} is also m-accretive in H. As in the case where $q \geq 2$, we can show that for $u \in D(S)$ and $n \in \mathbf{N}$,

$$|\mathrm{Im}(Su, (B^{\varepsilon})_n u)| \leq c(q)\mathrm{Re}(Su, (B^{\varepsilon})_n u),$$

where $(B^{\varepsilon})_n$ is the Yosida approximation of B^{ε}. Letting $\varepsilon \downarrow 0$ and noting that $(B^{\varepsilon})_n u \to B_n u$ ($\varepsilon \downarrow 0$) in H, we can obtain (2.9). □

Remark 3. Inequality (2.9) is also derived from a similar one proved in [14, Lemma 3.2]:

$$(2.10) \qquad |\mathrm{Im}(S_n u, Bu)| \leq c(q)\mathrm{Re}(S_n u, Bu), \quad u \in D(B), \quad n \in \mathbf{N}.$$

In fact, denoting by J_m the resolvent of B, we see that for $v \in H$ and $m, n \in \mathbf{N}$,

$$\begin{aligned}(S_n v, B_m v) &= (S_n(1 - J_m)v, B_m v) + (S_n(J_m v), B_m v)\\ &= m^{-1}(S_n(B_m v), B_m v) + (S_n(J_m v), B(J_m v)).\end{aligned}$$

Noting that S_n is also nonnegative selfadjoint in H and applying (2.10) with $u = J_m v \in D(B)$ to the second term on the right-hand side, we have

$$(2.11) \qquad |\mathrm{Im}(S_n v, B_m v)| \leq c(q)\mathrm{Re}(S_n v, B_m v), \quad v \in H, \quad m, n \in \mathbf{N}.$$

Letting $n \to \infty$ in (2.11) for $v \in D(S)$ yields (2.9). Conversely, we can derive (2.10) from (2.9) by using Lemma 2.1.

In the rest part of this section we give existence results on two kinds of abstract evolution equations. First we consider the following abstract Cauchy problem in H:

$$(2.12) \qquad \begin{cases} \dfrac{du}{dt} + (\mu + i\nu)Su + Fu = 0, \\ u(0) = u_0, \end{cases}$$

where $\mu \geq 0$, $\nu \in \mathbf{R}$ are constants and F is a Lipschitz continuous operator on H.

Lemma 2.5. *Let H be a complex Hilbert space. Let S be a nonnegative selfadjoint operator in H and F a Lipschitz continuous operator on H.*

(i) *If $\mu \geq 0$ and $u_0 \in D(S)$, then (2.12) has a unique C^1-solution $u(\cdot) \in C^1([0,\infty); H) \cap C([0,\infty); D(S))$.*

(ii) *If* $\mu > 0$ *and* $u_0 \in D(S^{1/2})$, *then* (2.12) *has a unique* C^1*-solution* $u(\cdot) \in C^1((0,\infty); H) \cap C((0,\infty); D(S)) \cap C([0,\infty); D(S^{1/2}))$.

(iii) *If* $\mu > 0$ *and* $u_0 \in H$, *then* (2.12) *has a unique* C^1*-solution* $u(\cdot) \in C^1((0,\infty); H) \cap C((0,\infty); D(S)) \cap C([0,\infty); H)$.

Proof. (i) Since $(\mu + i\nu)S$ is a linear m-accretive operator in H for $\mu \geq 0$, we can obtain the assertion by virtue of Pazy [20, Theorem 6.1.6] (see also Cazenave-Haraux [3, Proposition 4.3.9]).

(ii), (iii) If $\mu > 0$ and $u_0 \in D(S^{1/2})$ (resp. $u_0 \in H$), then (2.12) has a unique strong solution $u(\cdot) \in C([0,\infty); D(S^{1/2}))$ (resp. $u(\cdot) \in C([0,\infty); H)$) such that $u(t) \in D(S)$ for every $t > 0$ (see [16, Proposition 3.1]). Let $t_0 > 0$. Applying part (i) with u_0 replaced with $u(t_0) \in D(S)$, we see that $u(\cdot) \in C^1([t_0,\infty); H) \cap C([t_0,\infty); D(S))$. Since $t_0 > 0$ is arbitrary, we can obtain the assertion. □

We conclude this section with the next existence result on nonlinear evolution equations (see Kato [10] and Showalter [23, Theorem IV.4.1]).

Lemma 2.6. *Let* H *be a complex Hilbert space. Let* A *be a nonlinear operator such that* $A + \gamma$ *is* m*-accretive in* H *for some* $\gamma \in \mathbf{R}$. *Then for any* $u_0 \in D(A)$ *there exists a unique strong solution* $u(\cdot)$ *to the Cauchy problem*

$$(2.13) \qquad \begin{cases} \dfrac{du}{dt} + Au = 0, \\ u(0) = u_0, \end{cases}$$

in the following sense:

(a) $u(t) \in D(A)$ *and* $\|Au(t)\| \leq e^{\gamma t}\|Au_0\|$ *for every* $t \geq 0$;

(b) $u(\cdot) : [0,T] \to H$ *is Lipschitz continuous for every* $T > 0$ *so that* $u(\cdot)$ *is strongly differentiable a.e. on* $(0,\infty)$;

(c) $u(\cdot)$ *satisfies the equation in* (2.13) *a.e. on* $(0,\infty)$ *as well as the initial condition.*

3 Proof of Theorem 1.1

The proof of Theorem 1.1 is divided into four steps.

Step 1: approximate problems. Let $n \in \mathbf{N}$, $\lambda \geq 0$, $\kappa \geq 0$ with $\lambda\kappa = 0$ and $\alpha, \beta, \gamma \in \mathbf{R}$. Let S, B, ϕ, ψ, J_n, B_n and ψ_n be as in (2.1)–(2.7), respectively. Given $u_0 \in H := L^2(\Omega)$, we consider the following approximate problem:

$$(\mathrm{P})_n \qquad \begin{cases} \dfrac{du_n}{dt} + \Big(\dfrac{1}{n} + \lambda + i\alpha\Big)Su_n + (\kappa + i\beta)B_n u_n - \gamma u_n = 0, \\ u_n(0) = u_0. \end{cases}$$

Since $\mathrm{Re}(n^{-1} + \lambda + i\alpha) > 0$ and B_n is Lipschitz continuous on H, it follows from Lemma 2.5 (iii) that $(\mathrm{P})_n$ has a unique C^1-solution $u(\cdot) \in$

$C^1((0,\infty);H)\cap C((0,\infty);D(S))\cap C([0,\infty);H)$. Furthermore, we note that for $t>0$,

$$\frac{d}{dt}\phi(u_n(t)) = \mathrm{Re}\Big(Su_n(t), \frac{du_n}{dt}(t)\Big)_{L^2}, \tag{3.1}$$

$$\frac{d}{dt}\psi_n(u_n(t)) = \mathrm{Re}\Big(B_n u_n(t), \frac{du_n}{dt}(t)\Big)_{L^2} \tag{3.2}$$

(cf. [23, Lemma IV.4.3]).

Step 2: estimates. We prepare some estimates for approximate solutions.

Lemma 3.1. *Let $u_n(\cdot)$ be a unique C^1-solution to $(\mathrm{P})_n$ with $u_0\in H$. Then*

$$\|u_n(t)\|_{L^2} \le e^{\gamma t}\|u_0\|_{L^2} \quad \forall t\ge 0, \tag{3.3}$$

$$\lambda\int_0^T \|\nabla u_n(t)\|_{L^2}^2\,dt + \kappa\int_0^T \|J_n u_n(t)\|_{L^q}^q\,dt \le \frac{e^{2\gamma_+ T}}{2}\|u_0\|_{L^2}^2 \quad \forall T>0, \tag{3.4}$$

where $\gamma_+ := \max\{\gamma, 0\}$.

Proof. Making the inner product of the equation in $(\mathrm{P})_n$ with $u_n(\cdot)$, we see from (2.8) that

$$\frac{1}{2}\frac{d}{dt}\|u_n\|_{L^2}^2 + \lambda\|\nabla u_n\|_{L^2}^2 + \kappa\|J_n u_n\|_{L_q}^q \le \gamma\|u_n\|_{L^2}^2.$$

Integrating this inequality, we can obtain (3.3) and (3.4). □

We need more estimates for approximate solutions because (3.4) is useless when $\lambda=\kappa=0$.

Lemma 3.2. *Let $u_n(\cdot)$ be a unique C^1-solution to $(\mathrm{P})_n$ with $u_0\in H$.*

(i) *If $u_0\in H_0^1(\Omega)\cap L^q(\Omega)$ and $\alpha\beta\ge 0$, then for a.a. $t\ge 0$,*

$$\frac{1}{2}\|\nabla u_n(t)\|_{L^2}^2 + \frac{\delta}{q}\|J_n u_n(t)\|_{L^q}^q \le e^{M\gamma_+ t}\Big(\frac{1}{2}\|\nabla u_0\|_{L^2}^2 + \frac{\delta}{q}\|u_0\|_{L^q}^q\Big), \tag{3.5}$$

where $M := \max\{q,2\}$ and δ is a positive constant depending only on $\lambda+i\alpha$, $\kappa+i\beta$ and q.

(ii) *If $u_0\in H_0^1(\Omega)$, $\kappa>0$ and $|\beta|/\kappa\le c(q)^{-1}$, then*

$$\|\nabla u_n(t)\|_{L^2} \le e^{\gamma t}\|\nabla u_0\|_{L^2} \quad \forall t\ge 0. \tag{3.6}$$

(iii) *If $u_0\in L^q(\Omega)\cap H$, $\lambda>0$ and $|\alpha|/\lambda\le c(q)^{-1}$, then*

$$\|u_n(t)\|_{L^q} \le e^{\gamma_+ t}\|u_0\|_{L^q} \quad \forall t\ge 0. \tag{3.7}$$

Proof. (i) First, setting $Z_n := (Su_n, B_n u_n)_{L^2}$, we have

$$\frac{d}{dt}\phi(u_n) + \kappa \mathrm{Re} Z_n + \beta \mathrm{Im} Z_n \le 2\gamma\phi(u_n), \tag{3.8}$$

$$\frac{d}{dt}\psi_n(u_n) + \lambda \mathrm{Re} Z_n - \alpha \mathrm{Im} Z_n \le q\gamma_+\psi_n(u_n). \tag{3.9}$$

In fact, making the inner product of the equation in $(\mathrm{P})_n$ with Su_n or $B_n u_n$, we can obtain (3.8) and (3.9) by virtue of (3.1) and (3.2).

Next, adding (3.8) and (3.9) multiplied by $\delta > 0$, we see by (2.9) that

$$\begin{aligned} &M\gamma_+(\phi(u_n) + \delta\psi_n(u_n)) \\ &\ge \frac{d}{dt}(\phi(u_n) + \delta\psi_n(u_n)) + (\kappa + \lambda\delta)\mathrm{Re} Z_n + (\beta - \alpha\delta)\mathrm{Im} Z_n \\ &\ge \frac{d}{dt}(\phi(u_n) + \delta\psi_n(u_n)) + (c(q)^{-1}(\kappa + \lambda\delta) - |\beta - \alpha\delta|)|\mathrm{Im} Z_n|. \end{aligned}$$

Setting

$$\delta := \begin{cases} \beta/\alpha & (\alpha\beta > 0), \\ c(q)(|\beta|/\lambda) & (\alpha = 0, \beta \ne 0), \\ c(q)^{-1}(\kappa/|\alpha|) & (\alpha \ne 0, \beta = 0), \\ \text{any positive number} & (\alpha = \beta = 0), \end{cases}$$

we see that $c(q)^{-1}(\kappa + \lambda\delta) - |\beta - \alpha\delta| \ge 0$ and hence

$$\frac{d}{dt}(\phi(u_n) + \delta\psi_n(u_n)) \le M\gamma_+(\phi(u_n) + \delta\psi_n(u_n)).$$

Integrating this inequality gives

$$\phi(u_n(t)) + \delta\psi_n(u_n(t)) \le e^{M\gamma_+ t}(\phi(u_0) + \delta\psi_n(u_0)).$$

Since $\psi(J_n u) \le \psi_n(u) \le \psi(u)$ by (2.7), we obtain (3.5).

(ii) It follows from (2.9) that if $\kappa > 0$ and $|\beta|/\kappa \le c(q)^{-1}$ then

$$\kappa \mathrm{Re} Z_n + \beta \mathrm{Im} Z_n \ge (c(q)^{-1}\kappa - |\beta|)|\mathrm{Im} Z_n| \ge 0.$$

Applying this inequality to the left-hand side of (3.8), we have

$$\frac{d}{dt}\phi(u_n(t)) \le 2\gamma\phi(u_n(t)).$$

Integrating this inequality yields (3.6).

(iii) Let $\lambda > 0$ and $|\alpha|/\lambda \le c(q)^{-1}$. As in the proof of part (ii), we can obtain (3.7) by using (2.9) and (3.9). □

Step 3: convergence. It follows from Lemmas 3.1 and 3.2 that $\{u_n(\cdot)\}$ and $\{J_n u_n(\cdot)\}$ are bounded in $L^2(0,T;H_0^1(\Omega))$ and $L^q(0,T;L^q(\Omega))$, respectively, under the assumption of Theorem 1.1. Therefore, in the same way as in the proof of [16, Lemma A.3], we can obtain the following lemma by using compactness methods.

Lemma 3.3. *Let* $u_n(\cdot)$ *be as in* Lemma 3.2. *Set* $Q := (0,T) \times \Omega$. *Then there exist a subsequence* $\{u_{n_k}(\cdot)\}$ *of* $\{u_n(\cdot)\}$ *and a function* $u \in L^2(Q)$ *such that*

$$u_{n_k} \to u \ \ (k \to \infty) \ \ \text{in} \ \ L^2(Q) \ \ \text{and a.e. on} \ Q.$$

The next lemma is a consequence of Lemmas 3.1–3.3. Since the proof is just the same as that of [16, Lemma A.4], we omit it.

Lemma 3.4. *Let* $u_{n_k}(\cdot)$ *and* $u(\cdot)$ *be as in* Lemma 3.3. *Then*

$$\begin{aligned}
&u(\cdot) \in L^2(0,T;H_0^1(\Omega)) \cap L^q(0,T;L^q(\Omega)),\\
&u_{n_k}(\cdot) \to u(\cdot) \ \ (k \to \infty) \ \ \text{weakly in} \ \ L^2(0,T;H_0^1(\Omega)),\\
&J_{n_k}u_{n_k}(\cdot) \to u(\cdot) \ \ (k \to \infty) \ \ \text{weakly in} \ \ L^q(0,T;L^q(\Omega)),\\
&B_{n_k}u_{n_k}(\cdot) \to |u|^{q-2}u \ \ (k \to \infty) \ \ \text{weakly in} \ \ L^{q'}(0,T;L^{q'}(\Omega)).
\end{aligned}$$

Step 4: conclusion. Now we can complete

Proof of Theorem 1.1. Let $\varphi(\cdot) \in C^1([0,T];H_0^1(\Omega) \cap L^q(\Omega))$ with $\varphi(T) = 0$. Then it follows from $(\mathrm{P})_n$ that

$$\begin{aligned}
&-\int_0^T (u_n(t),\varphi'(t))_{L^2}\,dt + \Big(\frac{1}{n} + \lambda + i\alpha\Big)\int_0^T (\nabla u_n(t),\nabla\varphi(t))_{L^2}\,dt\\
&+(\kappa + i\beta)\int_0^T \langle B_n u_n(t),\varphi(t)\rangle_{L^{q'},L^q}\,dt - \gamma\int_0^T (u_n(t),\varphi(t))_{L^2}\,dt\\
&= (u_0,\varphi(0))_{L^2}.
\end{aligned}$$

Setting $n = n_k$ and letting $k \to \infty$, we see by Lemmas 3.3 and 3.4 that

$$\begin{aligned}
&-\int_0^T (u(t),\varphi'(t))_{L^2}\,dt + (\lambda + i\alpha)\int_0^T (\nabla u(t),\nabla\varphi(t))_{L^2}\,dt\\
&+(\kappa + i\beta)\int_0^T \langle |u(t)|^{q-2}u(t),\varphi(t)\rangle_{L^{q'},L^q}\,dt - \gamma\int_0^T (u(t),\varphi(t))_{L^2}\,dt\\
&= (u_0,\varphi(0))_{L^2}.
\end{aligned}$$

This implies that $u(\cdot)$ is a weak solution to (P); note that $u(\cdot)$ belongs to the Banach space

$$W := \{u(\cdot) \in L^r(0,T;H_0^1(\Omega) \cap L^q(\Omega));\ u'(\cdot) \in L^{r'}(0,T;H^{-1}(\Omega) + L^{q'}(\Omega))\}$$

($r := \min\{q,2\}$) which is continuously embedded in $C([0,T];H)$ (see, e.g., [23, Proposition III.1.2]). Moreover, we can obtain (1.5)–(1.8), (1.10) and

(1.11) by virtue of (3.3)–(3.7). It remains to prove (1.9). Lemma 2.1 yields that for $u, v \in L^q(\Omega)$,

$$|\mathrm{Im}\langle |u|^{q-2}u - |v|^{q-2}v, u - v\rangle_{L^{q'},L^q}| \le c(q)\mathrm{Re}\langle |u|^{q-2}u - |v|^{q-2}v, u - v\rangle_{L^{q'},L^q}.$$

Hence, if $\kappa > 0$ and $|\beta|/\kappa \le c(q)^{-1}$, then

$$\begin{aligned} &\mathrm{Re}(\kappa + i\beta)\langle |u|^{q-2}u - |v|^{q-2}v, u - v\rangle_{L^{q'},L^q} \\ &\ge (c(q)^{-1}\kappa - |\beta|)|\mathrm{Im}\langle |u|^{q-2}u - |v|^{q-2}v, u - v\rangle_{L^{q'},L^q}| \ge 0. \end{aligned} \tag{3.10}$$

Now we prove (1.9). Let $u(\cdot)$ and $v(\cdot)$ be weak solutions to (P) with $u(0) = u_0 \in H_0^1(\Omega)$ and $v(0) = v_0 \in H_0^1(\Omega)$, respectively. Then it follows from (3.10) that if $\kappa > 0$ and $|\beta|/\kappa \le c(q)^{-1}$, then

$$\begin{aligned} \frac{1}{2}\frac{d}{dt}\|u - v\|_{L^2}^2 &= \mathrm{Re}\langle du/dt - dv/dt, u - v\rangle \\ &\le -\mathrm{Re}(\kappa + i\beta)\langle |u|^{q-2}u - |v|^{q-2}v, u - v\rangle_{L^{q'},L^q} + \gamma\|u - v\|_{L^2}^2 \\ &\le \gamma\|u - v\|_{L^2}^2. \end{aligned}$$

Integrating this inequality, we obtain (1.9). □

4 Proof of Theorem 1.2

To prove Theorem 1.2 we shall apply Lemma 2.6 to the following operator in the complex Hilbert space $H := L^2(\Omega)$ with inner product $(\cdot,\cdot) := (\cdot,\cdot)_{L^2}$ and norm $\|\cdot\| := \|\cdot\|_{L^2}$:

$$A := i\alpha S + (\kappa + i\beta)B - \gamma \quad \text{with } D(A) := D(S) \cap D(B),$$

where S and B are the same as in (2.1) and (2.2).

Lemma 4.1. *Let $\kappa > 0$, $q \ge 2$ and $\alpha, \beta \in \mathbf{R}$ be as in* Theorem 1.2. *Let B_n be the Yosida approximation of B. Then for $u \in D(S)$ and $n \in \mathbf{N}$,*

$$\|B_n u\| \le K_2\|i\alpha Su + (\kappa + i\beta)B_n u\|, \tag{4.1}$$

where $K_2 := (1 + c(q))/(\kappa - \frac{(\alpha\beta)_-}{|\alpha|}c(q))$ and $(\alpha\beta)_- := -\min\{\alpha\beta, 0\}$.

Proof. It follows from (2.9) that for $u \in D(S)$ and $n \in \mathbf{N}$,

$$\begin{aligned} &\|i\alpha Su + (\kappa + i\beta)B_n u\|\cdot\|B_n u\| \\ &\ge \mathrm{Re}(i\alpha Su + (\kappa + i\beta)B_n u, B_n u) \\ &\ge -|\alpha|\cdot|\mathrm{Im}(Su, B_n u)| + \kappa\|B_n u\|^2 \\ &\ge -|\alpha|c(q)\,\mathrm{Re}(Su, B_n u) + \kappa\|B_n u\|^2 \\ &= -|\alpha|c(q)\,\mathrm{Re}(Su + \frac{\kappa + i\beta}{i\alpha}B_n u, B_n u) + (\kappa + \frac{\alpha\beta}{\alpha^2}|\alpha|c(q))\|B_n u\|^2 \\ &\ge -c(q)\|i\alpha Su + (\kappa + i\beta)B_n u\|\cdot\|B_n u\| + (\kappa - \frac{(\alpha\beta)_-}{|\alpha|}c(q))\|B_n u\|^2. \end{aligned}$$

Therefore we can obtain (4.1). □

Lemma 4.2. *Let $\kappa > 0$, $q \geq 2$ and $\alpha, \beta \in \mathbf{R}$ be as in* Theorem 1.2. *Then $A + \gamma = i\alpha S + (\kappa + i\beta)B$ is m-accretive in H.*

Proof. It follows from Lemma 2.1 that $(\kappa+i\beta)B_n$ is accretive and Lipschitz continuous on H: for $u, v \in H$,

$$\begin{aligned}&\mathrm{Re}((\kappa + i\beta)(B_n u - B_n v), u - v)\\&= \mathrm{Re}((\kappa + i\beta)(B(J_n u) - B(J_n v)), J_n u - J_n v) + n^{-1}\kappa\|B_n u - B_n v\|^2\\&\geq n^{-1}\kappa\|B_n u - B_n v\|^2\end{aligned}$$

(cf. (3.10)). Hence we see from [23, Lemma IV.2.1] that $i\alpha S + (\kappa+i\beta)B_n$ is m-accretive in H so that for $f \in H$ and $n \in \mathbf{N}$ there exists a unique solution $u_n \in D(S)$ of the equation

$$u_n + i\alpha S u_n + (\kappa + i\beta)B_n u_n = f. \tag{4.2}$$

Since $D(S) \cap D(B) \neq \emptyset$, it follows that $\{\|u_n\|\}$ is bounded as $n \to \infty$ (see [23, Lemma IV.2.2]). Noting that $\|B_n u_n\| \leq K_2\|f - u_n\|$ by (4.1) and (4.2), we see that $\{\|B_n u_n\|\}$ is also bounded as $n \to \infty$. Therefore the assertion follows from a complex space version of the Brézis-Crandall-Pazy theorem (see [23, Proposition IV.2.1]). □

As a consequence of Lemma 4.2 we can complete

Proof of Theorem 1.2. It follows from Lemmas 2.6 and 4.2 that for any $u_0 \in D(A)$ there exists a unique strong solution $u(\cdot) \in C([0,\infty); H)$ to (P) and $u(\cdot)$ satisfies (1.12) and

$$\|Au(t)\| \leq e^{\gamma t}\|Au_0\| \quad \forall t \geq 0. \tag{4.3}$$

Letting $n \to \infty$ in (4.1) with $u = u(t) \in D(S) \cap D(B)$, we see from (1.9) with $v \equiv 0$ and (4.3) that for $t \geq 0$,

$$\begin{aligned}\|Bu(t)\| &\leq K_2\|(A + \gamma)u(t)\| \leq K_2 e^{\gamma t}(\|Au_0\| + |\gamma|\cdot\|u_0\|),\\ \|Su(t)\| &= |\alpha|^{-1}\|(A + \gamma - (\kappa + i\beta)B)u(t)\|\\ &\leq |\alpha|^{-1}(1 + K_2|\kappa + i\beta|)e^{\gamma t}(\|Au_0\| + |\gamma|\cdot\|u_0\|),\end{aligned}$$

which are nothing but (1.16) and (1.15), respectively. So we obtain (1.14). Noting that for $u, v \in D(S) \cap D(B)$,

$$\begin{aligned}&\|\nabla u - \nabla v\|_{L^2}^2 \leq (\|\Delta u\|_{L^2} + \|\Delta v\|_{L^2})\|u - v\|_{L^2},\\&\|u - v\|_{L^q}^q \leq 2^{q-2}(\|u\|_{L^{2(q-1)}}^{q-1} + \|v\|_{L^{2(q-1)}}^{q-1})\|u - v\|_{L^2},\end{aligned}$$

we can obtain (1.13), (1.17) and (1.18) by using (1.9), (1.12), (1.15) and (1.16). □

5 Proof of Theorem 1.3

First we establish the local existence theorem for C^1-solutions to (P).

Theorem 5.1. *Let $N = 1, 2, 3$, $\lambda \geq 0$, $\kappa, \alpha, \beta, \gamma \in \mathbf{R}$ and $q \geq 2$. Then for any $u_0 \in H^2(\Omega) \cap H_0^1(\Omega)$ there exist a maximal time $T_m \in (0, \infty]$ and a unique C^1-solution $u(\cdot)$ to* (P) *such that*

$$u(\cdot) \in C^1([0, T_m); L^2(\Omega)) \cap C([0, T_m); H^2(\Omega) \cap H_0^1(\Omega)).$$

Furthermore, if $T_m < \infty$, then $\limsup_{t \uparrow T_m} \|u(t)\|_{L^\infty} = \infty$.

Proof. We prove Theorem 5.1 by the same way as in Ôtani [19, Theorem 8]. Let S and B be as in (2.1) and (2.2) and let $M > 0$. Setting

$$(P_M u)(x) := \begin{cases} u(x) & \text{if } |u(x)| \leq M, \\ M e^{i \arg u(x)} & \text{if } |u(x)| \geq M, \end{cases}$$

we define an operator B_M on $H := L^2(\Omega)$ as follows:

$$B_M u := B(P_M u) \quad \text{for } u \in H.$$

Then Lemma 2.2 implies that B_M is Lipschitz continuous on H: for $u, v \in H$,

$$\begin{aligned} \|B_M u - B_M v\|_{L^2} &\leq (q-1) M^{q-2} \|P_M u - P_M v\|_{L^2} \\ &\leq (q-1) M^{q-2} \|u - v\|_{L^2}. \end{aligned}$$

Hence Lemma 2.5 (i) assures that for $M > 0$ and $u_0 \in D(S)$ there exists a unique C^1-solution $u_M(\cdot) \in C^1([0, \infty); H) \cap C([0, \infty); D(S))$ to

$$(\mathrm{P})_M \qquad \begin{cases} \dfrac{du_M}{dt} + (\lambda + i\alpha) S u_M + (\kappa + i\beta) B_M u_M - \gamma u_M = 0, \\ u_M(0) = u_0. \end{cases}$$

Now let $N \leq 3$ and choose $M := \|u_0\|_{L^\infty} + 1$; note that $D(S) \subset L^\infty(\Omega)$. Then by continuity we see that there exists $T_0 > 0$ such that

$$\|u_M(t)\|_{L^\infty} \leq \|u_0\|_{L^\infty} + 1 = M \quad \forall t \in [0, T_0].$$

This implies that $B_M u_M(\cdot) = B(P_M u_M(\cdot)) = B u_M(\cdot)$ on $[0, T_0]$. Therefore we conclude that $u_M(\cdot)$ is a unique C^1-solution to (P) on $[0, T_0]$.

Next we show that if $T_m < \infty$ then $\limsup_{t \uparrow T_m} \|u(t)\|_{L^\infty} = \infty$, where T_m is the maximal time such that there exists a C^1-solution $u(\cdot)$ to (P) on $[0, T_m)$. To this end we argue by contradiction, assuming that $T_m < \infty$ and $\limsup_{t \uparrow T_m} \|u(t)\|_{L^\infty} < \infty$. Setting $M_0 := \sup_{0 \leq t < T_m} \|u(t)\|_{L^\infty} < \infty$, we see that $u(\cdot)$ is a C^1-solution to $(\mathrm{P})_{M_0}$ on $[0, T_m)$. It follows from the uniqueness of C^1-solutions to $(\mathrm{P})_{M_0}$ that $u(\cdot) = u_{M_0}(\cdot)$ on $[0, T_m)$, where

$u_{M_0}(\cdot) \in C^1([0,\infty); H) \cap C([0,\infty); D(S))$ is a unique "global" C^1-solution to $(\mathrm{P})_{M_0}$. Repeating the same argument as in the first half of the proof with M and u_0 replaced with $M_0 + 1$ and $u_{M_0}(T_m)$, respectively, we see that $u(\cdot)$ can be extended as a C^1-solution to (P) beyond T_m. This contradicts the definition of T_m. □

In Theorem 5.1 we have constructed a unique *local* C^1-solution $u(\cdot)$ to (P). The last assertion of Theorem 5.1 means that if $\limsup_{t \uparrow T_m} \|u(t)\|_{L^\infty} < \infty$, then the solution $u(\cdot)$ is a unique *global* C^1-solution to (P).

Now we can prove Theorem 1.3 by using the estimates for weak and strong solutions to (P).

Proof of Theorem 1.3. Let T_m and $u(\cdot)$ be as in Theorem 5.1.

(i) It follows from (1.5) and (1.6) that $\limsup_{t \uparrow T_m} \|u(t)\|_{H^1} < \infty$. Since $H^1(\Omega) \hookrightarrow L^\infty(\Omega)$ when $N = 1$, we see that $\limsup_{t \uparrow T_m} \|u(t)\|_{L^\infty} < \infty$. Therefore the assertion follows from Theorem 5.1.

(ii), (iii) In view of (1.5), (1.9) with $v \equiv 0$ and (1.15) we see that $\limsup_{t \uparrow T_m} \|u(t)\|_{H^2} < \infty$. Hence $\limsup_{t \uparrow T_m} \|u(t)\|_{L^\infty} < \infty$; note that $H^2(\Omega) \hookrightarrow L^\infty(\Omega)$ when $N \leq 3$. Therefore we can obtain the assertion by virtue of Theorem 5.1. □

Acknowledgements

The author expresses his gratitude to Professor N. Okazawa for his valuable comments and suggestions on the subject.

References

[1] D. Bahuguna and V. Raghavendra, Application of Rothe's method to nonlinear Schrödinger type equations, *Applicable Analysis,* **31** (1988), 149–160.

[2] P. Bechouche and A. Jüngel, Inviscid limits of the complex Ginzburg-Landau equation, *Comm. Math. Phys.,* **214** (2000), 201–226.

[3] T. Cazenave and A. Haraux, An introduction to semilinear evolution equations, Oxford Lecture Series in Mathematics and its Applications, Vol. 13, *The Clarendon Press, Oxford University Press*, New York, 1998.

[4] J. Ginibre and G. Velo, The Cauchy problem in local spaces for the complex Ginzburg-Landau equation. I. Compactness methods, *Physica D,* **95** (1996), 191–228.

[5] J. Ginibre and G. Velo, The Cauchy problem in local spaces for the complex Ginzburg-Landau equation. II. Contraction methods, *Comm. Math. Phys.,* **187** (1997), 45–79.

[6] N. Hayashi, E.I. Kaikina and P.I. Naumkin, Large-time behavior of solutions to the dissipative nonlinear Schrödinger equation, *Proc. Roy. Soc. Edinburgh Sect. A,* **130** (2000), 1029–1043.

[7] N. Hayashi, E.I. Kaikina and P.I. Naumkin, Large time behavior of solutions to the Landau-Ginzburg type equations, *Funkcialaj Ekvacioj,* **44** (2001), 171–200.

[8] N. Hayashi, E.I. Kaikina and P.I. Naumkin, Asymptotic expansion of small solutions to the Landau-Ginzburg type equations, *Asymptotic Anal.*, to appear.

[9] N. Hayashi, E.I. Kaikina and P.I. Naumkin, Global existence and time decay of small solutions to the Landau-Ginzburg type equations, *J. Analyse Math.*, to appear.

[10] T. Kato, Nonlinear semigroups and evolution equations, *J. Math. Soc. Japan,* **19** (1967), 508–520.

[11] C.D. Levermore and M. Oliver, The complex Ginzburg-Landau equation as a model problem, Dynamical Systems and Probabilistic Methods in Partial Differential Equations (Berkeley, 1994), 141–190, Lectures in Appl. Math., Vol. 31, *Amer. Math. Soc.*, Providence, RI, 1996.

[12] J.L. Lions, Quelques Méthodes de Résolution des Problèmes aux Limites Non Linéaires, *Dunod*, Paris, 1969.

[13] V.A. Liskevich and M.A. Perelmuter, Analyticity of submarkovian semigroups, *Proc. Amer. Math. Soc.,* **123** (1995), 1097–1104.

[14] N. Okazawa and T. Yokota, Monotonicity method applied to the complex Ginzburg-Landau and related equations, *J. Math. Anal. Appl.,* **267** (2002), 247–263.

[15] N. Okazawa and T. Yokota, Perturbation theory for m-accretive operators and generalized complex Ginzburg-Landau equations, *J. Math. Soc. Japan,* **54** (2002), 1–19.

[16] N. Okazawa and T. Yokota, Global existence and smoothing effect for the complex Ginzburg-Landau equation with p-Laplacian, *J. Differential Equations,* **182** (2002), 541-576.

[17] N. Okazawa and T. Yokota, Smoothing effect for generalized complex Ginzburg-Landau equations in unbounded domains, Dynamical Systems and Differential Equations (Kennesaw, GA, 2000), *Discrete Contin. Dynam. Systems,* **2001**, Added Volume, 280–288.

[18] N. Okazawa and T. Yokota, Monotonicity method for the complex Ginzburg-Landau equation, including smoothing effect, The Third World Congress of Nonlinear Analysts (Catania, Sicily, 2000), *Nonlinear Anal.,* **47** (2001), 79–88.

[19] M. Ôtani, An introduction to nonlinear evolution equations, *Summer Seminar Notes*, 1983 [Japanese].

[20] A. Pazy, Semigroups of linear operators and applications to partial differential equations, Applied Mathematical Sciences, Vol. 44, *Springer-Verlag*, New York, 1983.

[21] H. Pecher and W. von Wahl, Time dependent nonlinear Schrödinger equations, *Manuscripta Math.*, **27** (1979), 125–157.

[22] T. Shigeta, A characterization of m-accretivity and an application to nonlinear Schrödinger type equations, *Nonlinear Anal.*, **10** (1986), 823–838.

[23] R.E. Showalter, Monotone Operators in Banach Space and Nonlinear Partial Differential Equations, Math. Surv. Mono. Vol. 49, *Amer. Math. Soc.*, Providence, RI, 1997.

[24] R. Temam, Infinite-Dimensional Dynamical Systems in Mechanics and Physics, Applied Math. Sci., Vol. 68, *Springer-Verlag*, Berlin and New York, 1988; 2nd ed., 1997.